AF765248

ELECTROCHEMICAL WATER AND WASTEWATER TREATMENT

ELECTROCHEMICAL WATER AND WASTEWATER TREATMENT

Edited by

CARLOS ALBERTO MARTÍNEZ-HUITLE

Federal University of Rio Grande do Norte, Natal, Brazil

MANUEL ANDRÉS RODRIGO

University of Castilla La Mancha, Ciudad Real, Spain

ONOFRIO SCIALDONE

University of Palermo, Palermo, Italy

ELSEVIER

Butterworth-Heinemann
An imprint of Elsevier

Butterworth-Heinemann is an imprint of Elsevier
The Boulevard, Langford Lane, Kidlington, Oxford OX5 1GB, United Kingdom
50 Hampshire Street, 5th Floor, Cambridge, MA 02139, United States

Notices
Knowledge and best practice in this field are constantly changing. As new research and experience broaden our understanding, changes in research methods, professional practices, or medical treatment may become necessary.

Practitioners and researchers must always rely on their own experience and knowledge in evaluating and using any information, methods, compounds, or experiments described herein. In using such information or methods they should be mindful of their own safety and the safety of others, including parties for whom they have a professional responsibility.

To the fullest extent of the law, neither the Publisher nor the authors, contributors, or editors, assume any liability for any injury and/or damage to persons or property as a matter of products liability, negligence or otherwise, or from any use or operation of any methods, products, instructions, or ideas contained in the material herein.

Library of Congress Cataloging-in-Publication Data
A catalog record for this book is available from the Library of Congress

British Library Cataloguing-in-Publication Data
A catalogue record for this book is available from the British Library

ISBN: 978-0-12-813160-2

Publisher: Matthew Deans
Acquisition Editor: Ken McCombs
Editorial Project Manager: Charlotte Rowley
Production Project Manager: Kamesh Ramajogi
Cover Designer: Greg Harris

Typeset by SPi Global, India

Contents

11. Hybrid and Sequential Chemical and Electrochemical Processes for Water Decontamination

12. Microbial Fuel Cells and Wastewater Treatment

II

APPLICATIONS OF ELECTROCHEMICAL TECHNOLOGIES FOR DECONTAMINATION AND DISINFECTION OF WATER

13. Fine Chemical Industry, Pulp and Paper Industry, Petrochemical Industry and Pharmaceutical Industry

14. Application of Electrochemical Processes for Treating Effluents From Hydrocarbon Industries

Contributors

Jonathan Albo University of Cantabria, Santander, Spain

Manuel Alvarez-Guerra University of Cantabria, Santander, Spain

Sidney Aquino Neto Chemistry Department, Faculty of Philosophy, Sciences and Letters at Ribeirão Preto, Universidade de São Paulo, Ribeirão Preto, SP, Brazil

Patricia Balderas-Hernández Joint Center for Research in Sustainable Chemistry, UAEM-UNAM, Toluca, México

Carlos E. Barrera-Díaz Joint Center for Research in Sustainable Chemistry, UAEM-UNAM, Toluca, México

Katlin I. Barrios Eguiluz Universidade Tiradentes, Aracaju, SE, Brazil

Bryan Bilyeu Xavier University of Louisiana, New Orleans, LA, United States

Enric Brillas University of Barcelona, Barcelona, Spain

Brian P. Chaplin University of Illinois at Chicago, Chicago, IL, United States

Simona Corgiolu University of Cagliari, Cagliari, Italy

Adalgisa R. De Andrade Chemistry Department, Faculty of Philosophy, Sciences and Letters at Ribeirão Preto, Universidade de São Paulo, Ribeirão Preto, SP, Brazil

Achille De Battisti University of Ferrara, Ferrara, Italy

Marilia M. de Salles Pupo Universidade Tiradentes, Aracaju, SE, Brazil

Antonio Dominguez-Ramos University of Cantabria, Santander, Spain

Elisama V. dos Santos Federal University of Rio Grande do Norte, Natal, Brazil

Carolina Espinoza-Cisternas University of Santiago of Chile, Santiago, Chile

Eduardo Expósito University of Alicante, Alicante, Spain

Ana S. Fajardo Federal University of Rio Grande do Norte, Natal, Brazil

Francisco Gallud University of Alicante, Alicante, Spain

Vicente García-García University of Alicante, Alicante, Spain

Karine Groenen Serrano Toulouse University, Toulouse, France

Raquel Ibañez University of Cantabria, Santander, Spain

Jesús Iniesta University of Alicante, Alicante, Spain

Angel Irabien University of Cantabria, Santander, Spain

Cristina Locatelli University of Milan, Milan, Italy

Florencio Lugo Autonomous National University of Mexico, Mexico City, Mexico

Laura Mais University of Cagliari, Cagliari, Italy

Carlos A. Martínez-Huitle Federal University of Rio Grande do Norte, Natal, Brazil

Michele Mascia University of Cagliari, Cagliari, Italy

Yunny Meas Center for Research and Technological Development in Electrochemistry, Santiago de Querétaro, Mexico

Alejandro Medel Center for Research and Technological Development in Electrochemistry, Santiago de Querétaro, Mexico

Alessandro Minguzzi University of Milan, Milan, Italy

Vicente Montiel University of Alicante, Alicante, Spain

Inmaculada Ortiz University of Cantabria, Santander, Spain

Nihal Oturan Université Paris-Est, Laboratoire Géomatériaux et Environnement (LGE), Marne-la-Vallée, France

Mehmet A. Oturan Université Paris-Est, Laboratoire Géomatériaux et Environnement (LGE), Marne-la-Vallée, France

Martin A. Pacheco-Álvarez University of Guanajuato, México

Simonetta Palmas University of Cagliari, Cagliari, Italy

Marco Panizza University of Genoa, Genoa, Italy

Juan M. Peralta-Hernández University of Guanajuato, México

Elisabetta Petrucci University of Rome, La Sapienza, Roma, Italy

Valéria Reginatto Chemistry Department, Faculty of Philosophy, Sciences and Letters at Ribeirão Preto, Universidade de São Paulo, Ribeirão Preto, SP, Brazil

María J. Rivero University of Cantabria, Santander, Spain

Manuel Andrés Rodrigo University of Castilla La Mancha, Ciudad Real, Spain

Oscar Rodríguez-Narváez University of Guanajuato, México

Sandra Rondinini University of Milan, Milan, Italy

Cristina Sáez University of Castilla La Mancha, Ciudad Real, Spain

Ricardo Salazar University of Santiago of Chile, Santiago, Chile

Giancarlo R. Salazar Banda Universidade Tiradentes, Aracaju, SE, Brazil

Géssica de O. Santiago Santos Universidade Tiradentes, Aracaju, SE, Brazil

Onofrio Scialdone University of Palermo, Palermo, Italy

Ignasi Sirés University of Barcelona, Barcelona, Spain

Ane M. Urtiaga University of Cantabria, Santander, Spain

Annalisa Vacca University of Cagliari, Cagliari, Italy

David Valero University of Alicante, Alicante, Spain

Vanessa M. Vasconcelos Universidade Tiradentes, Aracaju, SE, Brazil

Alberto Vertova University of Milan, Milan, Italy

Shraddha Vijay University of Guanajuato, México

Preface

Over the last two decades, a very important effort has been made by many research groups all around the world in order to develop electrochemically based alternatives to conventional water and wastewater-treatment technologies. The advances in this interdisciplinary area have recently encouraged a close collaboration between chemists, electrochemists, engineers, and other scientists, particularly in the use of these alternative technologies for decontamination and disinfection of many industrial and domestic effluents. This book tries to summarize the current development status of these technologies in 19 chapters, written by several of today's most recognized researchers, pointing out the strengths and weaknesses of these promising technologies.

The book starts with the description of the *Historical Developments and Fundamentals* of electrochemical water and wastewater treatment technologies. This part deals with processes such as electrochemical reduction, electrocoagulation, electroflotation, and electrochemical oxidation, emphasizing the fundamentals and new approaches developed. In this description, a large portion of the attention is focused on electrodes, the key component of the electrochemical cell. Different types of electrodes that have been developed in recent years are described in terms of preparation methods, properties and performance in the electrochemical treatment of wastewater. Recent advances in electrocatalysis related to generation of the powerful hydroxyl radical, as well as other strong oxidants (persulfate, ozone, hydrogen peroxide, and ferrate), are also extensively examined. Likewise, the fast destruction of pollutants mediated by electrogenerated active chlorine active chlorine, by electro-Fenton or by photo assisted electrochemical methods is also a matter of interest. An important section is also devoted to novel electrochemical technologies (hybrid and sequential processes as well as microbial fuel cells) that can be used for treating water and wastewaters.

Once the fundamentals are described, the second part of the book (*Applications of Electrochemical Technologies for Decontamination and Disinfection of Water*) deals with innovations sought in research labs, that is, how today's researchers aim to extend the range of applicability for electrochemical technologies. Special emphasis is placed on production of strong oxidizing species which can promote the fast mineralization of organics contained in wastewater and enable water disinfection. Likewise, attention is focused on the direct use of sunlight, with a description of emerging solar photo-assisted processes. In addition, the main aspects of other novel, combined technologies involving photocatalysis, adsorption, nanofiltration, microwaves, and ultrasounds, among others, are also pointed out. Finally, the very novel and promising bioelectrochemical technologies are discussed as well.

The third part of the book deals with the *Chemical and Technological Advantages and Disadvantages of Electrochemical Approaches' Applicability*, summarizing the research and development in the area of electrochemical advanced oxidation processes for water and wastewater treatment. Specific topics covered include a discussion of stable anode materials, advantages and recent innovations, challenges associated with electrochemical advanced oxidation processes' implementation, and future research and engineering needs. Examples that aim to use electrochemical technologies as new alternatives for treatment of real effluents were also described, as were future perspectives for these approaches.

The last part of the book deals with the possibilities of combining *Renewable Energies with Electrochemical Technologies* towards wastewater treatment. The nature and characteristics of a solar panel, along with the connections of solar plants to a wide variety of electrochemical processes, are described, pointing out the principal research directions for each application. Some remarkable examples of the use of other renewable energies are also taken into account.

Based on the wide gamut of subjects considered in the contents of this book, it should be of interest for a large number of diverse readers such as researchers, industry experts, public administrators, and students who are interested in both wastewater treatment and electrochemical processes. For researchers and professors involved in various scientific sectors (for example, chemistry, electrochemistry, engineering, materials science, and biotechnology) this book offers the opportunity to examine the fundamentals, recent innovations, and applications of electrochemical advanced oxidation processes.

Currently, the successful treatment of wastewater requires the selection and utilization of various technologies, the choice of which depends upon the specific case at hand, and the knowledge of electrochemical advanced oxidation processes can be a very useful aid in finding an effective solution for treating many kinds of wastewater. Indeed, electrochemical advanced oxidation processes are considered one of the most promising areas for wastewater treatment, and they could be potentially applied in a large number of decontamination processes. Hence, experts working in the environmental field, both in the industry and in public administration, could be strongly interested in the book. In addition, this book could be used for chemistry, environmental engineering, and chemical engineering courses focused on advanced oxidation processes, in order to gain full knowledge of electrochemical advanced oxidation processes.

We strongly believe that this book will greatly promote research in this key field in the forthcoming years to the benefit of our society.

Editors

Manuel Andrés Rodrigo

UCLM UNIVERSIDAD DE CASTILLA-LA MANCHA

Onofrio Scialdone

UNIVERSITÀ DEGLI STUDI DI PALERMO

Carlos Alberto Martínez-Huitle

UFRN UNIVERSIDADE FEDERAL DO RIO GRANDE DO NORTE

Editors Biography

Carlos A. Martínez-Huitle graduated in chemistry at Universidad de las Américas-Puebla (México) under the supervision of Dr. Marco Antonio Quiroz Alfaro. After a work experience in Ciba—Specialty Chemicals (currently owned by the German chemical company BASF), he moved to Ferrara, Italy, where he received his PhD in chemical sciences at the University of Ferrara under supervision of Prof. Achille De Battisti. During the same period, he worked as visiting scientist in the group of Prof. Christos Comninellis at the EPFL Institute, Switzerland. From 2005 to 2008, he served as a faculty member in the Department of Chemistry at the University of Milan. In 2008, he also moved to Brazil, where he is currently an associate professor in the Institute of Chemistry at the Federal University of Rio Grande do Norte. He was awarded the "Oronzio and Niccolò De Nora Foundation Prize" by the Italian Chemical Society (2005) and the "Oronzio and Niccolò De Nora Foundation Prize on Environmental Electrochemistry" by the International Society of Electrochemistry (2009). The Mexican scientist was also recognized by the German government in 2009 with the "Green Talent Award" for his contributions in the field of water disinfection treatment using new electrochemical technologies. In 2017, he received a Humboldt fellowship to work as senior scientist in the group of Prof. Siegfried Waldvogel at the Mainz University in Germany. Over the years, he has studied and worked in Mexico, conducted research in Switzerland, and both taught and researched in Italy, Chile, Germany, Spain, and Brazil. He is author of more than 170 publications in international journals, 6 chapters of books, and is a co-editor of the book entitled *Synthetic Diamond Films*, and has contributed to national and international meetings. Also, he participates in the Editorial Board of Applied Catalysis B: Environmental, RSC Advances and Scientific Reports journals. His research interests include electrochemical technologies for water treatment, electrocatalytic materials, electrocatalysis, and electroanalysis.

Manuel A. Rodrigo obtained his PhD from the University of Valencia in 1997, with research focused on the modeling and automation of biological nutrient removal processes. In 1997, he joined the University of Castilla La Mancha and began researching electrochemical engineering. In this first postdoctoral stage, his research focused on electrolysis of industrial wastewater. After a first postdoctoral training in the lab of Prof. Christos

Comninellis (EPFL, Switzerland), he started working with diamond electrodes, one of the key topics in his research. In 2000, he was appointed an associate professor at the University of Castilla La Mancha, and began researching electrocoagulation and high-temperature PEM fuel cells. Oxidants production, microbial fuel cells, and soil electroremediation have also focused his research attention. In 2009, he became a full professor in chemical engineering at the University of Castilla La Mancha. He maintains strong consultant collaboration with many companies in energy and environmental engineering. He is the author of more than 300 papers in referenced journals and books (h-index 53), more than 75 technical reports for companies, 5 patents, and he has supervised 14 doctoral theses. He has been invited professor at the universities Paris-Est Marne-la-Vallée (France) and Politécnica de Valencia (Spain). He has served as a chairman of the Working Party of Electrochemical Engineering of the European Federation of Chemical Engineering since 2011. He is a member of the editorial board of Journal of Hazardous Materials, Separation and Purification Technology, and Journal of Applied Electrochemistry.

Onofrio Scialdone studied chemical engineering at the University of Palermo. He obtained his PhD in electrochemical engineering from the Politecnico di Milano in 1999 and a Master of Economics at the Scuola Mattei of ENI (2000). He is a professor of industrial chemistry at Università degli Studi di Palermo and he leads the research activities on electrochemistry of the Laboratory of Chemical and Electrochemical Technologies there.

His main research interests are in electrochemical reaction engineering, with particular emphasis on synthesis of fine chemicals, environmental protection, and electrical energy generation. He has actively contributed on both direct and indirect oxidation processes for the electrochemical treatment of wastewater and on their modeling, and, lately, on particular kinds of cells such as microfluidic cells. In recent years, he studied the effect of pressure on various electrochemical processes and the cathodic conversion of carbon dioxide to various products. He is also involved in research on microbial fuel cells, synthesis and modification of polymers, utilization of $scCO_2$, and preparation of biofuels. He is a member of the editorial board of ChemElectroChem, and he has been involved in various research programs funded by the European community, private companies, and the Italian government.

PART I

HISTORICAL DEVELOPMENTS AND FUNDAMENTALS

CHAPTER

1

Electroreduction

Sandra Rondinini, Cristina Locatelli, Alessandro Minguzzi, Alberto Vertova
University of Milan, Milan, Italy

OUTLINE

INTRODUCTION

Why Electroreduction Methods?

Considering the extremely wide variety of air, soil, and water pollutants and all their possible combinations in a variety of different matrixes, electrochemical remediation methods are particularly convenient thanks to their flexibility, versatility, and effectiveness.

https://doi.org/10.1016/B978-0-12-813160-2.00001-8

Generally, the core of direct electrochemical treatments is the electron transfer reaction, which modifies the chemical nature of the pollutant by direct electro-reduction or -oxidation. This usually also implies a drastic cutback in the usage of chemicals and the adoption of mild physical reaction conditions, so that electrochemical treatments are safely listed among the most sustainable methods, especially in a perspective synergism with renewable energy sources. Still, their successful application has to take into account and prevent possible drawbacks, the main one being the formation of undesired by-products, which deactivate the electrode reaction or lead to substances which have a higher toxicity than the ones started with. These aspects are particularly evident when dealing with oxidation methods, which in treating halogenated compounds, or other organic molecules in the presence of halide radicals, can lead to high toxicity and persistency species such as dioxins [1–3]. Under these perspectives, direct electroreductive processes can effectively tune reaction conditions in order to avoid the main drawbacks and to provide working solutions for the detoxification of wastewaters, airborne streams, and underground waters from the ubiquitous presence of organic halides in industrial and civil processes.

In principle, electroreduction fulfills these needs, being capable of transforming any organic halide in the correspondent hydrocarbon:

$$R{-}X + H^+ + 2e^- \rightarrow RH + X^- \qquad X = F, Cl, Br, I \tag{1}$$

As will be discussed here, reaction (1) actually occurs under certain well defined conditions, while, in some cases, other pathways (e.g., dimerization or other types of substitution) can prevail.

Electroreduction of organic halides for remediation purposes has been deeply studied in the last 20 years [4], and the reader will see how important progress has been made so far in terms of (i) the understanding of electrode reaction mechanisms, (ii) electrode materials, and (iii) process approaches. After a brief introduction on the high environmental impact of organic halides, these three key issues will be presented and discussed. Finally, the reader can take advantage of a conclusions and perspectives paragraph where we attempt to frame the actual level of technology, according to the needs and means of today's society.

Organic Halides

Organic halides are widely adopted in industrial and civil processes as solvents, intermediates, pesticides, fungicides, insecticides, dielectrics and coolants, plasticizers (PCBs), and drugs. The main environmental issue bound to their wide use is due to their (i) high persistency in the environment, since they either remain unaltered or evolve toward more toxic species; and (ii) high volatility, and consequently a tendency of traveling over wide distances. Therefore, a single country's effort is not sufficient for their limitation or elimination, leading to their effective remediation.

These aspects are at the bases of the environmental treaties or conventions on pollutants and climate changes, which have been produced starting from the second half of the 20th century. In all these documents, organic halides are always listed as one of the main (if not the main) category of persistent pollutants whose use has to be limited and controlled. The well-known United Nations Framework Convention on Climate Change (1992) and the Kyoto Protocol, entered into force in 2005, commit State Parties to reduce greenhouse gas emissions, which include hydrofluorocarbons and perfluorocarbons.

The most important treaty for the elimination and limitation of dispersion of persistent organic pollutants (POPs) is the Stockholm Convention, entered into force in 2004 [5] and amended in five Conference of the Parties meetings, the last in 2015.

Looking at the classes of POPs considered in the Stockholm Convention, one immediately realizes that most of them are organic halides. The convention classifies POPs in three categories: to be eliminated (Annex A), to be restricted (Annex B), or to reduce unintentional production (Annex C).

In Annex A, several pesticides/insecticides such as Aldrin, Chlordane, or Lindane are listed, whereas Annex 2 includes Dichlorodiphenyltrichloroethane (DDT) that has currently no affordable replacements in malaria-prone countries. In Annex C the convention lists substances that are not meant to be dispersed in their applications, such as Polychlorinated biphenyls (PCBs), which are used for the production of electronic and electrical parts (capacitors, transformers, voltage regulators), hydraulic systems, adhesives, and plastics.

Mechanistic Aspects

The elucidation of the electroreduction pathways that lead to the (A) C—X bond cleavage (an example of dissociative electron transfers, DET) and to the (B) sequence of electrochemical or chemical reactions that transform the starting molecule(s) in the detected product(s) has been, and still is, discussed in several authoritative studies, and is briefly summarized here.

First, it has to be emphasized that the overall step (A) can, in turn, consist either in (i) a single electron transfer-bond cleavage step (see reaction 2) or (ii) a sequence of two separate steps, the first leading to an activated radical anion (reaction 3) and the second representing the bond cleavage (reaction 4)

$$\mathrm{R{-}X} + e^- \rightarrow \mathrm{R}^\bullet + \mathrm{X}^- \tag{2}$$

$$R{-}X + e^- \rightarrow R{-}X^{\bullet -} \tag{3}$$

$$R{-}X^{\bullet -} \rightarrow R^{\bullet} + X^- \tag{4}$$

In turn, the $R^{\bullet}$ radical may undergo further reduction to give the corresponding anion

$$R^{\bullet} + e^- \rightarrow R^- \tag{5}$$

The branching ratio between (2) and (3) + (4) is crucial and, on inert electrodes, is strongly dependent on the nature of the organic moiety of the halide.

From an experimental point of view, the elucidation of step (B) mainly requires preparative electrolyses and various analytical methods for the qualitative/quantitative determination of intermediates and products, while (A) likely represents one of the most studied electron transfer + bond cleavage processes in electrochemistry, which provided the test bed for cardinal mechanistic models. Indeed, reactions (2)–(4) are here denoted according to the dissociative electron transfer (DET) model, whose assumptions and outcomes are summarized later on.

In this chapter we will focus our attention on (A), while for (B) the reader is invited to refer to notable review works [4,6], where several case studies are well described and commented, or to more specific literature.

The organic halide electroreduction processes have also been devoted to electro-synthetic purposes to yield dimers (reaction 6), alkanes (by hydrodehalogenation, as in reaction 1), or olefins (reaction 7):

$$2RX + 2e^- \rightarrow R_2 + 2X^- \tag{6}$$

$$2R{-}CH_2{-}CH_2{-}X + 2e^- \rightarrow R{-}CH_2{-}CH_3 + R{-}CH{=}CH_2 + 2X^- \tag{7}$$

This chapter is limited to treating environmentally-oriented works, while important contributions reviewing electrosynthetic approaches can be found in the literature [6,7]. Still, the historical role of the electrosynthetic approach cannot be neglected, since it prompted deep mechanistic researches devoted to better understanding the DET reactions. The R—X case, in particular, has been adopted by Andrieux et al. [8] as a simple case for investigating the kinetics of DETs, which represent a class of reactions at the bases of several natural and artificial phenomena and were not described by the original Marcus-Hush theory.

According to this model, the reductive cleavage of a R—X bond can occur through a stepwise sequence of electron insertion (3) and bond breaking (4), or by a concerted combination of the two (2). The process may end with the insertion of the second electron, leading to the anion R^- (5). In most cases, the overpotential, at which the insertion of the first electron is observed, is quite high and the second electron transfer is fast.

It is intuitive to assess that the stepwise mechanism requires a smaller energy barrier (and thus a smaller driving force) than the concerted one, the latter including the bond dissociation energy and thus leading to peak potentials significantly more negative than for the former. Under the conditions that guarantee outer sphere DET, that is on an inert electrode, the two mechanisms can be easily distinguished by verifying the effect of the driving force on the energy barrier [9], which can change in dependence on available electronic energy levels in the electron-accepting molecule. Generally, aromatic compounds may easily accommodate electrons in available π^* orbitals, thus leading to a stepwise mechanism, while aliphatic molecules undergo a concerted mechanism because the radical anion obtained by an electron inserted in σ^* orbitals is energetically less favorable.

The parameter that better describes this branching is α, easily accessed via the dependence of the cathodic peak potential on the potential scan rate [9]. For concerted pathways, α assumes values <0.5 [9–11], while for stepwise mechanisms α is larger than 0.5.

As mentioned, the approach is valid only for inert electrodes; that is, where no specific interactions of reactants/intermediates/products of the dehalogenation process are established with the electrode surface. Unfortunately, a real inert electrode does not really exist (see below), but glassy carbon (GC) is generally accepted as the best example of it.

If interaction exists, modeling the system becomes much more complicated and the analysis of the nature of adsorbed species is crucial. This is the approach adopted in the case of benzylchloride on Ag [12,13], silver being one of the most active (if not the most active) electrocatalyst for organic halide electroreductions (see "Materials" section), where surface-enhanced operando Raman spectroscopy has been adopted. The works have been paralleled by a study based on the combination of quantitative experimental cyclic voltammetries, recorded over a wide range of scan rates, and theoretical investigations. This approach allowed the establishment of a complex reduction reaction mechanism and introduced and demonstrated the validity of an extended Laviron's mechanism that takes into account an interplay between solution and surface (adsorption) steps [14]. The main results point to the significant contribution of benzylchloride adsorption on Ag, following the substrate's concerted reductive cleavage.

How inert is the nature of an "inert" electrode? GC, considered as the benchmark material for outer sphere DET, presents surface oxygen functions (e.g., quinonic, acidic) that might chemically interact with the reactant, thus promoting the ET reactions [15]. In studying the electrochemical performance of GC, several authors evidenced that the presence of edge-plane defects on the surface leads to their facile oxidation, thus forming oxygen-containing surface functional groups, like carbonyl, hydroxyl, and carboxyl groups. Moreover, these authors underlined that the composition and quantity of such oxo-groups is influenced by the treatments used to prepare the surface: cleaning or activation processes. They showed that the electroreduction of five alkyl halides followed a concerted mechanism, while for the electroreduction of eight aromatic halides the stepwise mechanism must be invoked, concluding that the O/C ratio, on GC surface, mostly influences DET for alkyl halides.

The presence of oxo-groups has been considered responsible for the improved activity of Ag nanoparticles (NPs) supported on "activated carbon" (i.e., enriched of acidic surface groups) for hydrodehalogenation reduction [16].

MATERIALS

Silver

Since the 1980s, silver was reported to have a strong electrocatalytic behavior toward dehalogenation of organic compounds. The electroreduction processes and the C—X bond cleavage have been widely studied on GC, without any detection of catalytic effect; on mercury, with the possibility to form the more toxic metal-organic compounds [17]; and also on silver [18–21], but without any attempt of defining a rationale for its behavior.

Only starting from the original works of Rondinini et al. [22–31], the strong electrocatalytic activity of silver for the reductive dehalogenation of organic halides was demonstrated. The authors also elaborated a mechanistic rationale for this effect, suggesting a "cage effect" involving silver, the organic moiety, and the halide, which act as a bridge between the electrode and the organic compound, thus allowing for a remarkable reduction of the energy required for the C—X bond cleavage. This behavior is modulated by the molecular structure of organic halides, the morphology of electrode surfaces, the halogen atom [32–36], and the nature of Ag—X interaction, which can affect the reaction mechanism, as discussed by Isse et al. [37]. In particular, Isse et al. discuss the relation between the two main mechanisms already presented by Saveant, concerted and stepwise, and the electrocatalytic effect of cathode material, evidencing a strong connection between electrocatalysis and DET mechanism. At silver cathodes, in the presence of different organic halides, see Scheme 1 in [37], authors found that Ag can act as a strong electrocatalyst when two main conditions occur: the DET mechanism is concerted and the organic moiety and the halide atom interact with the silver cathode; see Fig. 1.

In this case, an inner-sphere reaction, due to the adsorption of the two part of the organic halide, leads to the C—X bond weakening and to a lowering of the activation-free energy of the intermediate complex. The reduction pathway changes due to the electrocatalytic behavior of Ag.

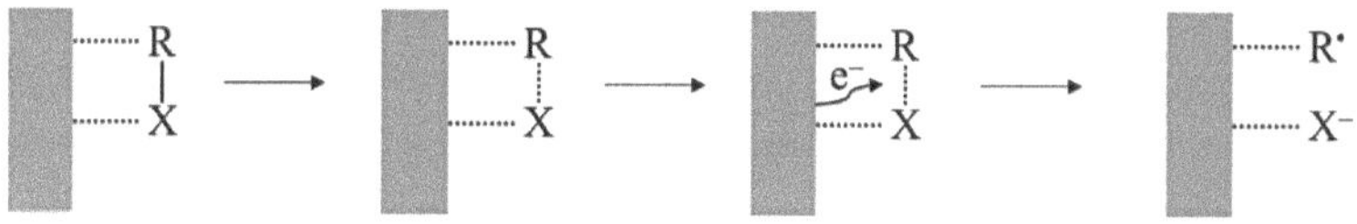

FIG. 1 Scheme of reaction (2). *Modified from A.A. Isse, L. Falciola, P.R. Mussini, A. Gennaro, Relevance of electron transfer mechanism in electrocatalysis: the reduction of organic halides at silver electrodes, Chem. Commun. 1 (2006) 344–346.*

In this field, silver has been used to dehalogenate a huge number of very different organic compounds, ranging from aliphatic long and short linear-chain molecules to aromatic rings.

The strong electrocatalytic activity of Ag can be even stronger when the metal NPs are supported on a carbonaceous matrix, appropriately modified with surface oxo-groups [16]. As evidenced in Fig. 2, the presence of oxide, hydroxide, and acidic groups on the supporting surface allows not only for a better clamping of Ag NPs, thus increasing their stability, durability, and loading, but also imparts hydrophilic character to the electrode. Now, considering the whole hydrodehalogenation process, as schematized in Fig. 2, the presence of mobile protons close to the electrocatalytic Ag site dramatically increases the reaction rate.

Massive silver has also been tested in room temperature ionic liquid [38], RTIL, to check its electrocatalytic activity for the C—X bond cleavage in the presence of the so-called "sticky DET" [39] reaction (8); see Fig. 3. This is a variant of concerted mechanism in which $R^{\bullet}$ and X^{-}, formed after

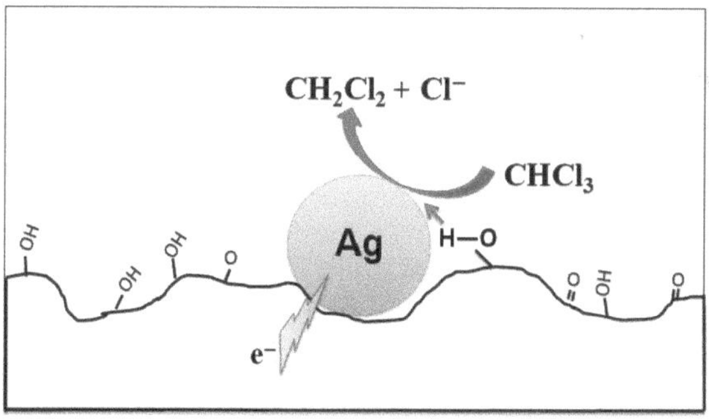

FIG. 2 Scheme for $CHCl_3$ hydrodehalogenation. *Reprinted with permission from A. Minguzzi, O. Lugaresi, G. Aricci, S. Rondinini, A. Vertova, Silver nanoparticles for hydrodehalogenation reduction: evidence of a synergistic effect between catalyst and support, Electrochem. Commun. 22 (2012) 25–28. Copyright 2012, Elsevier.*

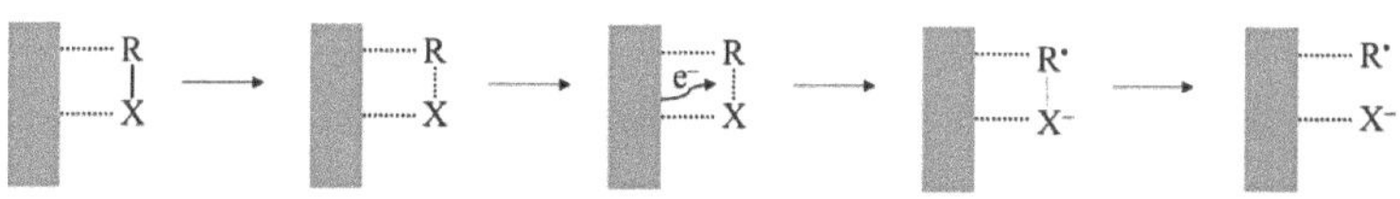

FIG. 3 Scheme of "sticky DET." *This scheme is described in Ref. 39.*

the bond breaking, present a ion-dipole interaction before leaving the electrode surface, thus influencing, in dependence on the interaction energy between the two leaving moieties, the DET. Isse et al. demonstrated that the DET of organic chloride in RTIL follows the same rules as in water or in organic solvent: a concerted reaction, involving alkyl- and benzyl-chloride, can be catalyzed by the usage of silver or copper as cathode materials, while a stepwise reaction, involving aromatic chloride, is completely independent on the cathode materials. For the concerted reaction, silver behaves better than copper, but both of them present a less electrocatalytic activity if compared with their usage in organic or water solvent.

$$R{-}X + e^- \rightarrow R^{\bullet} \cdots X^- \rightarrow R^{\bullet} + X^- \tag{8}$$

Finally, it can be worthwhile to highlight the use of silver to treat volatile organic halides (VOH) to reduce their toxicity. VOH are a family of organic compounds present in a large variety of waste: emulsion, organic solvent, airborne streams, and dilute aqueous mixtures, and they must be mostly removed from the gas phase. In this context, the use of gas diffusion electrodes (GDE) can allow the direct treatment of the gaseous streams. The cell setup comprises a working GDE cathode, consisting in a graphite fiber-cloth support loaded with Ag micro- or nanoparticles, in contact with a $KClO_4$ aqueous solution equipped with a Pt anode. This apparatus was used for the hydrodehalogenation of VOH [40]. The model molecule, trichloromethane, was fed under a N_2 stream inside the reactor closed at one hand with a GDE disk (see Fig. 4). This setup allowed for 100% CH_3Cl to CH_4 conversion.

In general, it has been demonstrated [16,41,42] that not only silver NPs present similar or even better electrocatalytic activity toward hydrodehalogenation of organic halides than massive silver, but also that other shapes of silver NPs can exhibit electrocatalytic activity. In this context, the possibility of a rapid screening of new composite materials for the preparation of cathodes to be employed in electroreductive dehalogenation processes has been also investigated and described by Rondinini et al. [41,43,44].

Recently, Brzózka et al. [45] discussed the electrocatalytic activity of silver nanowire/nanohemisphere array cathodes for organic halides dehalogenation. Using an anodic aluminium oxide template, the authors prepared silver nanowires and nanohemispheres by electrodeposition of Ag from $10\,g/dm^3$ $AgNO_3$ and $10\,g/dm^3$ H_3BO_3 under a constant current density of $1\,mA/cm^2$ for 60 min, see Fig. 5. The template was removed using a chemical etching, 0.5 M H_3PO_4, thus leaving free-standing nanowire or nanohemisphere array electrodes, which were tested for the electro-hydrodehalogenation reaction in the presence of $CHCl_3$ as a model molecule, using cyclic voltammetry and differential pulse voltammetry in 0.1 M NaOH solution. Their performance was compared with that of an Ag-rod electrode. The authors concluded that not only silver nanowires show about 200 mV less overpotential than an Ag-rod, while silver

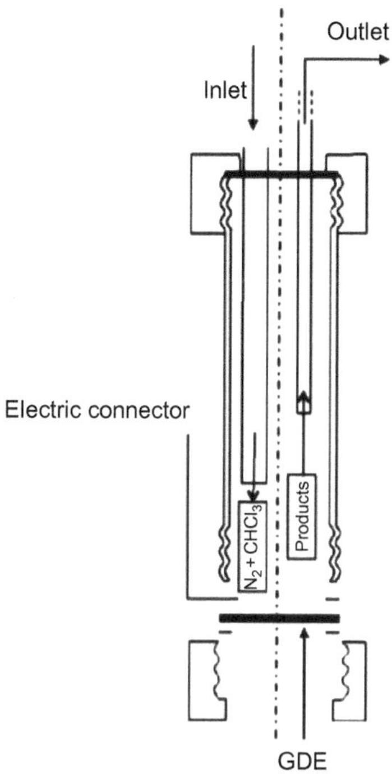

FIG. 4 Scheme of GDE to be immersed in aqueous solution. *Reprinted with permission from O. Lugaresi, H. Encontre, C. Locatelli, A. Minguzzi, A. Vertova, S. Rondinini, C. Comninellis, Gas-phase volatile organic chloride electroreduction: a versatile experimental setup for electrolytic dechlorination and voltammetric analysis, Electrochem. Commun. 44 (2014) 63–65. Copyright 2014, Elsevier.*

nanohemispheres show only 100 mV, but these nano-shaped materials also pose a good long-term stability.

In 2015 Wang et al. [46] studied dendritic silver NP cathodes for electroreductive dehalogenation of 5-bromo-2-chlorobenzoic acid in NaOH aqueous solution. The authors, using galvanic replacement reaction of $AgNO_3$ and Cu, prepared a dendritic Ag/Cu cathode, which showed increased electrocatalytic activity in comparison with massive Cu and Ag. Cyclic voltammetry in NaOH of dendritic Ag/Cu NPs, in the presence of 5-bromo-2-chlorobenzoic acid, showed a cathodic peak due to the dehalogenation process, less negative of about 150 mV and about 250 mV in comparison with Cu and Ag, respectively. In conclusion, silver can be diluted with a less expensive nonnoble metal, maintaining its electrocatalytic activity toward dehalogenation of organic halides.

Silver has been also used as a cathode for electroreductive dehalogenation reactions after an oxidation-reduction cycle process to transform the surface, creating a porous film formed by dendritic and spherical particles [47], thus reaching a roughness factor of about 64. In cycling an Ag plate from −0.4 V to 0.4 V vs SCE at 10 mV/s in an aqueous solution containing

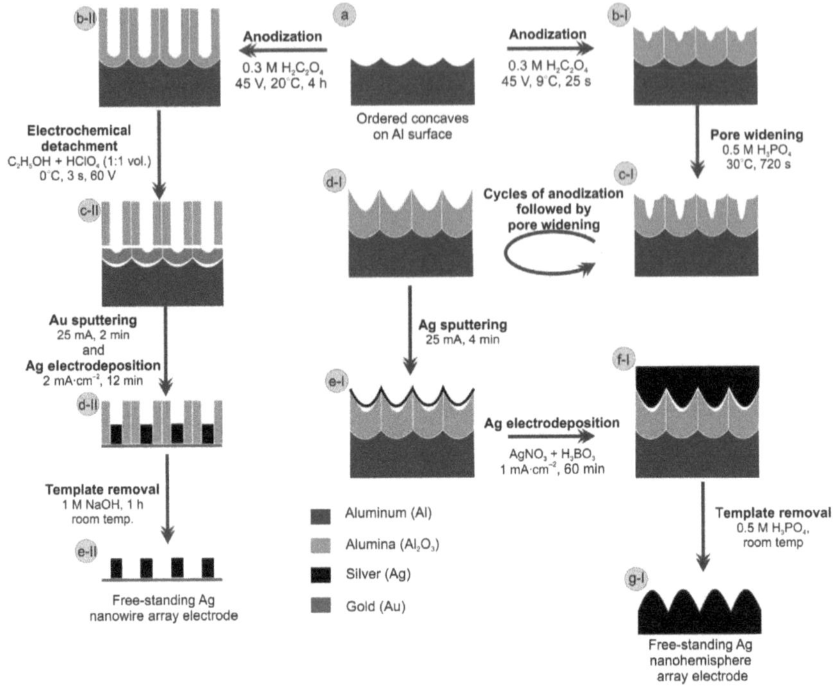

FIG. 5 Schematic representation of synthesis of the free-standing Ag nanohemisphere and nanowire array electrodes for $CHCl_3$ detection. *Reprinted with permission from A. Brzózka, A. Jeleń, A.M. Brudzisz, M.M. Marzec, G.D. Sulka, Electrocatalytic reduction of chloroform at nanostructured silver electrodes, Electrochim. Acta 225 (2017) 574–583. Copyright 2017, Elsevier.*

0.5 M NaCl, active metal area can be dramatically increased. In Ref. 47 authors report that this "activated" silver cathode reduces the overpotential by about 200 mV in comparison with a silver plate, for the treatment of chloroacetic acids and polyhalogenated phenols.

It should be noted that silver NPs supported on reduced graphene oxide were employed for nonelectrochemical reductive dehalogenation of a large number of pesticides and aliphatic halocarbons, such as chlorpyrifos, endosulfan, dichlorodiphenyldichloroethylene, carbon tetrachloride, chloroform, dichloromethane, and some fluorocarbons in water [48]. Koushik suggests a synergistic effect between the functional groups present on the reduced graphene oxide surface and the silver NPs, which account for the removal of these organic halides from the water stream, by a two-step mechanism: the degradation of organic molecules by means of silver NPs, producing AgCl, followed by the adsorption of dehalogenated products on graphene oxide surfaces by π-π interaction. Reduced graphene oxide can be chemically regenerated using NH_3, thus removing the AgCl, then refunctionalized with Ag NPs as previously described, and reused for other treatments.

Other Metals

Many other metals [49] have been investigated as possible electrocatalytic cathodes for reductive hydrodehalogenations of toxic organic halides, mainly aromatic chlorides, such as Pd [50–54], Pt [55–57], Ru [58], Rh [59], Ni [60], and Cu [61–63]. In general, they are used as massive cathodes or synthetized in micro- or nanosize, supported on carbonaceous materials.

Palladium

Among these metal substrates, Pd presents particular features that are worthwhile to discuss.

In the past, several authors [64–67] investigated the possibility of using zero valent iron to hydrodehalogenate different organic halides. In particular, Gillham [66] discussed the use of Fe^0 to treat trichloroethene in a flow-through reactor, mainly obtaining ethene and ethane as dehalogenated products, while the partially chlorinated compounds were only a minor fraction. These authors also underlined that the strong adsorption of trichloroethene on an iron surface, which leads to a tight contact between an organic halide and iron, is the reason for the very fast dehalogenation reaction if compared with abiotic treatments. In fact, the mechanism is an electron transfer reaction between zero valent iron, which is oxidized, and the organic halide, which is reduced. These outcomes caused a growing interest toward zero valent iron [68–71]. In this context, the possibility to increase the interaction between organic halides and iron has been exploited by different authors, assessing that bimetallic particles can present an enhanced catalytic and electrocatalytic activity for hydrodehalogenation processes, due to the formation of a short-circuit galvanic cell between the two metallic elements: iron and a more noble metal, especially Pd [72].

Different authors [73–75] described the treatment of toxic organic halides in water using palladized iron, leading to rapid non-electrochemical dechlorination. In a Pd/Fe bimetallic system, the zero valent iron can be oxidized in water to Fe^{2+} accompanied by production of molecular hydrogen, which in turn can be absorbed into Pd, thus forming the active chemical species PdH, which likely acts as a source of atomic hydrogen. The hydrodehalogenation reaction is:

$$R{-}X + 2H{-}Pd \rightarrow RH + H^+ + X^- \tag{9}$$

The rate of reaction is mainly a function of the surface area of metallic iron (consumed during the reaction) and palladium (not consumed during the reaction), the Pd/Fe ratio at the iron surface, and pH. Zha and coworkers demonstrated that the deposition of a more electropositive metallic element on iron increases the zero valent iron corrosion rate for the formation of a short-circuit galvanic cell. This, in turn, leads to a reaction rate increase for the dehalogenation process. The mechanism of this reaction, in presence of different Me/Fe systems, is still an open debate, but it is well ascertained that for Pd-like metal the strong affinity with hydrogen leads to a hydrodehalogenation process in which the hydrogenation reaction is the main process. In the case of Ag-like metals the β elimination reaction is the main mechanism, due also to the affinity of Ag for the leaving group. The authors easily prepared the bimetallic particles by a replacement reaction in an acidic deoxygenated solution and tested them in a batch reactor containing tribromoacetic acid. It must be underlined that iron is thermodynamically very active for a dehalogenation reduction reaction, due to its strong reductive potential, but the formation of an oxide surface film rapidly deactivates the metal. The presence of a more noble metal, Pd or Cu, allows for the formation of a short-circuited galvanic cell, Fe acting as an anode and Pd or Cu as a cathode. This leads to an easier dissolution of iron with the formation of the very active atomic hydrogen, as previously mentioned, due to the presence of the noble metal, especially when Pd is used for its electrocatalytic activity in the hydrogen evolution reaction. The authors discussed in detail the mechanism of the C—Br bond breaking, as presented in Fig. 6, for which bromide is substituted with hydrogen by means of a surface-mediated mechanism that explains why Pd/Fe particles are more active than Cu/Fe ones. Moreover, the greater activity of Pd/Fe than Cu/Fe particles is also due to the higher potential difference between Pd and Fe in comparison with Cu and Fe.

On the bases of these outcomes, the synergistic presence of Pd and atomic H, produced by zero valent iron, Cheng et al. [73] used palladized iron, carbon cloth, and graphite rod cathodes to perform the electroreductive dehalogenation of 4-chlorophenol in an aqueous solution. The authors demonstrated how these type of electrodes behave differently. At palladized graphite cathodes, the production of hydrogen, and hence the active species PdH, is observed together with the adsorption of 4-chlorophenol on the support, near the Pd "islands." This evidences a synergistic effect of the co-presence of Pd and C, that yield a pseudo-first order rate constant of about 10^{-3} min^{-1}; see Fig. 7C. Palladized iron and palladized carbon cloth showed one order of magnitude lower rate constants (see Fig. 7B), while plain carbon cloth cathodes presented very low reaction rates (Fig. 7A). The graphite rod evidenced no dehalogenation reaction.

Finally, Pd has been also used together with graphite to prepare cathodes for the electrochemical remediation of contaminated groundwater or sediments. Al-Abed [50] proposed the use of Pd-modified granular graphite as a cathode for in-situ PCB abatement. A cationic membrane divided reactor has been used to investigate the electrocatalytic activity of pure

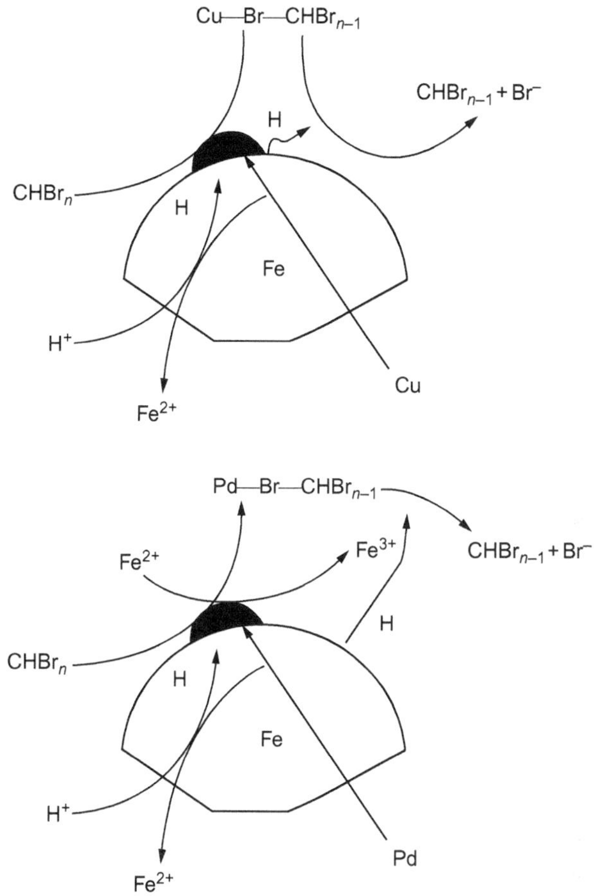

FIG. 6 Mechanisms of reductive dehalogenation by Cu/Fe and Pd/Fe bimetallic particles. *Reprinted with permission from X. Zha, L. Ma, Y. Liu, Reductive dehalogenation of brominated disinfection byproducts by iron based bimetallic systems, RSC Adv. 6 (2016) 16323–16330. Copyright 2016, Royal Society of Chemistry.*

and Pd-modified granular graphite, with the possibility to deposit Pd on the graphite surface during the dechlorination electrolysis, thus increasing the stability and durability of the cathodes.

Copper

Copper has been largely investigated as a possible low cost substitute for silver in the electroreductive dehalogenation processes [38,62,76]. Cu is reported to be a good electrocatalytic metal, which leads to a complete dechlorination of polychloromethanes, even if it is less active than Ag. In addition, Cu is about 100 times cheaper than Ag, largely cheaper than

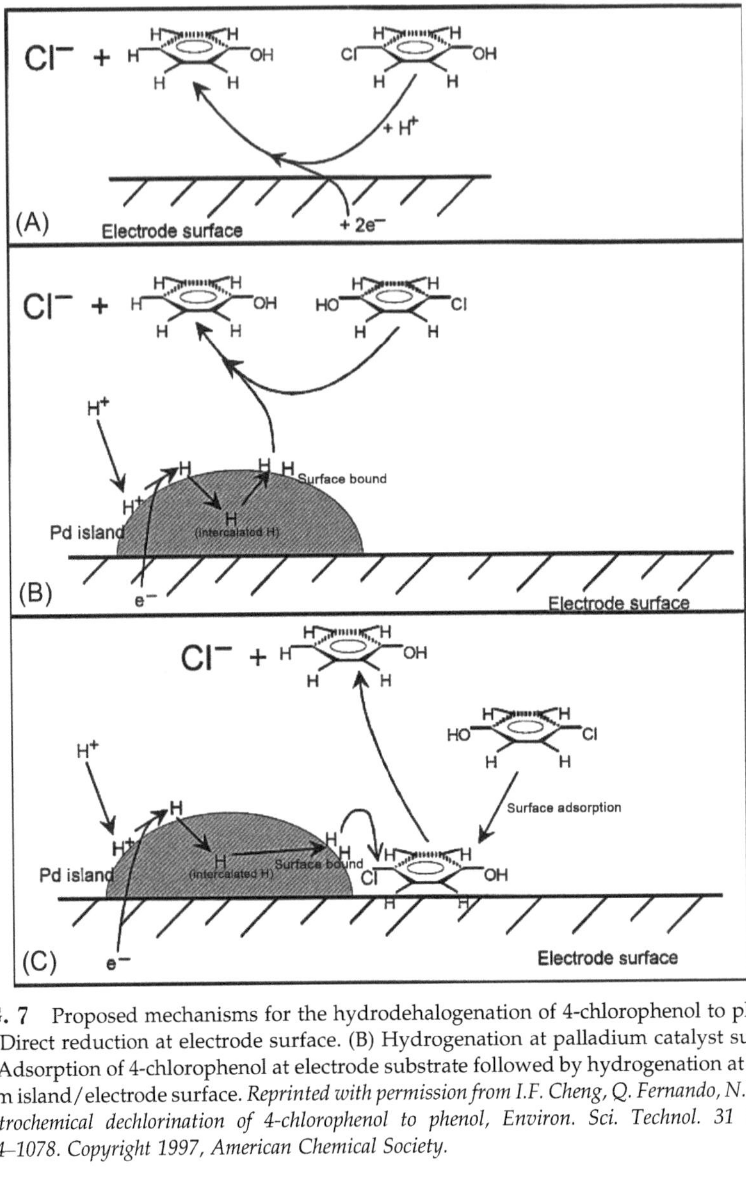

FIG. 7 Proposed mechanisms for the hydrodehalogenation of 4-chlorophenol to phenol. (A) Direct reduction at electrode surface. (B) Hydrogenation at palladium catalyst surface. (C) Adsorption of 4-chlorophenol at electrode substrate followed by hydrogenation at palladium island/electrode surface. *Reprinted with permission from I.F. Cheng, Q. Fernando, N. Korte, Electrochemical dechlorination of 4-chlorophenol to phenol, Environ. Sci. Technol. 31 (1997) 1074–1078. Copyright 1997, American Chemical Society.*

Pd and very abundant. Thus, bimetallic cathodes Ag/Cu or Pd/Cu have been prepared and studied for electrochemical dehalogenation of different organic halides [46,77]. These composite materials show an electrocatalytic activity comparable with pure Ag or Pd, avoiding the higher costs of the pure noble metal. Recently, Durante et al. discussed the possibility of using Pd/Cu [78] and Ag/Cu [79] decorated NPs for activating the C—X bond to lead to the hydrodehalogenation reaction. Nanomaterials allow not only for a further reduction of expensive noble metal usage, but also for the possibility of increasing stability, durability, and selectivity of the electrocatalyst by acting on the morphology and chemical composition of the NPs, via the appropriate synthetic route. In general, these composite materials are dispersed on a carbonaceous support, which acts both as an electronic conductor and as possible active material for the reaction of interest. In fact, on these cathodes, by tuning the quantity and dimension of NPs, it is possible to obtain the hydrodehalogenation products or the products of the C—C addition. Moreover, due to the high surface-to-volume ratio of NPs grafted on carbonaceous matrices, cathodes show a large increase of current density and efficiency, even if the noble metal loading has been reduced.

From an experimental point of view, bimetallic NPs can be easily prepared in two steps: first, on a carbonaceous support the Cu NPs are formed by electrodeposition procedure in water solvent. Modifying and/or cycling the potential and varying the electrolyte concentration of the deposition solution, it is possible to tune the dimension, shape, and dispersion of the particles on the support. Pd or Ag are added after the formation of core Cu NPs, by electroless displacement of Cu by Pd or Ag in water containing the specific metal salt. NPs are in the range of 10–40nm, in dependence on copper deposition time, and, after a thermal treatment, bimetallic particles show an enrichment of the noble metal on the surface, thus leaving Cu as a massive core material with a thin crust of costlier noble metals.

In conclusion, both bimetallic NPs, Pd/Cu and Ag/Cu, have demonstrated a good, in some cases even better than the massive metals, catalytic activity for the electroreductive dehalogenation processes, when the C—X bond cleavage follows a concerted mechanism. Finally, these cathodes are more stable than one formed by only Cu NPs, which can be easily oxidized at room temperature.

Mediator Catalyst

Transformation of a C—X bond into a C—H bond has been deeply investigated also in the case of homogeneous catalytic systems, in which organometallic compounds, formed by Pd, Ru, Rh, Fe, and other transition metals, are the catalysts [80–88]. These organometallic clusters largely employ phosphine or other very toxic ligands to be effective in the process, thus reducing the possibility of an environmental friendly reaction. Recently, new catalysts have been presented, based on a Ru(II) complex,

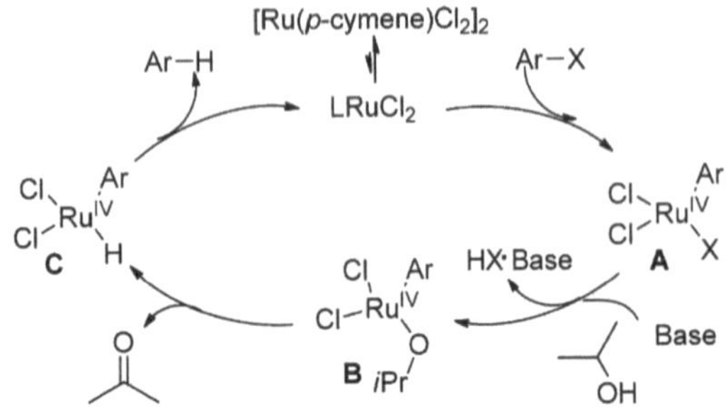

FIG. 8 Scheme of reaction mechanism of Ru(II) organometallic catalyst for the hydrodehalogenation of aromatic halides. *Reprinted with permission from T. You, Z. Wang, J. Chen, Y. Xia, Transfer hydro-dehalogenation of organic halides catalyzed by ruthenium(II) complex, J. Org. Chem. 82 (2017) 1340–1346. Copyright 2017, American Chemical Society.*

which are more environmentally compatible. Starting from the works of several groups [89–91], which described the use of a Ru(II) complex to catalyze the halogenation reaction on some aromatic compounds, thus forming a C—X bond from a C—H bond, Xia and coworkers [92] discussed the usage of commercially available $[RuCl_2(p\text{-cymene})]_2$ to catalyze the dehalogenation of aromatic, ester, and amide halides in isopropanol, used also as a hydrogen donor, and the possible reaction mechanism; see Fig. 8. They found that many different halocompounds can be hydrodehalogenated in the presence of the Ru(II) complex in a mild condition.

REACTORS AND REACTION SCHEMES

As evidenced in the previous paragraphs, the research efforts are mainly devoted to the study of nanostructured electrocatalysts, chiefly in the form of NPs. This is due not only to their higher activity and to lower costs with respect to "massive" electrodes, but also to their adaptability to the preparation of composite electrodes. For example, NPs can be supported on different types of carbonaceous substrates, like carbon cloths or carbon papers as in fuel cell-type electrodes, with adequate binders, for example, ionomers, assuring both ion permeation through the layer and good adhesion. This optimal assembling for gas-phase treatments can be easily extended to any kind of plate-and-frame cells. The general goal of increasing the electrode active surface while maintaining a compact geometry has prompted the development of metal foams [93,94] and 3-D composites or Ni complexes immobilized onto graphite felt [95].

Reactor design and engineering is then a key stage for the successful application of organic halide electroreductions, to take advantage of the newly developed electrode materials and compositions, and to attain the optimized reaction conditions of the laboratory-scale devices.

Nonetheless, most of the studies on organic halide electroreductions have been carried out in laboratory cells designed according to the type of experiment and the ancillary product analyses. In some cases, preparative electrolyses aim at analyzing in-line the reaction products by, for example, chromatographic techniques, possibly requiring the separation between anode and cathode chambers. To this end, Rondinini et al. reported the usage of an Ag/AgBr sacrificial anode and 2-compartment cell setup to allow for the easy recovery of reaction products [31,36]. In all cases, a suitable compartment for lodging the reference electrode is designed and a Luggin capillary is provided to compensate for the electrolyte ohmic resistances, especially in nonaqueous solvents.

The very popular plate-and-frame is used for a large variety of applications, as schematized in Fig. 9.

An innovative approach consists in microfluidic cells [96], which, thanks to the micrometric interelectrode distance (see Fig. 10), are characterized by acceptable ohmic drops without the addition of a supporting electrolyte. This feature presents the great advantage of directly treating low conductivity streams, thus avoiding the costs of electrolyte addition and recovery. This further aspect contributes to balance the higher pumping costs.

The use of nonaqueous media in wastewater treatments is not frequent. Nonetheless, examples of the adoption of organic solvents as reaction media have been proposed in the hypothesis of treating molecules previously extracted by means of activated carbon or by air-stripping [62].

For the treatment of gaseous streams, different strategies have been proposed, generally based on the preparation of GDEs which include electrocatalytic particles.

The first approach [97] is based on the use of Pd, Au, Cu, and Ag cathodes deposited onto a cation-exchange membrane that is in contact with the gaseous stream. The electrolyzer is completed by a Pt anode and was tested for the reduction of CCl_2F_2.

A second approach is based on a fuel-cell-like design and allowed the reduction of various chloromethanes and ethanes at a GDE cathode, while H_2 is oxidized at the anode [98]. The "core" of the cell consisted in a membrane electrode assembly (MEA) made of a cation-permeable membrane and two porous carbon-based electrodes hot pressed to both faces of the membrane. The electrode contained Pt at the anode and various metals at the cathode. The organic halides were fed at the cathode as a saturated N_2 stream.

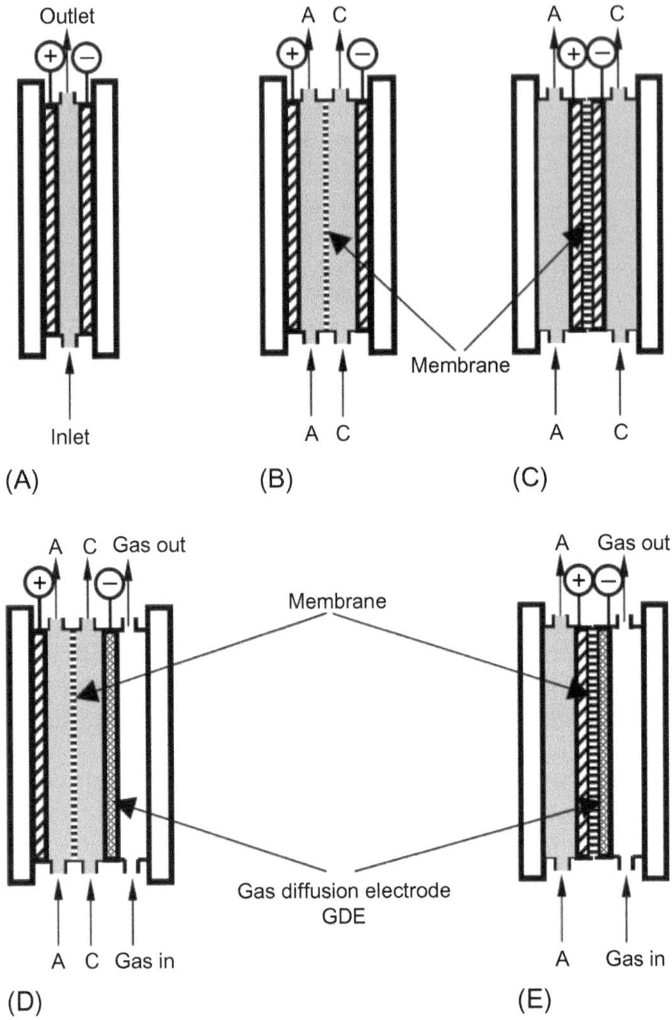

FIG. 9 Plate-and-frame cell schemes (single repetitive unit): (A) undivided cell; (B) 2-compartment membrane or separator cell; (C) solid polymer electrolyte reactor (SPE); (D) 2-compartment membrane cell with GDE; (E) SPE-GDE reactor. Liquid compartments are denoted in gray. *Reprinted with permission from S. Rondinini, A. Vertova, in: C. Comninellis, G. Chen (Eds.), Electrochemistry for the Environment, Springer New York, New York, NY, 2010. Copyright 2010, Springer Nature.*

One of the problems of the previous approaches is the formation of HCl as reduction by-product. In this optics, a possible solution is to separate the gas diffusion electrode from the membrane by means of a liquid compartment that can be recirculated and treated separately (Fig. 11). This setup was adopted to reduce $CHCl_3$ to CH_4 on GDEs made of various Ag NPs [42].

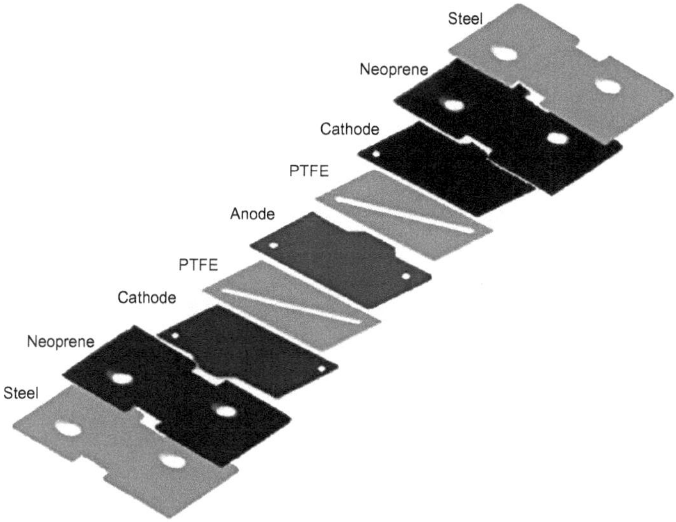

FIG. 10 Scheme for a microfluidic stack cell, proposed for the reduction of dichloroacetic acid. *Reprinted with permission from O. Scialdone, A. Galia, S. Sabatino, D. Mira, C. Amatore, Electrochemical conversion of dichloroacetic acid to chloroacetic acid in a microfluidic stack and in a series of microfluidic reactors, ChemElectroChem 2 (2015) 684–690. Copyright 2015, John Wiley & Sons, Inc., New Jersey, USA.*

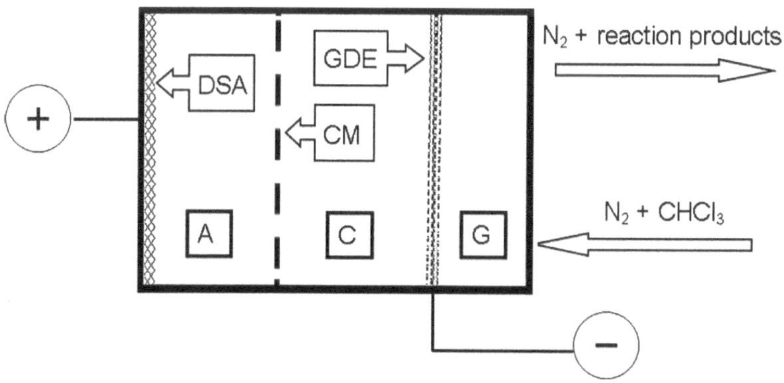

FIG. 11 Scheme of an electrolysis flow cell for gas-phase organic halide reduction. A: anolyte compartment; C: catholyte compartment; G: cathodic gas compartment. *CM*, cationic membrane; *DSA*, dimensionally stable anode (DSA anode); *GDE*, gas-diffusion cathode. Arrows denote gas in and gas out in the cathodic gas compartment.

CONCLUSIONS AND PERSPECTIVES

As for many other classes of pollutants, organic halides are at the bases of unprecedented difficult pollution situations, whose solution requires a step further in the research of new remediation strategies. A clear example is dense nonaqueous phase liquids (DNAPL), caused by the presence of substances of higher density than water, that accumulate at the bottom in groundwaters and are typically associated to chlorinated solvents.

The present chapter evidenced that the current research in organic halide remediation by electroreductive methods made important steps in the understanding of reaction mechanisms and in the development of active electrodes, but the scale-up and application of the relevant processes/reactors has still to be exploited, when compared to the electrooxidative ones.

Organic halide electroreduction has some clear limits, the main one being the competition with two other reactions (water reduction to hydrogen and oxygen reduction) that heavily limit the current efficiency and make the process costs rise accordingly. Future research on electrocatalysts will likely focus on increasing the selectivity toward the target reactions; for example, by applying high throughout methods for the rapid screening of alloys and new composite materials.

At the same time, one has to take into account that organic halide electroreduction:

- may lead to dehalogenated value-added hydrocarbons;
- is a flexible process, as generally assessed for electrochemical methods, easily added in line with conventional remediation processes. For example, after dehalogenation the substance may easily undergo complete mineralization by biotechnological methods. Extending the remediation capabilities to gas-phase pollution, the treatment of incinerator exhausts can be foreseen;
- can be applied in-situ: for example, for the treatment of DNAPLs;
- can be simultaneously combined with electrooxidation methods; for example, in the electroreduction of organic chlorides, the produced Cl^- ions can be the reactants for mediated oxidation at the anode.

It is the authors' opinion that one major obstacle to the wide applications of the electroreductive treatment of organic halides is (once again, as for many electrochemistry-based processes) the still-limited familiarity that scientists and engineers have for this type of approach. Under this perspective, modern methods, like ecodesign and Life Cycle Assessment studies, can be of great help to affirm the effectiveness and the economic competitiveness of these processes.

References

[1] B.G. Svensson, L. Barregård, G. Sällsten, A. Nilsson, M. Hasson, C. Rappe, Exposure to polychlorinated dioxins (PCDD) and dibenzofurans (PCDF) from graphite electrodes in a chloralkali plant, Chemosphere 27 (1993) 259–262.
[2] M. Vallejo, M.F. San Román, I. Ortiz, Quantitative assessment of the formation of polychlorinated derivatives, PCDD/Fs, in the electrochemical oxidation of 2-chlorophenol as function of the electrolyte type, Environ. Sci. Technol. 47 (2013) 12400–12408.
[3] R. Weber, Relevance of PCDD/PCDF formation for the evaluation of POPs destruction technologies—review on current status and assessment gaps, Chemosphere 67 (2007) 109–117.
[4] E.T. Martin, C.M. McGuire, M.S. Mubarak, D.G. Peters, Electroreductive remediation of halogenated environmental pollutants, Chem. Rev. 116 (2016) 15198–15234.
[5] EU, Regulation (EC) No 850/2004 of the European Parliament and of the Council of 29 April 2004 on persistent organic pollutants and amending Directive 79/117/EEC, Off. J. Eur. Commun. L158/7 (2004) 7–49.
[6] O. Hammerich, B. Speiser, Organic Electrochemistry, Fifth Edition: Revised and Expanded, fifth ed., CRC Press—Taylor and Francis Group, Boca Raton, FL, 2016.
[7] H. Lund, M.M. Baizer, Organic Electrochemistry and Introduction and a Guide, third ed., Marcel Dekker Inc., New York, 1991.
[8] C.P. Andrieux, A. Merz, J.M. Saveant, Dissociative electron transfer. autocatalysis and kinetic characteristics of the electrochemical reductive cleavage of the carbon halogen bond in alkyl halides giving rise to stable radicals and carbanions, J. Am. Chem. Soc. 107 (1985) 6097–6103.
[9] C.P. Andrieux, A.L. Gorande, J.M. Savéant, Electron transfer and bond breaking. Examples of passage from a sequential to a concerted mechanism in the electrochemical reductive cleavage of arylmethyl halides, J. Am. Chem. Soc. 114 (1992) 6892–6904.
[10] A. Cardinale, A.A. Isse, A. Gennaro, M. Robert, J.M. Savéant, Dissociative electron transfer to haloacetonitriles. An example of the dependency of in-cage ion-radical interactions upon the leaving group, J. Am. Chem. Soc. 124 (2002) 13533–13539.
[11] J.M. Savéant, Electron transfer, bond breaking and bond formation, Adv. Phys. Org. Chem. 35 (2000) 117–192.
[12] Y.F. Huang, D.Y. Wu, A. Wang, B. Ren, S. Rondinini, Z.Q. Tian, C. Amatore, Bridging the gap between electrochemical and organometallic activation: benzyl chloride reduction at silver cathodes, J. Am. Chem. Soc. 132 (2010) 17199–17210.
[13] A. Wang, Y.F. Huang, U.K. Sur, D.Y. Wu, B. Ren, S. Rondinini, Z.Q. Tian, In situ identification of intermediates of benzyl chloride reduction at a silver electrode by SERS coupled with DFT calculations, J. Am. Chem. Soc. 132 (2010) 9534–9536.
[14] O.V. Klymenko, O. Buriez, E. Labbé, D.P. Zhan, S. Rondinini, Z.Q. Tian, I. Svir, C. Amatore, Uncovering the missing link between molecular electrochemistry and electrocatalysis: mechanism of the reduction of benzyl chloride at silver cathodes, ChemElectroChem 1 (2014) 227–240.
[15] A. Gennaro, A.A. Isse, C.L. Bianchi, P.R. Mussini, M. Rossi, Is glassy carbon a really inert electrode material for the reduction of carbon-halogen bonds? Electrochem. Commun. 11 (2009) 1932–1935.
[16] A. Minguzzi, O. Lugaresi, G. Aricci, S. Rondinini, A. Vertova, Silver nanoparticles for hydrodehalogenation reduction: evidence of a synergistic effect between catalyst and support, Electrochem. Commun. 22 (2012) 25–28.
[17] M.R. Rifi, Organic electrochemistry, in: M.M. Baizer (Ed.), Organic Electrochemistry, Marcel Dekker Inc., New York, 1973, pp. 279–315.
[18] F. Y. Edamura, D. Kyriacou, J. Love, U.S. Pat. 4217185. U.S. Pat. 4217185 (1980).
[19] A. Inesi, L. Rossi, M. Feroci, M. Rizzuto, Electrochemically induced Favorskii rearrangement. α,β-unsaturated amides and esters in the electrochemical reduction of polyhaloketones, New J. Chem. 22 (1998) 57–61.

[20] R.M. Justice, D.A. Hall, Electrochemical reduction of 3-chlorobenzo[b]thiophenes, US Patent: US4588484, 1986.

[21] D. Kyriacou, Highly active silver cathode, preparation of same and use to make 2,3,5-trichloropyridine and/or 2,3,5,6-tetrachloropyridine, European Patent: EP0018069 A1, 1979.

[22] S. Ardizzone, G. Cappelletti, P.R. Mussini, S. Rondinini, L.M. Doubova, Adsorption competition effects in the electrocatalytic reduction of organic halides on silver, J. Electroanal. Chem. 532 (2002) 285–293.

[23] M. Benedetto, G. Miglierini, P.R. Mussini, F. Pelizzoni, S. Rondinini, G. Sello, Diastereoselective synthesis of aryl O-glycosides by electrochemical reduction of Glycosyl Halidesitle, Carbohydr. Lett. 1 (1995) 321–328.

[24] G. Fiori, P.R. Mussini, S. Rondinini, A. Vertova, Selective electrodehalogenation of bromophenols on Ag electrocatalyst, Anal. Chim. 92 (2002) 963–972.

[25] M. Guerrini, P. Mussini, S. Rondinini, G. Torri, E. Vismara, Electrochemical reduction of halogenosugars on silver: a new approach to C-disaccharide-like mimics, Chem. Commun. (15) (1998) 1575–1576.

[26] S. Rondinini, P.R. Mussini, F. Crippa, G. Sello, Electrocatalytic potentialities of silver as a cathode for organic halide reductions, Electrochem. Commun. 2 (2000) 491–496.

[27] S. Rondinini, P.R. Mussini, V. Ferzetti, D. Monti, Electrochemical reduction of α-D-glycopyranosyl bromides on a mercury cathode, Electrochim. Acta 36 (1991) 1095–1098.

[28] S. Rondinini, P.R. Mussini, G. Sello, E. Vismara, Glycosyl halides as building blocks for the electrosynthesis of glycosides, J. Electrochem. Soc. 145 (1998) 1108–1112.

[29] S. Rondinini, P.R. Mussini, G. Cantù, G. Sello, Cathode and medium effects on the electroreductive glucosidation of phenols, PCCP 1 (1999) 2989–2995.

[30] S. Rondinini, P.R. Mussini, M. Specchia, A. Vertova, The electrocatalytic performance of silver in the reductive dehalogenation of bromophenols, J. Electrochem. Soc. 148 (2001) D102–D107.

[31] S. Rondinini, A. Vertova, Electrocatalysis on silver and silver alloys for dichloromethane and trichloromethane dehalogenation, Electrochim. Acta 49 (2004) 4035–4046.

[32] S. Ardizzone, G. Cappelletti, P.R. Mussini, S. Rondinini, L.M. Doubova, Electrodeposited polycrystalline silver electrodes: surface control for electrocatalysis studies, Russ. J. Electrochem. 39 (2003) 170–176.

[33] S. Ardizzone, G. Cappelletti, L.M. Doubova, P.R. Mussini, S.M. Passeri, S. Rondinini, The role of surface morphology on the electrocatalytic reduction of organic halides on mono- and polycrystalline silver, Electrochim. Acta 48 (2003) 3789–3796.

[34] M. Fedurco, C.J. Sartoretti, J. Augustynski, Reductive cleavage of the carbon-halogen bond in simple methyl and methylene halides. reactions of the methyl radical and carbene at the polarized electrode/aqueous solution interface, Langmuir 17 (2001) 2380–2387.

[35] P.R. Mussini, S. Ardizzone, G. Cappelletti, M. Longhi, S. Rondinini, L.M. Doubova, Surface screening effects by specifically adsorbed halide anions in the electrocatalytic reduction of a model organic halide at mono- and polycrystalline silver in acetonitrile, J. Electroanal. Chem. 552 (2003) 213–221.

[36] S. Rondinini, P.R. Mussini, P. Muttini, G. Sello, Silver as a powerful electrocatalyst for organic halide reduction: the critical role of molecular structure, Electrochim. Acta 46 (2001) 3245–3258.

[37] A.A. Isse, L. Falciola, P.R. Mussini, A. Gennaro, Relevance of electron transfer mechanism in electrocatalysis: the reduction of organic halides at silver electrodes, Chem. Commun. 1 (2006) 344–346.

[38] A.A. Isse, L. Scarpa, C. Durante, A. Gennaro, Reductive cleavage of carbon–chlorine bonds at catalytic and non-catalytic electrodes in 1-butyl-3-methylimidazolium tetrafluoroborate, Phys. Chem. Chem. Phys. 17 (2015) 31228–31236.

[39] C. Costentin, M. Robert, J.M. Savéant, Successive removal of chloride ions from organic polychloride pollutants. Mechanisms of reductive electrochemical elimination in

aliphatic gem-polychlorides, α,β-polychloroalkenes, and α,β-polychloroalkanes in mildly protic medium, J. Am. Chem. Soc. 125 (2003) 10729–10739.
[40] O. Lugaresi, H. Encontre, C. Locatelli, A. Minguzzi, A. Vertova, S. Rondinini, C. Comninellis, Gas-phase volatile organic chloride electroreduction: a versatile experimental setup for electrolytic dechlorination and voltammetric analysis, Electrochem. Commun. 44 (2014) 63–65.
[41] O. Lugaresi, J.V. Perales-Rondón, A. Minguzzi, J. Solla-Gullón, S. Rondinini, J.M. Feliu, C.M. Sánchez-Sánchez, Rapid screening of silver nanoparticles for the catalytic degradation of chlorinated pollutants in water, Appl. Catal. B Environ. 163 (2015) 554–563.
[42] S. Rondinini, G. Aricci, Z. Krpetić, C. Locatelli, A. Minguzzi, F. Porta, A. Vertova, Electroreductions on silver-based electrocatalysts: the use of Ag nanoparticles for $CHCl_3$ to CH_4 conversion, Fuel Cells 9 (2009) 253–263.
[43] A. Minguzzi, C. Locatelli, O. Lugaresi, A. Vertova, S. Rondinini, Au-based/electrochemically etched cavity-microelectrodes as optimal tool for quantita, Electrochim. Acta 114 (2013) 637–642.
[44] A. Vertova, R. Barhdadi, C. Cachet-Vivier, C. Locatelli, A. Minguzzi, J.Y. Nedelec, S. Rondinini, Cavity microelectrodes for the voltammetric investigation of electrocatalysts: the electroreduction of volatile organic halides on micro-sized silver powders, J. Appl. Electrochem. 38 (2008) 965–971.
[45] A. Brzózka, A. Jeleń, A.M. Brudzisz, M.M. Marzec, G.D. Sulka, Electrocatalytic reduction of chloroform at nanostructured silver electrodes, Electrochim. Acta 225 (2017) 574–583.
[46] F. Wang, Z.Q. Fan, X. Wu, M.C. Li, C.A. Ma, The electroreductive dehalogenation reaction of 5-bromo-2-chlorobenzoic acid on dendritic Ag/Cu, Int. J. Electrochem. Sci. 10 (2015) 5101–5111.
[47] Y.H. Xu, H. Zhang, C.P. Chu, C.A. Ma, Dechlorination of chloroacetic acids by electrocatalytic reduction using activated silver electrodes in aqueous solutions of different pH, J. Electroanal. Chem. 664 (2012) 39–45.
[48] D. Koushik, S.S. Gupta, S.M. Maliyekkal, T. Pradeep, Rapid dehalogenation of pesticides and organics at the interface of reduced graphene oxide-silver nanocomposite, J. Hazard. Mater. 308 (2016) 192–198.
[49] B. Huang, A.A. Isse, C. Durante, C. Wei, A. Gennaro, Electrocatalytic properties of transition metals toward reductive dechlorination of polychloroethanes, Electrochim. Acta 70 (2012) 50–61.
[50] Y. Fang, S.R. Al-Abed, Alladium-facilitated electrolytic dechlorination of 2-chlorobiphenyl using a granular-graphite electrode, Chemosphere 66 (2007) 226–233.
[51] J.B. Hoke, G.A. Gramiccioni, E.N. Balko, Catalytic hydrodechlorination of chlorophenols, Appl. Catal. B Environ. 1 (1992) 285–296.
[52] M. Kuroboshi, H. Idei, N. Hara, Y. Kokui, H. Tanaka, Electroreduction of aryl halides loaded on palladium-immobilized activated carbon, Electrochemistry 81 (2013) 359–361.
[53] C. Schüth, M. Reinhard, Hydrodechlorination and hydrogenation of aromatic compounds over palladium on alumina in hydrogen-saturated water, Appl. Catal. B Environ. 18 (1998) 215–221.
[54] X. Xu, H. Zhou, P. He, D. Wang, Catalytic dechlorination kinetics of p-dichlorobenzene over Pd/Fe catalysts, Chemosphere 58 (2005) 1135–1140.
[55] G. Chen, E.A. Betterton, R.G. Arnold, W.P. Ela, Electrolytic reduction of trichloroethylene and chloroform at a Pt- or Pd-coated ceramic cathode, J. Appl. Electrochem. 33 (2003) 161–169.
[56] L. Liu, G. Zhao, M. Wu, Y. Lei, R. Geng, Electrochemical degradation of chlorobenzene on boron-doped diamond and platinum electrodes, J. Hazard. Mater. 168 (2009) 179–186.

[57] W.B. Wu, M.L. Li, J.M. Huang, Electrochemical hydrodefluorination of fluoroaromatic compounds, Tetrahedron Lett. 56 (2015) 1520–1523.
[58] K. Miyoshi, Y. Kamegaya, M. Matsumura, Electrochemical reduction of organohalogen compound by noble metal sintered electrode, Chemosphere 56 (2004) 187–193.
[59] Y. Xu, H. Ma, T. Ge, Y. Chu, C.A. Ma, Rhodium-catalyzed electrochemical hydrodefluorination: a mild approach for the degradation of fluoroaromatic pollutants, Electrochem. Commun. 66 (2016) 16–20.
[60] Z. Liu, R.G. Arnold, E.A. Betterton, K.D. Festa, Electrolytic reduction of CCl_4—effects of cathode material and potential on kinetics, selectivity, and product stoichiometry, Environ. Eng. Sci. 16 (1999) 1–13.
[61] J. He, A.E. Sáez, W.P. Ela, E.A. Betterton, R.G. Arnold, Destruction of aqueous-phase carbon tetrachloride in an electrochemical reactor with a porous cathode, Ind. Eng. Chem. Res. 43 (2004) 913–923.
[62] A.A. Isse, B. Huang, C. Durante, A. Gennaro, Electrocatalytic dechlorination of volatile organic compounds at a copper cathode. Part I: polychloromethanes, Appl. Catal. B Environ. 126 (2012) 347–354.
[63] X. Wen-ying, G. Ting-yao, Dechlorination of carbon tetrachloride by the catalyzed Fe-Cu process, J. Environ. Sci. 19 (2007) 792–799.
[64] T. Li, J. Farrell, Reductive dechlorination of trichloroethene and carbon tetrachloride using iron and palladized-iron cathodes, Environ. Sci. Technol. 34 (2000) 173–179.
[65] L.J. Matheson, P.G. Tratnyek, Reductive dehalogenation of chlorinated methanes by iron metal, Environ. Sci. Technol. 28 (1994) 2045–2053.
[66] W.S. Orth, R.W. Gillham, Dechlorination of trichloroethene in aqueous solution using Fe^0, Environ. Sci. Technol. 30 (1996) 66–71.
[67] L. Zhang, W.A. Arnold, R.M. Hozalski, Kinetics of haloacetic acid reactions with Fe(0), Environ. Sci. Technol. 38 (2004) 6881–6889.
[68] C.M. Dominguez, J. Parchão, S. Rodriguez, D. Lorenzo, A. Romero, A. Santos, Kinetics of lindane dechlorination by zerovalent iron microparticles: effect of different salts and stability study, Ind. Eng. Chem. Res. 55 (2016) 12776–12785.
[69] S. Giri, M. Bhaumik, R. Das, V.K. Gupta, A. Maity, Dehalogenation of aromatic halides by polyaniline/zero-valent iron composite nanofiber: kinetics and mechanisms, Appl. Catal. B Environ. 202 (2017) 207–216.
[70] P. Rao, T. Liu, X. Liang, G. Ding, Dual effects of humic acid in trichloroethylene removal from groundwater by zero-valent iron: Hydrophobic partition and surface adsorption, Water Air Soil Pollut. 227 (2016) 137.
[71] S. Wang, S. Chen, Y. Wang, A. Low, Q. Lu, R. Qiu, Integration of organohalide-respiring bacteria and nanoscale zero-valent iron (Bio-nZVI-RD): a perfect marriage for the remediation of organohalide pollutants? Biotechnol. Adv. 34 (2016) 1384–1395.
[72] Y. Han, C. Liu, J. Horita, W. Yan, Trichloroethene hydrodechlorination by Pd-Fe bimetallic nanoparticles: solute-induced catalyst deactivation analyzed by carbon isotope fractionation, Appl. Catal. B Environ. 188 (2016) 77–86.
[73] I.F. Cheng, Q. Fernando, N. Korte, Electrochemical dechlorination of 4-chlorophenol to phenol, Environ. Sci. Technol. 31 (1997) 1074–1078.
[74] C. Grittini, M. Malcomson, Q. Fernando, N. Korte, Rapid dechlorination of polychlorinated biphenyls on the surface of a Pd/Fe bimetallic system, Environ. Sci. Technol. 29 (1995) 2898–2900.
[75] X. Zha, L. Ma, Y. Liu, Reductive dehalogenation of brominated disinfection byproducts by iron based bimetallic systems, RSC Adv. 6 (2016) 16323–16330.
[76] C. Durante, B. Huang, A.A. Isse, A. Gennaro, Electrocatalytic dechlorination of volatile organic compounds at copper cathode. Part II: polychloroethanes, Appl. Catal., B 126 (2012) 355–362.

[77] V. Jouikov, P. Poizot, J. Simonet, Reactivity of long chain α, ω-dibromoalkanes at palladized solid cathodes: homocoupling vs oligomerization, J. Electrochem. Soc. 156 (2009) E171–E178.
[78] C. Durante, V. Perazzolo, A.A. Isse, M. Favaro, G. Granozzi, A. Gennaro, Electrochemical activation of carbon-halogen bonds: electrocatalysis at palladium-copper nanoparticles, ChemElectroChem 1 (2014) 1370–1381.
[79] C. Durante, V. Perazzolo, L. Perini, M. Favaro, G. Granozzi, A. Gennaro, Electrochemical activation of carbon–halogen bonds: electrocatalysis at silver/copper nanoparticles, Appl. Catal. B Environ. 158–159 (2014) 286–295.
[80] K.S. Chan, C.R. Liu, K.L. Wong, Cobalt porphyrin catalyzed hydrodehalogenation of aryl bromides with KOH, Tetrahedron Lett. 56 (2015) 2728–2731.
[81] B. Cirera, J. Björk, R. Otero, J.M. Gallego, R. Miranda, D. Ecija, Efficient lanthanide catalyzed debromination and oligomeric length-controlled Ullmann coupling of aryl halides, J. Phys. Chem. C 121 (2017) 8033–8041.
[82] M.E. Cucullu, S.P. Nolan, T.R. Belderrain, R.H. Grubbs, Catalytic dehalogenation of aryl chlorides mediated by ruthenium(II) phosphine complexes, Organometallics 18 (1999) 1299–1304.
[83] W.M. Czaplik, S. Grupe, M. Mayer, A.J. von Wangelin, Practical iron-catalyzed dehalogenation of aryl halides, Chem. Commun. 46 (2010) 6350–6352.
[84] M. Fabien, S. Raphael, F. Yves, I.C. Sandra, T. Olivier, Synergistic effect in bimetallic Ni-Al clusters. Application to efficient catalytic reductive dehalogenation of polychlorinated arenes, Tetrahedron 56 (2000) 4765–4768.
[85] Y. Huang, S. Liu, Z. Lin, W. Li, X. Li, R. Cao, Facile synthesis of palladium nanoparticles encapsulated in amine-functionalized mesoporous metal–organic frameworks and catalytic for dehalogenation of aryl chlorides, J. Catal. 292 (2012) 111–117.
[86] G. Lázaro, V. Polo, F.J. Fernández-Alvarez, P. García-Orduña, F.J. Lahoz, M. Iglesias, L.A. Oro, Catalytic hydrodechlorination of benzyl chloride promoted by Rh-N -heterocyclic carbene catalysts, ChemSusChem 8 (2015) 495–503.
[87] A. Pyo, S. Kim, M.R. Kumar, A. Byeun, M.S. Eom, M.S. Han, S. Lee, Palladium-catalyzed hydrodehalogenation of aryl halides using paraformaldehyde as the hydride source: high-throughput screening by paper-based colorimetric iodide sensor, Tetrahedron Lett. 54 (2013) 5207–5210.
[88] S. Sabater, J.A. Mata, E. Peris, Hydrodefluorination of carbon–fluorine bonds by the synergistic action of a ruthenium–palladium catalyst, Nat. Commun. 4 (2013).
[89] C.J. Teskey, A.Y.W. Lui, M.F. Greaney, Ruthenium-catalyzed meta-selective C—H bromination, Angew. Chem. Int. Ed. 54 (2015) 11677–11680.
[90] L. Voskressensky, N. Golantsov, A. Maharramov, Recent advances in bromination of aromatic and heteroaromatic compounds, Synthesis 48 (2016) 615–643.
[91] Q. Yu, L. Hu, S. Zheng, J. Huang, Directed meta-selective bromination of arenes with ruthenium catalysts, Angew. Chem. Int. Ed. 54 (2015) 15284–15288.
[92] T. You, Z. Wang, J. Chen, Y. Xia, Transfer hydro-dehalogenation of organic halides catalyzed by ruthenium(II) complex, J. Organomet. Chem. 82 (2017) 1340–1346.
[93] E. Verlato, W. He, A. Amrane, S. Barison, D. Floner, F. Fourcade, F. Geneste, M. Musiani, R. Seraglia, Preparation of silver-modified nickel foams by galvanic displacement and their use as cathodes for the reductive dechlorination of herbicides, ChemElectroChem 3 (2016) 2084–2092.
[94] E. Verlato, S. Cattarin, N. Comisso, L. Mattarozzi, L. Vázquez-Gómez, Electrochemical impedance spectroscopy study of the preparation of electrocatalysts through galvanic displacement reactions, J. Electroanal. Chem. 737 (2015) 100–107.
[95] J.M. Fontmorin, W.Y. He, D. Floner, F. Fourcade, A. Amrane, F. Geneste, Reductive dehalogenation of 1,3-dichloropropane by a [Ni(tetramethylcyclam)]Br2-Nafion® modified electrode, Electrochim. Acta 137 (2014) 511–517.

[96] O. Scialdone, A. Galia, S. Sabatino, D. Mira, C. Amatore, Electrochemical conversion of dichloroacetic acid to chloroacetic acid in a microfluidic stack and in a series of microfluidic reactors, ChemElectroChem 2 (2015) 684–690.
[97] E. Delli, G. Kyriacou, C. Lambrou, Electrochemical reduction of CCl2F2 on Nafion solid polymer electrolyte composite electrodes, Chem. Commun. 2 (1998) 1693–1694.
[98] Z. Liu, R.G. Arnold, E.A. Betterton, E. Smotkin, Reductive dehalogenation of gas-phase chlorinated solvents using a modified fuel cell, Environ. Sci. Technol. 35 (2001) 4320–4426.

Further Reading

[1] S. Rondinini, A. Vertova, in: C. Comninellis, G. Chen (Eds.), Electrochemistry for the Environment, Springer New York, New York, NY, 2010.

CHAPTER

2

Electrochemical Conversion of CO_2 to Value-Added Products

Angel Irabien, Manuel Alvarez-Guerra, Jonathan Albo, Antonio Dominguez-Ramos

University of Cantabria, Santander, Spain

OUTLINE

Electrochemical Water and Wastewater Treatment

https://doi.org/10.1016/B978-0-12-813160-2.00002-X

INTRODUCTION

Carbon dioxide (CO_2) is nontoxic, nonflammable, and available in high amounts. Undoubtedly, these are interesting properties to develop an industrial chemistry based off this feedstock for making commodity chemicals, fuels, polymers, and specialties. However, there are two main drawbacks: first, CO_2 is very stable, so it takes extra effort to activate the molecule to react; and, second, the huge amount of unwanted greenhouse gas escaping into the atmosphere demands high volume products in order to reach a significant impact.

Global carbon emissions from fossil fuels and industry have increased every decade from an average of $3.1 \pm 0.2\,Gt\,year^{-1}$ C in the 1960s to an average of $9.3 \pm 0.5\,Gt\,year^{-1}$ C during the period from 2006 to 2015, thus reaching 34.1 Gt CO_2 [1]. Fossil fuel emissions (including cement production) accounted for about 91% of total CO_2 emissions from human sources in 2014. This portion of emissions originates from several sources such as coal (42%), oil (33%), gas (19%), cement (6%), and gas flaring (1%) [2].

To solve the aforementioned problems, numerous studies have focused on energy- and environmental-related topics. During the past several decades, fossil fuels have been partly replaced by clean and renewable energy sources (e.g., wind, tide, and solar) to mitigate CO_2 emissions. Despite the growth of renewable energy at increasing rates, the percentage of these renewable sources is still very low (<5%) in today's overall energy consumption. In addition, most of these renewable sources are geographical, seasonal, and intermittent. Consequently, the full use of renewable energy is often concerned with energy conversion and storage technologies. Specifically, energy conversion and storage devices such as supercapacitors and rechargeable batteries (e.g., lithium-ion batteries) are supposed to level the electricity output. However, considering their low energy densities and high costs, the large-scale integration of these energy-storage devices into the grid can hardly be realized.

Apart from the development of renewable energy and related technologies, carbon capture and sequestration (CCS) has also been adopted to prevent the release of large quantities of CO_2 into the atmosphere. Nevertheless, CCS technology is energy-consuming and expensive. Moreover, the risky leakage of stored CO_2 is a major concern with CCS, preventing large-scale CCS deployment from being commercialized. Accordingly, carbon capture and utilization (CCU) is still a great challenge for human beings worldwide.

In fact, an ideal solution is to convert atmospheric CO_2 into small organic molecules with improved energy density, such as carbon monoxide (CO), formic acid (HCOOH), methanol (CH_3OH), methane (CH_4), etc., using renewable energy. Such a strategy, apart from reducing the accumulation of CO_2 in the atmosphere, also allows producing fuels and useful industrial chemicals, thus relieving our dependency on conventional fossil resources. To this end, various CO_2 reduction approaches, including electrochemical, photochemical, biochemical, and thermochemical methods, have been proposed and intensively studied in the past decades. Among these methods, reducing CO_2 with renewable electricity is particularly appealing due to its moderate efficiency, controllable selectivity, simple reaction units, and enormous potential for real industrial applications. Moreover, CO_2 electroreduction can also be regarded as a convenient way to store the renewable energies above in chemical forms. In order for CO_2 to be electrochemically reduced to industrially viable products, four requirements have to be considered: (i) the overpotential for CO_2 reduction should be low enough to be energy efficient; (ii) the Faradaic efficiency (FE) toward the target product (product selectivity) should be quite high in order to save energy and to avoid costly separation processes; (iii) the rate of CO_2 reduction (determined as current) should be high enough to be economically viable; and (iv) the electrochemical cell should be scaled up to match other industrial processes.

As previously mentioned, CO_2 is a very chemically stable molecule, indeed one of the most stable found in nature, so its reactivity tends to be very unfavorable. Table 1 lists the equilibrium potentials for CO_2 reduction reactions at neutral pH [3,4]. As can be observed, its equilibrium potential for reduction is not very different than the equilibrium potential for H_2 evolution from H_2O, which is established at −0.31 V vs a standard hydrogen electrode (SHE).

TABLE 1 Equilibrium Electrode Potentials Against the Standard Hydrogen Electrode (SHE) for CO_2 Reduction to Different Products

CO_2 reduction reaction, pH 7	**Equilibrium potential/V vs SHE**
$CO_2 + H_2O + 2e^- \leftrightharpoons CO + 2OH^-$	−0.52
$CO_2 + H_2O + 2e^- \leftrightharpoons HCOO^- + OH^-$	−0.43
$CO_2 + 5H_2O + 6e^- \leftrightharpoons CH_3OH + 6OH^-$	−0.39
$CO_2 + 8H_2O + 12e^- \leftrightharpoons C_2H_4 + 12OH^-$	−0.34
$CO_2 + 9H_2O + 12e^- \leftrightharpoons C_2H_5OH + 12OH^-$	−0.33
$CO_2 + 13H_2O + 18e^- \leftrightharpoons C_3H_7OH + 18OH^-$	−0.32
$CO_2 + 6H_2O + 8e^- \leftrightharpoons CH_4 + 8OH^-$	−0.25
Water reduction reaction, pH 7	**Equilibrium potential/V vs SHE**
$H_2O + 2e^- \leftrightharpoons H_2 + 2OH^-$	−0.31

Since the H_2 evolution is dominant in acidic media, and molecular CO_2 does not exist in alkaline media, most CO_2 reduction occurs in near-neutral media competing with the H_2 evolution from an equilibrium point of view.

Attending to these potentials, the production of formate, methanol, and other hydrocarbons by electroreduction then appears to be a promising way of carbon dioxide utilization. However, due to the previously mentioned low reactivity, it is clear that using renewable energy for the process requires the assessment of its environmental sustainability. Among the available tools to perform so, life cycle assessment (LCA) is one of the most popular approaches. ISO series 14040 describes rigorously the procedure to overcome the different interrelated stages of an LCA study [5,6]. In these studies, global warming is used as a metric of reference, as it is clear that net difference between inputs and outputs of CO_2 in the whole supply chain will play a key role. Therefore, LCA can be used to claim objectively the reduction of the CO_2 equivalent emissions from the production of a mass unit of the electro-reduced CH_3OH or HCOOH against its corresponding large-scale, coal/gas sourced equivalents. The application of LCA to the utilization of CO_2 must address some very specific points to avoid methodological mistakes [7].

The environmental rationality behind the electrochemical reduction of CO_2 to added-value products lays on the fact that renewable energy is available. In this sense, excess electricity, from photovoltaic solar or wind farms that otherwise would be curtailed, can be used to transfer electrons to the CO_2 molecule for further use as a chemical or fuel. In this way, CO_2 acts as an energy reservoir. Even if these products are later burned, thus returning CO_2 to the atmosphere, they can potentially substitute the corresponding production of coal/gas-based products. Consequently, as long as the electroreduction products present an overall better environmental profile, they can be seen as an overall better environmental alternative, as they avoid an environmentally worse product.

In this context, a low product concentration can hinder the potential benefits derived from the rationality of connecting electroreduction with photovoltaic solar or other low carbon energy sources. The steam demand for distillation of the water-diluted electroreduced products usually becomes the critical step, because they must be upgraded to commercial purity. The concentration of CH_3OH or HCOOH is currently at least one order of magnitude away from the level required for commercialization. LCA can therefore recommend that electroreduction developments must tackle different key issues such as long lifetime for the electrodes and to essentially reduce steam demand for purification processes. The last one can be accomplished either by increasing selectivity, thus concentration of the desired product, or by the use of alternative nonaqueous electrolytes such as ionic liquids [8].

State of the Art

Chemical stability is undoubtedly the most critical issue regarding the reduction of the CO_2 molecule. Using electrochemistry for its reduction requires the application of larger negative potentials than those shown in Table 1, close to the water reduction equilibrium potential. There is a consensus regarding the fact that the electroreduction of CO_2 implies the formation of intermediate species, via single electron transfer:

$$CO_2 + e^- \leftrightarrows CO_2^{\bullet -}$$

Electrochemical reduction of CO_2 to CO is the first step in the synthesis of more complex carbon-based fuels and feedstocks using renewable electricity [3,4,9–12]. Not only certain metals, but also transition-metal oxides, transition-metal chalcogenides, and materials whose inner structures are based on carbon are the most widely recognized heterogeneous inorganic solid catalysts when classified by chemical group. The wide-industrial application of CO_2 electroreduction is still hindered even if noticeable progress and advances have been made for this process in the last decades [13].

The mechanism behind the CO_2 electroreduction using electrodes with catalytic properties is usually considered to involve four individual steps, including the formation of charged intermediate species: (i) CO_2 adsorption on the active sites of the electrode; (ii) CO_2 activation, leading to charged intermediates such as $CO_2^{\bullet -}$ or $HCO_2^{\bullet -}$; (iii) C—O bond dissociation involving protons and single or multiple transfer of electrons involving protons; and (iv) desorption of electroreduced products from the active sites of the electrode [11].

As previously mentioned, this second step corresponding to the chemical activation of CO_2 is clearly a main barrier in the development of CO_2 electroreduction due to: (i) the relatively high overpotential requested, and (ii) the competition of this activation with other parallel parasitic reactions such as the hydrogen evolution reaction (HER). Consequently, if electrons are derived for other reduction reactions rather than activation, the specific energy consumption for the CO_2 electroreduced processes and their selectivity decreases, losing the attractive features of producing chemicals and fuels so far described. Developments in this field need to be focused on the bottleneck represented by the activation reaction [14].

A number of high-quality review articles covering homogeneous and heterogeneous catalysis and membrane reactors for CO_2 electrochemical reduction have been published [13,15,16]. Apart from traditional metal electrocatalysts, which have been treated as typical electrodes for CO_2 electroreduction, a variety of new inorganic heterogeneous electrocatalysts have been reported in recent years with modest electrochemical results, in virtue of the rapid development of advanced nanotechnology and the progress in computational methods.

Taking into account that the CO_2 molecule has a high chemical stability and is completely oxidized, the search for efficient and robust electrocatalysts is crucial to promote this kinetically slow reduction reaction, as previously mentioned. The chosen electrocatalyst does determine the actual performance of the CO_2 electroreduction: different properties will provide different specific energy consumptions and selectivities. Electroreduction of the CO_2 molecule can be carried out by means of heterogeneous or homogeneous reactions. Therefore, both homogeneous and heterogeneous catalysts can be used as electrocatalysts for CO_2 reduction.

Organics or metal-organic molecules posing unique active sites are selected as homogeneous CO_2 electrocatalysts. This very unique feature provides extraordinary selectivity values and therefore explains the large momentum toward its investigation since many decades ago. However, even if the capacity to bind the CO_2 molecule is of great interest, relevant problems of homogeneous catalysts can be identified as well, mainly related to energy (difficult subsequent separation process), economy (high cost per unit of mass of catalyst), and environment (ecotoxicity issues), which hamper their practical application. More recently, inorganic heterogeneous electrocatalysts have gathered notable interest in the scientific community due to their easier scale-up, better efficiency, minimum environmental interaction, and production [13].

The key parameters typically used for the evaluation and comparison of the performance of CO_2 electroreduction are no different from other electrochemical processes: (i) cathode reduction potential, (ii) FE, related with process selectivity, (iii) energetic efficiency (EE), (iv) current density, and (v) Tafel plot. The well-known Tafel plot is used to connect the overpotential with the logarithm of the current density, which is very useful in evaluating the performance of electrocatalysts. In general, a better catalytic performance is revealed by a smaller Tafel slope. Particularly for CO_2 electroreduction, the Tafel slope provides insight regarding the reaction mechanism. A value of $59\,mV\,dec^{-1}$ for the Tafel slope is indicative of a single electron equilibrium step before a later rate-limiting chemical step. Twice that value ($118\,mV\,dec^{-1}$) is representative of the fact that the rate-determining step is the activation of the CO_2 molecule, as previously shown; thus, the charged intermediate $CO_2^{\bullet -}$ is the product of the single electron transfer step [17].

Many metal species such as Cu, Ag, Au, and Sn have been used extensively as catalysts for CO_2 electroreduction. As the electroreduction demands the charged intermediate $CO_2^{\bullet -}$ to participate, the developments are mainly oriented to new electrodes that accomplish a better stabilization of that charged molecule. The different tendency of metal electrodes to temporarily bind intermediates and products to the electrocatalyst surface is typically used to classify them into three classes. The first class groups the metals for which the most stable bonds are found for the intermediate to the surface. The most representative metal of this class is Cu. Due to the stability of the bond of the charged intermediate $CO_2^{\bullet -}$ to the electrode surface, very reduced products can be obtained such as alcohols (e.g., CH_3OH) and hydrocarbons (e.g., CH_4). The second class is the set of metals that create a weaker bond than in Cu, a bond which is not enough to promote very reduced products and therefore yields CO as the main product. Noble metals such as Pd, Au, and Ag, and others such as Zn belong to the second class. Finally, the third class is the set of metals that has $HCOO^-$ as the main product due to the weak bind of the intermediate to the catalyst surface. This class includes metals such as Sn, In, Pb, and Hg.

In general, the major products generated by these three classes of catalysts are higher value-added products such as hydrocarbons or alcohols, CO, and, $HCOO^-$ (or HCOOH depending on the acidity). As mentioned above, the competitive process due to the HER should not be forgotten in the case of CO_2 electroreduction in aqueous solutions. Since the partially reduced intermediate $CO_2^{\bullet -}$ is adsorbed strongly enough to exclude further reduction of CO_2, in the first class of metals (Cu) the HER almost exclusively tends to occur on metals from the two other classes, such as Pt, Ti, Fe, and Ni. Thus, tuning the binding energies of the key reaction intermediates is critical for the selectivity of the desired products and the specific energy consumption of CO_2 electroreduction.

The improvement of the previous metal electrodes for the aforementioned CO_2 electroreduction has been attempted by using combinations of metal oxides and metal electrodes. For the third class of metals corresponding to weak binding of the intermediate $CO_2^{\bullet -}$, the electrolytical deposition of a SnO_x layer over a Sn electrode showed improvement of four times in the FE for CO_2 reduction [18]. As discussed later, gas-diffusion electrodes (GDEs) have been tested using a similar strategy with commercial Cu_2O and Cu_2O-ZnO mixtures deposited onto carbon papers and evaluated for the continuous CO_2 gas phase electroreduction in a filter-press electrochemical cell. The process mainly produces methanol, as well as small quantities of ethanol and *n*-propanol [19].

Transition-metal chalcogenides (TMCs) are chemical species that received their name due to the combination of a chalcogen anion, which can be at different oxidation states, and a transition metal. This kind of combination has gathered noticeable interest, since different structures can be observed when the atomic ratio transition-metal to chalcogen anion goes from 1:1, as in ZnS, to 1:4 in VS_4. Even if different oxidation states can be of interest, the application of these TMCs for CO_2 electroreduction has only been reported for MoS_2. Additional research work directed toward other atomic ratio 1:2 TMCs, such as $MoSe_2$ and WS_2, would be desirable in order to find new catalysts based on available raw materials.

In addition to metals and TMCs as electrocatalysts for CO_2 reduction, renewable metal-free catalysts, especially nanocarbon-based catalysts, have become increasingly attractive because of their distinct properties over conventional CO_2 electrocatalysts. One of the most prominent features of these catalysts is the ability of carbon atoms to assemble into a variety of nanocarbon materials with different dimensions and structures; for example, one-dimensional carbon nanotubes (CNTs) and carbon nanofibers (CNFs), and two-dimensional graphene. With respect to CNTs, $HCOO^-$ has been produced by using polyethylenimine-functionalized N-doped graphenated carbon nanotubes, thanks to the three-dimensional hierarchical structure [20]. Regarding CNFs, the reduction of CO_2 to CO is possible when polyacrylonitrile-based heteroatomic CNFs are used [21]. Not only CNTs and CNFs from carbon-based materials can be used. Boron doped diamond (BDD) can also be utilized to produce HCHO thank to the sp^3 orbitals of the carbon atoms in the diamond structure. These orbitals seem to have a more pronounced effect in the electroreduction process than the sp^2 orbitals. Another great advantage of using BBD as an electrocatalytic material is derived from its unique window for the HER, improving specific energy consumption and selectivity since undesired reactions are avoided [22]. Metal-organic frameworks (MOFs) have been also studied as electrodes, but results do not show higher efficiencies [23] than other electrocatalysts.

Focusing on formate/formic acid and methanol as the two main chemicals within all the plethora of substances that can be obtained from CO_2, reduction is abundantly justified in terms of their techno-economic potential as recently pointed out [24].

ELECTROCHEMICAL CO_2 REDUCTION TO FORMATE/FORMIC ACID

The conversion of CO_2 to formic acid HCOOH or formate $HCOO^-$ (depending on the pH value) can be achieved by an electrochemical pathway according to the following two-electron reduction:

$$CO_2 + H_2O + 2e^- \leftrightarrows HCOO^- + OH^-$$

where the equilibrium potential is reported to be −0.43 V against the SHE, as shown in Table 1. However, the actual reaction could be affected by large overpotentials, which are highly dependent on the particular system used for the electrochemical conversion (e.g., nature of catalyst, configuration of electrode, or reaction conditions) [25]. Different alternative pathways of $HCOO^-$ formation from the electroreduction of CO_2 and several mechanisms have been suggested and can be found in the literature [26]. The reduction of CO_2 to formic acid or formate is completed by a complementary oxidation at the anode, which is usually the oxygen evolution reaction.

Formic acid/formate are considered very important chemicals that can serve as a precursor for other value-added chemicals and feedstocks for fuels. The traditional industrial uses of these chemicals include silage preservation, additive in animal feeds, textile finishing, anti-icing agents, and as intermediates in the chemical and pharmaceutical industries. In 2013, the global demand for formic acid was 579 kt, of which 34% was attributed to animal feed, 32% to leather tanning, and 13% to textile dyeing [24]. But, above all, formic acid is one of the most promising materials for H_2 storage today [27], and formic acid/formate are recently attracting more attention as fuels in direct formic acid/formate fuel cells. It is widely accepted that H_2 as an energy vector could be a solution for storing and delivering energy from renewable sources, but the storage of H_2 in high volumetric and gravimetric capacities has been considered as one of the most critical practical issues that prevent the large-scale implementation of H_2 fuel cells [28]. Research on direct H_2 storage and transportation has not revealed economically usable possibilities, so attention is now focused on alternative chemical H_2 storage. Recently, formic acid has been suggested as one of the most promising, convenient, and safe compounds for chemical H_2 storage because it has a high volumetric ($53\,g\,L^{-1}$ H_2) and a moderate gravimetric (4.4 wt.%) H_2 storage capacity at room temperature and atmospheric pressure; at ambient conditions, formic acid stores 580 times more H_2 than the same volume of gas, and it can be easily decomposed into H_2 and CO_2 with the appropriate catalyst. However, currently formic acid is mainly produced by thermochemical processes based on the carbonylation of methanol or by the oxidation of hydrocarbons, which have negative environmental impacts and are relatively expensive production processes that mainly rely on fossil feedstocks.

Accounting for problems that still need to be solved (as discussed later), the CO_2 electroreduction to formic acid/formate is currently one of the most promising reactions for an industrial scale, and it has been considered to have the best chance for the practical development of technical and economically viable processes [29]. Recent gross-margin models that were developed to evaluate the techno-economic feasibility of producing different chemicals through CO_2 electroreduction also pointed out formic acid among the most economically viable products [30]. This section will briefly summarize relevant studies on the electrocatalytic conversion of CO_2 to formic acid/formate.

Applied Electrocatalytic Materials

The electroreduction of CO_2 to formic acid/formate has been mainly studied using metals of a different nature as electrocatalytic materials, such as Pb, In, Pd, and, particularly, Sn.

Several studies have been focused on the use of Pb as an electrocatalyst, such as [31–33]. Some references can be found reporting the performance of In [34–37]. Recently, Pd electrodes have shown promising results in allowing the CO_2 reduction to formate at lower overpotentials [38,39]. It is also noteworthy that Co-based catalysts in the form of partially oxidized atomic layers exhibit higher intrinsic activity and selectivity for formate production than the Co bulk material [40]. However, Sn-based catalysts are clearly the most studied materials for the selective CO_2 electroreduction to formic acid/formate, as can be checked in the studies discussed in the following sections.

Types of Electrochemical Cells

Different types of electrochemical reactors have been used in the CO_2 electroreduction to formic acid/formate. This section summarizes representative examples of most relevant types of cells.

Undivided Cells

There are some references in the literature where the CO_2 electroreduction to formic acid/formate has been studied in lab-scale undivided electrochemical cells. In this type of cells, both the cathode and the anode (and usually also a reference electrode, in a typical 3-electrode assembly) are in the same compartment (Fig. 1A). Most of the studies in undivided cells have been carried out to provide a fundamental understanding of the CO_2 electroreduction to formate, for example, [38,41].

Fixed-Bed Reactors

With the aim of extending the electrode surface as much as possible within a relatively small electrochemical cell volume, Köleli et al. [42] studied the reduction of CO_2 using Pb or Sn granules as a cathode in an undivided fixed-bed glass reactor (Fig. 1B), which they claimed to be the first time described in the literature. The maximum FE for HCOOH was 90%, although the observed current density was only $0.79\,mA\,cm^{-2}$, and after 120 min the value of FE fell to 30%, which was attributed to the consumption of the formic acid produced during electrolysis. Considering these results, Köleli and Balun [43] later studied the performance of a stainless steel fixed-bed reactor divided in anode and cathode compartments with a Nafion 417 membrane. Although increasing the applied CO_2 pressure resulted in increases of both the current density and the FE for formic acid, the maximum current density was still a low value of $1.5\,mA\,cm^{-2}$.

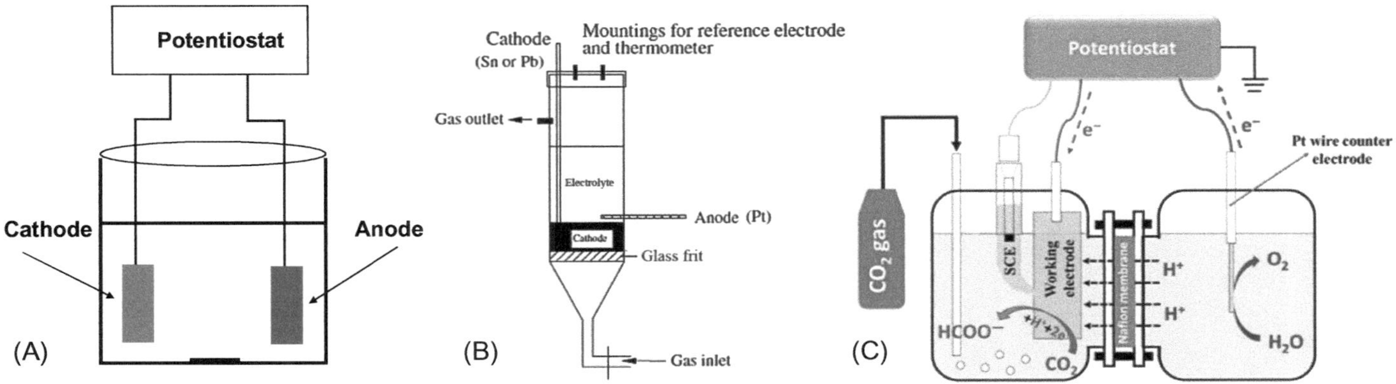

FIG. 1 Different types of electrochemical cells used in the CO_2 electroreduction to formic acid/formate: (A) undivided cell; (B) fixed-bed reactor; (C) H-type cell. *(B) Reproduced from F. Köleli, T. Atilan, N. Palamut, A.M. Gizir, R. Aydin, C.H. Hamann, Electrochemical reduction of CO_2 at Pb- and Sn-electrodes in a fixed-bed reactor in aqueous K_2CO_3 and $KHCO_3$ media, J. Appl. Electrochem. 33 (2003) 447–450. Copyright 2003. Springer. (C) From C. Zhao, J. Wang, Electrochemical reduction of CO_2 to formate in aqueous solution using electro-deposited Sn catalysts, Chem. Eng. J. 293 (2016) 161–170. Copyright 2016. Elsevier.*

H-Type and Two-Compartment Cells

An "H-type cell" is a divided electrochemical reactor whose name derives from the typical "H" form of the cell. These cells consist of a cathodic and an anodic compartment separated by a diaphragm, glass frit, or ion exchange membrane (Fig. 1C). A reactor with a two-compartment configuration prevents the oxidation in the anode of the desired products obtained from CO_2 (in this case, formic acid/formate) in the cathode (e.g., [44,45]).

Narayanan et al. [35] studied the performance of a two-compartment cell with a membrane-electrode assembly (MEA) configuration, using In as a powder metal catalyst, which was deposited on an alkaline polymer electrolyte membrane. They performed experiments at a constant current density of 40 mA cm^{-2}, but low FEs of formate production were obtained. An interesting divided cell with a buffer layer of aqueous electrolyte circulating between the cation exchange membrane and the cathode has been designed in the University of South Carolina (see e.g., [46,47]). Scialdone et al. [48] investigated the effect of many operating parameters at a Sn cathode in both batch divided and undivided cells. Interestingly, their experiments carried out in a wide range of CO_2 pressures showed that the increase of the pressure (up to 30 bar) resulted in an enhancement of the process performance. The utilization of a single divided glass cell with a Nafion membrane has also been proposed for the simultaneous anodic oxidation of organic pollutants in wastewater and the reduction of CO_2 to formic acid at a Sn cathode, with the aim of improving the overall economic figures [49].

Flow Reactors and Filter-Press Type Cells

Interesting studies focused on the electrochemical conversion of CO_2 to formic acid/formate have been carried out with flow or filter-press type reactors. These reactors are divided electrochemical cells in which the catholyte and the anolyte flow through their own compartments that are separated by an ion exchange membrane (usually a Nafion cation exchange membrane) placed between both chambers (Fig. 2A).

Microfluidic flow reactors: Some attempts using very small flows of electrolytes in microfluidic flow cells should be emphasized. Whipple et al. [50] used an electrochemical cell with microfluidic configuration (modification of a H_2 fuel cell), in which two GDEs were separated by a flowing 0.5 M KCl electrolyte stream. More recently, a dual electrolyte microfluidic reactor based on a laminar flow membrane-less architecture has been reported, which enhanced both reactivity and FE of the system [51].

Flow reactors in batch mode: Studies using flow cells at laboratory scale working in a batch mode have also been reported [32,52]. Innocent et al. [32] investigated the reduction of CO_2 to formate in a filter-press

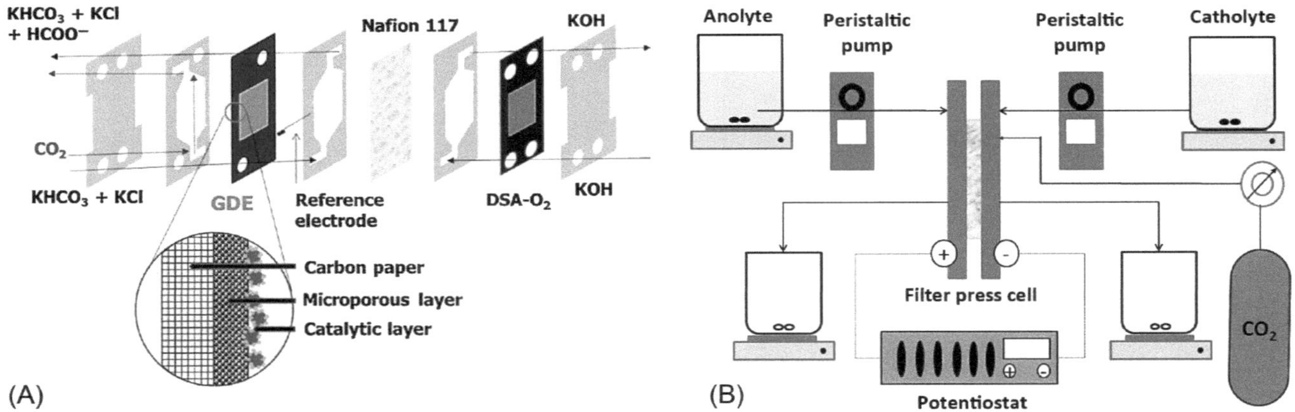

FIG. 2 (A) Filter-press electrochemical cell with cathodic and anodic compartment separated by a membrane, using a GDE as the cathode and (B) scheme for operation in continuous mode. *(A) Reproduced from A. Del Castillo, M. Alvarez-Guerra, J. Solla-Gullón, A. Sáez, V. Montiel, A. Irabien, Sn nanoparticles on gas diffusion electrodes: synthesis, characterization and use for continuous CO_2 electroreduction to formate, J. CO_2 Util. 18 (2017), 222–228. Copyright 2017. Elsevier. (B) Reproduced from A. Del Castillo, M. Alvarez-Guerra, A. Irabien, Continuous Electroreduction of CO_2 to Formate Using Sn Gas Diffusion Electrodes, AIChE J. 60 (2014) 3557–3564. Copyright 2014. John Wiley & Sons.*

reactor using a Pb plate cathode in noncontinuous operation of 4h. More recently, Irtem et al. [52] also carried out experiments in a filter-press type electrochemical cell, with a Sn catalyst electrodeposited on carbon fibers as the cathode, operating in batch mode by recirculation of the anolyte and catholyte (50mL) to the cell for 6h.

Flow reactors in continuous operation: However, continuous operation can be desirable, from an industrial point of view, for the further development of the types of processes for obtaining formic acid/formate at an industrial scale. Akahori et al. [34] are apparently the first to report continuous operation for the electroreduction of CO_2 to formic acid/formate in a cell using filter-press arrangement. With an In-impregnated Pb wire cathode, at ambient conditions, they obtained almost 100% in formate current efficiency at $2\,mA\,cm^{-2}$ and a flow rate of $1.4\,mL\,min^{-1}$. Then, a filter-press type flow reactor with a Pb-plated stainless steel-woven mesh cathode of a geometrical area of $45\,cm^2$ was designed and fabricated [31]. The optimum conditions were $3.2\,mL\,min^{-1}$ and $2.0\,mA\,cm^{-2}$, where a formate current efficiency of 93% and a product concentration of 1.5×10^{-2} $mol\,dm^{-3}$ were achieved. Although efficiencies for formate were very high in both [31,34], current densities were very small.

Li and Oloman (University of British Columbia, UBC, Canada) have performed extensive work on the development of a flow-by continuous electrochemical reactor for CO_2 conversion in formate at higher current densities and flow rates. In a preliminary investigation [53], they described the configuration of a laboratory bench-scale continuous reactor with a tinned-copper cathode ($45\,cm^2$), in which they carried out factorial and parametric experiments. Among the results reported, current efficiency for formate decreased with operating time, which was attributed to the progressive loss of tin from the cathode surface [53]. In a latter publication, part 1 of a two-part paper, Li and Oloman [54] extended their study using the same reactor configuration described previously with only minor changes. The best results reported in this work were a formate current efficiency of 86% and a formate concentration of 0.08M at a current density of $1.3\,kA\,m^{-2}$, obtained after 10min of operating time [54]. In part 2 of their study, Li and Oloman presented further experimental work employing two flow-by type reactors: reactor A (whose configuration and size, $45\,cm^2$, are essentially those described in their previous papers) and reactor B (tested as a seven-fold scaled-up version of reactor A). Although formate concentrations of up to 1M were obtained from a single pass of catholyte in B at $3.1\,kA\,m^{-2}$ with 63% current, notable problems, especially related to the degeneration of the tin cathode, were also reported [55]. Mantra Venture Group acquired the intellectual property from UBC in 2008 and is reported to be on a path toward commercialization [56], finishing the engineering work on a pilot plant project to produce $35\,t\,year^{-1}$ of formic acid [24].

Det Norske Veritas (DNV, United States) has developed since 2008 a continuous process with a flow cell electrochemical reactor using Sn as the base catalyst, referred to as ECFORM (electrochemical reduction of carbon dioxide to formate/formic acid). ECFORM was demonstrated on an experimental and semipilot scale (with a capability of around 1 kg of CO_2 converted per day) powered by solar panels in a mobile trailer [57]. Using this continuous flow cell previously reported by DNV [29], the performance of 3D Sn-carbon fiber paper electrodes prepared by a new electroplating process has also been tested, reporting current densities of 75 mA cm^{-2} and a FE of 78%, but at a relatively high cell potential of 3.75 V [58]. The degradation and long-term performance of Sn catalysts was investigated by using a rotating disk electrode in an aqueous KCl solution similar to the one used in the bench flow cell reactor [45].

Interesting experiments in a continuous mode of operation using a microstructured electrochemical flow-cell have been recently performed [59]. Moreover, approaches developed in the University of Cantabria (Spain) with flow reactors operating in continuous mode with only one pass of the catholyte through the cell (Fig. 2B) have also been reported [33,60–62]. These studies will be discussed in the next section, since the evolution in the improvement of the performance by using electrodes with different configurations is particularly noteworthy.

Types of Electrode Configurations

As already discussed, the nature of the metal used as an electrocatalyst in the cathode is a key aspect, but a certain metal can be used in electrodes with different forms, such as wires, plates, granules, powders, or nanoparticles deposited on porous supports. The configuration of the electrode also has a great influence on the performance of the CO_2 electroreduction process. An illustrative example can be found in the above-mentioned series of studies carried out in UC in which the performance of continuous processes for CO_2 electroreduction to formate in a flow reactor is progressively improved by using Sn in electrodes with different configurations [60–62].

Sn plate electrodes were initially used, achieving 70% FE but at current densities around 12 mA cm^{-2}, which resulted in a maximum formate rate of 4.4×10^{-4} mol m^{-2} s^{-1} and only 140 ppm of concentration [62]. Since this low performance was attributed to the small electrocatalytic surface of the plate electrode and the mass transport limitations, Sn particles were then deposited on carbon paper to manufacture particulate electrodes. In these electrodes, gaseous CO_2 is directly fed to the cell and flows through the electrode. Commercial Sn particles of 150 μm [62] gave 70% FE working at 40 mA cm^{-2}, with a formate concentration of 1.3 g L^{-1}. Reducing the particle size to 150 nm, it was possible to keep a 70% FE with a current density of 90 mA cm^{-2}, thus increasing the concentration to 1.5 g L^{-1} [61]. Further improvement was achieved using gas diffusion electrodes (GDEs), which consist of a carbonaceous support, a microporous layer, and a catalytic layer formed by nanoparticles of metal (see Fig. 2A). GDEs enhance the three-phase boundary area, avoiding mass transfer limitations, and allow working at higher current densities with high FEs. The use of GDEs with carbon-supported Sn nanoparticles outperformed the results obtained with particulate electrodes, since a 70% FE at a current density of 150 mA cm^{-2} was achieved, giving a formate concentration of 2.5 g L^{-1} [60]. By increasing the current density to 200 mA cm^{-2} and with a low catholyte flow, concentrations over 16 g L^{-1} were achieved, but this was at the expense of a 42% FE.

Several other recent studies also report encouraging results using nanocatalysts (especially Sn-based) in GDEs (e.g., [47,59,63]). For example, Kopljar et al. [59] have prepared GDEs using a dry deposition technique and loaded with carbon-supported SnO_2 nanoparticles, which allowed current densities of ~130 mA cm^{-2} with a FE toward formate of 80% to be achieved in a continuous mode of operation. Novel microstructured tin oxide catalyst-coated GDEs have been very recently reported to be very stable for at least 12 h of continuous operation, although maximum FE to formate is only 62% at a low current density in the order of 12 mA cm^{-2} [63].

Reaction Medium

The vast majority of studies of CO_2 electroreduction for formic acid/formate have been carried out using aqueous electrolytes, mainly solutions of carbonates, bicarbonates, or phosphates. However, some interesting attempts in other types of media have also been reported, especially related to the use of ionic liquids (ILs). ILs are a family of compounds (organic salts that are liquid below 100°C) with unique properties, which have allowed some remarkable improvements in CO_2 electroreduction [64]. Particularly, promising studies in which systems with ILs allow CO_2 electroreduction to formate at low overpotential and hence high efficiency have been published [65,66]. Very recently [66], the combination of the nanodendritic porous structure of a Cu-based material and the use of an IL (BMIM BF_4)/water-mixed electrolyte favors CO_2 reduction to formate over H_2 evolution at Cu electrodes, while more complex mixtures are obtained when using aqueous electrolytes, illustrating the importance of the reaction medium.

ELECTROCHEMICAL CO_2 REDUCTION TO METHANOL

Methanol can be considered as a green liquid fuel with a volumetric energy density (15.6 MJ L^{-1}) comparable to mostly used gasoline fuel (34.2 MJ L^{-1}), which make it an ideal alternative for combustion and transportation. Moreover, CH_3OH finds application as a primary feedstock for many organic compounds, as well as a building block for several bulk chemicals that are integral parts of our daily lives, such as plastics, paints, or adhesives.

At present, most of the commercial CH_3OH is synthesized from methane through syngas ($CO+H_2$) by steam reforming with optimized Cu/ZnO/Al_2O_3 catalysts at high temperatures and pressures (up to 300°C and 100 bar). This conventional process, however, might be replaced with CO_2-based processes, which would then be feasible to produce CH_3OH and simultaneously save diminishing natural fossil fuel resources. The so-called "methanol economy" first proposed by George Olah (Nobel Prize in 1994), can no longer be considered an unrealistic dream [67]. In fact, CH_3OH for renewable fuel (blended with gasoline) to be used in cars is currently being produced from CO_2 emissions and geothermal power in the Carbon Recycling International's George Olah Renewable Methanol Plant in Svartsengi, Iceland (production of 5×10^6 L $year^{-1}$, recycling 5.5×10^3 t $year^{-1}$ CO_2). This plant uses electricity to make H_2, which is then converted into CH_3OH in a catalytic reaction with CO_2. Nevertheless, using H_2O as a source of protons and renewable energy as a source of electrons required to reduce CO_2 could be a more promising route to produce CH_3OH, which also has a high market potential. The overall reaction of CH_3OH formation is a combination of a reduction and oxidation reaction at the cathode and anode of the system, respectively. Thermodynamically, it is feasible to electrochemically reduce CO_2 to CH_3OH. The reduction potential of CO_2, however, is only 20 mV positive with respect to the potential for water reduction, which results in the consumption of energy in the HER as shown in Table 1. Consequently, the CO_2-to-CH_3OH is a very challenging reaction, and it seems there is still a long way to go before the practical application of this technology can be done, although recent literature on the topic offers a strong start-up for future improvements [68]. The reaction not only requires highly active, durable, and affordable electrocatalytic materials, but also needs efficient reactor designs to minimize mass transfer limitations, reoxidation of reaction products, and the presence of dark zones, as well as appropriate reaction mediums. This section covers several fundamental aspects and opportunities to make the electrocatalytic CO_2-to-CH_3OH technology feasible for real application in the near future.

Applied Electrocatalytic Materials

The electrochemical transformation of CO_2 to CH_3OH requires 6 electrons (Table 1), which makes it kinetically slower. Therefore, it is required to apply electrode materials able to reach high selectivities for CH_3OH formation (represented by FE) with also high production rates, avoiding the undesirable HER. The literature offers the two most probable reaction sequences in the electroreduction of CO_2 to CH_3OH, that is, through CO or HCOOH to formaldehyde intermediates [68]. The other main by-products that are expected to be formed with CH_3OH are CO, HCOOH, and CH_2O, together with CH_4, depending mainly on the electrocatalyst employed and the potential applied. Among the wide range of applied materials, this subsection summarizes the most active electrodes for the electrochemical CO_2 transformation to CH_3OH.

Cu-Based Electrodes

Most of the reported studies for the electrochemical reduction of CO_2 to CH_3OH focused on the use of Cu as an active material. The reason behind this is the ability of this metal to directly produce alcohols at relatively high current efficiencies, in comparison to other metals that tend to be selective toward CO (i.e., Pd, Au, Ag, Zn, or Ga), $HCOO^-$ (i.e., Pb, Sn, Hg, In, or Cd) or H_2 (i.e., Ni, Fe, Pt, or Ti), as reviewed by Qiao et al. in 2014 [11].

Cu and Cu oxides: In 1991, Frese et al. first reported the application of oxidized Cu surfaces for the electrochemical reduction of CO_2 to CH_3OH, obtaining extraordinary high FEs (>100%) at $33\,mA\,cm^{-2}$. These results inspired further studies in the use of oxide surfaces [69]. In fact, in 2009, Chang et al. [70] synthesized and applied Cu_2O-catalyzed carbon clothes, confirming CH_3OH as the predominant product and revealing the notable electrochemical stability and catalytic ability of this material for CO_2 reduction. In the same line, Le et al. [71] reported the transformation of CO_2 to CH_3OH at cuprous thin film electrodes ($r = 1.2 \times 10^{-4}\,mol\,m^{-2}\,s^{-1}$, FE = 38%), suggesting the critical role of Cu(I) species in electrode activity and selectivity. Recently, our group airbrushed Cu_2O and ZnO particles (at different catalytic loadings and weight ratios) onto porous carbon papers. The results denoted the enhanced selectivity of the electrodes containing ZnO, in contrast to the strong deactivation with time observed in Cu_2O-deposited carbon papers. Thus, Cu_2O-ZnO mixtures were recommended for the reaction due to their stable catalytic activity ($r = 3.17 \times 10^{-5}\,mol\,m^{-2}\,s^{-1}$, FE = 17.7%) at $-1.3\,V$ vs Ag/AgCl [72].

Cu alloys: In an attempt to enhance the electrochemical properties for CO_2 reduction to CH_3OH, the research community has also focused on alloying Cu with other metals. Initially, some authors tested Cu in combination with other metals (e.g., Ni, Sn, Pb, Zn, Ag, and Cd), demonstrating that Cu-based alloys yielded CH_3OH at less negative potentials in aqueous solutions [68]. Lately, in 2012, Schizodimou and Kyriacou [73], evaluated a $Cu_{88}Sn_6Pb_6$ alloy material. The best CO_2 to CH_3OH performance ($r = 1.8 \times 10^{-5}$ mol m^{-2} s^{-1}, FE = 36.3%) was obtained at −0.70 V versus SCE in an acid solution. Recently, Jia et al. [74] used nanostructured $Cu_{36.9}Au_{36.1}$ alloys for CO_2 conversion. The results demonstrated that this material catalyzed not only CO_2 reduction, but also CO reduction, which contributed to enhance the conversion of CO_2 to alcohols (FE = 15.9%). Other reports demonstrated that CH_3OH production was favorable on Cu_3Pd and Cu_3Pt surfaces, although still they showed high overpotential (~0.7 V) [75]. Alloying Cu with Au could also bring enhancements based on CO binding energy, OH binding energy, and H binding energy criteria [76].

Cu nanostructures and organic frameworks: In 2013, Andrews et al. [77] used Cu nanoclusters on single crystal ZnO electrodes in an aqueous electrolyte for the electroreduction of CO_2. It was found that the selectivity toward CH_3OH was improved by at least an order of magnitude using the nanostructured Cu, in comparison to a Cu electrode at −1.45 vs SCE, in agreement with earlier reports on the use of crystal perovskite structures [78]. A recent report used Cu-loaded carbon nanotubes (CNTs). The material was able to reduce CO_2 at a high current density in a potential range of 0 V vs SCE to −3 V vs SCE, with a maximum activity for CH_3OH formation (FE = 38.5%) for CNTs, including 20 wt.% of Cu [79]. Additionally, MOFs are particularly suitable for electrochemical reactions due to their combined large surface area, high porosity, and shape-selective character, as well as their high electronic conductivities. Our group recently dealt, for the first time, with the use of Cu-based MOFs as electrocatalysts for the reduction of CO_2. The results showed the favorable electrocatalytic reduction of CO_2 to CH_3OH and C_2H_6O at 10 mA cm^{-2} (FE = 15.9%), due to the high surface area, and accessibility and exposure of Cu catalytic centers. The electrodes also showed activity for up to 17 h [23].

Ruthenium (Ru) and Molybdenum (Mo)

The formation of CH_3OH was first observed on Mo and several types of Ru electrodes with a FE near 100% in some cases at low current densities [68]. Based on this, Bandi [80] used a number of conductive oxide mixtures (e.g., MoO_2, RuO_2, or TiO_2, among others). They observed a high current efficiency for CH_3OH formation at RuO_2/TiO_2-based electrodes at very low current density levels (<100 μA cm^{-2}), before the onset potential of HER. In 2014, Li et al. [81] demonstrated that a regular structure of the catalyst was beneficial for electron transfer and catalytic activity when using MoS_2-rods/TiO_2 nanotubes for CO_2 conversion to CH_3OH (FE = 42%, yielding a concentration value of 202.2 mg L^{-1}). Unfortunately, the performance of the system gradually decreased as the reaction proceeded. A recent study [82] showed that by using Mo-Bi bimetallic chalcogenides supported on carbon paper, a FE = 71.2% could be reached at 12.1 mA cm^{-2}. They suggested that the superior performance of the electrode resulted from the excellent synergistic effect of Mo and Bi for producing CH_3OH.

Iron (Fe)

Modest efficiencies for CH_3OH formation have been also obtained when using Fe/Co complexes of porphyrin coated at plate electrodes. The first tests probed the positive performance of an Everitt's salt, ES ($K_2Fe(II)[Fe(II)(CN)_6]$), supported onto a Pt plate for the formation of CH_3OH (FE = 80%) [68]. Oher studies confirmed that CO_2 can be reduced to CH_3OH in a continuous electrochemical process if Prussian blue (PB, $KFe(III)[Fe(II)(CN)_6]$) from ES is continuously reduced back to ES, with no degradation of the complex. Despite the fact that Fe-based metal complexes can accelerate the rate for CO_2 reduction, high overpotentials are still needed because of low electrochemical efficiencies. Other attempts applied FeS_2/NiS nanocomposites, resulting in an unprecedented overpotential of 280 mV and a FE to CH_3OH of 64%, with a stable current of 3.1 mA cm^{-2} over 4 h. The high selectivity toward CO_2 electroreduction to CH_3OH was attributed to the special ladder structure of the FeS_2/NiS nanocomposite that can effectively suppress the side HER [83].

Other Metals

Pt, Pd, Pb, and Hg have also been tested for the electroreduction of CO_2 to CH_3OH, although their performance was limited. The literature also showed that Ti on its own possessed no significant activity for CO_2 electroreduction, although it could be used as a substrate for other metals to achieve enhanced formations of CH_3OH. Other reports suggested that Ga and In- based electrodes might present potential, alone or in combination with other metals, for the reduction of CO_2 to CH_3OH [68].

Design of the Electroreduction Cell

Types of Electrochemical Cells

The literature shows that there is no standard electrolyzer or methodology to conduct the CO_2-to-CH_3OH electrochemical reaction.

Undivided cells: Traditionally, undivided three-electrode cells have been applied with different electrocatalyst materials and electrolyte composition. This simple configuration, however, facilitates the reoxidation of reduced species in the anode, and thus productivity and selectivity of the reaction tend to be low [68]. To solve this important limitation, two-compartment electrochemical cells have been widely used in the literature.

Two-compartment electrolyzers: In these divided systems, the cathode and anode compartments are separated by ion exchange membranes, where Nafion (a cation exchange membrane, CEM) is the most common material applied. The membrane-based divided cell configurations are similar to those described in the previous section for the electroreduction of CO_2 to $HCOO^-$. Other reports applied divided electrolyzers separated by a Na_2SO_4, agar bridge, or glass frit [68].

Membrane electrode assemblies (MEAs): In this case, the working and counter electrodes are uniquely separated by an ion exchange membrane, in a so-called electrode-membrane assembly. Although only a few studies reported their use for the electrochemical production of CH_3OH from CO_2 [3], with this cell configuration the contact and transport of species between the electrodes can be enhanced, overcoming important limitations of the process [16]. For example, a novel Cu-activated carbon/Sterion/IrO_2 MEA was developed for the gas phase electrocatalytic conversion of CO_2. The high CO_2 electrocatalytic activity of the material was explained as the high surface area available and the large dispersion of Cu particles, leading to CH_3OH and C_2H_4O as the main reduction products with less energy requirements [84].

Structure of the Electrode

Maximizing electrode performance requires a deep study and then optimization of all transport processes occurring on the electrode structure.

Bulk metals: The first reports on the electrochemical reduction of CO_2 to CH_3OH focused on the application of metallic plates or foils, although this electrode configuration was soon replaced by other structures. Frese analyzed Cu foil electrodes, namely anodized, thermally oxidized, and air-oxidized Cu foils, reaching high FEs in 0.5 M $KHCO_3$ at −1.9 V vs SCE. Subsequently, different planar electrodes (e.g., Cu, Zn, Pd, Ga, Ti, or Pt) were evaluated for the reaction at low current densities ($5\,mA\,cm^{-2}$) [68]. The performance of these bulk metals was in fact very limited mainly due to the low availability of active surfaces, which led to the consideration of other configurations.

Metals electrodeposited in metallic or carbon supports: This configuration comprises an active catalytic layer and a porous support. The latter serves first to deliver CO_2 from flow channels to the active sites, transport of reduced species from the catalytic sites, and conduction of electrons with low resistance. Le et al. [71] demonstrated the remarkably higher conversion rates and efficiencies obtained at electrodeposited Cu_2O thin film electrodes in comparison with anodized or air-oxidized Cu electrodes for the transformation of CO_2 to CH_3OH in a three-electrode cell. They also explored the relationship between surface chemistry and reaction behavior. Moreover, Chang et al. brush-painted carbon clothes with Cu_2O particles for their evaluation in a CO_2-saturated 0.5 M NaOH solution contained in a three-electrode undivided cell. Cyclic voltammetry analyses and potentiostatic measurements denoted the notable catalytic ability of the material for CH_3OH formation [70].

Metals supported in gas diffusion electrodes (GDEs): Recent research efforts seem to be concentrated on the use of GDEs, due to the high current densities that can be obtained under mild (200–600 mA cm^{-2}) or pressurized conditions (300–900 mA cm^{-2}) [68]. These porous composite electrodes are normally composed of polymer-bonded catalyst particles and a carbon support. Schwartz et al. first studied CuO-based perovskite typo crystal structures including La, Pr, Gd, Sr, and Th for the preparation of GDEs. The current efficiency of the process was up to 40% for CO_2 reduction to CH_3OH, together with other by-products (C_2H_6O and C_3H_8O) [78]. Our group recently was able to improve CO_2 conversion efficiencies to alcohols (CH_3OH, C_2H_6O, and C_3H_8O) by using Cu_2O/ZnO GDEs (FE = 31.4% at −1.16 V vs Ag/AgCl), overcoming mass transfer limitations in the cell (Fig. 3) [19].

The results for the use of GDEs are promising so far, but further research is required in order to optimize the composition of the catalytic layer (e.g., pore size and distribution, binder material, etc.), as well as the

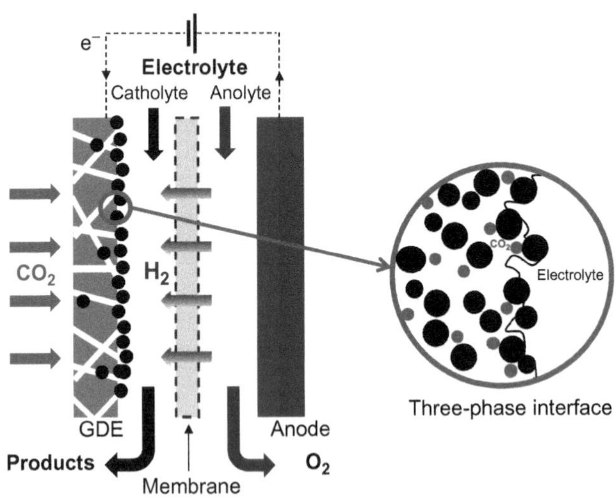

FIG. 3 Schematic diagram of the electrolyte cell configuration for the electroreduction of CO_2 supplied directly from gas phase. *Reproduced from J. Albo, A. Irabien, Cu_2O-loaded gas diffusion electrodes for the continuous electrochemical reduction of CO_2 to methanol, J. Catal. 343 (2016) 232–239. Copyright 2016. Elsevier.*

porous supporting material (e.g., thickness, porosity, etc.), that will impact the transport of CO_2 and the generated products.

Reaction Medium

Previous literature showed a great effect of different electrolytes and operation conditions on current, product selectivity, reaction rates, and CO_2 reduction energetic efficiency, even for the same metal and structure [68].

Aqueous Electrolytes

Commonly, CO_2 reduction experiments have been conducting in close to neutral solutions employing electrolytes based on alkali cations (e.g., K^+, Na^+) in combination with various anions (e.g., Cl^-, HCO_3^-, or OH^-). These inorganic salts possess high conductivity in water, providing the protons required for the reaction [68]. In 2012, the literature demonstrated that CO_2 reduction performance and product distribution could be controlled by simply varying the composition of the electrolyte and acidity of the solution. Additionally, it showed that the rate for CO_2 electroreduction increases with the increase of the surface charge of the cation of the supporting electrolyte in this order: $Na^+ < Mg^{2+} < Ca^{2+} < Ba^{2+} < Al^{3+} < Zr^{4+} < Nd^{3+} < La^{3+}$ [73]. Spataru et al. [85] showed before that a pH value c. 3.9 ensured the highest efficiency for the electroreduction of CO_2 to several products at RuO_2-coated diamond electrodes, although at pH 5.9 the selectivity and productivity for CH_3OH formation could be enhanced (FE = 8.12%, 31.31 ppm) at −0.5 V versus SHE. The results suggested that in a very acidic media, CO_2 reduction to CH_3OH is hindered by a strong hydrogen discharge, supporting the conclusion that a higher pH ensures favorable conditions for a CO_2-to-CH_3OH reduction reaction.

Pyridine (Py)-Based Molecular Catalysts

In order to help overcoming the overpotential limitations, organic molecules have been also applied. Specifically, nitrogen-containing heterocycles such as Py have been proven to lower reaction barriers (overpotentials) and increase the selectivity toward CO_2 to CH_3OH, as largely demonstrated by Bocarsly's group [86] where Py was found to act as an electron shuttle, implying the formation of a carbamate radical as intermediate during the electron transfer to CO_2 to CH_3OH. They also observed an optimum at pH 5, suggesting that both Py and proton accessibility (free or as pyridinium) are important for the overall process. Based on these inspiring results, further developments in the use of a Py-based molecular catalyst for CO_2 reduction to CH_3OH have been conducted. For example,

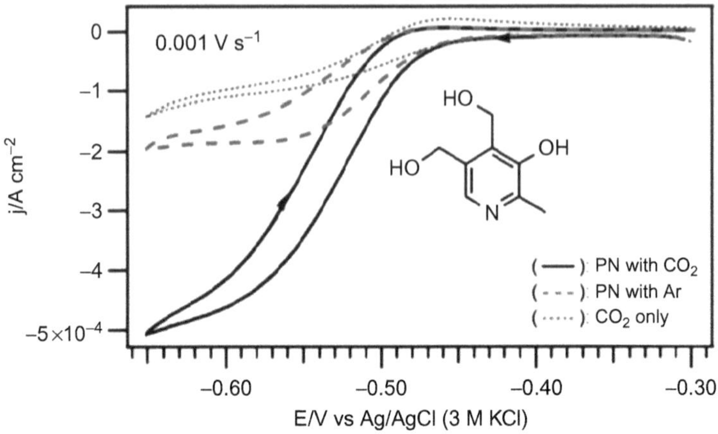

FIG. 4 Variable scan rate cyclic voltammograms obtained using a Pt electrode at pH 5 in the presence/absence of pyridoxine (PN). *Reproduced from J.H.Q. Lee, S.J.L. Lauw, R.D. Webster, The electrochemical reduction of carbon dioxide (CO_2) to methanol in the presence of pyridoxine (vitamin B6), Electrochem. Commun. 64 (2016) 69–73. Copyright 2016. Elsevier.*

Yang et al. [87] were able to entrap and stabilize the Py derivative within Cu-Pd and Pd (as heterogeneous support). A FE of 35% was obtained at only −0.6 V versus SCE in the case of [PYD]-Pd, while [PYD]-Cu/Pd composites led to a FE to CH_3OH of 26%. Recently, our group tested Py-based molecular catalysts containing electron-releasing or electron-withdrawing groups for CH_3OH synthesis at Cu_2O/ZnO-based surfaces at a low current density level (1 mA cm^{-2}). The reduction response occurred at around 200 mV lower overpotential in the presence of 2-methylpyridine, with a maximum performance at pH 5 ($r = 4.42$ μmol m^{-2} s^{-1} and FE = 25.6% [88]. Other studies focused on the potential utility of pyridoxine (vitamin B6 family) as an alternative reagent to a more toxic Py. The results at a Pt cathode showed a FE = 5% below 200 mV overpotential (Fig. 4) [89].

Ionic Liquids (ILs)

Replacing the conventional aqueous systems with ILs is advantageous for several reasons: (i) the HER can be suppressed; (ii) there is a higher solubility of CO_2, reducing mass transport limitations; (iii) there is a higher conductivity of the reaction medium; and (iv) the energy required for CO_2 activation can be reduced. There are, however, scarce analyses for the use of ILs in the CO_2-to-CH_3OH reaction. Carlesi et al. [90] synthesized and applied [H_2N-C_3mim][Br] IL, due to its good absorption capacity, high ionic conductivity, and high chemical-electrochemical stability. The IL was found to act as a CO_2-stabilizer, enabling the electrochemical reduction of absorbed CO_2 at a reduced potential. Xiaofu et al. recently reported [82] the performance of Mo-Bi electrodes in a [bmim]

[BF_4]-MeCN electrolyte at ambient conditions. They suggested that a complex [bmim-CO_2]$^+$ can form quickly, reducing the reaction barrier for electron transfer to CO_2. The results showed a FE for CH_3OH of 72.2%.

Solid Polymer Electrolytes (SPEs)

In order to suppress the HER, the application of SPE-based processes (and SPE in combination with GDEs) has been proposed in the literature [16]. In these systems, CO_2 is directly provided as a gas in the cathodic compartment and the SPE is responsible for the conduction of protons, separation of gaseous products, and the electrical insulation of the electrodes. This includes the use of CEMs and anionic exchange membranes (AEMs) that normally are easy to handle and present good tensile strength, increasing the electrode life. For the specific case of CH_3OH formation, Aeshala et al. [91] reported in 2012 the use of a Nafion/SPEEK/Amberlist SPE for CO_2 reduction to CH_3OH at modest current efficiencies with a Cu-coated carbon paper cathode. The material was characterized by thickness, ionic conductivity, thermal stability, and mechanical strength. The same group reported [92] the performance of electrodeposited Cu_2O/carbon papers using cationic and anionic SPEs. The main products formed were CH_3OH, CH_4, and C_2H_4O, with a cumulative FE of 45% at 5.4 mA cm^{-2}, concluding that the SPEs alleviated CO_2 transport limitations in the cell.

CONCLUSIONS

This chapter has provided a brief overview of the state of the art and challenges for the development of CO_2 electroreduction processes to obtain value-added products, with particular emphasis on two products of great interest: formic acid/formate and methanol.

In recent years, a rapidly growing number of research studies have allowed significant advances on the CO_2 electroreduction to formic acid/formate. A few approaches at a larger scale than laboratory plants have also been reported. However, there are still some challenges that have to be overcome for the practical application of this technology. Despite the encouraging results achieved, improvements are definitely necessary to combine high FE, high rates of formate production, and low overpotential in order to ensure a commercially viable process. The durability of the electrocatalysts is another important issue which has barely been studied. Moreover, a key challenge, which is usually neglected, is the concentration of product formic acid/formate obtained, still far from 85 wt.%, the most common concentration in the market. Purification of diluted solutions is energy intensive, and implies an economic penalty. Therefore, further research is required for improving the CO_2

electroreduction processes and enabling that the feasible production of formic acid/formate is concentrated enough to be marketed, ensuring at the same time that this approach is environmentally sustainable.

With respect to the electrocatalytic reduction of CO_2 to methanol, this appears to be an environmentally sustainable option that offers the possibility of closing the carbon loop, generating useful products. Despite all the efforts put into this reaction, the production of methanol from CO_2 and water seems to be still far from adequate, and conversion rates should be boosted by order of magnitudes to reach productions rates that can be practically used. In order to reach industrial-scale implementation, a breakthrough in catalyst materials for activating the molecule of CO_2 at lower overpotentials is definitively needed. These achievements will probably come with the development of new complex catalysts and catalysts assemblies, including hybrid metal catalysts and innovative nanostructures, along with the use of metal organic frameworks. It is important to focus on the precise tuning of electrode active areas and morphology, composition of multimetal catalysts, and porosity of the support layer, which facilitate the diffusion pathways and lead to faster kinetics. Low catalyst stability produced by blocking and poisoning of catalyst actives sites seems also to be one of the major limitations at present. Besides, the efforts in fundamental CO_2 reduction mechanistic studies are still insufficient, which in turn will guide the development of new cathode materials and the optimization of operating conditions. Moreover, a fine-tuning of electrolyte composition for a given catalyst offers an opportunity for great performance enhancements by reducing the overpotentials required and suppressing the formation of hydrogen. In this regard, the use of pyridine solutions or nonaqueous media, such as ionic liquids, present a great potential for conversion improvements.

References

[1] C. Le Quéré, R.M. Andrew, J.G. Canadell, S. Sitch, J.I. Korsbakken, G.P. Peters, A.C. Manning, T.A. Boden, P.P. Tans, R.A. Houghton, R.F. Keeling, S. Alin, O.D. Andrews, P. Anthoni, L. Barbero, L. Bopp, F. Chevallier, L.P. Chini, P. Ciais, K. Currie, C. Delire, S.C. Doney, P. Friedlingstein, T. Gkritzalis, I. Harris, J. Hauck, V. Haverd, M. Hoppema, K.K. Goldewijk, A.K. Jain, E. Kato, A. Körtzinger, P. Landschützer, N. Lefèvre, A. Lenton, S. Lienert, D. Lombardozzi, J.R. Melton, N. Metzl, F. Millero, P.M.S. Monteiro, D.R. Munro, J.E.M.S. Nabel, S. Nakaoka, K. O'Brien, A. Olsen, A.M. Omar, T. Ono, D. Pierrot, B. Poulter, C. Rödenbeck, J. Salisbury, U. Schuster, J. Schwinger, R. Séférian, I. Skjelvan, B.D. Stocker, A.J. Sutton, T. Takahashi, H. Tian, B. Tilbrook, I.T. van der Laan-Luijkx, G.R. van der Werf, N. Viovy, A.P. Walker, A.J. Wiltshire, Global carbon budget 2016, Earth Syst. Sci. Data 8 (2016) 605–649.

[2] The Intergovernmental Panel on Climate Change (IPCC), Climate Change 2014: Mitigation of Climate Change, 2014. Work. Gr. III Contrib. to Fifth Assess. Rep. Intergov. Panel Clim. Chang.
[3] Y. Hori, Electrochemical CO_2 reduction on metal electrodes, in: C.G. Vayenas, R.E. White, M.E. Gamboa-Aldeco (Eds.), Modern Aspects of Electrochemistry, Springer, New York, NY, 2008, pp. 89–189.
[4] C. Costentin, S. Drouet, M. Robert, J.-M. Savéant, A local proton source enhances CO_2 electroreduction to CO by a molecular Fe catalyst, Science 338 (2012) 90–94.
[5] ISO, ISO 14040:2006 Environmental Management — Life Cycle Assessment — Principles and Framework, 2006. Int. Organ. Stand.
[6] ISO, ISO 14044:2006-Environmental Management — Life Cycle Assessment — Requirements and Guidelines, 2006. Int. Organ. Stand.
[7] N. von der Assen, J. Jung, A. Bardow, Life-cycle assessment of carbon dioxide capture and utilization: avoiding the pitfalls, Energy Environ. Sci. 6 (2013) 2721–2734.
[8] A. Dominguez-Ramos, B. Singh, X. Zhang, E.G. Hertwich, A. Irabien, Global warming footprint of the electrochemical reduction of carbon dioxide to formate, J. Clean. Prod. 104 (2015) 148–155.
[9] S. Lin, C.S. Diercks, Y.-B. Zhang, N. Kornienko, E.M. Nichols, Y. Zhao, A.R. Paris, D. Kim, P. Yang, O.M. Yaghi, C.J. Chang, Covalent organic frameworks comprising cobalt porphyrins for catalytic CO_2 reduction in water, Science 349 (2015) 1208–1213.
[10] B.A. Rosen, A. Salehi-Khojin, M.R. Thorson, W. Zhu, D.T. Whipple, P.J. Kenis, R.I. Masel, Ionic liquid-mediated selective conversion of CO_2 to CO at low overpotentials, Science 334 (2011) 643–644.
[11] J. Qiao, Y. Liu, F. Hong, J. Zhang, A review of catalysts for the electroreduction of carbon dioxide to produce low-carbon fuels, Chem. Soc. Rev. 43 (2014) 631–675.
[12] M. Liu, Y. Pang, B. Zhang, P. De Luna, O. Voznyy, J. Xu, X. Zheng, C.T. Dinh, F. Fan, C. Cao, F.P.G. Arquer, T.S. Safaei, A. Mepham, A. Klinkova, E. Kumacheva, T. Filleter, D. Sinton, S.O. Kelley, E.H. Sargent, Enhanced electrocatalytic CO_2 reduction via field-induced reagent concentration, Nature 537 (2016) 382–386.
[13] D.D. Zhu, J.L. Liu, S.Z. Qiao, Recent advances in inorganic heterogeneous electrocatalysts for reduction of carbon dioxide, Adv. Mater. 28 (2016) 3423–3452.
[14] S. Gao, Z. Sun, W. Liu, X. Jiao, X. Zu, Q. Hu, T. Yao, W. Zhang, S. Wei, Y. Xie, Atomic layer confined vacancies for atomic-level insights into carbon dioxide electroreduction, Nat. Commun. 8 (2017) 1–9.
[15] Q. Lu, F. Jiao, Electrochemical CO_2 reduction: electrocatalyst, reaction mechanism, and process engineering, Nano Energy 29 (2016) 439–456.
[16] I. Merino-Garcia, E. Alvarez-Guerra, J. Albo, A. Irabien, Electrochemical membrane reactors for the utilisation of carbon dioxide, Chem. Eng. J. 305 (2016) 104–120.
[17] H. Wang, Y. Chen, X. Hou, C. Ma, T. Tan, Nitrogen-doped graphenes as efficient electrocatalysts for the selective reduction of carbon dioxide to formate in aqueous solution, Green Chem. 18 (2016) 3250–3256.
[18] Y. Chen, M.W. Kanan, Tin oxide dependence of the CO_2 reduction efficiency on tin electrodes and enhanced activity for tin/tin oxide thin-film catalysts, J. Am. Chem. Soc. 134 (2012) 1986–1989.
[19] J. Albo, A. Irabien, Cu_2O-loaded gas diffusion electrodes for the continuous electrochemical reduction of CO_2 to methanol, J. Catal. 343 (2016) 232–239.
[20] S. Zhang, P. Kang, S. Ubnoske, M.K. Brennaman, N. Song, R.L. House, J.T. Glass, T.J. Meyer, Polyethylenimine enhanced electrocatalytic reduction of CO_2 to formate at nitrogen doped carbon nanomaterials, J. Am. Chem. Soc. 136 (2014) 1–18.
[21] B. Kumar, M. Asadi, D. Pisasale, S. Sinha-Ray, B.A. Rosen, R. Haasch, J. Abiade, A.L. Yarin, A. Salehi-Khojin, Renewable and metal-free carbon nanofibre catalysts for carbon dioxide reduction, Nat. Commun. 4 (2013) 1–9.

[22] K. Nakata, T. Ozaki, C. Terashima, A. Fujishima, Y. Einaga, High-yield electrochemical production of formaldehyde from CO_2 and seawater, Angew. Chem. Int. Ed. 53 (2014) 871–874.
[23] J. Albo, D. Vallejo, G. Beobide, O. Castillo, P. Castaño, A. Irabien, Copper-based metal-organic porous materials for CO_2 electrocatalytic reduction to alcohols, ChemSusChem 9 (2016) 1–11.
[24] M. Pérez-Fortes, E. Tzimas, Techno-Economic and Environmental Assessment of CO_2 Utilisation for Fuel Production, Joint Research Center (JRC) Science for Policy Report, European Union, Luxembourg, 2016.
[25] X. Lu, D.Y.C. Leung, H. Wang, M.K.H. Leung, J. Xuan, Electrochemical reduction of carbon dioxide to formic acid, ChemElectroChem 1 (2014) 836–849.
[26] H. Huo, J. Zhang, D. Ren, Y. Liu, Electrochemical reduction of carbon dioxide, in: J. Qiao, Y. Liu, J. Zhang (Eds.), Fundamentals and Technologies, CRC Press, Taylor & Francis Group, Boca Raton, FL, 2016, pp. 47–102.
[27] D. Mellmann, P. Sponholz, H. Junge, M. Beller, Formic acid as a hydrogen storage material – development of homogeneous catalysts for selective hydrogen release, Chem. Soc. Rev. 45 (2016) 3954–3988.
[28] J.S. Yoo, R. Christensen, T. Vegge, J.K. Nørskov, F. Studt, Theoretical insight into the trends that guide the electrochemical reduction of carbon dioxide to formic acid, ChemSusChem 9 (2016) 358–363.
[29] A.S. Agarwal, Y. Zhai, D. Hill, N. Sridhar, The electrochemical reduction of carbon dioxide to formate/formic acid: engineering and economic feasibility, ChemSusChem 4 (2011) 1301–1310.
[30] S. Verma, B. Kim, H.R.M. Jhong, S. Ma, P.J.A. Kenis, A gross-margin model for defining technoeconomic benchmarks in the electroreduction of CO_2, ChemSusChem 9 (2016) 1–9.
[31] K. Subramanian, K. Asokan, D. Jeevarathinam, M. Chandrasekaran, Electrochemical membrane reactor for the reduction of carbondioxide to formate, J. Appl. Electrochem. 37 (2007) 255–260.
[32] B. Innocent, D. Liaigre, D. Pasquier, F. Ropital, J.M. Léger, K.B. Kokoh, Electro-reduction of carbon dioxide to formate on lead electrode in aqueous medium, J. Appl. Electrochem. 39 (2009) 227–232.
[33] M. Alvarez-Guerra, S. Quintanilla, A. Irabien, Conversion of carbon dioxide into formate using a continuous electrochemical reduction process in a lead cathode, Chem. Eng. J. 207–208 (2012) 278–284.
[34] Y. Akahori, N. Iwanaga, Y. Kato, O. Hamamoto, M. Ishii, New electrochemical process for CO_2 reduction to from formic acid from combustion flue gases, Electrochemistry 72 (2004) 266–270.
[35] S.R. Narayanan, B. Haines, J. Soler, T.I. Valdez, Electrochemical conversion of carbon dioxide to formate in alkaline polymer electrolyte membrane cells, J. Electrochem. Soc. 158 (2011) A167–A173.
[36] Z.M. Detweiler, J.L. White, S.L. Bernasek, A.B. Bocarsly, Anodized indium metal electrodes for enhanced carbon dioxide reduction in aqueous electrolyte, Langmuir 30 (2014) 7593–7600.
[37] Z. Bitar, A. Fecant, E. Trela-Baudot, S. Chardon-Noblat, D. Pasquier, Electrocatalytic reduction of carbon dioxide on indium coated gas diffusion electrodes – comparison with indium foil, Appl. Catal. B Environ. 189 (2016) 172–180.
[38] R. Kortlever, C. Balemans, Y. Kwon, M.T.M. Koper, Electrochemical CO_2 reduction to formic acid on a Pd-based formic acid oxidation catalyst, Catal. Today 244 (2015) 58–62.
[39] X. Min, M.W. Kanan, Pd-catalyzed electrohydrogenation of carbon dioxide to formate: high mass activity at low overpotential and identification of the deactivation pathway, J. Am. Chem. Soc. 137 (2015) 4701–4708.

[40] S. Gao, Y. Lin, X. Jiao, Y. Sun, Q. Luo, W. Zhang, D. Li, J. Yang, Y. Xie, Partially oxidized atomic cobalt layers for carbon dioxide electroreduction to liquid fuel, Nature 529 (2016) 68–71.
[41] J. Wu, F.G. Risalvato, F.-S. Ke, P.J. Pellechia, X.-D. Zhou, Electrochemical reduction of carbon dioxide I. Effects of the electrolyte on the selectivity and activity with Sn electrode, J. Electrochem. Soc. 159 (2012) F353–F359.
[42] F. Köleli, T. Atilan, N. Palamut, A.M. Gizir, R. Aydin, C.H. Hamann, Electrochemical reduction of CO_2 at Pb- and Sn-electrodes in a fixed-bed reactor in aqueous K_2CO_3 and $KHCO_3$ media, J. Appl. Electrochem. 33 (2003) 447–450.
[43] F. Köleli, D. Balun, Reduction of CO_2 under high pressure and high temperature on Pb-granule electrodes in a fixed-bed reactor in aqueous medium, Appl. Catal. A Gen. 274 (2004) 237–242.
[44] C. Zhao, J. Wang, Electrochemical reduction of CO_2 to formate in aqueous solution using electro-deposited Sn catalysts, Chem. Eng. J. 293 (2016) 161–170.
[45] Anawati, G.S. Frankel, A. Agarwal, N. Sridhar, Degradation and deactivation of Sn catalyst used for CO_2 reduction as function of overpotential, Electrochim. Acta 133 (2014) 188–196.
[46] J. Wu, F.G. Risalvato, P.P. Sharma, P.J. Pellechia, F.-S. Ke, X.-D. Zhou, Electrochemical reduction of carbon dioxide: II. Design, assembly, and performance of low temperature full electrochemical cells, J. Electrochem. Soc. 160 (2013) 953–957.
[47] J. Wu, S.G. Sun, X.D. Zhou, Origin of the performance degradation and implementation of stable tin electrodes for the conversion of CO_2 to fuels, Nano Energy 27 (2016) 225–229.
[48] O. Scialdone, A. Galia, G. Lo Nero, F. Proietto, S. Sabatino, B. Schiavo, Electrochemical reduction of carbon dioxide to formic acid at a tin cathode in divided and undivided cells: effect of carbon dioxide pressure and other operating parameters, Electrochim. Acta 199 (2015) 332–341.
[49] S. Sabatino, A. Galia, G. Saracco, O. Scialdone, Development of an electrochemical process for the simultaneous treatment of wastewater and the conversion of carbon dioxide to higher value products, ChemElectroChem 4 (2016) 150–159.
[50] D.T. Whipple, E.C. Finke, P.J.A. Kenis, Microfluidic reactor for the electrochemical reduction of carbon dioxide: the effect of pH, Electrochem. Solid-State Lett. 13 (2010) B109–B111.
[51] X. Lu, D.Y.C. Leung, H. Wang, M.M. Maroto-Valer, J. Xuan, A pH-differential dual-electrolyte microfluidic electrochemical cells for CO_2 utilization, Renew. Energy 95 (2016) 277–285.
[52] E. Irtem, T. Andreu, A. Parra, M.D. Hernandez-Alonso, S. García-Rodríguez, J.M. Riesco García, G. Penelas, J.R. Morante, Low-energy formate production from CO_2 electroreduction using electrodeposited tin on GDE, J. Mater. Chem. A 4 (2016) 13582–13588.
[53] H. Li, C. Oloman, The electro-reduction of carbon dioxide in a continuous reactor, J. Appl. Electrochem. 35 (2005) 955–965.
[54] H. Li, C. Oloman, Development of a continuous reactor for the electro-reduction of carbon dioxide to formate—Part 1: process variables, J. Appl. Electrochem. 36 (2006) 1105–1115.
[55] H. Li, C. Oloman, Development of a continuous reactor for the electro-reduction of carbon dioxide to formate—Part 2: scale-up, J. Appl. Electrochem. 37 (2007) 1107–1117.
[56] R. Masel, Z. Liu, D.I. Zhao, Q. Chen, CO_2 conversion to chemicals with emphasis on using renewable energy/resources to drive the conversion, in: S.W. Snyder (Ed.), Commercializing Biobased Products: Opportunities, Challenges, Benefits, and Risks, The Royal Society of Chemistry, Cambridge, UK, 2016, pp. 215–257.

[57] C. Georgopoulou, S. Jain, A. Agarwal, E. Rode, G. Dimopoulos, N. Sridhar, N. Kakalis, On the modelling of multidisciplinary electrochemical systems with application on the electrochemical conversion of CO_2 to formate/formic acid, Comput. Chem. Eng. 93 (2016) 160–170.
[58] S. Guan, A. Agarwal, E. Rode, D. Hill, N. Sridhar, 3-D tin-carbon fiber paper electrodes for electrochemically converting CO_2, in: J. Matyás, T. Ohji, X. Liu, M.P. Paranthaman, R. Devanatham, K. Fox, M. Singh, W. Wong-Ng (Eds.), Advances in Materials Science for Environmental and Energy Technologies II, John Wiley & Sons, Inc., Hoboken, New Jersey, 2013, pp. 231–243.
[59] D. Kopljar, N. Wagner, E. Klemm, Transferring electrochemical CO_2 reduction from semi-batch into continuous operation mode using gas diffusion electrodes, Chem. Eng. Technol. 39 (2016) 2042–2050.
[60] A. Del Castillo, M. Alvarez-Guerra, J. Solla-Gullón, A. Sáez, V. Montiel, A. Irabien, Sn nanoparticles on gas diffusion electrodes: synthesis, characterization and use for continuous CO_2 electroreduction to formate, J. CO_2 Util. 18 (2017) 222–228.
[61] A. Del Castillo, M. Alvarez-Guerra, J. Solla-Gullón, A. Sáez, V. Montiel, A. Irabien, Electrocatalytic reduction of CO_2 to formate using particulate Sn electrodes: effect of metal loading and particle size, Appl. Energy 157 (2015) 165–173.
[62] M. Alvarez-Guerra, A. Del Castillo, A. Irabien, Continuous electrochemical reduction of carbon dioxide into formate using a tin cathode: comparison with lead cathode, Chem. Eng. Res. Des. 92 (2014) 692–701.
[63] Y. Fu, Y. Li, X. Zhang, Y. Liu, J. Qiao, J. Zhang, D.P. Wilkinson, Novel hierarchical SnO_2 microsphere catalyst coated on gas diffusion electrode for enhancing energy efficiency of CO_2 reduction to formate fuel, Appl. Energy 175 (2016) 536–544.
[64] M. Alvarez-Guerra, J. Albo, E. Alvarez-Guerra, A. Irabien, Ionic liquids in the electrochemical valorisation of CO_2, Energy Environ. Sci. 8 (2015) 2574–2599.
[65] N. Hollingsworth, S.F.R. Taylor, M.T. Galante, J. Jacquemin, C. Longo, K.B. Holt, N. H. De Leeuw, C. Hardacre, Reduction of carbon dioxide to formate at low overpotential using a superbase ionic liquid, Angew. Chem. Int. Ed. 54 (2015) 14164–14168.
[66] T.N. Huan, P. Simon, G. Rousse, I. Génois, V. Artero, M. Fontecave, Porous dendritic copper: an electrocatalyst for highly selective CO_2 reduction to formate in water/ionic liquid electrolyte, Chem. Sci. 8 (2017) 742–747.
[67] G.A. Olah, A. Goeppert, G.K.S. Prakash, Beyond Oil and Gas: The Methanol Economy, second ed., John Wiley & Sons, Inc., Hoboken, New Jersey, 2009, pp. 1–334.
[68] J. Albo, M. Alvarez-Guerra, P. Castaño, A. Irabien, Towards the electrochemical conversion of carbon dioxide into methanol, Green Chem. 17 (2015) 2304–2324.
[69] K.W. Frese, Electrochemical reduction of CO_2 at intentionally oxidized copper electrodes, J. Electrochem. Soc. 138 (1991) 3338–3344.
[70] T.Y. Chang, R.M. Liang, P.W. Wu, J.Y. Chen, Y.C. Hsieh, Electrochemical reduction of CO_2 by Cu_2O-catalyzed carbon clothes, Mater. Lett. 63 (2009) 1001–1003.
[71] M. Le, M. Ren, Z. Zhang, P.T. Sprunger, R.L. Kurtz, J.C. Flake, Electrochemical reduction of CO_2 to CH_3OH at copper oxide surfaces, J. Electrochem. Soc. 158 (2011) E45–E49.
[72] J. Albo, A. Sáez, J. Solla-Gullón, V. Montiel, A. Irabien, Production of methanol from CO_2 electroreduction at Cu_2O and Cu_2O/ZnO-based electrodes in aqueous solution, Appl. Catal. B Environ. 176–177 (2015) 709–717.
[73] A. Schizodimou, G. Kyriacou, Acceleration of the reduction of carbon dioxide in the presence of multivalent cations, Electrochim. Acta 78 (2012) 171–176.
[74] F. Jia, X. Yu, L. Zhang, Enhanced selectivity for the electrochemical reduction of CO_2 to alcohols in aqueous solution with nanostructured Cu-Au alloy as catalyst, J. Power Sources 252 (2014) 85–89.
[75] P. Hirunsit, W. Soodsawang, J. Limtrakul, CO_2 electrochemical reduction to methane and methanol on copper-based alloys: theoretical insight, J. Phys. Chem. C 119 (2015) 8238–8249.

[76] A. Javier, J.H. Baricuatro, Y.G. Kim, M.P. Soriaga, Overlayer Au-on-W near-surface alloy for the selective electrochemical reduction of CO_2 to methanol: empirical (DEMS) corroboration of a computational (DFT) prediction, Electrocatalysis 6 (2015) 493–497.
[77] E. Andrews, M. Ren, F. Wang, Z. Zhang, P. Sprunger, R. Kurtz, J. Flake, Electrochemical reduction of CO_2 at Cu nanocluster (100) ZnO electrodes, J. Electrochem. Soc. 160 (2013) H841–H846.
[78] M. Schwartz, R.L. Cook, V.M. Kehoe, R.C. MacDuff, J. Patel, A.F. Sammells, Carbon dioxide reduction to alcohols using perovskite-type electrocatalysts, J. Electrochem. Soc. 140 (1993) 614.
[79] S. Safdar Hossain, S.U. Rahman, S. Ahmed, Electrochemical reduction of carbon dioxide over CNT-supported nanoscale copper electrocatalysts, J. Nanomater. 2014 (2014) 1–10.
[80] A. Bandi, H. Kiihne, Electrochemical reduction of carbon dioxide in water: analysis of reaction mechanism on ruthenium-titanium-oxide, J. Electrochem. Soc. 139 (1992) 1605–1610.
[81] P. Li, H. Hu, J. Xu, H. Jing, H. Peng, J. Lu, C. Wu, S. Ai, New insights into the photo-enhanced electrocatalytic reduction of carbon dioxide on MoS_2-rods/TiO_2 NTs with unmatched energy band, Appl. Catal. B Environ. 147 (2014) 912–919.
[82] X. Sun, Q. Zhu, X. Kang, H. Liu, Q. Qian, Z. Zhang, B. Han, Molybdenum-bismuth bimetallic chalcogenide nanosheets for highly efficient electrocatalytic reduction of carbon dioxide to methanol, Angew. Chem. Int. Ed. 128 (2016) 6883–6887.
[83] S. Zhao, S. Guo, C. Zhu, J. Gao, H. Li, H. Huang, Y. Liu, Z. Kang, Achieving electroreduction of CO_2 to CH_3OH with high selectivity using a pyrite–nickel sulfide nanocomposite, RSC Adv. 7 (2017) 1376–1381.
[84] N. Gutierrez-Guerra, L. Moreno-Lopez, J.C. Serrano-Ruiz, J.L. Valverde, A. de Lucas-Consuegra, Gas phase electrocatalytic conversion of CO_2 to syn-fuels on Cu based catalysts-electrodes, Appl. Catal. B Environ. 188 (2016) 272–282.
[85] N. Spataru, K. Tokuhiro, C. Terashima, T.N. Rao, A. Fujishima, Electrochemical reduction of carbon dioxide at ruthenium dioxide deposited on boron-doped diamond, J. Appl. Electrochem. 33 (2003) 1205–1210.
[86] G. Seshadri, C. Lin, A.B. Bocarsly, A new homogeneous electrocatalyst for the reduction of carbon dioxide to methanol at low overpotential, J. Electroanal. Chem. 372 (1994) 145–150.
[87] H.-P. Yang, S. Qin, H. Wang, J.-X. Lu, Organically doped palladium: a highly efficient catalyst for electroreduction of CO_2 to methanol, Green Chem. 17 (2015) 5144–5148.
[88] J. Albo, G. Beobide, P. Castaño, A. Irabien, Methanol electrosynthesis from CO_2 at Cu_2O/ZnO prompted by pyridine-based aqueous solutions, J. CO_2 Util. 18 (2017) 164–172.
[89] J.H.Q. Lee, S.J.L. Lauw, R.D. Webster, The electrochemical reduction of carbon dioxide (CO_2) to methanol in the presence of pyridoxine (vitamin B_6), Electrochem. Commun. 64 (2016) 69–73.
[90] C. Carlesi, D. Carvajal, D. Vasquez, R.S. Arratia, Analysis of carbon dioxide-to-methanol direct electrochemical conversion mediated by an ionic liquid, Chem. Eng. Process. Process Intensif. 85 (2014) 48–56.
[91] L.M. Aeshala, S.U. Rahman, A. Verma, Effect of solid polymer electrolyte on electrochemical reduction of CO_2, Sep. Purif. Technol. 94 (2012) 131–137.
[92] L.M. Aeshala, R.G. Uppaluri, A. Verma, Effect of cationic and anionic solid polymer electrolyte on direct electrochemical reduction of gaseous CO_2 to fuel, J. CO_2 Util. 3–4 (2013) 49–55.

CHAPTER

3

Electrocoagulation: Fundamentals and Prospectives

Carlos E. Barrera-Díaz, Patricia Balderas-Hernández*, Bryan Bilyeu†*

*Joint Center for Research in Sustainable Chemistry, UAEM-UNAM, Toluca, México

†Xavier University of Louisiana, New Orleans, LA, United States

OUTLINE

COAGULATION VERSUS ELECTROCOAGULATION

Chemical coagulation is an essential part of drinking water and wastewater treatment. In drinking water, it is used for the clarification of water, using coagulant agents mainly for municipal water treatment.

https://doi.org/10.1016/B978-0-12-813160-2.00003-1

Physical separation techniques for treating colloidal wastewater, such as sedimentation, are not adequate since colloidal systems contain particles from 0.1 μm to 1 nm. This makes the settling velocity range from 0.3 to 3 m per year, and implies that the process time will be extremely long. In many cases, wastewater that contains colloids does not agglomerate and is stable. The main reason is that the particles have an electric charge which makes the ions that surround them create a double electric layer, so colloidal particles will repel, preventing contact and not agglomerating. Therefore, Fe or Al cations are introduced into the solution to neutralize the charge and allow the particles to join, forming flocs that can sediment [1–3].

Traditional chemical coagulation uses aluminum and iron coagulants. The most common aluminum coagulants are aluminum sulfate, aluminum chloride, and sodium aluminate. Iron coagulants include ferric sulfate, ferrous sulfate, ferric chloride, and ferric chloride sulfate [4].

Coagulant effectiveness relies on its ability to form multicharged polynuclear complexes with enhanced adsorption characteristics. The nature of the complexes formed may be controlled by the pH of the system (for a more detailed description, see the predominant chemical species distribution in an aqueous solution during the electrocoagulation process section). Coagulation effectiveness depends on the following process variables: mixing, pH, and coagulant dose, and it is common practice to use a laboratory jar test to determine optimized values [5].

For practical wastewater applications, chemical treatment (Fig. 1) may require up to three chemical reagents and pH adjustments for effective treatment, resulting in an expensive and labor-intensive process. On the other hand, electrocoagulation doesn't require external chemicals and removes any size of suspended solids, oil, and grease, as well as heavy metals.

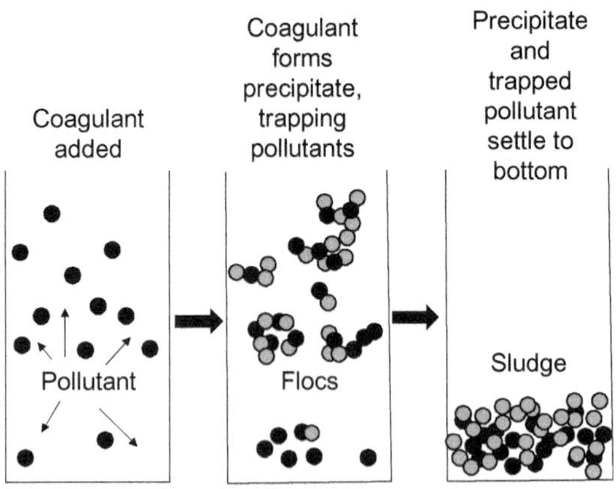

FIG. 1 Illustration of the chemical coagulation process to remove colloidal pollutants.

One of the main drawbacks of using chemical coagulation is the large amount of residual sludge formed at the end of the process. Electrocoagulation produces about half the sludge of chemical coagulation. Rigorous legislation is now in effect in several countries to limit the impact of such residuals on the environment. Wastewater sludge is generally characterized by BOD concentrations in the hundreds and COD concentrations in the thousands of mg L^{-1} [6].

ELECTROCOAGULATION FUNDAMENTALS

Electrocoagulation is an electrochemical water and wastewater treatment which uses an electrochemical cell in which direct current voltage is applied to sacrificial electrodes, commonly iron or aluminum.

Commonly, electrocoagulation takes place when electrochemical reactions take place on the anode and cathode, but solution reactions also play an important role in the process, as shown in reactions (1)–(3).

Anodic reaction

$$Al^0 \leftrightharpoons Al^{3+} + 3e^- \quad (1)$$

Cathodic reaction

$$2H_2O + 2e^- \leftrightharpoons 2OH^- + H_2{}^0 \quad (2)$$

Solution reaction

$$Al^{3+} + 3OH^- \leftrightharpoons Al(OH)_3 \quad (3)$$

The following steps can be identified: (1) anode oxidation, in which the metal cations (Al or Fe) are generated; (2) water is electrolyzed in the cathode, producing small hydrogen bubbles and hydroxide; (3) solution reactions, in which metal ions react with hydroxide to form hydroxy complexes, which adsorb the pollutants, form coagulants, and can then be separated by coagulation/flocculation processes [7,8].

Fig. 2 shows the electrochemical reactions that take place on the electrodes (A) and the chemical reactions in the aqueous solution (B).

As seen in Fig. 2, the electrochemical dissolution of Al and the hydrogen production (bubble generation) takes place at the same time in electrocoagulation.

Thus, the electrocoagulation reactor is primarily operating as an electrochemical dosing reactor in which the Al or Fe addition is controlled and the metal ions produce colloidal aggregation that results in increased size and floc formation. In the case of the bubble generation, if the applied current is low the number of bubbles is small, so this favors the sedimentation of the flocs. With higher applied currents, the bubble generation

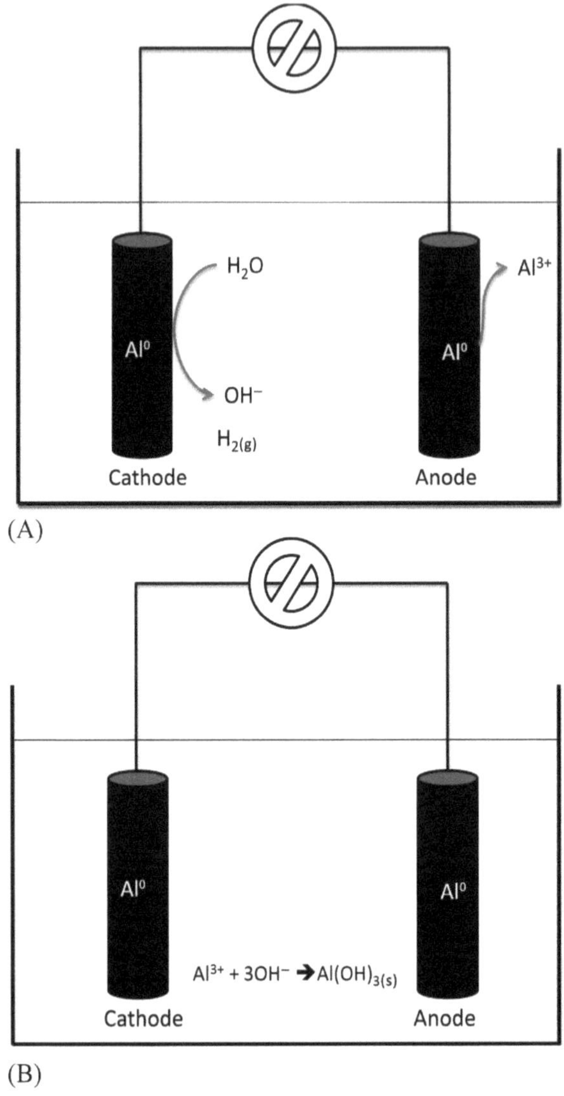

FIG. 2 (A) Electrolytic reactions and (B) solution reactions in the electrocoagulation process.

increases. Thus, the combination of increased aggregation and bubble production makes flotation the primary removal mechanism [9–11].

It has been reported that the pH also affects the bubble size. Typical bubble sizes in electrocoagulation are in the range of 20–70 μm. The size of the bubbles is smaller than those observed in conventional air-assisted flotation, which provides both sufficient surface area for gas-liquid-solid interfaces and mixing efficiency to favor the aggregation of tiny destabilized particles. In particular, hydrogen bubbles usually obey a log normal size distribution, known to be the smallest at neutral pH [12].

THE FARADAY LAW AND ITS IMPORTANCE IN ELECTROCOAGULATION

Current density is a key factor in electrocoagulation, since it is the only operational parameter that can be controlled directly. Indeed, current density controls the anodic and cathodic reactions and also the chemical rate. The importance of the Faraday law in electrocoagulation is that it describes the relationship of the current density and the amount of metallic electrodes dissolved [13].

The Faraday law allows calculation of the mass of Fe or Al ions electrochemically produced, as indicated in Eq. (4), [14,15]:

$$n = \frac{it}{zF} \tag{4}$$

Where n is the number of moles of metal dissolved, i is the current in amperes, t is the electrolysis time in seconds, F is the Faraday constant ($F = 96{,}500\ \mathrm{C\ mol^{-1}}$), and z is the charge of the cation ($z = 3+$ for Al or 2 + for Fe).

SUPERFARADIC EFFICIENCY

As previously described, the concentration of metal ions in a solution using electrocoagulation can be determined by the Faraday law; thus, this concentration should correspond to the maximum theoretical quantity of electrochemical Al or Fe ions produced and the current efficiency should be 100%. However, experimental data suggests that the ion concentration can be higher than the one predicted by the Faraday law [16]. This phenomenon has been called "superfaradiac efficiency."

Superfaradaic efficiency implies that there can be more Fe and Al ions in a solution than expected. This effect can be beneficial for the process since less energy or process time is required to obtain a desired coagulant dose. The question that arises is: where does the added concentration come from? There are two answers, one electrochemical and one chemical. The electrochemical answer can be found in a detailed revision of the cathodic reaction. In principle, this reaction corresponds to the hydrogen generation as shown in Fig. 2; but from the thermodynamics point of view, it is possible that an electrochemical process takes place at the cathode promoting Al or Fe dissolution. Therefore, two simultaneous reactions are taking place in the cathode: hydrogen evolution and at some extent Al of Fe dissolution.

Another effect is that electrochemical oxidation and reduction of water can modify the pH on the anode and cathode surfaces with respect to the bulk pH. Thus, a chemical reaction contributes to the chemical dissolution of the electrode [17–19].

PREDOMINANT CHEMICAL SPECIES DISTRIBUTION IN AQUEOUS SOLUTION DURING ELECTROCOAGULATION PROCESS

As previously stated, many pollutants contained in wastewater are in colloidal form. Colloids normally have a negative charge around them, and when metal cations are released into a solution this charge is neutralized (measured as Zeta potential). Thus, the colloids stop the repulsion and start to form flocs, causing sedimentation and separation from the water. Traditionally, the electrodes used in electrocoagulation are Fe and Al. However, recent research indicate that Cu and Zn can be used. In Figs. 3 and 4, the predominant chemical species of these metals are shown. It can be observed that there are different chemical species as a function of pH. In particular, these diagrams indicate the existence and abundance of some chemical species according to the conditions in an aqueous solution. For instance, in Fig. 3A Fe^{2+} is the predominant chemical specie up to a pH of 8, and then $Fe(OH)_{2(s)}$ starts to prevail. The same behavior is observed with Al, Cu, and Zn. In acidic conditions the cation is ready available for a reaction, and when the pH is raised hydroxocomplexes start to appear. This implies that solutions reactions will have different reaction rates and the process efficacy will be reduced.

Note that in all cases at low pH values, the simple cation form of the metal is present. However, once the pH is increased the hydro-complexes of the cations start to appear. In the case of iron, Fe(II) has a larger pH window in which it can be a free cation; on the other hand, Fe(III) on has a small range of pH values in which it can be present as the cation. This implies that redox reactions in an aqueous solution could affect the effectiveness of the pollutant removal process. Aluminum presents three clear zones; the first one is related with the free ion, and the other two are related with hydro-complexes. Cu and Zn species diagrams are quite similar, presenting two zones for the free and complex forms according to the pH values.

Recent reports of using Al, Cu, Fe, and Zn anodes are provided in Table 1. Note that in some cases the use of Al or Fe provide better pollutant removal results; i.e., for pulp and paper wastewater Al works better; on the other hand, for carwash wastewater Fe provides superior results. Note

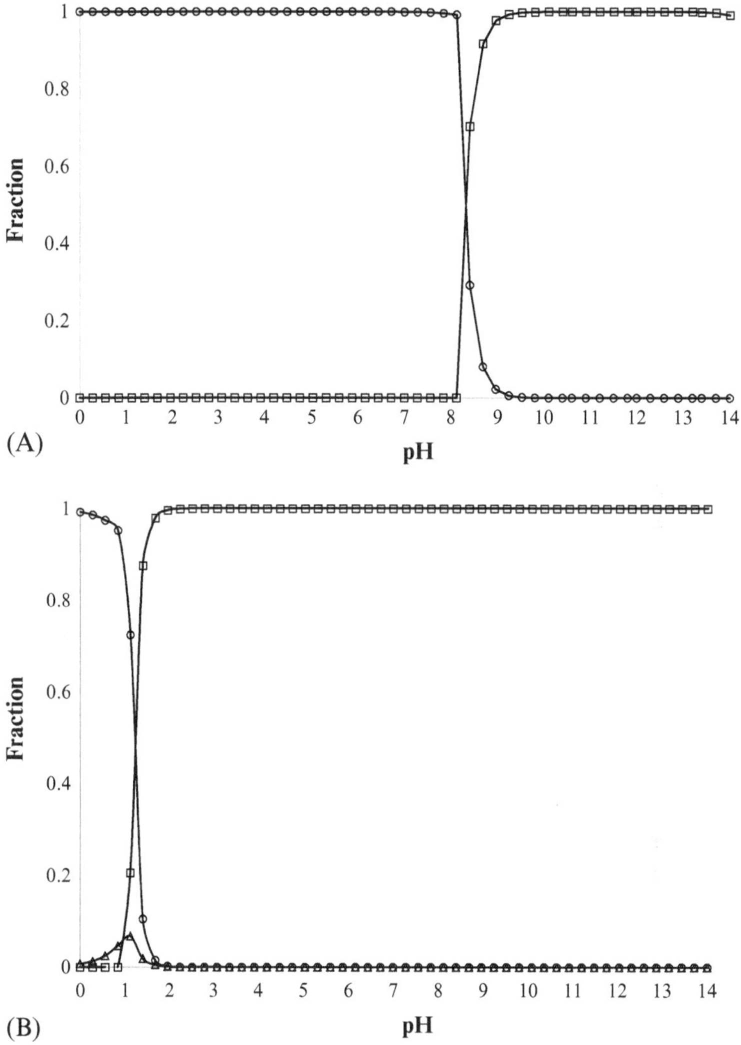

FIG. 3 Chemical species distribution diagrams of (A) Fe(II) [○, Fe^{2+}; □, $Fe(OH)_{2(s)}$] and (B) Fe(III) [○, Fe^{3+}; □, Fe_2O_3(cr); △, $Fe(OH)^{2+}$] species in an aqueous solution.

that although traditional Al and Fe electrodes have been extensively used, other metals can be used as sacrificial anodes. In the case of Cu and Zn, high pollutant removal efficiencies can be achieved for different types of wastewater. Furthermore, some alloys such as Al-Zn-In and Mg-Al-Zn present high removal rates, indicating that not only pure anodes can be used in electrocoagulation.

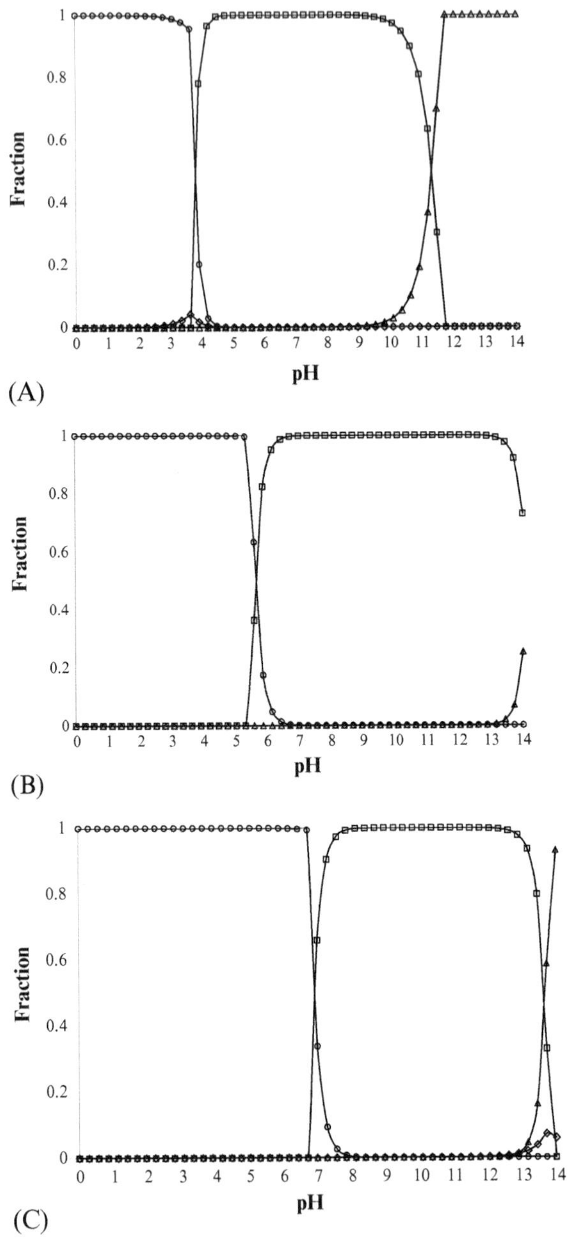

FIG. 4 Chemical species diagram of (A) Al [○, Al^{3+}; □, $Al(OH)_3$(cr); △, $Al(OH)_4^-$; ◇, $Al(OH)^{2+}$], (B) Cu(II) [○, Cu^{2+}; □, $Cu(OH)_4^{2-}$; △, CuO(cr)] and (C) Zn(II) [○, Zn^{2+}; □, ZnO(cr); △, $Zn(OH)_4^{2-}$; ◇, $Zn(OH)_3^-$] species in an aqueous solution.

TABLE 1 Electrocoagulation Using Al, Cu, Fe, and Zn as Anodes

Wastewater	Process conditions	Results	Ref
Pulp and paper mill wastewater treatment	Fe and Al anodes, current density of 20 Am^{-2}	Al electrode system showed an increase by 31% in COD removal and a decrease by 48% in energy consumption compared with Fe electrode system	[20]
Removal of the Reactive Black 5 (RB5) dye from aqueous solutions	Current density of 16 $A\ m^{-2}$, initial pH of 6, 100 $mg\ L^{-1}$ of RB5, and Al anodes	76% and 97% of decolorization after 45 and 120 min of operation	[21]
Electrochemical treatment of carwash wastewater using Fe and Al electrode	Al and Fe electrodes, pH: 8, current density: 3 $A\ m^{-2}$, operating time: 30	Fe electrode presents higher COD and oil-grease removals from carwash wastewaters	[22]
Removal of ciprofloxacin from hospital wastewater	Al electrodes pH7.78, interelectrode distance 1 cm, reaction time 20 min, current density 12.5 $mA\ cm^{-2}$, and electrolyte dose of 0.07 M NaCl	Removal efficiency of 89%	[23]
Treatment of wastewater from sugarcane process industry	Al electrode optimum pH of 7.5, electrode distance of 20 mm, current density of 178 $A\ m^{-2}$, and electrolyte concentration of 0.5 M	84% of COD reduction and 88% of color removal	[24]
Soft drink wastewater treatment	Boron-doped diamond-Cu electrodes and a current density of 30 $A\ m^{-2}$	Removal efficiencies were COD 75% and TOC 85%	[25]
Treatment of a synthetic phenolic mixture	Zn anode pH = 3.0, current intensity = 0.4 A, [NaCl] = 1.5 $g\ L^{-1}$, intergap distance = 1.0 cm	Removal of 86% of the total phenolic content and 49% of chemical oxygen demand	[26]
Removal of chromium from drinking water	Al-Zn-In-alloy anode at a current density of 0.2 $A\ dm^{-2}$, at a pH of 7.0	Removal efficiency of 98.2%	[27]
Removal of fluoride from drinking water	Mg-Al-Zn alloy as an anode and stainless steel as cathode at a current density of 0.2 $A\ dm^{-2}$ and a pH of 7.0	Removal efficiency of 96%	[28]

Continued

TABLE 1 Electrocoagulation Using Al, Cu, Fe, and Zn as Anodes—cont'd

Wastewater	Process conditions	Results	Ref
Arsenic removal from underground water	Zn, Cu-Zn, and Fe anodes, with a current density of 1.5 mA cm^{-2}	Removal efficient, with the followed tendency using anodes of Fe (>93%) Zn (>93%) > Cu-Zn (>73%)	[29]
Reduction of pollutants and disinfection of industrial wastewater	Copper electrocoagulation and hydrogen peroxide	Reduction of BOD_5 by 81%, and color by 97%; under the same conditions, the total coliforms, fecal coliforms, and bacteria are all reduced by 99%	[30]
Removal of organic pollutants in industrial wastewater	Cu anodes for electrocoagulation and BDD anodes for H_2O_2 production, optimal conditions of pH 2.8, and 14.2 mA cm^{-2} of current density	Integrated electrochemical process reduces the COD by 78% and fecal coliforms by 99.9%	[31]
Removal of Perfluoro-alkyl Acids	Zn anodes for electrocoagulation	Reduction of 90% of perfluoroalkyl acids	[32]
Removal of boron from water	Zn anodes using a current density of 0.2 A dm^{-2}, at pH of 7.0	Reduction of 93% of boron	[33]

COUPLED ELECTROCOAGULATION-OZONE

The use of combined processes results in higher pollutant removal, reduced process time, and decreased sludge generation. In this sense, these processes tend to be more efficient and sustainable. Previous research dealing with ozone-coupled methods indicates that the ozonation of anaerobically pretreated waste significantly enhances the organic removal in comparison to the ozonation of unpretreated waste, and substrate conversions in the range of 40%–67% can be reached. In the case of electrocoagulation-ozone coupled process, the ozone is applied at the same time as electrocoagulation takes place, as shown in Fig. 5.

Ozone is introduced in the bottom of the reactor through a performed plate which promotes small bubbles. The process requires the ozone molecules to transfer from the gas phase to the aqueous phase, where the attack on the organic molecules occurs. Ozone is a well-known strong oxidant that interacts with organic pollutants in two ways: direct oxidation of the organic matter and an indirect process that relies on the generation of free-radical intermediates such as the •OH radical, which is a powerful, effective, and nonselective oxidizing agent.

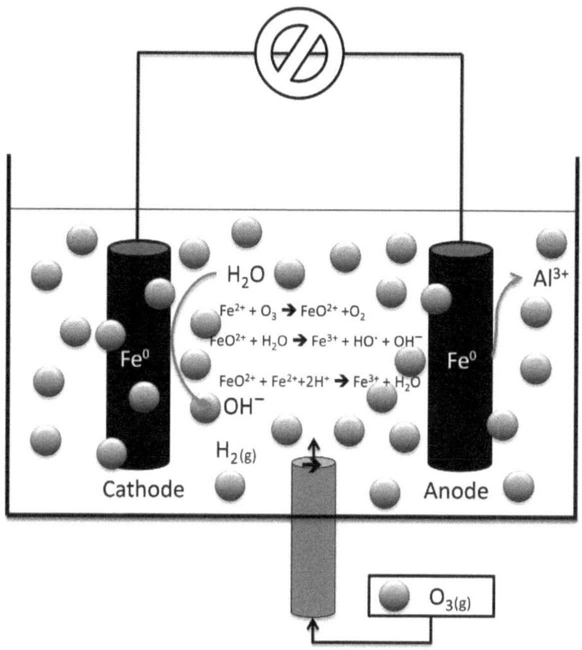

FIG. 5 Ozone addition in an electrocoagulation reactor.

In the coupled ozone-electrocoagulation treatment, the following simultaneous process takes place:

(a) Ozone oxidizes the organic molecules that are dissolved in the solution and that cannot be eliminated via electrocoagulation.
(b) The dissolved iron ions provided by the electrochemical reaction promote hydroxide radical production because Fe^{2+} ions catalyze ozone decomposition to generate hydroxyl radicals as a Fenton-like process.
(c) The electrochemical reactions at the anode and cathode induce the aqueous solution reactions that promote the coagulation/flocculation process.
(d) Supplying ozone bubbles into the reactor promotes the mixing between the reactants, so the mass transfer is enhanced.
(e) The ozone action contributes to the reduction in the amount of sludge produced.

This process helps to explain the synergistic effect of the combination of both technologies and the resulting high efficiencies.

The main reactions in the combined process are the following:

$$Fe^{2+} + O_3 \rightarrow (FeO)^{2+} + O_2 \quad (5)$$

$$FeO^{2+} + H_2O \rightarrow Fe^{3+} + HO\cdot + OH^- \tag{6}$$

$$FeO^{2+} + Fe^{2+} + 2H^+ \rightarrow 2Fe^{3+} + H_2O \tag{7}$$

Table 2 shows some recent examples of wastewater treatment using a combined electrocoagulation-ozone process.

As shown in Table 2, the use of ozone during the electrocoagulation process can be applied to a wide variety of wastewaters, which indicates that this process offers the key advantage of versatility. In all cases, the use of only ozone is limited by the mass transfer of the gas/liquid phase. However, the presence of electrochemically produced cations such as Fe or Al induces two simultaneous reactions: Fenton-like reactions that increase the oxidation of organic pollutants and floc formation for the sedimentation of colloids. An additional advantage of this process is that the sludge formation is less than electrocoagulation alone.

TABLE 2 Recent Wastewater Treatment Using a Coupled Electrocoagulation-Ozone Treatment

Wastewater	Process conditions	Results	Ref
Distillery effluent	3 A dm^{-2}, initial COD 3000 ppm, initial pH7, process time 5 h, ozone flow rate 15 L min^{-1}	Process reduced total chemical oxygen demand by 95%	[34]
Gray wastewater	Aluminum and graphite electrodes 0.9 A dm^{-2}, during 90 min pH8, and at 9.2 g h^{-1} of ozone dosage	Process reduced total chemical oxygen demand by 91% and total organic carbon by 85%	[35]
Removal of arsenic, phosphates, and ammonia from well water	Iron and aluminum electrode plates with simultaneous ozonation	Complete removal of arsenic, phosphates, color, turbidity, suspended solids, and ammonia	[36]
Wastewater that contains offset printing dye	Al electrocoagulation and ozonation	Color removal of 99.9% color and 99.4% COD	[37]
Dairy industry wastewater	Rectangular Al anodes and Fe cathodes were used in parallel using 5 mA cm^{-2} and ozone	80% COD removal	[38]
Wastewater containing hydrolyzed poly-acryamide	Fe electrocoagulation and ozone dosage 4.2 mg L^{-1}	Removal rates reached COD 82.2% and 94.4% HPAM	[39]

TABLE 2 Recent Wastewater Treatment Using a Coupled Electrocoagulation-Ozone Treatment—cont'd

Wastewater	Process conditions	Results	Ref
Winery wastewater	Electrocoagulation using stainless steel, iron, and aluminum electrode with ozonation	Removal of COD and sulfates was 77% and 62%, respectively	[40]
Electro-plating wastewater	Electrocoagulation with aluminum electrodes and ozonation	Removal efficiencies of Cr(VI), Fe, Ni, Cu, Zn, Pb, TOC, and COD were 99.9%, 100%, 95.8%, 98.6%, 99.9%, 96.8%, 93.2%, and 93.4%, respectively	[41]
Preparation of drinking water used in water supply systems	Electrocoagulation using iron and aluminum electrode plates with ozonation	Removal efficiencies for color, turbidity, suspended solids, total arsenic, total chromium, Ni(II), total copper, sulfates, fluorides, chemical oxygen demand, ammonia, nitrates, and nitrites were 100%	[42]
Dye removal in denim effluents		65% color removal, 76% turbidity removal, and 37% COD reduction	[43]

FUTURE PERSPECTIVES

Electrocoagulation wastewater treatment has received special attention as a high-performance and reliable process, due to its easy operation and low maintenance, along with its ability to be adapted to current systems in many industrial applications. One limitation of this process is anodic/cathodic surface passivation that takes place over time, reducing the efficiency of the process. Another limitation of the process is that it is only effective on colloidal and suspended contaminants. Thus, for dissolved substances other methods would need to be coupled with electrocoagulation.

References

[1] P. Cañizares, M. Carmona, J. Lobato, F. Martinez, M.A. Rodrigo, Electrodissolution of aluminum electrodes in electrocoagulation processes, Ind. Eng. Chem. Res. 44 (2005) 4178–4185.

[2] C.E. Barrera, M.E. Palomar, M. Romero, S. Martínez, Chemical and electrochemical considerations on the removal process of hexavalent chromium from aqueous media, J. Appl. Electrochem. 33 (2003) 61–71.

[3] H.S. Peavy, D.R. Rowe, G. Tchobanoglous, Environmental Engineering, McGraw-Hill, New York, 1985, p. 56.
[4] J. Bratby, Coagulation and Flocculation in Water and Wastewater Treatment, IWA Publishing, London, Seattle, 2006. 75 pp.
[5] M. Fanun, The Role of Colloidal Systems in Environmental Protection, first ed., Elsevier, UK, 2014, 692.
[6] W. David, Hendricks Fundamentals of Water Treatment Unit Processes: Physical, Chemical, and Biological, IWA Publishing, Netherlands, 2010. 110 pp.
[7] A.S. Koparal, U.B. Ogutveren, Removal of nitrate from water by electroreduction and electrocoagulation, J. Hazard. Mater. 89 (2002) 83–94.
[8] P.R. Kumar, S.C. Chaudhari, K.C. Khilar, S.P. Majahan, Removal of arsenic from wastewater by electrocoagulation, Chemosphere 55 (2004) 1245–1252.
[9] P.K. Holt, G.W. Barton, C.A. Mitchell, Deciphering the science behind electrocoagulation to remove suspended clay particles from water, Water Sci. Technol. 50 (2004) 177–184.
[10] C.Y. Hu, S.L. Lo, W.H. Kuan, Y.D. Lee, Removal of fluoride from semiconductor wastewater by electrocoagulation-flotation, Water Res. 39 (2005) 895–901.
[11] C.T. Souris, D.W. DePaoil, J.T. Shor, M.Z.C. Hu, T.Y. Ying, Electrocoagulation of magnetic seeding of colloidal particles, Coll. Surf. A: Phys. Chem. Asp. 177 (2001) 223–233.
[12] N. Adhoum, L. Monser, N. Bellakhal, J. Belgaied, Treatment of electroplating wastewater containing Cu2+, Zn2+ and Cr(VI) by electrocoagulation, J. Hazard. Mater. 112 (2004) 207–213.
[13] C. Comninellis, G. Chen, Electrochemistry for the Environment, Springer, New York, 2009. 566 pp.
[14] D. Wang, H. Tang, J. Gregory, Relative importance of charge neutralization and precipitation on coagulation of kaolin with PACl: effect of sulfate ion, Environ. Sci. Technol. 36 (2002) 1815–1820.
[15] M.I. Aguilar, J. Sáez, M. Lloréis, A. Soler, J.F. Ortuño, V. Meseguer, A. Fuentes, Improvement of coagulation-flocculation process using anionic polyacrylamide as coagulant aid, Chemosphere 58 (2005) 47–56.
[16] T. Picard, G. Cathalifaud-Feuillade, M. Mazet, C. Vandensteendam, Cathodic dissolution in the electrocoagulation process using aluminium electrodes, J. Environ. Monit. 2 (2000) 77–80.
[17] P. Cañizares, F. Martínez, M. Carmona, J. Lobato, M.A. Rodrigo, Continuous electrocoagulation of synthetic colloid-polluted wastes, Ind. Eng. Chem. Res. 44 (2005) 8171–8177.
[18] C.E. Barrera, N.F. Ureña, E. Campos, M. Palomar, M. Romero, A combined electrochemical-irradiation treatment of highly colored and polluted industrial wastewater, Radiat. Phys. Chem. 67 (2003) 657–663.
[19] C.E. Barrera, G. Roa, L. Ávila, T. Pavón, B. Bilyeu, Electrochemical treatment applied to food-processing industrial wastewater, Ind. Eng. Chem. Res. 45 (2006) 34–38.
[20] S. Camcioglu, B. Ozyurt, H. Hapoglu, Effect of process control on optimization of pulp and paper mill wastewater treatment by electrocoagulation, Process. Saf. Environ. Prot. 111 (2017) 300–319.
[21] A.S. Fajardo, R.C. Martins, D.R. Silva, C.A. Martínez-Huitle, R.M. Quinta-Ferreira, Dye wastewaters treatment using batch and recirculation flow electrocoagulation systems, J. Electroanal. Chem. 801 (2017) 30–37.
[22] Z.B. Gönder, G. Balcıoğlu, I. Vergili, Y. Kaya, Electrochemical treatment of carwash wastewater using Fe and Al electrode: techno-economic analysis and sludge characterization, J. Environ. Manag. 200 (2017) 380–390.
[23] S. Ahmadzadeh, A. Asadipour, M. Pournamdari, B. Behnam, H.R. Rahimi, M. Dolatabadi, Removal of ciprofloxacin from hospital wastewater using electrocoagulation technique by aluminum electrode: optimization and modelling through response surface methodology, Process. Saf. Environ. Prot. 109 (2017) 538–547.

[24] O. Sahu, D.G. Rao, R. Gopal, A. Tiwari, D. Pal, Treatment of wastewater from sugarcane process industry by electrochemical and chemical process: aluminum (metal and salt), J. Water Process. Eng. 17 (2017) 50–62.
[25] I. Linares, C.E. Barrera, M. Valdés, P.T. Almazán, M. Castañeda, V. Lugo, Soft drink wastewater treatment by electrocoagulation—electrooxidation processes, Environ. Technol. 38 (2017) 433–442.
[26] A.S. Fajardo, R.C. Martins, R.M. Quinta-Ferreira, Treatment of a synthetic phenolic mixture by electrocoagulation using Al, Cu, Fe, Pb, and Zn as anode materials, Ind. Eng. Chem. Res. 53 (2014) 18339–18345.
[27] S. Vasudevan, J. Lakshmi, G. Sozhan, Studies on the Al-Zn-In-alloy as anode material for the removal of chromium from drinking water in electrocoagulation process, Desalination 275 (2011) 260–268.
[28] S. Vasudevan, J. Lakshmi, G. Sozhan, Studies on a Mg-Al-Zn alloy as an anode for the removal of fluoride from drinking water in an electrocoagulation process, Clean - Soil, Air, Water 37 (2009) 372–378.
[29] C. Montero-Ocampo, A. Maldonado, O. Solorza-Feria, in: Arsenic removal from underground water by electrocoagulation with zinc, brass and iron (conference paper), 209th ECS meeting; Denver, CO; United States. ECS Transactions 2: 71–85, 2006.
[30] C.E. Barrera-Díaz, B.A. Frontana-Uribe, G. Roa-Morales, B.W. Bilyeu, Reduction of pollutants and disinfection of industrial wastewater by an integrated system of copper electrocoagulation and electrochemically generated hydrogen peroxide, J. Environ. Sci. Health - Part A Tox. Hazard. Subst. Environ. Eng. 50 (2015) 406–413.
[31] C.E. Barrera-Díaz, B. Frontana-Uribe, B.W. Bilyeu, Removal of organic pollutants in industrial wastewater with an integrated system of copper electrocoagulation and electrogenerated H_2O_2, Chemosphere 105 (2014) 160–164.
[32] H. Lin, Y. Wang, J. Niu, Z. Yue, Q. Huang, Efficient sorption and removal of Perfluoroalkyl acids (PFAAs) from aqueous solution by metal hydroxides generated in situ by electrocoagulation, Environ. Sci. Technol. 49 (2015) 10562–10569.
[33] S. Vasudevan, J. Lakshmi, G. Sozhan, Electrochemically assisted coagulation for the removal of boron from water using zinc anode, Desalination 310 (2013) 122–129.
[34] P. Asaithambi, A.R.A. Aziz, W.M.A.B.W. Daud, Integrated ozone—Electrocoagulation process for the removal of pollutant from industrial effluent: optimization through response surface methodology, Chem. Eng. Process. Process Intensif. 105 (2016) 92–102.
[35] R. Daghrir, A. Gherrou, I. Noel, B. Seyhi, Hybrid process combining electrocoagulation, electroreduction, and ozonation processes for the treatment of grey wastewater in batch mode, J. Environ. Eng. 142 (2016) 1123–1240.
[36] V. Orescanin, R. Kollar, K. Nad, I. Halkijevic, M. Kuspilic, S. Findri, Removal of arsenic, phosphates and ammonia from well water using electrochemical/chemical methods and advanced oxidation: a pilot plant approach, J. Environ. Sci. Health - Part A Tox. Hazard. Subst. Environ. Eng. 49 (2014) 1007–1014.
[37] G. Roa-Morales, C. Barrera-Díaz, P. Balderas-Hernández, F.Z. Aldumbide-Ortiz, H. Reyes Perez, B. Bilyeu, removal of color and chemical oxygen demand using a coupled coagulation-electrocoagulation-ozone treatment of industrial wastewater that contains offset printing dyes, J. Mex. Chem. Soc. 58 (2014) 362–368.
[38] A.L. Torres-Sánchez, S.J. López-Cervera, C. de la Rosa, M. Maldonado-Vega, M. Maldonado-Santoyo, J.M. Peralta-Hernández, Electrocoagulation process coupled with advance oxidation techniques to treatment of dairy industry wastewater, Int. J. Electrochem. Sci. 9 (2014) 6103–6112.
[39] B. Wang, J. Li, H. Ren, J. Li, C. Yue, Degradation process of hydrolyzed polyacryamide in polymer flooding produced wastewater by ozonation, Shinyou Huagong/Petr. Technol. 43 (2014) 832–837.

[40] V. Orescanin, R. Kollar, I.L. Mikelic, K. Nad, Electroplating wastewater treatment by the combined electrochemical and ozonation methods, J. Environ. Sci. Health - Part A. Tox. Hazard. Subst. Environ. Eng. 48 (2013) 1450–1455.
[41] V. Orescanin, R. Kollar, K. Nad, I.L. Mikelic, S.F. Gustek, Treatment of winery wastewater by electrochemical methods and advanced oxidation processes, J. Environ. Sci. Health - Part A Tox. Hazard. Subst. Environ. Eng. 48 (2013) 1543–1547.
[42] V. Orescanin, R. Kollar, K. Nad, N. Mikulic, Preparation of drinking water used in water supply systems of the towns Zrenjanin and Temerin by electrochemical methods, J. Environ. Sci.Health - Part A Tox. Hazard. Subst. Environ. Eng. 48 (2013) 437–445.
[43] M.A. García-Morales, G. Roa-Morales, C. Barrera-Díaz, V.M. Miranda, P.B. Hernández, T.B. Silva, Integrated advanced oxidation process (ozonation) and electrocoagulation treatments for dye removal in denim effluents, Int. J. Electrochem. Sci. 8 (2013) 8752–8763.

CHAPTER

4

Electroflotation

Géssica de O. Santiago Santos, Marilia M. de Salles Pupo, Vanessa M. Vasconcelos, Katlin I. Barrios Eguiluz, Giancarlo R. Salazar Banda

Universidade Tiradentes, Aracaju, SE, Brazil

OUTLINE

https://doi.org/10.1016/B978-0-12-813160-2.00004-3

INTRODUCTION AND HISTORICAL BACKGROUND

In the 20th century, flotation of small particles began to gain technological importance; several attempts were made to improve the beneficiation/purification of low grade ores and the treatment of industrial wastewater. The increasing need to develop technologies capable of separating small particles from aqueous effluents, which were not effectively removed with the usual methods applied at the time, led to the development of electroflotation (EF). EF was initially proposed by Elmore in 1904 for the flotation of valuable minerals from ores [1]; however, it only began to arouse the curiosity of many researchers in 1965 when it was first used in the field of wastewater treatment. The technique was then further studied, mostly in the Soviet Union and sparsely in Germany, Japan, and the United Kingdom [2].

EF is an electrochemical version of the typically applied flotation, which presents high efficiency in promoting separation of solid from liquid phases or one liquid phase from another. In EF even the smallest particles can be separated from a liquid, due to uniform and tiny bubbles of hydrogen and oxygen generated at the electrode's surfaces during electrolysis of an aqueous solution. This feature places the electrodes as the most important part of the system, being considered the heart of an EF unit [1–3].

The EF process differs from flotation technique mainly because of the bubble production mechanism. While for EF small bubbles are produced at the cathode and anode surfaces, for dispersed (induced) air flotation (IAF) bubbles are mechanically generated by the combination of a high speed mechanical mixer and air injection to the system. This technology uses centrifuge force in which the gas is inserted through the dispenser, forming several bubbles with diameters between 700 and 1500 μm. In the dissolved air flotation (DAF) process, bubbles are formed from the reduction of presaturated liquid pressure with air at pressures superior than atmospheric pressure. The saturated liquid is forced through needle valves or special orifices, producing clouds of bubbles with 30–1000 μm diameter [4].

One of EF's main features is the ability to produce very small bubbles, which enhance the flotation of small particles. Besides, EF presents some other advantages when compared to DAF or IAF, such as: (i) electrodes can be arranged to offer total coverage of the flotation tank surface, avoiding bypass; (ii) the electric field gradient between the electrodes helps on flocculation and flotation even in the presence of surfactant; (iii) the gas evolution, flotation time, and other operational conditions can be rapidly checked and are relatively easy to control; (iv) the management of the equipment is safe; and (v) ionic charge of the effluent is not a crucial variable, as is seen in DAF and IAF [5].

The EF process is very efficient for separating oil from water or oil emulsions, along with dyes and heavy metals found in water bodies and wastewater [4–6]. EF can also be applied as a preliminary treatment for water used in cooling and washing systems in industrial processes, recovery and separation of minerals, extraction of suspended compounds of heavy and nonferrous metals and emulsions of organic substances, biodiesel wastewater treatment, and restaurant wastewater treatment, among others [7]. In fact, the advantage of effectively removing small particles from a variety of wastewaters has been successfully studied by EF, and the technology has proven to be simple, fast, cost-effective, and very efficient, especially when treating highly toxic contaminants [8].

Many researches have combined electrocoagulation (EC) with EF into a single system capable of treating various effluents. The main difference between EC and EF is that in the EC process, considering the anode, the electrochemical generation of coagulant agents occurs due to the use of sacrificial anodes, such as Al and Fe anodes, which are consumed during electrolysis generation of Al^{3+} and Fe^{3+} ions, which are rapidly converted into hydroxide and polyhydroxide complexes. These are excellent coagulant agents, and adhere physical or chemically to the contaminants present in the effluent, which are removed afterward by EF [5].

FUNDAMENTALS

The main goal of EF is to promote the separation of suspended matter from aqueous solutions in order to produce a final effluent that is in agreement with environmental demands and that does not represent any hazardous threats. Furthermore, the suspended matter can also be applied as material for feed, fuels, or mineral recovery, or simply become more suitable for disposal [9].

Electrochemical Reactions (Oxygen and Hydrogen Evolution Reactions)

The fundamentals of EF are similar to flotation, differing only by the mechanisms of bubble generation, which are electrochemically generated in the water electrolysis process, where the electrodes used as cathodes and anodes become very relevant [9]. Many materials can be used as cathodes and anodes, and the most commonly used cathode is stainless steel [10–14], but other materials such as nickel [15] and titanium [16] can also be applied. As for anodes, graphite [17], PbO_2 [18], Pt [14,19], and dimensionally stable anodes (DSA) [20], among others, are used.

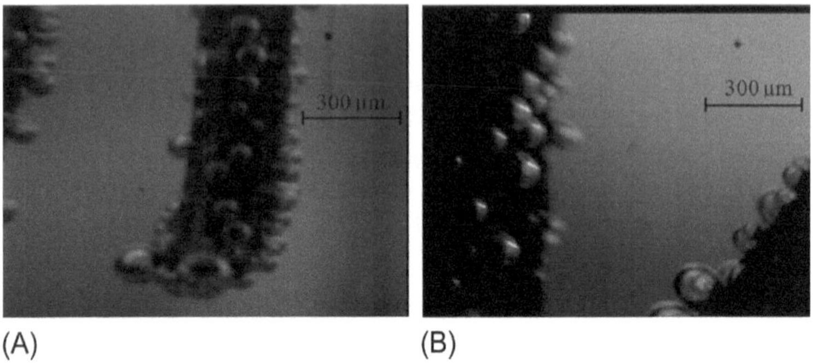

FIG. 1 Image of the hydrogen bubbles formed on the stainless-steel cathode (A) and of the oxygen bubbles formed on the anode (B), with a cell potential of 3.5 V, pH7, and 0.1 mol L^{-1} NaCl solution. Images obtained by laser diffraction. *Reproduced with permission from S.G. Da Cruz, A.J.B. Dutra, M.B.M. Monte, The influence of some parameters on bubble average diameter in an electroflotation cell by laser diffraction method, J. Environ. Chem. Eng. 4 (2016) 3681–3687, Copyright 2016. Elsevier.*

During aqueous solution electrolysis, electrochemical reactions take place, such as oxidation and reduction of water leading to evolution of oxygen and hydrogen at the surface of the anode and cathode electrodes, respectively (Reactions 1 and 2), as can be seen in Fig. 1 [10].

$$2H_2O \rightleftharpoons O_2 + 4H^+ + 4e^- \tag{1}$$

$$2H_2O + 2e^- \rightleftharpoons H_2 + 2OH^- \tag{2}$$

Bubble production by electrolysis occurs in three stages: nucleation, growth, and detachment [21,22]. The electronucleation step is directly influenced by the current density applied. The onset of bubble electronucleation is marked by the accumulation of dissolved gas on the surface of the electrode until a critical supersaturated concentration is reached, undergoing a physical phase change [10]. If the current density is low, the dissolved gas is removed from the electrode surface and migrates toward the liquid bulk by molecular diffusion and convection, and thus no gas phase is formed. This behavior is also observed at high current densities, where the dissolved gas reacts homogenously and quickly, and the electronucleation does not occur. Thus, the applied current density needs to be sufficiently high to form the gas bubbles at predetermined nucleation sites on the electrode surface [22].

According to Tawfik and Diez [23], the dissolved gas mass transfer rates command the bubbles' formation time on the electrode surface. At this point, two processes are competing: the first is the generation of dissolved gas molecules from the chemical reactions occurring on the electrode's surface, and the second is the molecule diffusion toward the bulk solution. The authors analyzed variations in current density with the dissolved gas concentrations on the electrode surface and the time of nucleation. They proposed a model based on Fick's second law, to predict the dissolved gas supersaturation concentration up to the onset of bubble nucleation at the electrode's surface (heterogeneous nucleation). The average H_2 concentration within the diffusion layer can be estimated by Eq. (3):

$$\tilde{C}_i\ (t) = \frac{1}{\delta}\int_0^{\delta} C_i(x,t)d_x \cong \frac{0.274Js}{\sqrt{D_{H_2}}}\sqrt{t} + C_i^{\infty} \tag{3}$$

where $\tilde{C}_i$ is the average concentration of the species in the diffusion layer, δ is the diffusion layer thickness, s is the resulting flux on the electrode's surface, and D_i is the diffusion coefficient of the species i.

According to Fick's first law, the molar mass flux, J_s, across an area A is given by Eq. (4), where I is the current, z is the valency number of the ionic species, and F is the Faraday constant.

$$-J_s = -J_i(0,t) = \frac{I}{zFA} = D_i\left(\frac{\partial C_i(x,t)}{\partial x}\right)_{x=0} \tag{4}$$

Therefore, the higher the current density, the higher the flux of dissolved gases on the electrode's surface; consequently, the average concentration of the dissolved gases within the diffusion layer increases, resulting in a shorter nucleation time and favoring a smaller bubble size [10,23].

The nucleation time is a function of the applied current flux and depends on the species concentrations at the electrode surface and the average concentration within the diffuse layer. The species concentrations at the electrode surface and the average concentration within the diffuse layer can be obtained from the diffusion equation, and are simply related by a constant. Therefore the nucleation time (t^*) is diffusion-controlled and related with current density by Eq. (5), which allows prediction of the onset of bubble nucleation [23].

$$t^* = 0.19\pi z^2 F^2 D\left(\frac{I}{A}\right)^{-1} \tag{5}$$

After nucleation, the bubbles adhered at the electrode grow due to three causes: (i) diffusion of dissolved gas into the bubble surface, (ii) coalescence with other bubbles, and (iii) due to the high internal pressure [10]. According to Venkatachalam [22], only bubbles with a critical radius grow in diameter, while bubbles smaller than the critical radius decrease. Bubble growing rates are given by Eq. (6); also, the author presents a statement that when current density is above 100 mA cm^{-2}, bubble coalescence occurs. This process happens very rapidly (± 0.10 s) and can occur by three routes: (i) bubbles that are close, and touch each other while growing promoting coalescence; (ii) bubbles with moderate size (40 μm) embrace smaller bubbles, growing around them; and (iii) a larger bubble embraces smaller bubbles while migrating along the electrode.

$$R = 2\beta(Dt)^{1/2} \tag{6}$$

where R is the bubble radius, D is diffusivity, t is time, and β is the saturation coefficient.

The last stage of the gas evolution is the bubble detachment, which occurs when the buoyancy or liquid shearing forces overcome the binding forces [21,22]. Bubble detachment efficiency is given by Eq. (7).

$$f_G = \frac{N_G}{N_D}, N_D = N_E + N_G \tag{7}$$

where N_D is total flow density, N_E is transported flow density of the dissolved gas toward the bulk, and N_G is flow density of the gas on the electrode. As current density is directly dependent on generated gas flow, the efficiency of bubble detachment depends linearly on current density; therefore, a higher bubble detachment is noted for higher current densities [10]. The suspended particles attach to the gas bubbles floating on the top of the fluid, and they can be collected as a sludge [9].

Bubble Size Influencing Factors

Electrolysis leads to the formation of uniform, extremely finely dispersed gas bubbles that rise very slowly, producing a slightly undesired convection [9]. The very fine bubbles produced improve the EF process, because the fine particles float easier with small bubbles and the smaller bubbles provide larger surface area for particle attachment [1,2,24]. In general, the bubbles formed during conventional flotation process present an average diameter from 0.9 to 2.0 mm, whereas bubbles formed during the EF process range from 8 to 15 μm, being therefore more homogeneous in size [2].

As the bubble size decreases, the probability of particle-bubble collision (and hence flotation separation) increases [24]. One of the main advantages of EF treatment is that the size and the distribution curve of the desired bubbles can be controlled by choosing the electrode material, electrode surface/morphological properties, the curvature, the treatment of the electrode surface, pH, and current density [21,23,25]. Among these parameters, the most important is the electrode material and its position in the electrochemical cell, which plays an important role in the size of bubbles. At fixed current density and pH, the bubble diameter is proportional to the wire diameter (electrode), and electrodes with larger diameters generate bubbles with larger diameters [19].

Effects of current density on the size of the bubbles has been previously reported [19]. Some researchers noticed the increase in hydrogen bubbles' diameters with increased current density due to the bubble coalescence. However, other studies have found the opposite effect when high current densities are applied [22]. Additionally, apparently there is a relation between the surface of electrodes and the influence of current density on the size of bubbles. For a smooth-surface electrode, the bubble size increased with an increase in current density, and for rough-surface electrodes the opposite trend was noted [19].

The pH is also an important parameter for bubble size. Hydrogen bubbles are larger in an acid medium (20–70 μm) than in a neutral (15–20 μm) or alkaline medium (15–30 μm). In a neutral or alkaline medium, the hydrogen bubble size is smallest indifferent of the cathode material; however, in an acid medium the cathode material influences the bubble size. The oxygen bubbles are larger in a neutral and alkaline medium than in an acid medium. For example, for platinum bubbles with 25 μm are formed in an acid medium, whereas in a neutral and alkaline medium the diameter increases to 30 and 55 μm, respectively. Therefore, the bubble size variation tends to follow a trend opposite to ion excess, and thus bubble size can be easily controlled by varying the pH value of the medium and the electrode materials. However, it is difficult to control pH inside the system due to the continuous release of OH^- and H^+ ions (Reactions 1 and 2), where the higher the current density, the higher the changes in the pH of the system [2].

Finally, bubble size depends on several parameters. For example, bubble position, the diameter of bubbles detached from an electrode, is different from bubble diameter in the solution bulk; this is because of the absorption of the dissolved gas from the surrounding liquid. With the ratio of dissolved-to-gaseous hydrogen formation, the bubble increase is due to the nucleation, as shown in Fig. 2. However, later the diameter of bubbles can also be reduced due to dissolution of the hydrogen from the interior of the bubble back into the liquid. Furthermore, bubble size depends on reaction time, the dissolved gas concentration profile, and the electrode surface treatment; this behavior explains why there is such wide variation in reported measurements across the literature [19].

The efficiency of the EF process can also be improved by an effective quick dispersion of the small bubbles produced electrochemically into the wastewater flow, because a quick dispersion of the bubbles favors the prevalence of tiny bubbles. Therefore, an EF system in which the anode and the cathode are in contact with the wastewater allows bubble dispersion at both electrodes, ensuring higher flotation efficiency than a system when only the screen cathode faces the wastewater. In the second system, the oxygen bubbles produced cannot be dispersed immediately into the

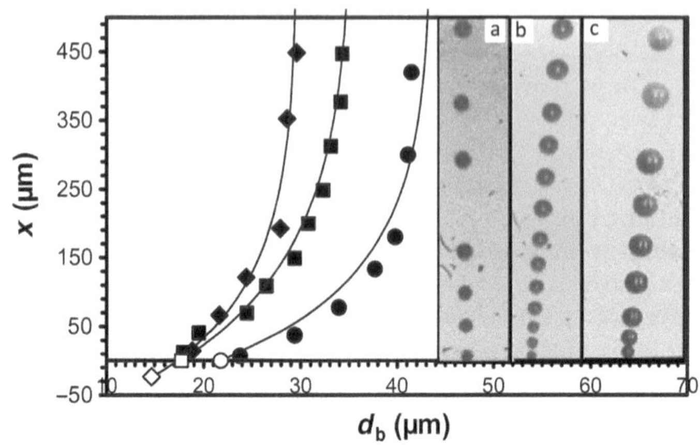

FIG. 2 Bubble diameter (d_b) versus distance from top of electrode (x) for different Pt wire diameter (D) [D=(◆) 90, (▪) 120, and (●) 190 μm]; current density = 354 A m^{-2}; photograph: D=(a) 90, (b) 120, and (c) 190 μm. *Reproduced with permission from M.S.K.A. Sarkar, G.M. Evans, S.W. Donne, Bubble size measurement in electroflotation, Miner. Eng. 23 (2010) 1058–1065, Copyright 2010. Elsevier.*

wastewater. This favors the coalescence and forming of large bubbles, thus decreasing the availability of the effective small bubbles and consequently affecting the flotation efficiency [1].

Electroflocculation Mechanisms

During EF treatment of contaminated waters, the pollutants float to the surface of a water body by attaching to the tiny bubbles of hydrogen and oxygen, promoting the separation of small particles. However, other mechanisms of electrochemical treatment are also possible, such as in the presence of chloride ions, where there is the formation of chlorine gas, hypochlorous acid, and hypochlorite (Reactions 8–10), which are strongly oxidizing species that can oxidize the organic pollutants forming CO_2, H_2O, and inorganic acids. In addition, the free atomic oxygen generated at the anode (Reaction 1) can be carried from the anode diffusion layer by convection toward the wastewater and act as an oxidizing agent of organic compounds. Likewise, hydrogen can promote reduction of the constituents of wastewater [24].

$$2Cl^- \rightarrow Cl_2 + 2e^- \quad (8)$$

$$Cl_2 + H_2O \rightarrow HOCl + Cl^- + H^+ \quad (9)$$

$$HOCl \rightarrow OCl^- + H^+ \quad (10)$$

Furthermore, when the effluent passes through the reactor between two electrodes with different charges, a positive (anode) and negative (cathode), the electric field built up between them due to the use of electrolyte (which offers adequate conductivity) promotes a preliminary coagulation of effluent particulate matter, even without the addition of chemical reagents, forming groups of negative and positive particle agglomerates [24]. It is also possible to generate the coagulants in situ by using a sacrificial electrode such as aluminum or iron electrodes that dissolve electrically, forming aluminum or iron ions at the anode. At the cathode, hydrogen gas is generated which favors the flotation of the flocculated particles. The treatment mechanism is therefore an initial electrocoagulation followed by an EF step [26].

A fluid treated by EF process must present some basic features, such as certain conductivity, in order to avoid addition of large amounts of electrolytes, and the suspended matter in the fluid must be floatable; if not, an additional step of coagulation will be required in order to enable the EF process [9,24]. Besides, in most studies EF behaves as a first-order kinetic model, except in initial times, illustrated by Eq. (11) [24].

$$\ln\left(\frac{C}{C_0}\right) = -kt \tag{11}$$

where k is the flotation rate coefficient, C is concentration in time $t = t$, and C_0 is concentration in $t = 0$.

Although the principles of EF are similar to flotation, the EF process offers certain additional advantages over other flotation techniques, such as that the operating conditions (gas evolution, flotation time) can be checked quickly and are relatively easily controlled; the equipment is reliable and safe in operation; the ionic strength of the effluent solution (slurry) to be treated is not a crucial parameter, as in the other flotation techniques; and by-product recovery is a potentially profitable aspect of EF treatment, so that proteins, fats, oil, and organics, along with toxic or valuable metals, could be recovered from waste streams, especially when no other chemical reagents (like surfactants) are added to help the pollutant separation [24].

Other advantages of the EF process are that transport efficiency is high, current density is a parameter that allows bubble size control and distribution, charged bubbles avoid coalescence, the absence of agitation induces a high floc stability, a high temperature is acceptable, electrochemistry can be added on, fewer chemicals are used, and anodic disinfection is possible depending only on the electrolyte used [9]. In addition, the EF process requires few accessories and uses simple and compact equipment [1].

On the other hand, the EF process presents the disadvantages of requiring DC power, effluents need to be conductive, tank heights must be <1 m, the electrode material may suffer corrosion or passivation, and poorer performance is obtained with high specific gravity materials [5,23].

ENGINEERING ASPECTS

Electrodes

As previously described, electroflotation systems are composed of metallic plates called electrodes arranged in pairs—anodes and cathodes. These electrodes consist of inert materials and can be placed in various configurations leading to arrangements with optimized conditions for EF application.

Cathodes

For the EF process, hydrogen bubble generation has a fundamental role; therefore, cathodic materials, usually metals, are selected for the promotion of a hydrogen evolution reaction.

The most common material used as cathodes in EF is stainless steel [10,12–14,27], chosen for its low cost and availability. However, other electrodes can also be applied, such as nickel [15], which presents low overpotential for a hydrogen evolution reaction, leading to lower energy consumption. Other material commonly applied is titanium [16], which is highly stable and thus is selected as a cathode material for cases of corrosive effluents.

Additionally, some studies combining electrocoagulation-electroflotation (ECEF) have applied aluminum and iron [25,26]. Majlesi et al. [28], for example, tested combinations of Al-Al, Fe-Fe, Fe-Al, and Al-Fe as anodes and cathodes.

Anodes

The anodic material is very influential on the EF process efficiency; therefore, in the literature many types of anodes have been tested, seeking to obtain the best outcomes in optimal conditions. In this context, lifetime and low overpotential for oxygen evolution reaction (OER) are the most important properties for these electrodes [26].

Graphite [17], PbO_2 [18], and Pt [14,19] are among the insoluble anodes most commonly used in EF. However, each of these presents certain limitations, such as: graphite anodes present low durability; lead dioxide anodes are also cheap and of easy fabrication but present high oxygen evolution overpotential and consequently low electroactivity for OER; PbO_2 can potentially generate Pb^{2+} species which are highly toxic, producing a secondary source of pollution; and, finally, platinum anodes are much more stable than lead or graphite electrodes, also presenting a low OER overpotential, but their high cost is a burden for their industrial application.

With the development of dimensionally stable anodes (DSA) by Beer [20], a new class of highly stable electrodes with high electroactivity for oxygen production was achieved which enabled industrial application of EF. Normally, the DSA coating is composed of a metallic oxide (such as RuO_2 and IrO_2), due to its high electrocatalytic activity, and by non-conductive metallic oxides (TiO_2, Ta_2O_5, ZrO_2, Nb_2O_5, etc.), which form a solution to recover the metallic support (Ti, Ta, Zr, W, Nb, Bi, etc.).

DSA electrodes covered with TiO_2-RuO_2 [10] present a high chloride evolution reaction tendency; however, a low service lifetime is seen for OER. Thus, DSA electrodes composed of IrO_x [27,29] have received great attention, since they significantly increase electrode service lifetime under these conditions. Also, studies with Ta_2O_5, TiO_2, and ZrO_2 [30,31] show how stabilizing or dispersing agents represent a good alternative to reducing costs and/or enhancing oxide film properties. The titanium electrodes coated with IrO_x-Ta_2O_5 were used with success as anodes in EF systems [32]. Furthermore, electrodes with ternary coatings containing CeO_2 [33] present enhanced stability.

On the other hand, some researchers have combined electrocoagulation with electroflotation in a single unit to treat wastewater. Hence, electrocoagulation occurs in an anodic site, being necessary for the use of consumable anodes (those that lose mass during their service lifetime) or the so-called sacrifice anodes. The most commonly used anodes for ECEF are aluminum and iron [28,34–36], which are responsible for providing ions in the effluent flow. Fe/Al is dissolved on the anode generating corresponding metallic ions, which rapidly form hydroxide or polyhydroxide complexes of iron and aluminum. These complexes are excellent coagulant agents; therefore, they adhere physical or chemically to the contaminants present in the effluent, and then the coagulated particles are removed by EF from O_2 and H_2 bubbles generated on the electrodes [28]. Unlike conventional coagulation, the coagulant agents Al^{3+} and Fe^{2+} are generated in situ by applying an electrical current to the process of ECEF. The main disadvantage of using a sacrifice anode is the dissolution of these in the aqueous medium, being required routine substitutions [8].

Typical Designs

EF reactors are usually composed of two electrodes, an energy supply and a slurry treatment unit, where electrodes are placed at the bottom of the reactor. The placing of electrodes is an important aspect of the EF process, since the configuration used is able to maximize the system efficiency. Therefore, some factors must be considered in electrode arrangement. Basically, ohmic drop and O_2 and H_2 bubble accumulation between the anode and cathode, which form a nucleus on the surface of the electrode, must be taken into account. These configurations should be minimized in order to guarantee that mass transfer limitations between the electrodes is the smallest possible. Therefore, electrodes used for EF present a series of arrangements that will be presented as: *conventional arrangements, vertical arrangement, and open arrangements.*

Conventional Arrangement

In EF reactors, usually, an anode is installed at the bottom of the reactor and a cathode is fixed around 10–50 mm distant from the anode; see Fig. 3 [1,7,37,38]. However, this arrangement does not contribute for direct interaction from bubbles produced at the bottom with the effluent flow. Thus, the bubbles dispersed on the bottom electrode are not well dispersed immediately in the effluent. Furthemore, the potential risk of flock ruptures is also present, affecting the EF process efficiency.

Additionally, if effluent conductivity is not adequate, a larger spacing between electrodes is necessary to avoid short-circuiting between a cathode and an anode, resulting in elevated energy consumption [26].

Vertical Arrangement

Another option of electrode arrangement is by vertically arranged plates; see Fig. 4. In this configuration, electrodes are placed in parallel and in an equal distance, and therefore can be easily placed. Their main disadvantage is that the bubbles generated tend to rise along the electrode; consequently, a rapid coalescence is seen and undesirably big bubbles are formed, which hinders the process efficiency.

Open Arrangements

See the schematic in Fig. 5, where the anode and cathode are placed in the same position; this is known as open arrangement. The rapid dispersion of small bubbles generated inside the effluent flow is responsible for the flotation efficiency. The fact that both the cathode and the anode are

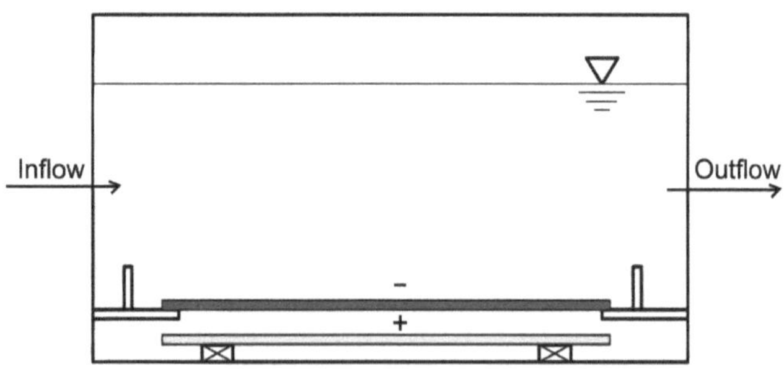

FIG. 3 Conventional electrodes arrangement. *Reproduced with permission from G. Chen, Electrochemical technologies in wastewater treatment, Sep. Purif. Technol. 38 (2004) 11–41, Copyright 2004. Elsevier.*

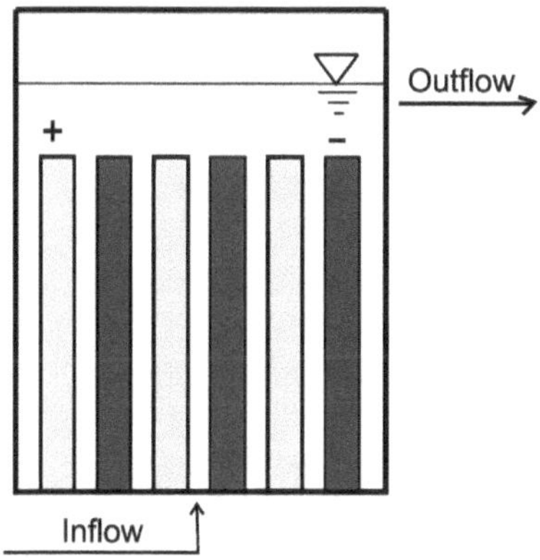

FIG. 4 Electrodes vertically arranged. *Reproduced with permission from M.Y.A. Mollah, P. Morkovsky, J.A.G. Gomes, M. Kesmez, J. Parga, D.L. Cocke, Fundamentals, present and future perspectives of electrocoagulation, J. Hazard. Mater. 114 (2004) 199–210, Copyright 2004. Elsevier.*

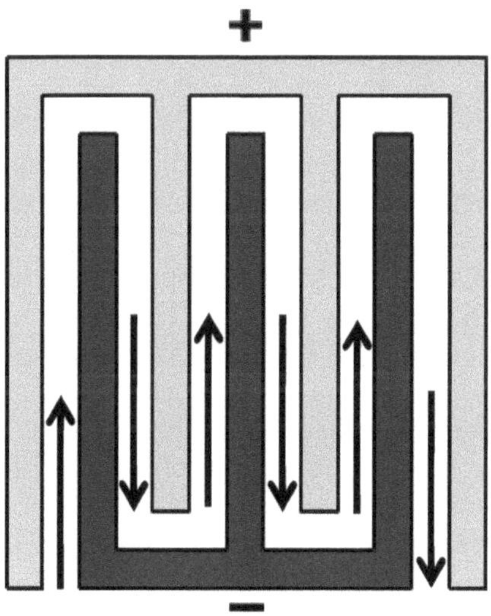

FIG. 5 Open electrodes arrangement with single channel flow for EF. *Reproduced with permission from X. Chen, G. Chen, P.L. Yue, Novel electrode system for electroflotation of wastewater, Environ. Sci. Technol. 36 (2002) 778–783, Copyright 2002. Elsevier.*

leveled, combined with the open configuration, leads to electrode direct interaction [1]. Therefore, bubbles generated and rapidly dispersed are fixed onto the flocks with high efficiency [26].

An alternative for open electrode configuration is shown in Fig. 6, which has proven high efficiencies for oil and suspended oil flotation [39]. The small gap between electrodes allows energy consumption reduction during electrolysis.

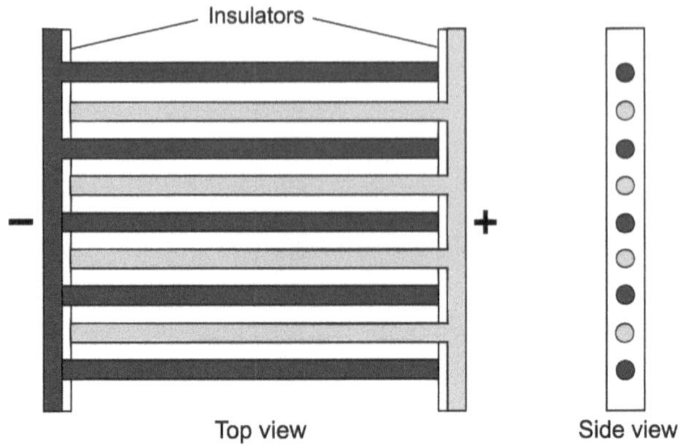

FIG. 6 Alternative open electrodes arrangement with multiple channels flow for EF. *Reproduced with permission from G. Chen, Electrochemical technologies in wastewater treatment, Sep. Purif. Technol. 38 (2004) 11–41, Copyright 2004. Elsevier.*

One-Chamber Reactors

An EF reactor is basically constituted of two pairs of electrodes, a power supply, and a sludge collector unit. Generally, the electrodes are located at the bottom or near the bottom of the reactor. As previously described, the electrodes can be arranged vertically [40] or horizontally [13,34,41,42]. Furthermore, EF reactors arranged as columns have also been commonly studied [38,43,44].

For most cases [45,46], a single EF unit operating with horizontal flow is a good choice, due to simplicity (Fig. 7A). However, an alternative for the cell project is the single vertical flow (Fig. 7B). This new configuration can enhance treatment efficiency due to uniform flow distribution [47].

Furthermore, it is known that the comprehension of fluid characteristics in certain reactors is an important factor for project design [24,48,49]. In this sense, the advantages of several operational modes in a continuous flow reactor have been studied [50,51]. Operation modes with flow in the co-current (same direction) or counter-current are shown

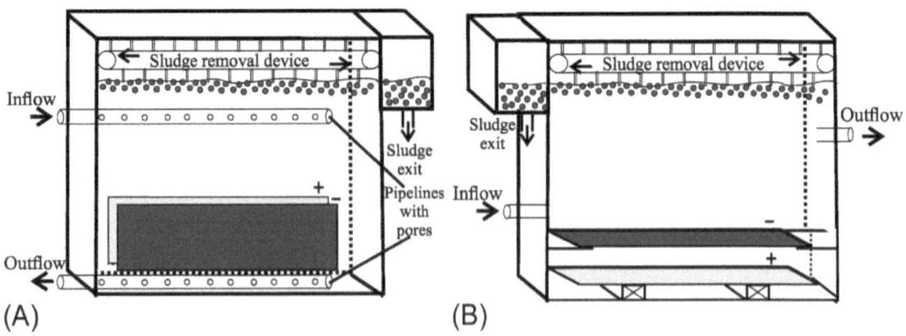

FIG. 7 Schematic of one chamber reactor with horizontal (A) and vertical (B) flow.

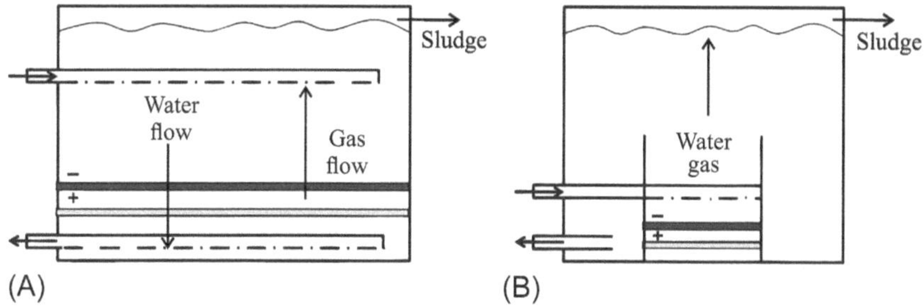

FIG. 8 Operation modes of flow in a EF unit: counter-current (A) and co-current (B). *Reproduced with permission from G. Chen, Electrochemical technologies in wastewater treatment, Sep. Purif. Technol. 38 (2004) 11–41, Copyright 2004. Elsevier.*

in Fig. 8. Comparatively, it was shown that the counter-current flow is more efficient, since in this configuration bubbles arise continuously through suspended particles that are in a descending flow, and therefore a considerable opportunity for bubbles to shock with particles is generated, increasing treatment efficiency. As for flow in the same direction, the inflow is close to the area forming bubbles and dead zones were noted [50,51], which does not happen in the counter-current configuration. Additionally, no bypass was noted in neither configuration [51].

Two-Chamber Reactors

The EF system proposed with the use of two chambers seeks to enhance treatment efficiency, since it ensures the removal of extremely small and disperse particles. Fig. 9 shows a typical two-chamber reactor used for ECEF. In the first chamber there is a rapid formation of gas bubbles due to high current, accelerating adherence and flotation of contaminant

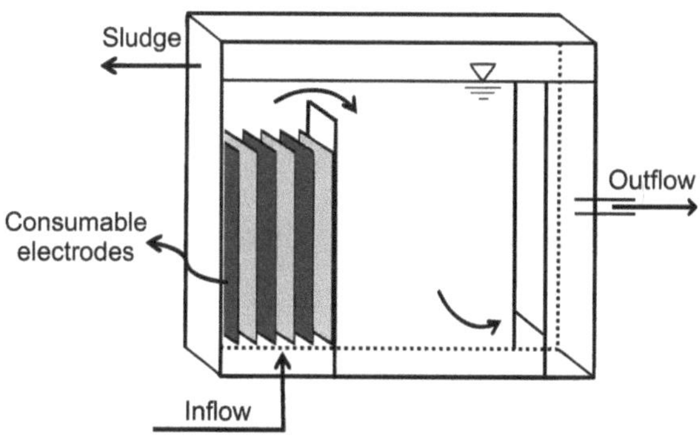

FIG. 9 Two-chamber reactor for ECEF. *Reproduced with permission from G. Chen, Electrochemical technologies in wastewater treatment, Sep. Purif. Technol. 38 (2004) 11–41, Copyright 2004. Elsevier.*

particles up to the surface, forming foam. In the second chamber, the final treatment occurs where current applied is chosen in order to produce bubbles at a laminar flow regimen, propitiating the removal of suspended impurities [1,46].

Operating Parameters

EF is easy to operate, which is one of the main attractive features of this technology. Automation and process control are tailored by electric current and voltage, enabling gas bubble production and concentration in the floating medium. Operation safety is also a concern in industrial scenarios, and due to low operation voltages (5–20 V) this technique is also a good industrial option, besides presenting low energy consumption [39]. As far as treatment efficiency, research on treatment efficiency of EF for wastewater and effluent treatments have been thoroughly carried out [6,40,47,52–55]. Thus, it is well known that EF efficiency is closely dependent on the uniform formation of small bubbles, since the smaller the bubbles, the higher superficial area they present, which can then bond with the pollutant [19,56].

Volumetric ratio among bubbles and particles in suspension, electrode (type, amounts, and sizes), electrolyte (type and concentration), current density, pH, and hydrodynamic regimen have been pointed out as important variables that affect EF treatment efficiency, being closely related to bubble size [7,8]. When considering EF in comparison with conventional flotation, besides dismissing the use of chemical flocculants, the promotion of smaller bubbles is highlighted, enabling more efficient treatment [57,58].

Electrode Type

It is known that the electrode material constitutes the most important part of the EF cell, since the production of small and uniform bubbles, responsible for treatment efficiency, is closely related to it.

Studies developed by Venkatachalam [22] in the early 1990s showed that the kind of cathode used in a neutral media had strong influence in hydrogen bubble size. The following scheme shows how electrogenerated bubbles change when considering electrode material in comparison with other material:

$$\xrightarrow{\text{Increase in hydrogen bubbles}} \\ \text{Pb, Sn, Fe, Ni, Pt} \\ \xleftarrow[\text{Increase in oxygen bubbles}]{}$$

Graphite and lead anodes are among the insoluble anodes most commonly used in EF; although they are cheap and easily found, they present high OER overpotential, which leads to treatment deficiency and increases the energy consumption of the process [17,18]. Platinum electrodes present lower OER overpotential; however, the material's cost is its main limitation. Still, aluminum, iron, and stainless steel are also commonly used. Studies developed by Arslan-Alaton et al. [59], comparing aluminum and stainless steel, show the complete color removal of Acid Blue 193, and partial chemical oxygen demand (COD) removal can be obtained by combining electrodes used. However, the authors observed that a higher degradation rate was seen for stainless steel, presenting lower electric energy consumption and iodine production (8 kWh m^{-3} and 700 g m^{-3}) when compared to aluminum (17 kWh m^{-3} and 8.2 g m^{-3}).

DSA are commonly used in EF due to their anodical insolubility and since they present high electrocatalytic activity for OER. Santos et al. [60] treated effluents with DSA anodes of TiO_2-RuO_2, which were much more reactive for organic mineralization in an effluent containing oil. However, although this electrode composition had good outcomes for chloride evolution reaction, it presented a short service lifetime for OER. Therefore, studies with iridium-based DSA receive great attention because these present a longer service lifetime than ruthenium electrodes, having as their main limitation the high cost associated with iridium [32]. Cho et al. [61] used a DSA anode of TiO_2-IrO_2, applied successfully in the EF, as a final clarification step in activated sludge treatment. The electrodes containing IrO_x-Sb_2O_5-SnO_2 with 10% iridium were studied as a viable alternative, since a little amount of iridium on the oxide film propitiated a long service lifetime for the composition and a good catalytic activity, where the bubbles formed were rapidly dispersed on the effluent and adhered to flocks formed in previous EC.

Majlesi et al. [28] studied the effect of anode material in a cathode application with several combinations (Fe-Fe, Fe-Al, Al-Fe, Al-Al) for nitrate removal by ECEF. The experimental results show that maximum nitrate removal was obtained for the Al-Al arrangement.

Electrode Amount

Depending on the EF reactor geometry, a pair of electrodes can be arranged vertically or horizontally [26]. However, the use of only two electrodes requires a large spacing between them, in a manner to avoid short-circuiting, besides avoiding energy consumption elevation. Moreover, treatment efficiency when comparing the arrangements with many electrodes [1] is lower, since for each pair of cathode and anode some disadvantages are clearly seen, such as low interaction of the bubbles produced with the bottom of the reactor, when using two parallel plates, and a bubble coalescence tendency for the vertical arrangement.

Aoudj et al. [36] studied several Fe and Al anode arrangements, besides altering the number of anodes used. 4 Al and 4 Fe plates in three hybrid configurations, 3 Al and 1 Fe, 2 Al and 2 Fe, 3 Fe and 1 Al plate, were used for simultaneous treatment of Cr^{6+} and fluoride. The optimal combination for Cr^{6+} and fluoride removal was of 2 Fe and 2 Al plates. However, most of the Cr^{6+} removal was seen for Fe plates, while most of the fluoride was removed by Al plates. Therefore, the behavior of hybrid anodes depends on the individual properties of Al and Fe when considering the pollutant treated [62].

Electrode Size

Bubble size depends on the geometry of the electrode. Bubble diameter increases with increasing thickness of the wire mesh. Experiments with wire electrodes have shown that the bubble diameter depends on the electrode curvature. Further studies by Jiménez et al. [56] showed that the electrode surface curvature influences the distribution of hydrogen and oxygen bubbles, enabling the control of production ratio and bubble size. Also, when analyzing the electrode material, comparing bubble production on stainless steel or titanium as a cathode and platinum as an anode, it was possible to conclude that stainless steel produced smaller bubbles than titanium. Finally, roughness was pointed out to present strong influence on the bubble production, where rougher surfaces were able to produce a higher amount of small bubbles, which favors EF treatment efficiency.

Sarkar et al. [12] evaluated bubble diameter for different wire diameters in a stainless steel mesh as a cathode. It was seen that hydrogen bubbles produced in the mesh with wholes presenting diameters of 0.190 mm were smaller than those produced in mesh with wholes presenting diameters of 0.400 mm. Similar results were also seen in the literature [19,63].

Gadd et al. [47] studied eight plate electrode configurations, with aluminum or iron plates as anodes and stainless steel plates as cathodes, and showed that treatment efficiency improvement was related to the increase in electrode area, which is directly related to the generation of bubbles and electrogenerated coagulants.

Electrolyte

In general, studies point out a linear increase in pollutant removal from various effluents as electrolyte concentration increases [64,65]. This effect can be associated with an increase in ionic strength resulting in a thinner electrical double layer, thus reducing bubble sizes; this is also known as electronic compression. This compression favors bubble detachment, decreasing bubble diameter [66]. However, very high saline concentrations caused by corrosion and presence of excessive ions can disturb the EF process, leading to a decrease in the removal rates [12].

Current Density

Current density applied to the electrodes constitutes an important variable which has significant effect on the treatment efficiency, being strongly associated to the number and size of bubbles [19]. Also, it represents a key parameter defining the determinant mechanism of pollutant removal (i.e., settling or flotation) [3].

It is well known that when a current density is applied to sacrificial anodes (typically Fe or Al), the metal dissolves, generating ions that are hydrolyzed almost immediately into their respective hydroxides or polyhydroxides. Therefore, coagulation occurs when these hydroxides and polyhydroxides combine with the pollutant forming flocks, which then can float or sediment [28]. Current density will determine which is the dominant mechanism. Low current densities produce a lower amount of bubbles, which does not stimulate ascending flow conditions, favoring sedimentation of the pollutant; on the other hand, when current density increases, the ascending flow increases due to a higher number of bubbles generated, stimulating pollutant flotation [3].

Da Cruz et al. [10] reported the influence of current density on the average bubble diameter. Bubbles with sizes of 35 and 31 μm were obtained for current density, varying from 14 to 77 mA cm^{-2}, respectively, using a mesh cathode and having a pH of 11. A decrease in bubble size was seen with the increase of current density, which was associated with dissolved gas flow generated at the electrode surface, having as a consequence the increase of average concentration of gases inside the diffusion layer and resulting in a short nucleation time, producing consequently smaller bubbles. Therefore, the effluent containing finely dispersed phosphate was treated at 77 mA cm^{-2} with bubbles around 31 μm, and 82% of the recovery was obtained. On the other hand, when 14 mA cm^{-2} was applied, the average bubble size was of 35 μm and a lower percentage of phosphate was recovered, around 77%. Studies for current densities over 100 mA cm^{-2} reported in the literature [22,25] show that very high current density is responsible for the promotion of bubble coalescence at the surface of the electrode, which increases bubble size.

Majlesi et al. [28] investigated nitrate removal from aqueous effluent, combining ECEF. The increase in current density from 0.4 to 3.2 mA cm^{-2} increased nitrate removal efficiency from 55% to 96% in optimal time and pH conditions, using Al-Al electrodes. This increase occurs because the increase in current density promotes a higher anode dissolution rate, which will increase Al^{3+} ion production. Therefore, it is expected, for the case of ECEF combined treatment, that higher current densities lead to higher treatment efficiencies. Similarly, Hooshmandfar et al. [41] studied dye removal (Acid Red 14) by ECEF, in which the current density varied from 10 to 60 mA cm^{-2}, reducing the treatment time for color removal of 90% from 480 to <90 min. This occurred due to an increase in Al^{3+} ions released from the aluminum anode, but also due to sludge generated in low current density, which does not float easily and returns to the effluent.

pH of the Media

For electronic process, due to its ionic nature pH is always highlighted as a strong determinant in treatment efficiency. By controlling the pH of the electrolyte, formation of determined co-products favored by a certain pH level (either dominant in acid or basic medium) can be controlled. As for EF, pH can determine the ratio of bubble formation, as well as bubble size. Hydrogen bubbles have a tendency to be smaller in neutral and alkaline pH, while oxygen bubbles present a tendency to gradually increase in size as pH increases [67].

Also, solution pH has a fundamental role in ECEF process when applying sacrificial electrodes, since the solubility of the metallic hydroxides formed, which are excellent coagulant agents, is affected. When Al and Fe electrodes are used, the $Fe^{3+}_{(aq)}$ or $Al^{3+}_{(aq)}$ generated undergo spontaneous reactions, producing their corresponding hydroxide and/or polyhydroxide [26].

Al ions, such as, Al^{3+}, $Al(OH)^{2+}$, $Al(OH)_2^+$, $Al(OH)_2^{4+}$, $Al(OH)_3$, and $Al_{13}(OH)_{32}^{7+}$ are formed in a wide pH gap (from 4 to 9). However, when $pH > 10$, $Al(OH)^{4-}$ is the dominant specie formed, and coagulation rates decrease rapidly. On the other hand, when $pH < 4$, Al^{3+} is predominantly formed; it also presents no coagulant property [3].

Similarly, for Fe ions, such as Fe^{2+} and Fe^{3+} formed during electrochemical oxidation, the corresponding hydroxides and polyhydroxides are also formed. In acid pH ($pH < 4$), the presence of Fe^{2+} and Fe^{3+} is favored. However, with the increase in pH from 4 to 14, iron ions undergo hydrolysis, forming their specific hydroxides [39].

Da Cruz et al. [10] investigated the influence of some parameters on the average bubble diameter in EF by laser diffraction method, conducing analysis in several pH media. Thus, the authors observed that the experiments carried out in various pH values presented a clear reduction in bubble size with the increase in pH. When applying a current density of 14 mA cm^{-2}, the average bubble size was 52.5 μm and phosphate removal percentage in an aqueous effluent was 35% with a pH of 4. On the other hand, when increasing pH values to 12, the average bubble size decreased to 39 μm and phosphate removal increased to 76.8%. Such increase in removal is seen because when the environment becomes more alkaline, H^+ species in solution reduce; with this, a lower amount of H_2 bubbles suffer nucleation, which reduces bubble diameter. Additionally, an average bubble size of O_2 generated in the anode increased with the increase of pH to 8, finding a plateau at 45 μm, a higher value than that found for the H_2 bubbles.

On the other hand, additional studies [65,68] analyzing the oil-water contaminated effluent treated by EF found that at neutral pH a higher removal efficiency is noted, probably due to the minimum bubble dimensions occurring in a neutral or slightly alkaline medium in these cases. Therefore, as the contact surface area between the oil droplets and gas bubbles increases, the percentage of removed oil also increases.

Hydrodynamic Regimen

Considering the delicate nature of the EF process, a very slight agitation is preferred, avoiding a Reynolds number of over 4000, and seeking to mix the gas in the liquid for enhanced dispersion, easing collision of the bubbles with the pollutant particles [44]. As for hydrodynamic behavior, operational parameters altering the Reynolds number in EF reactors show that the highest efficiency was reached when current density and liquid flow were kept in laminar regimen [25].

Yoon and Luttell [69] in the late 1980s had already developed a probability model to estimate collision and adherence between bubbles and particles according to the Reynolds number, bubble speed, bubble diameter, and particle diameter. Therefore, they demonstrated that for EF to occur, a minimal movement is necessary in order to promote contact between bubble and particle; besides, they noted that small particles tend to change their trajectory and follow flow lines that surround the bubble.

Kotti et al. [44] used a EF column to study hydrodynamic process, analyzing the effect of mechanical agitation in a EF reactor. They observed the influence of mass transport coefficient from the fluid flow (K_La) or different heterogeneous dynamic bubble regimens. As agitation increases, an increase in the Reynolds number is seen and the K_La increases progressively until stability is reached, when the hydrodynamic regimen changes from laminar to turbulent. Therefore, turbulence caused by the agitation is the limit where agitation will no longer affect mass transport.

Electrical Properties

Continuous current electric energy source is the most commonly applied electric configuration for electrolysis treatments. In EF, the electrode polarity can be inverted, seeking to reduce passivation effects on the cathode. The EF system can therefore be arranged in various electrode configurations, being continuous or plug-flow treatments. Thus, the electrodes can be arranged in a monopolar or bipolar configuration [26,70].

Fig. 10A shows the parallel monopolar configuration, where the electrodes are polarized interrelatedly and connected to external energy sources. In contrast, bipolar arrangement (Fig. 10B) shows that electrodes placed between the anode and cathode are placed without external or

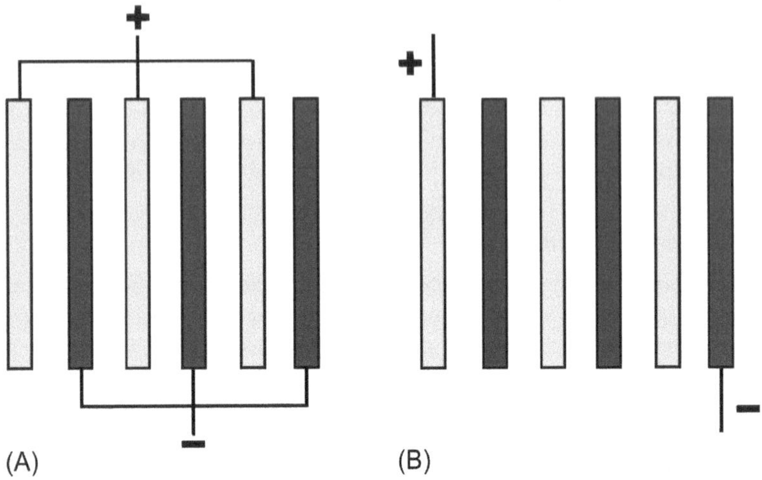

FIG. 10 Electrode connections in the EF process. (A) Monopolar and (B) bipolar. *Reproduced with permission from A.K. Golder, A.N. Samanta, S. Ray, Removal of Cr3+ by electrocoagulation with multiple electrodes: bipolar and monopolar configurations, J. Hazard. Mater. 141 (2007) 653–661, https://doi.org/10.1016/j.jhazmat.2006.07.025.*

internal electric interconnection; thus, the potential variation is equally distributed between the neighboring electrodes. After the electric current passes through the electrodes, the neutral sides of the conductive plate become charged positively or negatively. This arrangement allows a simple configuration, which eases the maintenance during use [28,39].

Cell voltage and current are easily controlled during EF treatment. The operational current in a EF cell should surpass equilibrium potential, anode and cathode overpotential, and solution ohmic potential in order to promote electrochemical reactions, as seen in Eq. (12) [39,71,72].

$$E_{cell} = E_{eq} + \eta_{a,a} + \eta_{a,c} + \eta_{a,p} + |\eta_{c,a}| + |n_{c,c}| + \eta_{ohm} \quad (12)$$

where E_{cell} is the electrolysis voltage (V), E_{eq} is the difference between equilibrium and water molecule breakage (V), $\eta_{a,a}$ is the anode activation overpotential (V), $\eta_{a,c}$ is the anode concentration overpotential (V), $\eta_{a,p}$ is the passive anode overpotential (V), $\eta_{c,a}$ is the cathode activation overpotential (V), $n_{c,c}$ is the cathode concentration overpotential (V), and η_{ohm} is ohmic resistance. The typical voltage components in an EF cell are shown in Fig. 11.

The direct estimation of voltage by Eq. (12) is still very difficult [1]. Normally, the overpotentials that include activation and concentration are related to several electrochemical properties of the electrodes, current density applied, and solution pH. Thus, the necessary voltage for an EF process is mostly given by the potential of the solution ohmic resistance drop.

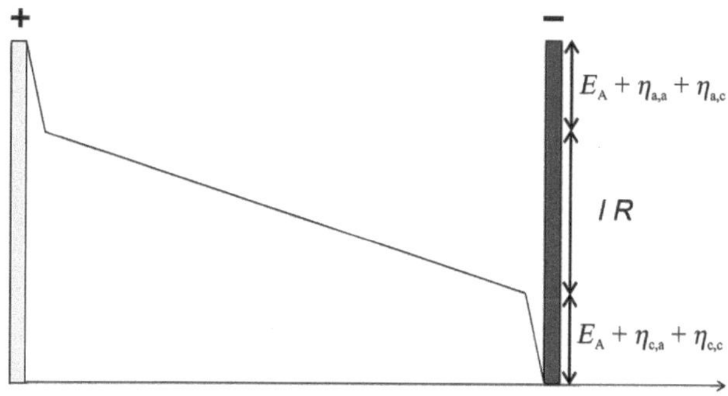

FIG. 11 Typical potential distribution between the electrodes. *Reproduced with permission from R. Alam, J.Q. Shang, Electrochemical model of electro-flotation, J. Water Process Eng. 12 (2016) 78–88, https://doi.org/10.1016/j.jwpe.2016.06.009.*

Energy Requirements

Nowadays, due to high costs associated with energy consumption, current efficiency has become one of the most important parameters to be considered in industrial applications. Energy consumption depends on reactor project, electrode material, current density, and solution conductivity, among others [41].

Energy consumption (EC) is given in kWh m^{-3}, and can be estimated from Eq. (13) [6,73].

$$EC = \frac{UIt}{V\,1000} \tag{13}$$

where U is the cell potential (V), I is the applied current (A), t is flotation time (h), V is the volume treated (m^3), and Q is the flowrate ($m^3\,h^{-1}$).

As for project design, the distance between electrodes is very important in order to reduce electric energy consumption, since the bigger the distance, the more ohmic resistant the solution will be.

Studies on the effect of current density and interelectrode distance on the removal of turbidity efficiency were carried out by Merzouk et al. [34]. It was seen that when the distance of two aluminum electrodes varied from 1 to 3 cm, the current density varied from 11.55 to 91.5 mA cm^{-2}, respectively. Still, the turbidity removal efficiency reached 89% with the smallest spacing, and reduced as the distance between electrodes increased.

Recent studies by Alam et al. [72] developed an electrochemical model to calculate total voltage necessary for electrolysis in an aqueous medium using EF in different conditions. Therefore, the variation in electrode distance was investigated and experiments altering spacing from 2.0 to 9.5 cm in an aqueous medium with conductivity of 2100 μS cm^{-1} were conducted. As a result, the resistance increases linearly with the increase in electrode spacing.

Nagai et al. [74] had already studied the effect of electrode spacing (0–20 mm), and concluded that if the electrodes are too close to one another, between 1 and 2 mm, the voltage can increase depending on the current density applied. This can occur due to the volume of hydrogen and oxygen bubbles increasing, which will further increase solution electric resistance. In the case studied, an increase in the voltage was observed when the current densities were above 0.6 A cm^{-2}, which was attributed to the large bubble layer created when current density is high. However, at lower current densities (0.1–0.5 A cm^{-2}) the voltage decreased as the electrode spacing became smaller (1–2 mm). Thus, optimal spacing found for current densities studied from 0.1 to 0.9 A cm^{-2} was of 1–2 mm applying 0.5 A cm^{-2}.

Gao et al. [62] showed a reduction in energy consumption when applying electrolytes with higher conductivity. However, it was noted that a limit of energy consumption reduction was reached after conductivity was increased over 1.5 μS cm^{-1}.

As for the effect of current density, recently Alam et al. [27] studied removal of bitumen, varying current density from 50 to 300 A m^{-2}, keeping the other variables constant. Bitumen percentage removal (%) increases with the increase of current density up to an optimal value of 150 A m^{-2}. However, when the bubble volume increases to a specific limit, a higher current density promotes bubble coalescence, and consequently the number of small bubbles is reduced, which decreases treatment efficiency. As a result, an optimal current density was found for the removal of bitumen, which also reduces energy cost when operating at nonoptimal conditions.

Shin et al. [29] carried out studies to evaluate the effect of applied current density on the *Scenedesmus obliquus* recovery, applying 0.25, 0.50, and 0.75 A. An increase in efficiency removal was seen when increasing the current, noting that outcomes of applying 0.50 and 0.75 A were very similar, reaching 95% of removal. The difference in energy consumption is of 0.909 Wh g^{-1} for 0.5 A, and 1.626 Wh g^{-1} for 0.75 A, which represents considerable economy when considering energy consumption analysis.

WATER AND WASTEWATER TREATMENT BY ELECTROFLOTATION

Potable Water Treatment

The need for proper water treatment is a continuously growing concern of modern society. Problems related to water safety coupled with continuous industrial pollution in water bodies and groundwater have been imposing the need for increased pretreatment in order to attain acceptable levels for human consumption. Furthermore, the pretreatment applied must guarantee the removal of potentially hazardous compounds with the addition of the smallest amount possible of additives. In this sense, EF emerges as a potential treatment due to its rapid procedure free of additives.

Regarding the removal of fluoride anions (associated with several adverse effects) from portable water, studies developed by Essadki et al. [75] analyzed reaction kinetics of EF combined with electrocoagulation treatment applying 12 mA cm^{-2} on Al electrodes for 30 min inside an external loop airlift reactor to treat typical Casablanca drinking water. Besides presenting several analyses on kinetic and retention time on a set of variations in reactor design, the study proved that by combining EC and EF it was possible to reduce fluoride levels to as low as 1.5 mg L^{-1}, the maximum acceptable requirement set by the World Health Organization (WHO). Thus, with no addition of chemicals, coagulants, or flocculants, the EF treatment is a potential alternative for portable water treatment.

Furthermore, considering optimization studies, Zuo et al. [76] carried out analysis on ECEF for fluoride removal adjusting parameters such as pH (3–8) and hydraulic retention time (20–60 min). They observed that the optimal influent pH is set in the range 6–7, combined with a current density of 22 A m^{-2}. This finding actually represents an additional advantage for the treatment process since it does not require further pH adjustment.

Domestic Wastewater Treatment

With the increase in population density in urban areas, and increasingly stricter environmental legislation, the typical wastewater treatment plants (WWTPs) have become ineffective, with sludge being a secondary source of pollution. Thus, flotation stands out as a means of fine solids separation from the solution bulk, replacing the typical sedimentation tank and leading to a final clarification of active sludge process. Thus, the combined treatment of a bioreactor with a final EF clarifier stage has been studied by Cho et al. [61], focusing on nitrogen removal reaching up to 74% removal with this combined treatment. Similar studies were developed by Chung et al. [77], comparing the outcomes for an EF procedure when compared to natural gravity settling, also known as sedimentation tank. The main findings of his work concluded that EF presented thickening of 2.6–9.2 folds higher than sedimentation, besides presenting a better quality subnatant. Both studies highlight the ability of EF as a secondary clarifier for solid-liquid separation independent of sludge properties.

Additionally, over the years urban wastewater has received an increasing variation in emission source and contamination index. For example, since they are mostly located near residential and commercial areas, restaurants are well known to dispose of considerable amounts of oils and grease into conventional WWTPs, which are not prepared to efficiently treat this wastewater. In face of the need to previously treat the wastewater produced, methods such as EF are very attractive, especially because of their limited space requirement. Ji et al. [78] analyzed real wastewater from a university cafeteria and studied the treatment efficiency in applying an ECEF procedure. After optimization studies, they reached a total oil and grease removal of 95% in 34 min of treatment applying 43 A m^{-2}, reinforcing the advantage of rapid treatments applying electronic methodologies.

Industrial Wastewater Treatment

A wide variety of pollutants are potentially found in industrial wastewater, which has degradation ease strongly associated to its composition. Among the main burdens found in industrial wastewater treatment are its high COD, biological oxygen demand (BOD), toxicity, and strong color, among others [5,79]. Thus, EF, due to its nonselective nature, can positively contribute to increase effluent degradability, enabling a posterior biological treatment or even leading to complete contaminant mineralization, ensuring water safety.

In this sense, several studies have been developed [7,80–84] analyzing various industrial wastewaters to understand EF applicability. Some specific cases will therefore be described.

Found as one of the main components of molasses (a sugarcane by-product), melanoidins are toxic to many organisms due to their antioxidant properties. Previous attempts for anaerobic-aerobic conventional treatment degradation reaches only up to 6% or 7% of mineralization [85]. Liakos et al. [86] addressed melanoidin degradation by EF treatment, comparing their outcomes to conventional coagulation-flocculation procedures and finding that ECEF presents as an advantage the reduction of chemical additives. However, for this particular study case, they saw a loss in conductivity found by the formation of a passivation layer, and this was solved by the addition of Cl^{-} to acidic and neutral solutions, being possible to attain up to 90% of melanoidin removal in 120 min of treatment applying 0.5 A to aluminum electrodes.

With the increase in the personal care products industry, as well as the higher industrialization of food, considerable amounts of surfactants are used. These materials are not biodegraded in anaerobic conditions, and represent a risk to aquatic life when associated to other chemicals. Furthermore, due to the amphiphilic property of surfactants, they are potentially treated because of their strong tendency to adsorb at air/liquid interface.

Thus, studies developed by Skender et al. [64] analyzed EF treatment by applying 8.42 mA cm^{-2} to two stainless steel plates in Na_2SO_4 electrolyte with pH at 7 for the removal of Tween 20, with concentrations ranging from 20 to 600 mg L^{-1} (nonionic surfactant). A maximum of 74.79% of Tween 20 removal was obtained after 30 min of reaction.

Some industrial activities can form very small particle size contaminants, as in the case of mining activity, where particle size can be as small as 8 μm, which therefore presents a very difficult extraction for conventional removal treatments. The small particle passes through typical wastewater grids and is not easily treated by flocculation or coagulation, especially because of their large specific areas and low mass. In this sense, Gonzales et al. [87] studied the EF of cassiterite, using a hydrophobic bacteria (*Rhodococcus opacus*) which contributed to bubble adhesion. They found that the higher current densities and bacterial concentration would form the smallest bubbles, promoting an efficient route to treat these nanoparticles.

Heavy Metals

Due to their toxic nature and high solubility, heavy metals are considered to present a strong environmental hazard [6,7,88]. Mainly found in nature as result of industrial discharge (such as oil well drilling, metal finishing, ceramics, and inorganic pigment [83,88,89]), heavy metals (e.g., lead, barium, zinc, nickel, copper, manganese, chromium) are known to be found in several contamination and industrial spilling points.

Several studies have focused on tailoring EF for wastewater contaminated with heavy metal treatments, highlighting the advantages of EF treatment specially when low concentrations are considered [6,54]. Due to the specifically high toxicity found for heavy metals, even the lowest concentrations can lead to serious environmental damage, and most wastewater treatment does not satisfactorily attend the proper removal of these hydrophilic species.

As far as heavy metal is understood, the contamination occurs normally with metallic ion associations, presenting more than one compound as the pollutant. This specific configuration tends to burden wastewater treatment, it being very difficult to simultaneously remove all contaminants. Hence, studies developed by Mota et al. [6], treating by EF a synthetic solution containing Pb, Ba, and Zn, analyzed variations in pH, electrolysis time, current density, and electrolyte used. With their study, they were able to remove simultaneously up to 97% of the pollutants when applying a current density of 350 A m^{-2}, an ionic strength of 3.2×10^{-3} mol L^{-1}, a pH=10, ethanol (as foamy), and collector, resulting in a heavy metal ratio of 3.1 with only 20 min of EF treatment.

Furthermore, in order to improve the EF efficiency treatment the addition of flocculants can be used, especially in cases where the contaminant has a very aggressive behavior to the environment [54]. As an example, consider the Cr^{3+} that is typically seen in wastewaters from ink, dye, leather, and textile industries, which represent a serious hazard to humans and natural biomes. Studies carried out by Perfileva et al. [90] attempting to improve removal of Cr^{3+} from synthetic solutions pointed out that when adding M-10 (commercial flocculants) to the solution, the Cr^{3+} compounds particle sizes increased from 13.5 to 90 μm, leading to 99% extraction. In such cases where the contaminants are highly toxic, the addition and cost associated with flocculant insertion is well accepted. The results obtained in the cited study consider an addition of strictly 2 mg L^{-1} of flocculants in order to enhance 150 mg L^{-1} of Cr^{3+} compound removal.

Additionally, concerning the use of flocculants in the EF treatment of effluents contaminated with heavy metals, studies developed by Brodskiy et al. [54] analyzed the effect of adding flocculants of anionic, cationic, and nonionic natures to treat effluents contaminated with copper, nickel, cobalt, and manganese, obtaining several outcomes. In this study, it was possible to note that the addition of flocculants contributed positively for heavy metal adhesion to electrolytic gas bubbles. However, when adding nonionic flocculants, due to the positioning of hydrophilic ends of the flocculants, the adhesion of the gas bubbles was hindered, lowering metal ion removal about twofold. Studies like these highlight the importance of analyzing proper conditions, especially considering the contaminant to be treated.

Textile Dyes

Textile industries are typically known for their high production of wastewater contaminated with dye residues which contain carcinogenic, mutagenic, or teratogenic toxics and are resistant to common wastewater treatment plants [91]. Due to their high toxicity, textile wastewaters tend to resist biological attack and treatment alternatives are strongly required. EF, due to its nonselectivity, tends to efficiently remove chemical and biological oxygen demand levels combined with a very effective color removal percentage [80].

Further considering the different ionic nature of dyes, a study developed by Balla et al. [92] analyzed the behavior of combined treatment of ECEF, analyzing energy costs associated with treating effluents contaminated with reactive, disperse, and the mixture of both effluents. Thus solutions of 890, 3570, and 920 mg L^{-1} of disperse, reactive, and mixture dyes, respectively, were prepared and treated analyzing the influences of anode, current density, and pH. Thus, in order to attain an 80% decolorization, 170, 120, and 50 kWh kg^{-1} for reactive, disperse, and mixed dyes were found, being strongly related to the initial conductivity of the material.

Organic and Inorganic Compounds Removal

The presence of phosphate ions in wastewater has been commonly associated with the development of bacteria, mushrooms, and algae, which compromises water potability. Studies developed by Ricordel et al. [93] treating raw surface waters by EF combined with electrocoagulation found that a mild decrease in phosphate, nitrate, hardness, and total suspended solids was seen. These findings corroborate with a reduction in biocide uses, improving groundwater safety as well.

As for nitrate removal in wastewater treatment units, due to its potential risk associated with methemoglobinoma, hypertension, and thyroid problems, the need for proper treatment has been a continuously increasing problem. The main route for treatment in case of nitrate pollution involves coagulation, which has limited use due to the formation of large amounts of sludge. Thus, the use of electrochemical routes for nitrate removal has been proposed in face of the reduction in chemicals used and the possibility of high efficiency and adjustments. Recent studies by Majlesi et al. [28] showed that efficient removal from nitrate solutions (100 mg L^{-1}), reaching up to 96% removal, is possible with ECEF treatments. Furthermore, it was also shown that by altering the pH in the system, the co-products produced can be controlled by stimulating reduction or oxidation reactions.

For cases of biologically refractory pollutants such as ethylenediamine tetraacetic acid (EDTA), cyanide, etc., the use of specific treatments to enable biodegradability of these materials is required. Studies of EDTA degradation show that the use of various electrolytes will considerably influence the degradation of the material. Therefore, studies developed by Khelifa et al. [88] analyzing EDTA degradation in sulfate and chloride medium noted that due to the presence of active chlorine species (which are powerful oxidizing agents), EDTA degradation reached significant extent in this medium. The specific condition of treatment applying chloride ions generated electrochemically is known as electrochlorination. Furthermore, analysis carried out by the same group studied the combination of electrochlorination with EF, showing that for effluents contaminated with heavy metals and EDTA (typically seen in metal finishing industries) the one stage process proposed of electrochlorination-EF was able to initially degrade EDTA (which was not possible with the EF process alone) and free the metal ions for proper EF outcomes.

Combined Treatments

Wastewater treatment, due to the complexity of contaminants emitted in water bodies, has been increasingly studied, and several proposals are seen in order to promote complete mineralization and decontamination of the effluents generated. Thus, it is understood that in the near future wastewater treatment will strongly depend on the combination of treatment routes that, when combined, can ensure the safety and sustainability of water bodies.

Combined With Electrocoagulation

Also considered a physical means for wastewater treatment, the combination of electrocoagulation and electroflotation is understood as an electrolytic treatment where a sacrificial anode (iron or aluminum) dissolves and produces coagulant agents (Ag^{3+} or Fe^{2+}), while the cathode yields hydrogen that is responsible for the flotation process [28,91,94]. The wastewater treatment therefore occurs from coagulation, adsorption, precipitation, and flotation. The main advantages of this combination are: less sludge formation, compact equipment, rapid treatment, energy efficiency, safety, and selectivity [28,75].

The advantages of combining ECEF have been well known and studied for more than a decade. Studies developed by Ge et al. [95] analyzing the application of ECEF for laundry wastewater treatments, using 8 Ti plates and 21 Al plates applying from 0.8 to 2.0 A of current, found removal efficiencies of over 70% for COD within 30 min for treatment, sustaining a relatively neutral pH during the process. However, the need for proper dimensioning and proposal of standard equipment design has hindered the industrial application of this method [91].

Furthermore, due to the wide variety of parameters potentially controlled in ECEF treatment, such as current density, anode material, electrolyte concentration, pH, and reactor dimensions, the system can be continuously improved to reach optimum treatment conditions [96]. Hence, studies developed by Majlesi et al. [28] analyzed current density (0.4–3.2 mA cm^{-2}), anode-cathode material (Al-Fe, Al-Al), and pH (7–11), reaching removal efficiencies of up to 96% for nitrate removal in very rapid treatments (30 min). The main advantage in this case is the low cost associated with the treatment, which for this study was 0.98 kWh g^{-1} of nitrate.

Considering the electrolytic oxidation of the anode as being responsible for the coagulant generation, a low consumption of alkali reagents is seen due to the limited variations in pH, and the process is suitable for portable water treatment units, as previously described [75,76].

Combined With Adsorption and Membranes

One of the main advantages of EF treatment is the ability to treat a large variation of disperse phase particle sizes, which is mostly compromised by the negative charge of the particle. So, if the particle to be treated presents a high negative charge, it will not adsorb the bubbles promoted by EF and remain dispersed in the solution. Thus, membranes which physically filter these materials are seen as potential after treatment removing 1%–5% of the disperse phase, which presents diameters of 1–10 μm [97].

The EF technique has also been combined with adsorption process, where certain features are favorable for an optimized system. EF is thereby dependent on the ionic behavior of the contaminant, and in some cases will present low removal efficiency due to bonding limitations. In such conditions, adsorption means tend to collaborate once they form more complex structures that will easily tend to dissociate from the solution bulk and be suitable for EF treatments. In this sense, Zouboulis et al. [83] analyzed the influence of adding activated sludge for cadmium removal in a simulated effluent, and the experimental conditions were: natural pH ± 7, 5 mg L^{-1} of initial cadmium concentration, 0.5 g L^{-1} of activated sludge, aluminum electrodes, and 130 A m^{-2} current density. Operational parameters were analyzed, such as pH $(2-12)$, current density (50–200 A m^{-2}), and biomass concentration (0.5–3 g L^{-1}). Treatment outcomes for combined activated sludge and EF, and EF alone, were compared, finding that total cadmium removal was possible in the combined treatment, while 25% of removal was seen for EF alone. Thus, by adding activated sludge as an adsorbent, the process is optimized fourfold without compromising operational costs or adding chemical flocculants.

Combined With Other Treatments

As the demand for treatment efficiency becomes stricter, and pollutants more complex, the variety of treatment routes are even more diversified. Furthermore, in the continuous seeking to promote wastewater treatment with environmentally friendly approaches, EF combined with other advanced oxidation processes (AOPs) is commonly seen. For instance, Yang et al. [98] carried out studies on combined EF with electrooxidation (EO) for sludge solubilization. These authors found that EF has a very short optimum operation time (10 min), applying 5 A in an experimental design containing 11 electrodes (5 anodes and 6 cathodes). Therefore, when a consecutive EO applying 10 mA cm^{-2} was carried out, COD solubilization was improved by 20% depending on electrode gap, current density, and potential applied.

Additionally, combination of EF with AOPs have been specially applied when the contaminated effluent is particularly complex. For example, Esfandyari et al. [99] combined peroxi-electrocoagulation and electrooxidation-electroflotation in the treatment of olive mill wastewater by using bipolar aluminum electrodes analyzing influence of pH, current density, NaCl concentration, and peroxide concentration. Due to the high toxicity of the contaminant, such efforts are becoming increasingly common in order to deal with the high COD, BOD, color, polyphenols, and such. For this specific study, optimal conditions were established at $H_2O_2 = 1000$ mg L^{-1}, NaCl = 1 g L^{-1}, pH = 4, current density = 40 mA cm^{-2}, and retention time of 30 min. All parameters analyzed presented a reduction of 90% in these conditions.

Furthermore, effluents that present high natural salinity are also good candidates for AOPs treatments since they do not require additional electrolytes. For example, Yunqing et al. [100] studied total ammonia nitrogen (TAN), nitrite, and COD removal from a marine recirculating aquaculture system by applying electrooxidation combined with further EF. Considering the high loads of animal excreting and noneaten decomposing feed, this wastewater represents a high burden on environmental impact. In this study, a correlation between TAN, nitrite and COD was clearly seen. The optimal operation conditions were: 25 A m^{-2} of current density and 40 min of retention time for 0.6 L of wastewater, reaching an overall 95% removal. Additionally, energy costs were estimated to be 1.75 kWh m^{-3} of wastewater treated.

A brief summary on the various applications of EF and combined proposals on synthetic and real wastewater treatments can be seen in Table 1.

COST ANALYSIS

The cost of the EF treatment involves the cost of the system apparatus (tank, electrodes, etc.), the installation, which is estimated to be 25% of the capital cost for the system, and the sludge treatment and disposal cost [89]. Besides the cost of electricity, once the electrochemical methods are applied for effluent treatment, an increase on energy consumption would lead to a subsequent increase in treatment cost, since current density is one of the most important variables in the process [5]. As discussed previously on the energy requirements topic, the energy consumption (Eq. 13) depends on the potential between anode and cathode electrodes, which is dependent on the current density, and it is mainly dependent on the solution resistance ohmic potential drop. In addition, the overpotentials are related to many factors, including the electrochemical properties of electrodes, the current density used, and the pH value of wastewater [1].

The EF process is versatile and competitive when compared to other flotation processes, such as DAF and IAF, since the EF system is small, compact, and requires low maintenance, which reduces costs when compared to other flotation processes. Also, EF presents three main advantages when compared to other flotation process: (i) the bubbles are extremely small and uniform, increasing the contact surface among the pollutant and gases; (ii) bubble distribution and size can be controlled by current density applied, increasing the possibility of collisions between the bubbles and the pollutant; and (iii) the process can be optimized,

TABLE 1 Summary of Main Synthetic and Real Wastewater Treatments Using EF as Innovative Treatment: Experimental Conditions and Efficiencies

Wastewater treatment with EF					
Anodes	**pH conditions**	**Current density**	**Pollutant**	**Treatment efficiency (%)**	**Ref.**
Al	10	350 A m^{-2}	Heavy metals	97	[6]
Ti-IrO_2	7	150 A m^{-2}	Bitumen	92.6	[11]
Ti/IrO_2-Sb_2O_5-SnO_2	5–9	2 F m^{-2}	Restaurant wastewater	84.3	[16]
Graphite-*Al*	7.5	1–3.6 A cm^{-2}	Emulsion and oil-bearing wastewater	94–95	[17]
PbO_2-Al	4.5	1.2 A	Oil-water emulsions	92	[18]
Al	7	8.42 mA cm^{-2}	Tween 20	74.79	[64]
Al	4–5	12 mA cm^{-2}	Fluor	100	[75]
Al	6–7	75 A m^{-2}	Fluor	79	[76]
Ti/IrO_2	7.5	100 A m^{-2}	Synthetic wastewater	94	[77]
Al	5–6	43 A m^{-2}	Cafeteria wastewater	>95	[78]
Al	8	20 V	Heavy metals	99	[80]
Al-Ti/$Ru_{0.34}Ti_{0.66}O_2$	4.5	19.40 A m^{-2}	Emulsified oil	95.90	[81]
Al	5	0.5 A m^{-2}	Melanoidins	95	[86]
Al-Fe	7.5	40 mA cm^{-2}	Dyes	98	[92]
Ti-Al	6–7	2.0 A m^{-2}	Laundry wastewater	70	[95]
Al-Fe	7–11	3.2 mA cm^{-2}	Nitrate	96	[28]

choosing adequate electrode material and electrolytes. Additionally, pollutant separation occurs faster in EF than in flotation [27]. Poon [89] reported that the cost of the EF treatment to remove nickel, lead, zinc, and copper/cyanide from polluted groundwater at a contaminated site was lower than for conventional treatment system using cyanide oxidation/alkaline precipitation/polymer-aided clarification, saving

43% by using the EF system. The EF system was composed by one reactor that replaced three tanks of the conventional treatment used in succession for cyanide oxidation, alkaline cadmium precipitation, and polymer-aided clarification, respectively. In the compact EF system, the three steps took place simultaneously in one reactor.

Hooshmandfar et al. [41] studied the optimization of energy consumption and material on the removal of Acid Red 14 by the ECEF process, concluding that the advantages of this method are low consumption of materials and energy and low production of sludge, which will generate economy associated with disposal costs.

Ulucan and Kurt [101] studied the treatability of bilge water treatment by ECEF (using aluminum and iron electrodes with a size of 50 mm × 200 mm) and by electro-Fenton (addition of H_2O_2). Bilge waters can be briefly defined as saline and greasy wastewater. Bilge waters that are stored at the bottom of ships are greasy wastewaters, including a number of pollutants with various characteristics. The ECEF with aluminum electrodes was more efficient and removed the pollutant in less time (10 min) than the process using iron. However, electro-Fenton presented higher removal efficiencies adapting the effluent to the standards applied by the International Convention on Preventing Pollution from Ships (MARPOL). Besides, the electro-Fenton process presented less energy cost, being 0.51 $\$ \, m^{-3}$, while ECEF for aluminum and iron cost 0.54 and 0.53 $\$ \, m^{-3}$, respectively.

Palomino-Romero et al. [102], after estimating the total cost of a pilot plant with the capacity to treat 7.92 m^3 of domestic wastewater per day, concluded that EF is very promising for large-scale treatment of extremely contaminated domestic wastewater. Aluminum electrodes were used as an anode and a cathode as parallel monopolar connections, the total geometric area was 5.6 m^2, and the electrodes were spaced in 2 cm; the EF reactor operated continuously. Costs were estimated as direct (electricity, electrodes, and reagents) and indirect (equipment, construction material, and reactor operation); the costs with safety were not included. The operational cost was calculated based on the Brazilian economy in 2011. The optimized condition (current density of 14.18 $A \, m^{-2}$, generating 4.25 V, flow of 0.33 $m \, h^{-1}$, and adding 0.5 g of NaCl) had its costs estimated in 0.12 $\$ \, m^{-3}$ (using the charges of the Company of Power and Light of São Paulo—*Companhia Paulista de Força e Luz*—CPFL). Therefore, the estimated cost to treat effluent was 0.37 $\$ \, m^{-3}$, considering electric energy and the value of adding NaCl.

In Hong Kong, businesses that generate domestic sewage above the pollution limits are charged an extra fee to discharge their effluents in public sewage. Qin et al. [16] observed that the previous local treatment (before the effluent is discharged in public sewage) by the ECEF in restaurants with high COD, many suspended solids, and a high level of oils, reduces the pollution up to 52%. This procedure enables the effluent to be discharged in public sewage free of charge.

CONCLUDING REMARKS AND PERSPECTIVES

Electroflotation is an attractive method in wastewater treatment because it presents an easy and safe operation, no harmful chemical reagents are added, automation and process control are possible, and low energy consumption is seen. Besides, it is well known that EF presents high efficiency in removal of pollutants since fine bubbles are produced by electrolysis. Furthermore, several variables affect EF treatment efficiency, such as electrode (type, treatment, form, and sizes), electrolyte (type and concentration), current density, pH, and hydrodynamic regimen, which makes it difficult to determine the effect of the variables on EF treatment efficiency; consequently, it is difficult to design efficient EF systems for fine particle recovery.

Before a viable EF process is developed, it is necessary to find the best electrode materials since the heart of the EF process is the electrode system that produces fine bubbles by electrolysis. Research has recently focused on material sciences to find electrodes with adequate and appropriate mechanical and geometrical structures, which have good electrical properties and an extended lifetime, which present low O_2 or H_2 evolution overpotentials, and which have an acceptable cost to make the electrolytic systems more competitive than the other processes (chemical, physical, and biological process). Another burden is related to the electrolytic reactors, which have to be designed to obtain excellent mass transfer, thus increasing the effectiveness of the treatment.

Due to the complexity of contaminants emitted in water bodies, combined treatment methods involving electrochemical techniques, including EF, have been developed and must be further developed to enhance the efficiency of biological, chemical, and physical processes and reduce energy consumption. Therefore, it is understood that in the near future wastewater treatment will strongly depend on the combination of treatment routes that when combined can ensure the safety and sustainability of water bodies. Additionally, due to the variety of pollutants in aqueous effluents, studies have shown that different types of effluent have different optimal conditions for operational parameters, such as current density, pH, and electrode type, which makes it difficult to establish a standard tendency for the process general efficiency. Therefore, research focusing on the effect of different effluents in the same operational conditions can be useful or even the next step when seeking to enable EF industrial application.

The energy consumption and mainly the cost operation of the EF process to treat water and wastewater also need further study by researchers to motivate the industrial application of the EF process. Another action that can motivate the industrial application of the EF process is the imposition of more rigid environmental laws for the disposal of wastewater. Although EF is an economically feasible technology to treat some of the most environmentally troublesome effluents [5,9], there is a big gap of unexplored industrial applications.

Therefore, perhaps with progress in electrochemistry, including suitable electrode materials and hence improved design for EF cells, the EF process can be an interesting technique which is economically feasible for industrial applications.

References

[1] X. Chen, G. Chen, P.L. Yue, Novel electrode system for electroflotation of wastewater, Environ. Sci. Technol. 36 (2002) 778–783, https://doi.org/10.1021/es011003u.
[2] G. Bhaskar Raju, P.R. Khangaonkar, Electroflotation-A critical review, Trans. Indian Inst. Metals 37 (1984) 59–66.
[3] X. Chen, G. Chen, Electroflotation, in: C. Comninellis, G. Chen (Eds.), Electrochemistry for the Environment, Vol. 1, (2010) 263–278.
[4] J. Rubio, R.W. Smith, Overview of flotation as a wastewater treatment technique, Miner. Eng. 15 (2002) 139–155.
[5] W.T. Mook, M.K. Aroua, G. Issabayeva, Prospective applications of renewable energy based electrochemical systems in wastewater treatment: a review, Renew. Sust. Energ. Rev. 38 (2014) 36–46, https://doi.org/10.1016/j.rser.2014.05.042.
[6] I. Mota, J.A. Castro, R. Casqueira, A.G. Oliveira Junior, Study of electroflotation method for treatment of wastewater from washing soil contaminated by heavy metals, J. Mater. Res. Technol. 4 (2015) 109–113, https://doi.org/10.1016/j.jmrt.2014.11.004.
[7] G.Z. Kyzas, K.A. Matis, Electroflotation process: a review, J. Mol. Liq. 220 (2016) 657–664, https://doi.org/10.1016/j.molliq.2016.04.128.
[8] M.M. Emamjomeh, M. Sivakumar, Review of pollutants removed by electrocoagulation and electrocoagulation/flotation processes, J. Environ. Manag. 90 (2009) 1663–1679, https://doi.org/10.1016/j.jenvman.2008.12.011.
[9] K. Muller, Electroflotation from the double layer to troubled waters, in: O.J. Murphy, S. Srinivasan, B.E. Conway (Eds.), Electrochemistry in Transition—From 20th to 21st Century, (1992) 702, https://doi.org/10.1002/anie.201306828.
[10] S.G. Da Cruz, A.J.B. Dutra, M.B.M. Monte, The influence of some parameters on bubble average diameter in an electroflotation cell by laser diffraction method, J. Environ. Chem. Eng. 4 (2016) 3681–3687, https://doi.org/10.1016/j.jece.2016.05.017.
[11] R. Alam, J.Q. Shang, Removal of bitumen from mature oil sands tailings slurries by electro-flotation, J. Water Process. Eng. (2016) 1–5, https://doi.org/10.1016/j.talanta.2011.11.033.
[12] M.S.K.A. Sarkar, S.W. Donne, G.M. Evans, Utilization of hydrogen in electroflotation of silica, Adv. Powder Technol. 22 (2011) 482–492, https://doi.org/10.1016/j.apt.2011.05.007.
[13] D. Ghernaout, C. Benblidia, F. Khemici, Microalgae removal from Ghrib dam (Ain Defla, Algeria) water by electroflotation using stainless steel electrodes, Desalin. Water Treat. 54 (2015) 3328–3337, https://doi.org/10.1080/19443994.2014.907749.

[14] R.G. Casqueira, M.L. Torem, H.M. Kohler, The removal of zinc from liquid streams by electroflotation, Miner. Eng. 19 (2006) 1388–1392, https://doi.org/10.1016/j.mineng.2006.02.001.
[15] L.J.J. Janssen, C.W.M.P. Sillen, E. Barendrecht, S.J.D. van Stralen, Bubble behaviour during oxygen and hydrogen evolution at transparent electrodes in KOH solution, Electrochim. Acta 29 (1984) 633–642, https://doi.org/10.1016/0013-4686(84)87122-4.
[16] X. Qin, B. Yang, F. Gao, G. Chen, Treatment of restaurant wastewater by pilot-scale electrocoagulation-electroflotation: optimization of operating conditions, J. Environ. Eng. (2013) 1004–1016, https://doi.org/10.1061/(ASCE)EE.1943-7870.0000687.
[17] E.A. Maksimov, A.A. Ostsemin, Intensifying the cleaning of emulsion- and oil-bearing waste water from rolled-product manufacturing by electroflotation, Metallurgist 58 (2015) 945–949, https://doi.org/10.1007/s11015-015-0022-8.
[18] A.Y. Hosny, Separating oil from oil-water emulsions by electroflotation technique, Sep. Technol. 6 (1996) 9–17, https://doi.org/10.1016/0956-9618(95)00136-0.
[19] M.S.K.A. Sarkar, G.M. Evans, S.W. Donne, Bubble size measurement in electroflotation, Miner. Eng. 23 (2010) 1058–1065, https://doi.org/10.1016/j.mineng.2010.08.015.
[20] H.B. Beer, Electrode and coating therefore. USA Patent 3,632,498, 1972.
[21] N.K. Khosla, S. Venkatachalam, P. Somasundaran, Pulsed electrogeneration of bubbles for electroflotation, J. Appl. Electrochem. 21 (1991) 986–990, https://doi.org/10.1007/BF01077584.
[22] S. Venkatachalam, Electrogenerated gas bubbles in flotation, Miner. Process. Extr. Metall. Rev. 8 (1992) 47–55, https://doi.org/10.1080/08827509208952677.
[23] M.E. Tawfik, F.J. Diez, On the relation between onset of bubble nucleation and gas supersaturation concentration, Electrochim. Acta 146 (2014) 792–797, https://doi.org/10.1016/j.electacta.2014.08.147.
[24] K.a. Matis, E.N. Peleka, Alternative flotation techniques for wastewater treatment: focus on electroflotation, Sep. Sci. Technol. 45 (2010) 2465–2474, https://doi.org/10.1080/01496395.2010.508065.
[25] L. Ben Mansour, S. Chalbi, I. Kesentini, Experimental study of hydrodynamic and bubble size distributions in electroflotation process, Indian J. Chem. Technol. 14 (2007) 253–257.
[26] G. Chen, Electrochemical technologies in wastewater treatment, Sep. Purif. Technol. 38 (2004) 11–41, https://doi.org/10.1016/j.seppur.2003.10.006.
[27] R. Alam, J.Q. Shang, Removal of bitumen from mature oil sands tailings slurries by electro-flotation, J. Water Process Eng. 15 (2017) 116–123, https://doi.org/10.1016/j.jwpe.2016.10.002.
[28] M. Majlesi, S.M. Mohseny, M. Sardar, S. Golmohammadi, A. Sheikhmohammadi, Improvement of aqueous nitrate removal by using continuous electrocoagulation/electroflotation unit with vertical monopolar electrodes, Sustain. Environ. Res. 26 (2016) 287–290, https://doi.org/10.1016/j.serj.2016.09.002.
[29] H. Shin, K. Kim, J. Jung, S.C. Bai, Y.K. Chang, J. Han, Harvesting of Scenedesmus obliquus cultivated in seawater using electro-flotation, Korean J. Chem. Eng. 34 (2017) 62–65, https://doi.org/10.1007/s11814-016-0251-y.
[30] F. Cardarelli, P. Taxil, A. Savall, Preparation of oxygen evolving electrodes with long service life under extreme conditions, J. Appl. Electrochem. 28 (1998) 245–250, https://doi.org/10.1023/A:1003251329958.
[31] G.P. Vercesi, J. Rolewicz, C. Comninellis, J. Hinder, Characterization of dsa-type oxygen evolving electrodes. choice of base metal, Thermochim. Acta 176 (1991) 31–47, https://doi.org/10.1016/0040-6031(91)80257-J.
[32] R. Mráz, J. Krýsa, Long service life IrO2/Ta2O5 electrodes for electroflotation, J. Appl. Electrochem. 24 (1994) 1262–1266, https://doi.org/10.1007/BF00249891.
[33] V.A. Alves, L.A. Da Silva, E. De Oliveira, J.F.C. Boodts, Investigation under conditions of accelerated anodic corrosion of the effect of TiO2 substitution by CeO2 on the stability of

Ir-based ceramic coatings, Mater. Sci. Forum 289–292 (1998) 655–666, https://doi.org/10.4028/www.scientific.net/MSF.289-292.655.
[34] B. Merzouk, B. Gourich, A. Sekki, K. Madani, M. Chibane, Removal turbidity and separation of heavy metals using electrocoagulation-electroflotation technique. A case study, J. Hazard. Mater. 164 (2009) 215–222, https://doi.org/10.1016/j.jhazmat.2008.07.144.
[35] H. Chenik, M. Elhafdi, A. Dassaa, A.H. Essadki, M. Azzi, Removal of real textile dyes by electrocoagulation/electroflotation in a pilot external-loop airlift reactor, Water Res. Prot. 5 (2013) 1000–1006, https://doi.org/10.4236/jwarp.2013.510104.
[36] S. Aoudj, A. Khelifa, N. Drouiche, R. Belkada, D. Miroud, Simultaneous removal of chromium(VI) and fluoride by electrocoagulation-electroflotation: application of a hybrid Fe-Al anode, Chem. Eng. J. 267 (2015) 153–162, https://doi.org/10.1016/j.cej.2014.12.081.
[37] C. Llerena, J.C.K. Ho, D.L. Piron, Effects of ph on electroflotation of sphalerite, Chem. Eng. Commun. 155 (1996) 217–228, https://doi.org/10.1080/00986449608936664.
[38] G. Montes-Atenas, F.J. Garcia-Garcia, R. Mermillod-Blondin, S. Montes, Effect of suspension chemistry onto voltage drop: application to electro-flotation, Powder Technol. 204 (2010) 1–10, https://doi.org/10.1016/j.powtec.2010.06.024.
[39] M.Y.A. Mollah, P. Morkovsky, J.A.G. Gomes, M. Kesmez, J. Parga, D.L. Cocke, Fundamentals, present and future perspectives of electrocoagulation, J. Hazard. Mater. 114 (2004) 199–210, https://doi.org/10.1016/j.jhazmat.2004.08.009.
[40] R.G. De Carvalho Neto, J.G. Da Silva Do, M.C.C. Nascimento, A.C. Lopes, E.F. Abdala Neto, C. Rossas Mota, A.B.D.S. Filho, Microalgae harvesting and cell disruption: a preliminary evaluation of the technology electroflotation by alternating current, Water Sci. Technol. 70 (2014) 315–320, https://doi.org/10.2166/wst.2014.220.
[41] A. Hooshmandfar, B. Ayati, A.K. Darban, Optimization of material and energy consumption for removal of acid red 14 by simultaneous electrocoagulation and electroflotation, Water Sci. Technol. 73 (2016) 192–202, https://doi.org/10.2166/wst.2015.477.
[42] S. Aoudj, A. Khelifa, N. Drouiche, M. Hecini, Removal of fluoride and turbidity from semiconductor industry wastewater by combined coagulation and electroflotation, Desalin. Water Treat. 57 (2016) 18398–18405, https://doi.org/10.1080/19443994.2015.1095120.
[43] M. Kotti, I. Ksentini, L. Ben Mansour, Impact of hydrodynamic regime on the capacity of oxygen transfer in a stirred electroflotation column, Desalin. Water Treat. 52 (2014) 1693–1698, https://doi.org/10.1080/19443994.2013.874136.
[44] I. Ksentini, M. Kotti, L. Ben Mansour, Effect of liquid phase physicochemical characteristics on hydrodynamics of an electroflotation column, Desalin. Water Treat. 52 (2013) 1–8, https://doi.org/10.1080/19443994.2013.800279.
[45] A. Pandey, C.R. Soccol, P. Nigam, V.T. Soccol, Biotechnological potential of agro-industrial residues. I: sugarcane bagasse, Bioresour. Technol. 74 (2000) 69–80, https://doi.org/10.1016/S0960-8524(99)00142-X.
[46] V.I. Il'in, V.A. Kolesnikov, Y.I. Parshina, Purification of highly concentrated industrial sewage from the porcelain and faience industry by the electric flotation method, Glas. Ceram. 59 (2002) 242–244, https://doi.org/10.1023/a:1020983329822.
[47] A.S. Gadd, D.R. Ryan, J.M. Kavanagh, G.W. Barton, Design development of an electrocoagulation reactor for molasses process wastewater treatment, Water Sci. Technol. 61 (2010) 3221–3227, https://doi.org/10.2166/wst.2010.086.
[48] I. Ksentini, M.L. Aouadi, H. BenBacha, L. BenMansour, Solar energy integration in the treatment of industrial effluent by coagulation-electroflotation, Desalin. Water Treat. 20 (2010) 60–65, https://doi.org/10.5004/dwt.2010.1398.
[49] I. Ksentini, L. Ben Mansour, Modeling the hydrodynamic of an electroflotation column for the treatment of industrial wastewaters, Desalin. Water Treat. 56 (2015) 1722–1727, https://doi.org/10.1080/19443994.2014.950989.

[50] K.A. Matis, G.P. Gallios, K.A. Kydros, Separation of fines by flotation techniques, Sep. Technol. 3 (1993) 76–90, https://doi.org/10.1016/0956-9618(93)80007-E.
[51] K.A. Matis, P. Mavros, Foam/froth flotation II. Removal of particle matter, Sep. Purif. Methods. 20 (1991) 163–198, https://doi.org/10.1080/03602549108021414.
[52] N.P. Gamage, J.D. Rimer, S. Chellam, Improvements in permeate flux by aluminum electroflotation pretreatment during microfiltration of surface water, J. Membr. Sci. 411–412 (2012) 45–53, https://doi.org/10.1016/j.memsci.2012.04.014.
[53] A.V. Kolesnikov, V.V. Kuznetsov, V.A. Kolesnikov, Y.I. Kapustin, The role of surfactants in the electroflotation extraction of copper, nickel, and zinc hydroxides and phosphates, Theor. Found. Chem. Eng. 49 (2015) 1–9, https://doi.org/10.1134/S0040579515010042.
[54] V.A. Brodskiy, V.A. Kolesnikov, V.I. Il'in, Effect of the physicochemical characteristics of the disperse phase of slightly soluble compounds of nonferrous metals on the efficiency of their electroflotation extraction from aqueous solutions, Theor. Found. Chem. Eng. 49 (2015) 138–144, https://doi.org/10.1134/S0040579515020037.
[55] F. Baierle, D.K. John, M.P. Souza, T.R. Bjerk, M.S.A. Moraes, M. Hoeltz, A.L.B. Rohlfes, M.E. Camargo, V.A. Corbellini, R.C.S. Schneider, Biomass from microalgae separation by electroflotation with iron and aluminum spiral electrodes, Chem. Eng. J. 267 (2015) 274–281, https://doi.org/10.1016/j.cej.2015.01.031.
[56] C. Jiménez, B. Talavera, C. Sáez, P. Cañizares, M.A. Rodrigo, Study of the production of hydrogen bubbles at low current densities for electroflotation processes, J. Chem. Technol. Biotechnol. 85 (2010) 1368–1373, https://doi.org/10.1002/jctb.2442.
[57] B. Lakghomi, Y. Lawryshyn, R. Hofmann, A model of particle removal in a dissolved air flotation tank: importance of stratified flow and bubble size, Water Res. 68 (2015) 262–272, https://doi.org/10.1016/j.watres.2014.09.053.
[58] L. Ren, Y. Zhang, W. Qin, S. Bao, P. Wang, C. Yang, Investigation of condition-induced bubble size and distribution in electroflotation using a high-speed camera, Int. J. Min. Sci. Technol. 24 (2014) 7–12, https://doi.org/10.1016/j.ijmst.2013.12.002.
[59] I. Arslan-Alaton, B.H. Gursoy, A. Akyol, M. Kobya, M. Bayramoglu, Modeling and optimization of acid dye manufacturing wastewater treatment with Fenton's reagent: comparison with electrocoagulation treatment results and effects on activated sludge inhibition, Water Sci. Technol. 62 (2010) 209–216, https://doi.org/10.2166/wst.2010.256.
[60] M.R.G. Santos, M.O.F. Goulart, J. Tonholo, C.L.P.S. Zanta, The application of electrochemical technology to the remediation of oily wastewater, Chemosphere 64 (2006) 393–399, https://doi.org/10.1016/j.chemosphere.2005.12.036.
[61] K. Cho, C.M. Chung, Y.J. Kim, M.R. Hoffmann, T.H. Chung, Electroflotation clarifier to enhance nitrogen removal in a two-stage alternating aeration bioreactor, Environ. Technol. 34 (2013) 2769–2776, https://doi.org/10.1080/09593330.2013.788072.
[62] P. Gao, X. Chen, F. Shen, G. Chen, Removal of chromium(VI) from wastewater by combined electrocoagulation- electroflotation without a filter, Sep. Purif. Technol. 43 (2005) 117–123, https://doi.org/10.1016/j.seppur.2004.10.008.
[63] M.S.K.A. Sarkar, S.W. Donne, G.M. Evans, Hydrogen bubble flotation of silica, Adv. Powder Technol. 21 (2010) 412–418, https://doi.org/10.1016/j.apt.2010.04.005.
[64] A. Skender, N. Moulai-Mostefa, M. Tir, Effects of operational parameters on the removal efficiency of non-ionic surfactant by electroflotation, Desalin. Water Treat. 13 (2010) 213–216, https://doi.org/10.5004/dwt.2010.992.
[65] M.Y. Ibrahim, S.R. Mostafa, M.F.M. Fahmy, A.I. Hafez, Utilization of Electroflotation in remediation of oily wastewater, Sep. Sci. Technol. 36 (2001) 3749–3762, https://doi.org/10.1081/SS-100108360.
[66] R.H. Rojas, M.L. Torem, A.G. Merma, J.G.S. Puelles, R. De Janeiro, R. De Janeiro-rj, in: Main factors affecting the size of hydrogen and oxygen bubbles, IMPC 2016 XXVIII International Mineral Processing Congress Proceedings, 2016, pp. 1–10.

[67] L.J.J. Janssen, J.G. Hoogland, The effect of electrolytically evolved gas bubbles on the thickness of the diffusion layer-II, Electrochim. Acta 18 (1973) 543–550, https://doi.org/10.1016/0013-4686(73)85016-9.
[68] R.M. Bande, B. Prasad, I.M. Mishra, K.L. Wasewar, Oil field effluent water treatment for safe disposal by electroflotation, Chem. Eng. J. 137 (2008) 503–509, https://doi.org/10.1016/j.cej.2007.05.003.
[69] R.H. Yoon, G.H. Luttrell, The effect of bubble size on fine particle flotation, Miner. Process. Extr. Metall. Rev. 5 (1989) 101–122, https://doi.org/10.1080/08827508908952646.
[70] A.K. Golder, A.N. Samanta, S. Ray, Removal of Cr3+ by electrocoagulation with multiple electrodes: bipolar and monopolar configurations, J. Hazard. Mater. 141 (2007) 653–661, https://doi.org/10.1016/j.jhazmat.2006.07.025.
[71] V.K. Scott, Electrochemical Process for Clean Technology, Royal Society of Chemistry, Cambridge, 1995.
[72] R. Alam, J.Q. Shang, Electrochemical model of electro-flotation, J. Water Process Eng. 12 (2016) 78–88, https://doi.org/10.1016/j.jwpe.2016.06.009.
[73] I. Heidmann, W. Calmano, Removal of Ni, Cu and Cr from a galvanic wastewater in an electrocoagulation system with Fe- and Al-electrodes, Sep. Purif. Technol. 71 (2010) 308–314, https://doi.org/10.1016/j.seppur.2009.12.016.
[74] N. Nagai, M. Takeuchi, T. Kimura, T. Oka, Existence of optimum space between electrodes on hydrogen production by water electrolysis, Int. J. Hydrog. Energy 28 (2003) 35–41, https://doi.org/10.1016/S0360-3199(02)00027-7.
[75] A.H. Essadki, B. Gourich, C. Vial, H. Delmas, M. Bennajah, Defluoridation of drinking water by electrocoagulation/electroflotation in a stirred tank reactor with a comparative performance to an external-loop airlift reactor, 168 (2009) 1325–1333. https://doi.org/10.1016/j.jhazmat.2009.03.021.
[76] Q. Zuo, X. Chen, W. Li, G. Chen, Combined electrocoagulation and electroflotation for removal of fluoride from drinking water, 159 (2008) 452–457. https://doi.org/10.1016/j.jhazmat.2008.02.039.
[77] C.M. Chung, K.W. Cho, S.W. Hong, Y.J. Kim, T.H. Chung, Feasibility of electroflotation to separate solids and liquid in an activated sludge process, Environ. Technol. 30 (2009) 1565–1573, https://doi.org/10.1080/09593330903313802.
[78] M. Ji, X. Jiang, F. Wang, A mechanistic approach and response surface optimization of the removal of oil and grease from restaurant wastewater by electrocoagulation and electroflotation, Desalin. Water Treat. 1 (2014) 37–41, https://doi.org/10.1080/19443994.2014.929034.
[79] D. Ghernaout, M.W. Naceur, B. Ghernaout, A review of electrocoagulation as a promising coagulation process for improved organic and inorganic matters removal by electrophoresis and electroflotation, Desalin. Water Treat. 28 (2011) 287–320, https://doi.org/10.5004/dwt.2011.1493.
[80] M. Belkacem, M. Khodir, S. Abdelkrim, Treatment characteristics of textile wastewater and removal of heavy metals using the electroflotation technique, Desalination 228 (2008) 245–254, https://doi.org/10.1016/j.desal.2007.10.013.
[81] F.N.B. Nahui, M.R. Nascimento, E.B. Cavalcanti, E.O. Vilar, Electroflotation of emulsified oil in industrial wastes evaluated with a full factorial design, Brazilian J. Chem. Eng. 25 (2008) 435–442, https://doi.org/10.1590/S0104-66322008000300001.
[82] M.F. García-Montoya, S. Gutiérrez-Granados, A. Alatorre-Ordaz, R. Galindo, R. Ornelas, J.M. Peralta-Hernández, Application of electrochemical/BDD process for the treatment wastewater effluents containing pharmaceutical compounds, J. Ind. Eng. Chem. 31 (2015) 238–243, https://doi.org/10.1016/j.jiec.2015.06.030.
[83] A.I. Zouboulis, K.A. Matis, Cadmium ion removal by electroflotation onto sewage sludge biomass, Int. J. Environ. Waste Manag. 9 (2012) 246–256.

[84] L. Ben Mansour, S. Chalbi, Removal of oil from oil/water emulsions using electroflotation process, J. Appl. Electrochem. 36 (2006) 577–581, https://doi.org/10.1007/s10800-005-9109-4.
[85] V.O. Ojijo, M.S. Onyango, A. Ochieng, Decolourization of Melanoidin containing wastewater using south African coal fly ash, Int. J. Civ. Environ. Eng. 2 (2010) 17–23.
[86] T.I. Liakos, N.K. Lazaridis, Melanoidins removal from simulated and real wastewaters by coagulation and electro-flotation, Chem. Eng. J. 242 (2014) 269–277, https://doi.org/10.1016/j.cej.2014.01.003.
[87] L. Gonzales, G. Pino, M.L. Torem, Electroflotation of cassiterite fines using a hydrophobic bacterium strain, Mining 66 (2013) 507–512.
[88] A. Khelifa, S. Aoudj, S. Moulay, M. De Petris-Wery, A one-step electrochlorination/electroflotation process for the treatment of heavy metals wastewater in presence of EDTA, Chem. Eng. Process. Process Intensif. 70 (2013) 110–116, https://doi.org/10.1016/j.cep.2013.04.013.
[89] C.P.C. Poon, Electroflotation for groundwater decontamination, J. Hazard. Mater. 55 (1997) 159–170, https://doi.org/10.1016/S0304-3894(97)00013-7.
[90] A.V. Perfileva, V.A. Brodskii, V.I. Il, V.A. Kolesnikov, Effect of the composition of the medium and electroflotation processing parameters on extraction efficiency of chromium (III) dispersed phase from aqueous solutions, Russ. J. Appl. Chem. 89 (2016) 388–393, https://doi.org/10.1134/S1070427216030071.
[91] B. Merzouk, K. Madani, A. Sekki, Using electrocoagulation-electroflotation technology to treat synthetic solution and textile wastewater, two case studies, Desalination 250 (2010) 573–577, https://doi.org/10.1016/j.desal.2009.09.026.
[92] W. Balla, A.H. Essadki, B. Gourich, A. Dassaa, H. Chenik, M. Azzi, Electrocoagulation/electroflotation of reactive, disperse and mixture dyes in an external-loop airlift reactor, J. Hazard. Mater. 184 (2010) 710–716, https://doi.org/10.1016/j.jhazmat.2010.08.097.
[93] C. Ricordel, A. Darchen, D. Hadjiev, Electrocoagulation-electroflotation as a surface water treatment for industrial uses, Sep. Purif. Technol. 74 (2010) 342–347, https://doi.org/10.1016/j.seppur.2010.06.024.
[94] H. Särkkä, A. Bhatnagar, M. Sillanpää, Recent developments of electro-oxidation in water treatment–a review, J. Electroanal. Chem. 754 (2015) 46–56, https://doi.org/10.1016/j.jelechem.2015.06.016.
[95] J. Ge, J. Qu, P. Lei, H. Liu, New bipolar electrocoagulation-electroflotation process for the treatment of laundry wastewater, Sep. Purif. Technol. 36 (2004) 33–39, https://doi.org/10.1016/S1383-5866(03)00150-3.
[96] A.H. Essadki, B. Gourich, M. Azzi, C. Vial, H. Delmas, Kinetic study of defluoridation of drinking water by electrocoagulation/electroflotation in a stirred tank reactor and in an external-loop airlift reactor, Chem. Eng. J. 164 (2010) 106–114, https://doi.org/10.1016/j.cej.2010.08.037.
[97] V.A. Kolesnikov, V.I. Ilin, V.A. Brodskii, T.V. Guseva, M.A. Vartanyan, Improvement of electroflotation treatment of waste waters from ceramic enterprises, Glas. Ceram. (English Transl. Steklo I Keramika). 71 (2015) 421–424, https://doi.org/10.1007/s10717-015-9701-y.
[98] H.G. Yang, H.Y. Chun, D. Pak, Improvement of sludge anaerobic degradability by combined electro-flotation and electro-oxidation treatment, Biochem. Eng. J. 90 (2014) 44–48, https://doi.org/10.1016/j.bej.2014.05.010.
[99] Y. Esfandyari, Y. Mahdavi, M. Seyedsalehi, M. Hoseini, Degradation and biodegradability improvement of the olive mill wastewater by peroxi-electrocoagulation/electrooxidation- electroflotation process with bipolar aluminum electrodes, Environ. Sci. Pollut. Res. 22 (2014) 6288–6297, https://doi.org/10.1007/s11356-014-3832-5.

[100] X. Yunqing, L. Jianwei, Application of electrochemical treatment for the effluent from marine recirculating aquaculture systems, Procedia Environ Sci 10 (2011) 2329–2335, https://doi.org/10.1016/j.proenv.2011.09.363.
[101] K. Ulucan, U. Kurt, Comparative study of electrochemical wastewater treatment processes for bilge water as oily wastewater: a kinetic approach, J. Electroanal. Chem. 747 (2015) 104–111, https://doi.org/10.1016/j.jelechem.2015.04.005.
[102] J.A.P. Romero, G.R. Salazar-Banda, M.O.O. Rezende, Treatment of sewage by electroflotation: a pilot study, Sep. Sci. Technol. 48 (2013) 192–198, https://doi.org/10.1080/01496395.2012.682616.

CHAPTER

5

Electrocatalysis in Wastewater Treatment

Achille De Battisti, Carlos A. Martínez-Huitle*†

*University of Ferrara, Ferrara, Italy

†Federal University of Rio Grande do Norte, Natal, Brazil

OUTLINE

FUNDAMENTALS OF THE ELECTROCHEMICAL OXIDATION

In the last four decades, electrochemical technologies have had significant advances for their use in environmental protection. In fact, the application of these technologies has been the topic of several books and authoritative reviews [1–13]. Nevertheless, the most important contribution has been for the abatement of organic and inorganic pollutants in wastewater and drinking water. The main electrochemical procedures

https://doi.org/10.1016/B978-0-12-813160-2.00005-5

utilized for the remediation of wastewater are electrocoagulation (EC), electroflotation, electrodialysis, electroreduction, electrochemical oxidation (EO), and indirect EO with active oxidants (frequently, active chlorine), the so-called mediated EO [13,14]. Recently, new treatments by emerging technologies have also received a great deal of attention, including electro-Fenton (EF) and photo-assisted systems such as photoelectro-Fenton (PEF), photoelectrocatalysis (PEC), and the solar-assisted electrochemical processes (solar photoelectrocatalysis (SPEC) and solar photoelectro-Fenton (SPEF)) [14]. Nevertheless, anodic oxidation, also called EO, continues to be the most popular electrochemical procedure among the electrochemical advanced oxidation processes (EAOPs) for removing organic pollutants from wastewater; many of the EAOPs are essentially based on the EO approach [13]. Fig. 1 summarizes the most important electrochemical technologies proposed until now.

From the beginning of research, pioneering studies by Dabrowski in the 1970s, Kirk, Stucki, Kotz, Chettiar, and Watkinson in the 1980s, and Johnson, De Battisti, and Comninellis in the 1990s, allowed the elucidation of

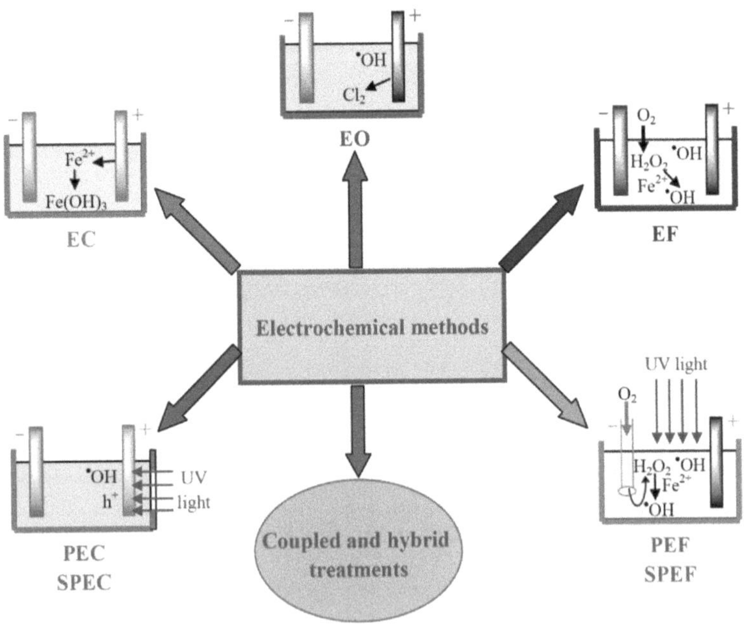

FIG. 1 The most important electrochemical technologies used for removing organic matter from wastewater and drinking water. *Reproduced with permission from E. Brillas, C.A. Martinez-Huitle, Decontamination of wastewaters containing synthetic organic dyes by electrochemical methods. An updated review, Appl. Catal. B Environ. 166–167 (2015) 603–643. Copyright 2015, Elsevier.*

the fundamentals of EO regarding its application as a promising alternative for treating wastewater [13]. After that, several research groups have studied the feasibility of EO for wastewater decontamination, some of them going from laboratory to pilot scale systems, contributing to the understanding of this technology [13].

In early studies, a mechanism for oxidizing organic pollutants was proposed by Feng and Johnson [15], based on the mediated action of reactive oxygen species formed at a high potential as intermediates of the oxygen evolution reaction (OER). This process involves the transfer of anodic oxygen from H_2O to the organics via adsorbed hydroxyl radicals generated by the water discharge:

$$S[\,] + H_2O \rightarrow S[^{\bullet}OH] + H^+ + e^- \tag{1}$$

$$S[^{\bullet}OH] + R \rightarrow S[\,] + RO + H^+ + e^- \tag{2}$$

where S[ ] represents the surface sites where the $^{\bullet}OH$ species can be adsorbed. Concomitantly, an inevitable and undesirable event takes place, the OER due to the oxidation of water:

$$S[^{\bullet}OH] + H_2O \rightarrow S[\,] + O_2 + 3H^+ + 3e^- \tag{3}$$

However, Feng and Johnson [15] indicated tentatively that the oxidation of organics could also be directly on the anode surface.

Later, Comninellis [16] proposed an initial scheme that described the mechanisms related to the electrochemical conversion/combustion of organics with simultaneous oxygen evolution, using different anode materials (Pt, Ti/IrO_2 and Ti/SnO_2). Based on the fact that a significant amount of hydroxyl radicals was formed at the anode surfaces and, consequently, these reactive oxygen species reacted with organic pollutants. The detection of $^{\bullet}OH$ radicals formed by water discharge at these anodes using spin trapping was the key parameter in these experiments to confirm the pathways proposed by Comninellis. Results clearly showed that at the surface of the Pt and IrO_2 anodes, the $^{\bullet}OH$ radical concentration was almost zero while at the surface of an SnO_2 anode, there was a much higher accumulation of $^{\bullet}OH$ radicals. Nevertheless, in 1996, Comninellis and De Battisti [17] published new evidence about the simplified mechanism reported in 1994 [16]. The use of iridium oxide-based electrodes confirmed this as a deciding factor, the electrocatalysis of oxidative electrochemical process that involves the formation of hydroxyl radicals as an intermediate step of the oxidation reaction and the oxygen-metal-ion interaction, controlling their surface concentration. Thus, according to the proposed mechanisms by Feng and Johnson [15] and Comninellis [16] as well as Comninellis and De Battisti [17], generally speaking, the EO approach offers two options for oxidizing organic compounds:

(i) Direct anodic oxidation (or direct electron transfer to the anode), which yields very poor decontamination.
(ii) Chemical reaction with electrogenerated species from water discharge at the anode, such as physically adsorbed "active oxygen" (physisorbed hydroxyl radical ($^{\bullet}OH$)) or chemisorbed "active oxygen" (oxygen in the lattice of a metal oxide (MO) anode). The action of these oxidizing species leads to total or partial decontamination, respectively.

Based on the existence of different heterogeneous species formed from water discharge, two main approaches have been proposed for wastewater pollution abatement by EO [14]:

(i) The electrochemical conversion, in which refractory organics are selectively transformed into biodegradable compounds under the action of chemisorbed "active oxygen."
(ii) The electrochemical combustion (i.e., electrochemical incineration), where organics are completely mineralized to H_2O, CO_2, and inorganic ions by physisorbed $^{\bullet}OH$ [16,17].

Both approaches require the application of high anode potentials, thus causing competition with the oxygen evolution reaction (OER). Conversely, direct EO is theoretically possible at low potential values, which are set before O_2 evolution, but the reaction rate is usually slow and depends strongly on the electrocatalytic activity of the anode [13,16–18].

Later on, a new generation of electrode materials was introduced in the EO method. The innovative material was the boron doped diamond (BDD) electrode, which is characterized by high reactivity toward organic oxidation via the formation of quasifree hydroxyl radicals on its surface [18]. These electrogenerated hydroxyl radicals on the BDD anode are very reactive (owing to a weak BDD-hydroxyl radical interaction) and thus can lead to organic combustion with a high current efficiency [5,6,8,13]. Thus, the results obtained by using a BDD anode allowed that the first simplified model was significantly improved [18] as well as allowing the generation of a classification of the different anode materials according to their oxidation power in acid media.

As mentioned above, the results obtained by several groups demonstrated that the nature of the electrode material influences the selectivity and the efficiency of an electrochemical process for the oxidation of organic compounds, especially, the figures reported by using BDD anodes. In this frame, the anode materials are divided into two types: (i) active and (ii) nonactive electrodes in the new model proposed by Comninellis in 2000 [18].

Active anodes (e.g., Pt, IrO_2, and RuO_2), which present low oxygen evolution overpotential, are good electrocatalysts for the OER and, consequently, lead to selective oxidation of the organic pollutants, also denominated, as partial oxidation of organics or electrochemical conversion [3–5,13]. This is due to the fact that the electrogenerated species from water discharge at the anode are present as chemisorbed "active oxygen" (oxygen in the lattice of a metal oxide (MO) anode), limiting the amount of free-hydroxyl radicals. Conversely, nonactive anodes (e.g., PbO_2, SnO_2, and BDD), which present high oxygen evolution overpotential, are poor electrocatalysts for the OER, and direct electrochemical oxidation is expected to occur for these electrodes. Also, they present no higher oxidation state available and the organic species are directly oxidized by nonadsorbed hydroxyl radicals, giving complete combustion (so-called electrochemical combustion or incineration) [3–5,13].

The complete proposed model (see Fig. 2) assumes that the initial reaction in both anode types (generically denoted as M) corresponds to the oxidation of water molecules leading to the formation of physisorbed hydroxyl radical (M($^\bullet$OH)) [18]:

$$M + H_2O \rightarrow M(^\bullet OH) + H^+ + e^- \quad (4)$$

Both the electrochemical and the chemical reactivity of heterogeneous M($^\bullet$OH) are dependent on the nature of the electrode material. The surface of active anodes interacts strongly with $^\bullet$OH and thus a so-called higher

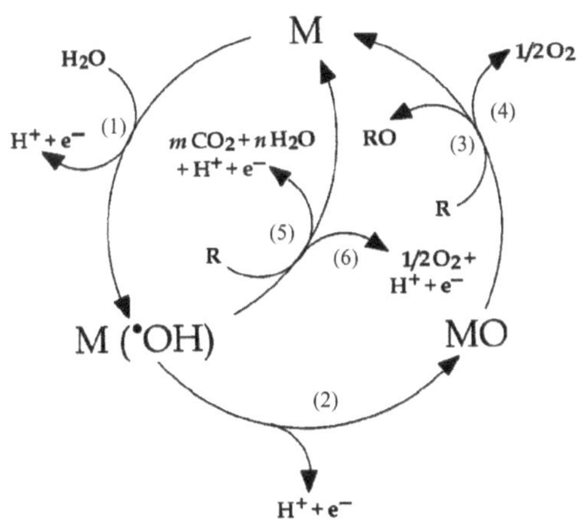

FIG. 2 Mechanistic scheme of anodic oxidation of organic compounds with simultaneous oxygen evolution on nonactive anodes and on active anodes. (1) water discharge to hydroxyl radicals, $^\bullet$OH; (2) formation of the higher metal oxide, MO; (3) partial (selective) oxidation of the organic compounds; (4) oxygen evolution by chemical decomposition of the higher metal oxide; (5) combustion of the organic compounds, R, via hydroxyl radicals; (6) oxygen evolution by electrochemical oxidation of hydroxyl radicals. *Reproduced with permission from B. Marselli, J. Garcia-Gomez, P.A. Michaud, M.A. Rodrigo, C. Comninellis, Electrogeneration of hydroxyl radicals on boron-doped diamond electrodes, J. Electrochem. Soc. 150 (2003) D79–D83. Copyright 2003, The Electrochemical Society.*

oxide or superoxide (MO) may be formed according to reaction (5). This may occur when higher oxidation states are available for a metal oxide anode, above the standard potential for oxygen evolution.

$$M(^{\bullet}OH) \rightarrow MO + H^{+} + e^{-} \quad (5)$$

The redox couple MO/M acts as a mediator in the organic oxidation according to reaction (6), which competes with the OER, a side reaction, via the chemical decomposition of the higher oxide species, as shown in reaction (7).

$$MO + R \rightarrow M + RO \quad (6)$$

$$MO \rightarrow M + \tfrac{1}{2}O_2 \quad (7)$$

In contrast, the surface of a nonactive anode interacts so weakly with $^{\bullet}OH$ that it allows the direct reaction of the organics with $M(^{\bullet}OH)$ to yield fully oxidized reaction products such as CO_2, as follows [18,19]:

$$aM(^{\bullet}OH) + R \rightarrow M + mCO_2 + nH_2O + xH^{+} + ye^{-} \quad (8)$$

where R is an organic compound with m carbon atoms and without any heteroatom, which needs a ($=2m + n$) oxygen atoms to be totally mineralized to CO_2. The oxidative reaction (6) involving the surface redox couple MO/M is much more selective than the mineralization reaction (8) involving the physisorbed heterogeneous hydroxyl radical [5,6,13,18,19]. The latter reaction also competes with the side reactions of $M(^{\bullet}OH)$ such as the direct oxidation to O_2 (reaction 9) or the indirect consumption through dimerization to hydrogen peroxide (reaction 10):

$$M(^{\bullet}OH) \rightarrow M + \tfrac{1}{2}O_2 + H^{+} + e^{-} \quad (9)$$

$$2M(^{\bullet}OH) \rightarrow 2M + H_2O_2 \quad (10)$$

A nonactive electrode does not participate in the direct anodic reaction of organics and does not provide any catalytic active site for their adsorption from aqueous media. It only acts as an inert substrate and as a sink for the removal of electrons. In principle, only outer-sphere reactions and water oxidation are possible with this kind of anode. Hydroxyl radicals produced from water discharge by reaction (4) are subsequently involved in the organic oxidation process. The above model presupposes that the electrochemical activity and the chemical reactivity of physisorbed $M(^{\bullet}OH)$ are strongly linked to the strength of the M-$^{\bullet}$OH interaction. Therefore, the BDD anode is the best nonactive electrode that presents this behavior [5,6,13,18,19], thus being proposed as the preferable anode for the treatment of organics by electrochemical oxidation.

It is important to note that the simplified model proposed by Comninellis has been modified or elaborated in different ways for other authors; however, it is clear that the mechanistic scheme reported in Fig. 2 [18] is the most clear form to comprehend and express the electrochemical reactions occurred by the EO approach, considering the nature of the electrode materials as well as the reactivity of hydroxyl radicals on their surfaces.

RECENT MECHANISM ADVANCES

The formation and reactivity of free hydroxyl radicals depends on the nature of the electrode material. In 2009, Kapalka and coworkers determined the profile of hydroxyl radicals in the vicinity of the BDD surface during an oxygen evolution reaction as well as the oxidation of organics (methanol and formic acid) in order to better understand the mechanisms involved in the oxidation processes [20].

Concentration Profile of $^{\bullet}OH$ During Oxygen Evolution

They have considered that, in the absence of organic compounds, electrogenerated free hydroxyl radicals can react with each other to form hydrogen peroxide, as given by the following equation:

$$HO^{\bullet} + HO^{\bullet} \rightarrow H_2O_2 \quad (11)$$

Hydrogen peroxide can be further oxidized to oxygen either by its direct discharge on the electrode surface (Eq. 12) or via the assistance of hydroxyl radicals (Eq. 13):

$$H_2O_2 \rightarrow O_2 + 2H^+ + 2e^- \quad (12)$$

$$H_2O_2 + 2HO^{\bullet} \rightarrow O_2 + 2H_2O \quad (13)$$

assuming that the first step (Eq. 11) is the rate determining step and that four hydroxyl radicals (four electrons) are needed for oxygen evolution. Then, the mass balance for hydroxyl radicals was considered by the authors by applying the one-dimensional Fick law referred to as the molecular diffusion in a stagnant layer and considering the diffusion coefficient of $HO^{\bullet}$, the rate constant of reaction (11) $k_{HO^{\bullet}}$ ($m^3\ mol^{-1}\ s^{-1}$), and the concentration of $HO^{\bullet}$ as well as assuming that far from the electrode ($x = \infty$), that concentration of hydroxyl radicals is 0 whereas at the electrode ($x = 0$) there is a surface concentration of hydroxyl radicals. The concentration profile of hydroxyl radicals, as a function of the distance x from the electrode surface during oxygen evolution, was determined by them.

Thus, the flux of hydroxyl radicals at the electrode surface was expressed in terms of the current density, allowing the determination of the concentration profile of hydroxyl radicals during oxygen evolution as a function of the distance from the electrode surface. From these results, at current density ($j = 300\,A\,m^{-2}$), the reaction layer thickness is about 1 μm whereas the maximum (surface) concentration of hydroxyl radicals reaches the value of several tenths of μM.

Concentration Profile of $HO^{\bullet}$ During Oxidation of Organics

In the presence of organic pollutants, it is possible to determine the concentration profile of hydroxyl radicals by analogous mathematical treatment. However, the hypotheses considered are:

- Oxidation of organic compounds R proceeds only via the assistance of hydroxyl radicals in the vicinity of the electrode surface:

$$R + zHO^{\bullet} \rightarrow \text{oxidation products} \tag{14}$$

- Concentration of organic compounds is high enough to be considered a constant in the reaction layer.
- Oxygen evolution via H_2O_2 oxidation is negligible.

Then, in the presence of organic compounds, the concentration of hydroxyl radicals decreases exponentially with the distance from the electrode surface. The thickness of the reaction layer (reaction cage) depends on the concentration of the organic compound, the rate constant of organic oxidation (via hydroxyl radicals), and the applied current density. The simulated $HO^{\bullet}$ concentration profile during the oxidation of formic acid (0.25–1 M) at $300\,A\,m^{-2}$ showed that the higher the formic acid concentration, the lower the surface concentration of $HO^{\bullet}$ and the smaller the thickness of the reaction layer. In this frame, the comparison of the $HO^{\bullet}$ concentration profile for 1 M formic acid, methanol, and ethanol showed that, depending on the rate constant, the thickness of the reaction layer varied between a few nanometers and tenths of nanometers [20].

MINERALIZATION OF ORGANICS BY MOLECULAR OXYGEN

The evidence in [21] showed in a study by using BDD anodes indicated that organic mineralization involves hydroxyl radicals but also molecular oxygen present in air/oxygen-saturated aqueous organic solutions. The direct evidence for this process was found during EO of an acetic acid (AA) solution saturated with isotopically labeled $^{18}O_2$, resulting in the formation of $C^{18}O_2$ and $C^{16}O^{18}O$. Oxygen, $^{16}O_2$, or isotopically labeled oxygen, $^{18}O_2$, was bubbled in the solution during measurements in order to understand the participation of the oxygen dissolved in the solution. Voltammetric profiles coupled to differential electrochemical mass spectrometry showed that the ionic current associated to $^{16}O_2$ decreased in parallel with the $C^{16}O_2$ evolution during acetic acid oxidation, for both aerated and $^{16}O_2$-saturated solutions. The decrease of the ionic current associated to $^{16}O_2$ indicates that molecular oxygen dissolved in the solution could participate in the oxidation process. In fact, when $^{18}O_2$-saturated solution was used for acetic acid oxidation, in parallel with the decrease of $^{18}O_2$, an increase of the ionic current associated to $^{16}O^{18}O$ (in $^{18}O_2$-saturated solution) was observed. Thus, this formation of $^{16}O^{18}O$ indicated that intermediates involved in oxygen evolution (probably $HO^{\bullet}$) are regenerated during chain reactions. It was also demonstrated by using isotopically labeled $C^{18}O_2$ and $C^{18}O^{16}O$, because these species are evolved in parallel with the $^{18}O_2$ decrease, confirming that oxygen dissolved in the solution is involved. These results allowed the proposition of an additional mechanism for the electrochemically induced oxidation of organic compounds via molecular oxygen dissolved in aerated solutions on BDD electrodes (Fig. 3).

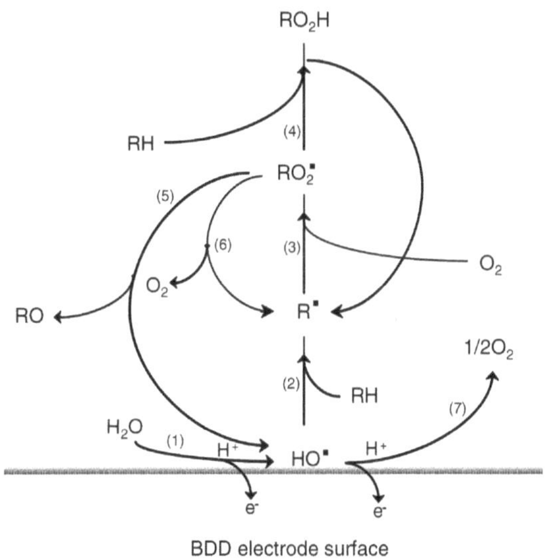

FIG. 3 (1) Water discharge to hydroxyl radical $HO^{\bullet}$; (2) dehydrogenation of organic compound RH via $HO^{\bullet}$ and formation of free organic radical $R^{\bullet}$; (3) addition of molecular oxygen to $R^{\bullet}$ resulting in the formation of an organic peroxy radical $RO^{\bullet}{}_2$; (4) dehydrogenation of RH via $RO^{\bullet}{}_2$ resulting in the formation of hydroperoxide ROOH and regeneration of $R^{\bullet}$; (5) decomposition of $RO^{\bullet}{}_2$ leading to the regeneration of $HO^{\bullet}$ and formation of RO; (6) decomposition of $RO^{\bullet}{}_2$ to $R^{\bullet}$; and (7) oxygen evolution, a side reaction. *Reproduced with permission from A. Kapałka, B. Lanova, H. Baltruschat, G. Fóti, C. Comninellis, Electrochemically induced mineralization of organics by molecular oxygen on boron-doped diamond electrode, Electrochem. Commun. 10 (2008) 1215–1218. Copyright 2008, Elsevier.*

This activation of the reaction with molecular oxygen proceeds most likely through the sequence of reactions that are initiated by $HO^{\bullet}$ formed on the electrode surface. This indicates that hydroxyl radicals, molecular oxygen, and organic species affect the rate of organic mineralization.

NEW APPROACHES FOR ELECTROCHEMICAL OXIDATION

Two new approaches for enhancing the efficiency of the EO of organic pollutants were developed. In the first [22], the electrochemical approach is performed under galvanostatic conditions with the potential "buffered" by the competing side reaction of OER. Meanwhile, the use of high pressures by increasing the oxygen dissolved in solution has been proposed as a second way.

The working potential is fixed by the nature of the electrode material and is buffered during organic oxidation by the side reaction of OER, in the former. This concept was used for the selective oxidation of secondary alcohols to the corresponding ketones and the oxidation of phenol to benzoquinone using the Ti/IrO_2 anode. Then, phenol was selectively oxidized with lower current efficiency (>1%) by using the approach described above, achieving the formation of hydroquinone, *p*-benzoquinone, and pyrocatechol. The fact that the total aromatic selectivity is close to one indicated that, at the working anodic potential (1.7 V) buffered by the side reaction of oxygen evolution on the Ti/IrO_2 anode, the oxidation products of phenol are stable toward further oxidation even after a quasicomplete conversion of phenol. On the other hand, the high region selectivity toward benzoquinone formation (*para* to *ortho* ratio: 3 to 1) indicated that there was a specific orientation of phenol on the IrO_2 surface during electrooxidation. In fact, in the absence of any interaction with the electrode, the main product is pyrocatechol (*para* to *ortho* ratio: 1 to 2).

Meanwhile, the use of thermal or pressure reactors to increase the temperature or the concentration of dissolved oxygen, respectively, allowed the increase of the production of oxidants, and consequently, enhanced the elimination of organic pollutants in the second case. The injection of oxygen at higher pressures will increase the dissolved oxygen to promote the effective conversion of it to hydrogen peroxide on the cathode surface, but the dissolved oxygen can be active at the anode surface to form reactive oxygen species. In both cases, the amount of oxidants in solution is considerably higher, increasing the homogeneous oxidative rate and, consequently, improving the current and removal efficiencies. As can be noted, this approach is related to the molecular oxygen mechanism proposed by Kapalka [21] or an improvement of the EF process when carbonaceous cathodes are used to produce H_2O_2 [6]. Then, this new approach can be a viable way to reduce energy consumption, increasing the efficiency for real industrial application. However, further studies are necessary in order to establish the operating conditions.

ADVANTAGES AND DISADVANTAGES OF THE EO APPROACH

High efficiency, mild operation conditions, ease of automation, versatility, and low costs, especially when powered by renewable energy from wind and solar sources, are some of the most important advantages of EO technology [3]. For this reason, several groups have tested a wide variety of synthetic wastewater types containing a diversity of target compounds in order to understand the effects, operating conditions, disadvantages, and potential efficiency of the EO approach [3–14]. However, the role of operation parameters for the efficient applicability of the EO approach drastically influenced the different routes involved in them. For this reason, its industrial application is still limited. The advantages and disadvantages are summarized in Table 1 [13].

TABLE 1 Main Characteristics of EO Technology

Abilities	Production of reactive oxygen species ($^{\bullet}OH$ radicals), production of strong oxidants from salts, microorganism inactivation, evolution versatility from 2D to 3D electrodes, no pH restrictions and effective degradation of different organic pollutants in wastewater: (i) Chemical industry (fine chemical industry, pulp and paper industry, petrochemical industry, pharmaceutical industry, textile industry, tannery industry, food industry, landfill leachate; (ii) Agroindustry (olive oil and dairy manure) and (iii) Urban and domestic wastewater
Advantages	Ambient temperatures and pressure requirements, robustness, versatility to treat large volumes, ease of automation, addition of nontoxic reagents for increasing conductivity, attractive compact technology, application as pretreatment or posttreatment of effluents in combination with other depuration technologies: biological treatment, Fenton oxidation, ion exchange, membrane filtration, membrane bioreactors, electrochemical technologies
Disadvantages/challenges to be further studied	Potential formation of halogenated byproducts, electrode fouling, and corrosion phenomena, high operation costs due to the high energy consumption (but coupling with renewable energy sources is possible), low conductivity of the effluent, optimization of reactor hydrodynamic conditions, expensive, high O_2 overpotential anodes (depending on the purpose)

Modified from C.A. Martínez-Huitle, M.A. Rodrigo, I. Sirés, O. Scialdone, Chem. Rev. 115 (2015) 13362–13407.

A short lifetime and/or high cost of some electrode materials and the low current efficiency under some conditions are important limitations. Some intrinsic drawbacks, such as mass-transport limitations, low space-time yield and surface area to volume ratio, and gradual temperature increase have not been satisfactorily solved yet. Also, the effect of pH, pollutant type and concentration, effluent conductivity, electrochemical reactor design and electrode arrangement, potential or current density distribution, and the occurrence of undesired reactions on process performance was not fully evaluated in all reports. Eventually, the studies on the reactors that can be used for all these processes have been reported, thus concluding that further studies are necessary in order to better identify the best solutions from the reactor perspective. Generally speaking, the EO approach appears mature enough to become a very promising tool for the treatment of wastewater contaminated by recalcitrant pollutants; however, further studies are necessary to develop these processes on an applicative scale.

CONCLUDING REMARKS

The future for the applicability of the EO process for wastewater and water treatment is bright because new EO approaches are used under specific operating conditions, emerging technologies are proposed, or novel electrocatalytical materials are prepared. Diamond electrodes are starting to make important contributions for their use on pilot scale plants or flow electrochemical reactors that have the capacity to treat high effluent volumes, putting in evidence the performances of the EO-based technologies. Several other complementary technologies are emerging and these can provide an enhancement or intensification on the decontamination of waters in short treatment times and lower energy consumption. Future developments will rely upon the close collaboration of analytical chemists, engineers, and electrochemists to ensure effective application and exploitation of new electrochemical technologies based on the EO approach or the direct use of the EO process coupled with physical-chemical methods such as pre- or post-treatment steps. Several areas of interest will give more support to the electrochemical technologies to develop new electrocatalytic materials, innovative reactors, original pilot scale processes, and novel method combinations in order to complete their effective industrial application, as discussed in the next chapters of this book. Finally, it is important to remark that, the fundamentals of EO approach are very important because other novel electrochemical technologies can be developed taking into consideration the scientific contributions published by the experts in the beginning of the research.

Acknowledgments

Carlos A. Martínez-Huitle acknowledges the funding provided by the Alexander von Humboldt Foundation (Germany) and Coordenação de Aperfeiçoamento de Pessoal de Nível Superior (Brazil) as a Humboldt fellowship for Experienced Researcher (88881.136108/2017-01) at the Johannes Gutenberg-Universität Mainz, Germany.

References

[1] K. Rajeshwar, J.G. Ibanez, G.M. Swain, J. Appl. Electrochem. 24 (1994) 1077.
[2] K. Rajeshwar, J.G. Ibanez, Environmental Electrochemistry: Fundamentals and Applications in Pollution Abatement, Academic Press, San Diego, CA, 1997.
[3] C. Comninellis, G. Chen, Electrochemistry for the Environment, Springer, New York, 2010.
[4] C.A. Martínez-Huitle, S. Ferro, Chem. Soc. Rev. 35 (2006) 1324–1340.
[5] M. Panizza, G. Cerisola, Chem. Rev. 109 (2009) 6541–6569.
[6] E. Brillas, I. Sirés, M.A. Oturan, Chem. Rev. 109 (2009) 6570–6631.
[7] M.M. Emamjomeh, M. Sivakumar, J. Environ. Manag. 90 (2009) 1663–1679.
[8] I. Sirés, E. Brillas, M.A. Oturan, M.A. Rodrigo, M. Panizza, Environ. Sci. Pollut. Res. Int. 21 (2014) 8336–8367.
[9] Á. Anglada, A. Urtiaga, I. Ortiz, J. Chem. Technol. Biotechnol. 84 (2009) 1747–1755.
[10] W. Wu, Z.H. Huang, T.T. Lim, Appl. Catal. A 480 (2014) 58–78.
[11] A.N.S. Rao, V.T. Venkatarangaiah, Environ. Sci. Pollut. Res. 21 (2014) 3197–3217.
[12] P. Cañizares, R. Paz, C. Sáez, M.A. Rodrigo, J. Environ. Manag. 90 (2009) 410–420.
[13] C.A. Martínez-Huitle, M.A. Rodrigo, I. Sirés, O. Scialdone, Chem. Rev. 115 (2015) 13362–13407.
[14] E. Brillas, C.A. Martinez-Huitle, Appl. Catal. B Environ. 166–167 (2015) 603–643.
[15] J. Feng, D.C. Johnson, J. Electrochem. Soc. 137 (1990) 507–510.
[16] Ch. Comninellis, Electrochim. Acta 39 (1994) 1857–1862.
[17] C. Comninellis, A. De Battisti, J. Chim. Phys. 93 (1996) 673–679.
[18] B. Marselli, J. Garcia-Gomez, P.A. Michaud, M.A. Rodrigo, C. Comninellis, J. Electrochem. Soc. 150 (2003) D79–D83.
[19] C. Comninellis, A. Kapalka, S. Malato, S.A. Parsons, I. Poulios, D. Mantzavinos, J. Chem. Technol. Biotechnol. 83 (2008) 769–776.
[20] A. Kapałka, G. Fóti, C. Comninellis, Electrochim. Acta 54 (2009) 2018–2023.
[21] A. Kapałka, B. Lanova, H. Baltruschat, G. Fóti, C. Comninellis, Electrochem. Commun. 10 (2008) 1215–1218.
[22] S. Fierro, E. Passas-Lagos, E. Chatzisymeon, D. Mantzavinos, C. Comninellis, Electrochem. Commun. 11 (2009) 1358–1361.

CHAPTER

6

Indirect Electrochemical Oxidation Using Hydroxyl Radical, Active Chlorine, and Peroxodisulfate

Karine Groenen Serrano
Toulouse University, Toulouse, France

OUTLINE

https://doi.org/10.1016/B978-0-12-813160-2.00006-7

INTRODUCTION

The most important method to decrease the chemical oxygen demand (COD) of wastewaters is based on biological treatment. It is well established and relatively cheap, with minimal operational and maintenance requirements. But, unfortunately, certain substances are not easily biodegradable. In particular, toxic substances (pesticides, pharmaceuticals, etc.) can inactivate microorganisms and slow down biological processes. A growing number of these compounds, called persistent organic pollutants (POPs), are present in wastewater because of the intensification of industrial activities and technological advances. They pass through the wastewater treatment plants (WWTP) and accumulate in the aquatic system.

Whether these POPs are present at high concentrations or at a trace level in aquatic systems, they continue to pose significant threats to the sustainability of the ecosystem and the safety of drinking water.

Consequently, considering the threat that POPs represent, a specific treatment at source should be developed to prevent their leakage into aquatic systems in order to limit risks.

As a response, a large variety of advanced oxidation processes (AOPs) have been developed to remove POPs from wastewater. AOPs are environmentally friendly methods, based on the in situ production of the hydroxyl radical ($^{\bullet}OH$), one of the strongest oxidants in aqueous solutions. Among these emerging AOPs, advanced electrochemical processes have significant advantages. Using only electrons as reactants, the degradation of organics into carbon dioxide and water can be achieved under particular conditions. The mechanism of electroxidation depends mainly on the electrocatalytic properties of the anodic material. As a matter of fact, since the POPs oxidize at high potentials, an anode with a high-oxygen overpotential is required.

Studies on electrochemical processes by oxidation for water treatment began in the 1970s. Progress was made thanks to the discoveries and the development in electrode materials. In the 1980s, much research was performed which developed metallic oxides [1–3]. Electrochemical water disinfection attained technical maturity in the 2000s, since sufficiently stable and efficient electrode materials were developed, in particular dimensionally stable anodes (DSA) based on mixed oxide iridium and/or ruthenium coated on Ti substrate. Nowadays, such electrode materials are widely used for the disinfection of private swimming pools by salt water chlorination. In addition, over the past 30 years more electrode material has appeared: boron doped diamond electrodes (BDD). The first electrochemical studies were reported by Pleskov et al. [4]. Carey et al. [5] patented the use of diamond electrodes for organic pollutant oxidation. The growing interest in BDD anodes is due to their unique properties: a high electrochemical window, excellent electrochemical stability, and chemical inertness in acid and caustic media. In the light of their extraordinary properties, the BDD electrodes allow the complete removal of the organic matter (mineralization) to be envisaged. The electrochemical oxidation of species in aqueous solutions can be carried out either by the transfer of electrons or by the transfer of oxygen atoms; alternatively, both pathways can be taken simultaneously. Direct electrochemical oxidation is heterogeneous: electron exchange occurs between organic species and the electrocatalytic electrode surface. Indirect oxidation uses an electrogenerated mediator. Indeed, a strong oxidant is generated at the anode and reacts chemically with the organics. The action of this oxidant can either occur close to the electrode surface if the oxidant is very reactive and not stable (the case of radicals such as hydroxyl radical), or in the bulk if the oxidant is sufficiently stable to diffuse into the solution (the case of active chlorine). The production of oxidants in situ presents various advantages: (i) in certain cases, complete mineralization can be reached and reuse of the solution can be considered; or (ii) in other cases, the action of the oxidant enables mass transfer limitations and the treatment of dilute solutions to be overcome.

This chapter is dedicated to indirect electrochemical oxidation with the electrogeneration of strong oxidants from salts (chlorides, sulfates, and phosphates) or from water. These salts are present in wastewater or added as supporting electrolytes to increase the ionic conductivity of the solution, in order to reduce the energetic cost of the process.

First, the oxidation of organics performed through the mediation of reactive oxygen species (ROS) will be presented. These species (ozone, hydrogen peroxide, and hydroxyl radicals) are formed at the electrode by water discharge or oxygen reduction. Next, the electrogeneration of oxidants from salts will be presented: active chlorine and peroxocompounds. For each case, several aspects will be discussed: (i) the reactive properties with organics, (ii) the operating conditions for the electrosynthesis, and (iii) some examples of coupling the electrosynthesis of strong oxidants through chemical reactions with organics.

REACTIVE OXYGEN SPECIES

Reactive oxygen species (ROS) are reactive chemical species containing oxygen such as hydrogen peroxide (H_2O_2), ozone (O_3), or hydroxyl radical ($^{\bullet}OH$). Their chemical reactivity is due to the oxygen molecule's unpaired electron.

The role of ROS (that include $^{\bullet}OH$, O_3, H_2O_2, and $^{\bullet}O$) has been underlined by their high disinfecting activity. Table 1 summarizes the redox potential of some ROS [6].

TABLE 1 Redox Potential of Some ROS [6]

Oxidant	Redox potential (V vs SHE)
$^{\bullet}OH$	2.8
$^{\bullet}O$	2.42
O_3	2.07
H_2O_2	1.78

The generation of such oxidants strongly depends on several key reaction parameters. These include electrode material, electrolyte composition, applied current (or voltage), pH, and temperature. Electrode material has been regarded as one of the most important parameters governing the electrogenerated species and yield of generated oxidants. For generating ROS, a high oxygen overpotential is required; otherwise, a large portion of the applied current will be wasted to produce oxygen as a side reaction, thereby reducing the efficiency of the electrochemical process. Several electrode materials such as PbO_2, SnO_2-based DSA, and BDD have been reported to exhibit high oxygen overpotential. These electrode materials are able to hinder the thermodynamically favored oxygen evolution reaction, thus improving the ROS generation efficiency [7–9]. This aspect will be discussed for the selected ROS: ozone and hydroxyl radical.

OZONE

Ozone is a strong oxidant used for water treatment and disinfection purposes ($E° = 2.1\,V$). For many applications it replaces chlorine; its primary advantage is that it prevents the formation of unwanted chlorinated by-products. An increasingly important O_3 application is its use in the treatment of drinking water to improve taste and color. O_3 is also used to disinfect water for the pharmaceutical industry as a disinfectant in purified water loops and electronic industries where extreme standards of purification are needed. However, ozone is unstable in water. Depending on the water quality, the half-life of ozone is in the range of seconds to hours. The major secondary oxidant formed from ozone decomposition in water is the hydroxyl radical [10].

$$O_3 + OH^- \rightarrow HO_2^- + O_2 \quad k = 70\,M^{-1}s^{-1} \tag{1}$$

$$O_3 + HO_2^- \rightarrow {}^{\bullet}OH + O_2^{-\bullet} + O_2 \quad k = 2.8 \times 10^6\,M^{-1}s^{-1} \tag{2}$$

However, for different applications of ozone the two species are of differing importance. While disinfection occurs dominantly through ozone, oxidation processes may occur through both oxidants, ozone and $^{\bullet}OH$. Ozone is a very selective oxidant. As shown in Table 2, $^{\bullet}OH$ radicals react fast with many dissolved compounds and the water matrix.

For most applications today, ozone is produced by corona discharge, which is based on passing an electric discharge through dry oxygen or air. The low concentrations of O_3 obtained have limited its viability as an oxidant for many potential applications. Therefore, the development of electrochemical ozone production (EOP) technology is of interest. The main advantages of EOP are that it is a low DC voltage technology and ozone can be produced directly and continuously in water. During water electrolysis, oxygen evolution is the main rival reaction to ozone production. Thermodynamically, oxygen evolution is strongly favored versus ozone production.

$$3H_2O \rightarrow O_3 + 6H^+ + 6e \quad E^\circ = 1.51\,V/ENH \tag{3}$$

$$2H_2O \rightarrow O_2 + 4H^+ + 4e \quad E^\circ = 1.23\,V/ENH \tag{4}$$

Therefore, ozone production is only possible if high oxygen overvoltage anode materials are used. Different materials have been tested such

TABLE 2 Reaction-Rate Constants of Selected Organics With Ozone and OH Radicals [10–12]

	Compound	kO_3 ($M^{-1}\,s^{-1}$)	$k^{\bullet}OH$ ($M^{-1}\,s^{-1}$)
Organics	Phenol	1.3×10^3	6.6×10^9
	Acetic acid	3×10^{-5}	1.6×10^7
	Maleic acid	1×10^3	1.2×10^{10}
Pesticides	Atrazine	6	3×10^9
	Alachlor	3.8	7×10^9
Solvents	Chloroethene	1.4×10^4	1.2×10^{10}
	Chlorobenzene	0.75	5.6×10^9
	Benzene	2	7.9×10^9
	Toluene	14	5.1×10^9
	Ethanol	0.37	1.9×10^9
Pharmaceuticals	Diclofenac	1×10^6	7.9×10^9
	Carbamazepine	3×10^5	8.8×10^9
	Sulfamethoxazole	2.5×10^6	5.5×10^9

as SnO_2, Pt, glassy carbon, IrO_2/Nb_2O_5, or IrO_2/Ta_2O_5. For all cases, the electrolysis must be performed at low temperatures, high current densities, and very low interfacial pH. Another important parameter is the composition of the solution; it has been shown that the addition of fluoride in the form of HBF_2 or KPF_6 allows the current efficiency [13,14] to be increased. However, the addition of such compounds may not be easily compatible with applications in the field of wastewater treatment.

A significant contribution to EOP technology was made by Stucki et al. [15]. These authors developed an electrochemical ozone generator that produces ozone directly into a stream of water at a porous PbO_2 anode in contact with a solid polymer electrolyte (SPE). No electrolyte is required, and ozone is generated at room temperature. The current efficiency was close to 14%. In conventional electrochemical reactors, O_3 is produced on BDD, with a low current efficiency of a few percent. Kraft et al. [16] combined diamond anodes formed on a niobium-expanded metal substrate and the SPE technology. A current efficiency reaching 24% was achieved with an applied current of 2 A and a $40\,L\,h^{-1}$ flow rate at 20°C. In parallel, Arihara et al. [17] developed a cell with a polymer electrolyte membrane positioned between a perforated conductive diamond plate and a Pt mesh cathode. They attained high current efficiency, reaching 47% at 20°C. This EOP technology is well-suited to small productions of ozone (until $50\,m^3\,h^{-1}$).

A combined process was presented by Santana et al. [18] to treat a solution containing a textile dye (Fig. 1). The electrochemical ozone reactor presented a current efficiency of 21 wt% with a specific power consumption of 70 Wh/g. The electrolyte was a 3 M H_2SO_4 solution containing KPF_6 at 0°C. The gas leaving the anodic compartment of the electrochemical reactor was introduced into the column bubble reactor. It was shown that total discoloration was rapidly reached, with a high rate of mineralization (>70%) after 1 h of ozonation.

HYDROGEN PEROXIDE

Hydrogen peroxide can be produced by the electrochemical reduction of oxygen on the cathode or by the decomposition of other oxidants in the bulk. To produce H_2O_2 electrochemically, an oxygen gas supply in the electrochemical cell is required.

$$O_2\,(g) + 2H^+ + 2e^- \rightarrow H_2O_2 \qquad (5)$$

Different materials can be used for hydrogen peroxide production: graphite, vitreous carbon, oxygen diffusion cathodes, or boron doped diamond.

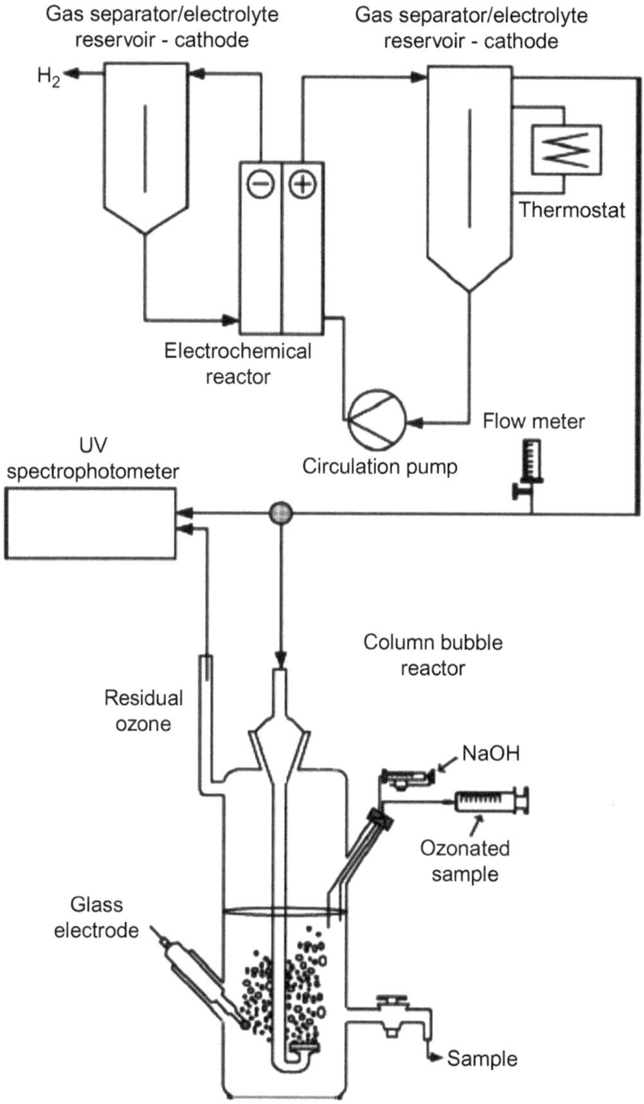

FIG. 1 Experimental setup for electrochemical ozone generation and ozonation under semi-batch conditions of aqueous solutions containing the RO122 textile dye. *Reproduced with permission from M.H.P. Santana, L.M. Da Silva, A.C. Freitas, J.F.C. Boodts, K.C. Fernandes, L.A. De Faria, Application of electrochemically generated ozone to the discoloration and degradation of solutions containing the dye Reactive Orange 122, J. Hazard. Mater. 164 (2009) 10–17. Copyright 2009. Elsevier.*

Although hydrogen peroxide is not a very strong oxidant, it is possible to activate it by combining it with ozone, which results in the formation of hydroxyl radicals:

$$H_2O_2 + 2O_3 \rightarrow 3O_2 + 2{}^{\bullet}OH \tag{6}$$

In the presence of iron or copper, hydrogen peroxide is decomposed into hydroxyl radicals:

$$H_2O_2 + Fe^{2+} \rightarrow Fe^{3+} + OH^- + {}^{\bullet}OH \tag{7}$$

This last reaction, called Fenton's reaction, takes place in acidic conditions (pH in the range of 2–4) to avoid the complexes of Fe(II) and Fe(III). The role of hydroxyl radicals is widely developed in the next paragraph. It is possible to use a sacrificial iron anode to supply Fe^{2+} ions to the aqueous solution. This process is called Electro-Fenton [19]. This part is discussed more fully in the next chapter.

HYDROXYL RADICALS

${}^{\bullet}OH$ radicals are the strongest oxidant in an aqueous solution. Table 3 illustrates the strong reactivity of hydroxyl radicals with organic compounds.

In the case of oxidation by oxygen transfer, the first step of the process is the generation of adsorbed hydroxyl radicals (see Eq. 8). The hydroxyl radicals are generated by water discharge (one electron per water molecule) [3,23,24]:

$$H_2O + M[\,] \rightarrow M[{}^{\bullet}OH] + H^+ + e \tag{8}$$

$$M[{}^{\bullet}OH] + R \rightarrow M[\,] + RO + H^+ + e \tag{9}$$

This reaction is the limiting step of the oxidation process for numerous materials, in particular those that have a low oxygen overpotential, such as metal oxides [3,21,25,26].

Electrochemical oxidation on a dimensionally-stable anode made of metal oxide was widely studied during the 1980s. Trasatti suggested a general scheme of oxidation pathway involving a first step of adsorbed radicals.

$$MO_x + H_2O \rightarrow MO_x({}^{\bullet}OH) + H^+ + e \tag{10}$$

These radicals can be physisorbed or chemisorbed. The physisorbed radicals react with organics following Eq. (11) until the end-stage of carbon dioxide.

$$R + MO_x({}^{\bullet}OH)_z \rightarrow CO_2 + zH^+ + ze + MO_x \tag{11}$$

TABLE 3 List of Various Organic Compounds and Their Rate Constant With Hydroxyl Radicals [12,20–22]

Chemicals	Hydroxyl radical constant rate ($L\,mol^{-1}\,s^{-1}$)
Acetic acid	1.6×10^{7}
Anisole	5.4×10^{9}
Benzene	8.0×10^{9}
Bromobenzene	4.4×10^{9}
Chlorobenzene	5.5×10^{9}
Ethanol	1.9×10^{9}
Formaldehyde	0.8×10^{9}
Formic acid	1.3×10^{8}
Maleic acid	6.0×10^{9}
Nitrobenzene	3.9×10^{9}
p-Nitrophenol	3.8×10^{9}
Phenol	6.6×10^{9}
Resorcinol	1.2×10^{10}
Styrene	6×10^{9}
Succinic acid	3.1×10^{8}
Toluene	5.1×10^{9}
Xylene-p	7.5×10^{9}

The chemisorbed radicals modify the electrode activity because they interact with the surface. This is the case of the anodes which are covered with metallic oxides, such as iridium oxides (IrO_3/IrO_2):

$$MO_x(^{\bullet}OH)\rightarrow MO_{x+1}+H^{+}+e \tag{12}$$

In this case, the chemisorbed active oxygen takes part in selective oxidation; it is not possible to reach complete mineralization. In the case of platinum, the oxygen–carbon bonds can stay in the adsorbed sites of the electrode, slowing the progression of oxidation of organic molecules [27].

$$R+MO_{x+1}\rightarrow RO+MO_x \tag{13}$$

In this case, the organic oxidation following Eqs. (12), (13) is in competition with oxygen formation (Eq. (14)):

$$MO_x(^{\bullet}OH)\rightarrow \tfrac{1}{2}O_2+H^{+}+e+MO_x \tag{14}$$

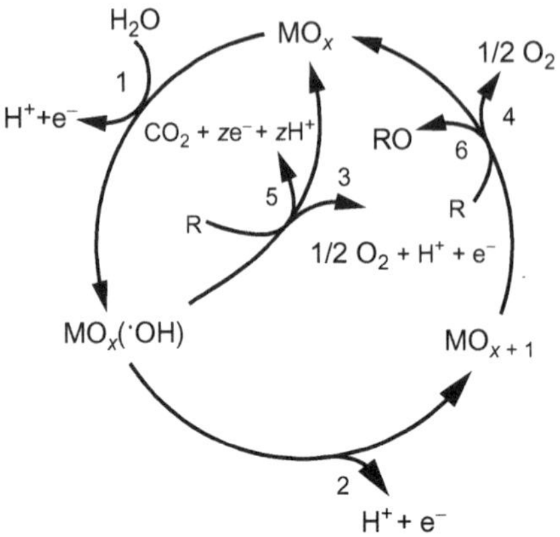

FIG. 2 Scheme of the electrochemical oxidation of organics on active anodes (6) and nonactive anodes (5); M is an active site at the electrode surface. *Reproduced with permission from C. Comninellis, Electrocatalysis in the electrochemical conversion/combustion of organic pollutants for waste water treatment, Electrochim. Acta 39 (11/12) (1994) 1857–1862. Copyright 1994. Elsevier.*

Consequently, the electrochemical oxidation depends strongly on the nature of the electrode material, which influences the selectivity and the efficiency of the process. To characterize the anode materials, Comninellis suggested that the electrode materials be classified in two types: "active" and "nonactive" electrodes [26]. Fig. 2 shows the different electrochemical pathways that can occur in acid media.

For "nonactive" electrodes there is a weak interaction between electrode M and hydroxyl radicals ($^\bullet OH$). In this case, the oxidation of organics is mediated by hydroxyl radicals. This reaction occurs in competition with oxygen evolution by hydroxyl radical discharge. This case is typical for boron doped diamond electrodes.

For "active" electrodes there is a strong interaction between electrode M and hydroxyl radicals. In this latter case, an oxygen transfer occurs between the hydroxyl radicals and the anode surface to form an oxide on the electrode surface. The couple MO/M can act as a mediator in the selective oxidation of organics, the surface redox. This is the case of IrO_2.

Although this terminology is presently highly used in the bibliography, it is difficult to describe the anodic behavior of materials by considering only two categories. IrO_2 and BDD electrodes represent two extreme cases of adsorption. It makes more sense to compare the strengths of adsorption of the various materials or the oxygen overpotentials. Table 4 gives different oxygen overpotentials for several anode materials.

The comparison of the oxygen overpotential shows that lead dioxide, tin dioxide, and BDD have electrochemical windows larger than that of platinum. This means that under anodic polarization in the region of

TABLE 4 Potentials for Oxygen Evolution Reactions on Different Anodes in H_2SO_4

Chemisorption	Anode	Value vs SHE	Conditions	Selectivity
↓	RuO_2	1.47	0.5M H_2SO_4	↓
	IrO_2	1.52	0.5M H_2SO_4	
	Pt	1.6	0.5M H_2SO_4	
	Oriented pyrolytic graphite	1.7	0.5M H_2SO_4	
	SnO_2	1.9	0.05M H_2SO_4	
	PbO_2	1.9	1M H_2SO_4	
Physisorption	Boron doped diamond (BDD)	2.3	0.5M H_2SO_4	Mineralization

Reproduced with modifications permission from M. Panizza, G. Cerisola, Direct and mediated anodic oxidation of organic pollutants, Chem. Rev. 109 (2009) 6541–6569. Copyright 2009. American Chemical Society.

water oxidation, the production of strong oxidative species that are weakly adsorbed on the electrode surface are particularly active for oxidation. From this results a remarkable electrocatalytic activity toward organic compounds. In other words, the oxygen overpotential increases as the adsorption strength of hydroxyl radicals on the electrode surface decreases. So, electrode materials that favor the chemisorption emphasize the selective oxidation, like platinum, IrO_2, or RuO_2. Conversely, BDD exhibits a lower adsorption capacity, and hydroxyl radicals which are quasi free on the electrode surface react very quickly and strongly, and favor complete mineralization.

$$BDD + H_2O \rightarrow BDD(^{\bullet}OH) + H^+ + e^- \quad (15)$$

$$BDD(^{\bullet}OH) + R \rightarrow BDD + pCO_2 + qH_2O \quad (16)$$

To illustrate this fact, a comparison of the performance of two electrode materials (PbO_2, BDD) used in the same electrolysis device exhibits the difference of activity of these materials toward phenol (Fig. 3). It appears that under mass transfer control, the rate of disappearance of phenol is the same on both materials. Furthermore, the number and amount of intermediates are higher with PbO_2; in particular, quinones produced by its oxidation are even more toxic than phenol. The detoxification of the solution is reached when all aromatics are removed, which leads to a required additional time using a PbO_2 anode.

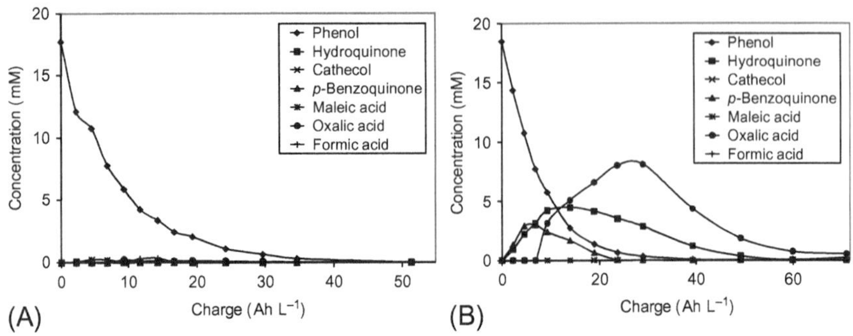

FIG. 3 Concentration variation of organic compounds during phenol oxidation. Electrolysis of 1 L of 0.02 M phenol solution in 0.1 M H_2SO_4. Flow rate = 200 L h^{-1}, $i = 142\,mA\,cm^{-2}$; $i_{lim} = 108\,mA\,cm^{-2}$. Reactor without separator; cathode: Zr; Anode: BDD (A) PbO_2 (B). *Reproduced with permission from E. Weiss, K. Groenen Serrano, A. Savall, A comparison study of electrochemical degradation of phenol on boron doped diamond and lead dioxide, J. Applied Electrochem, 38 (2008) 329–337. Copyright 2008. Springer.*

Thanks to the specific properties of nonselectivity of BDD, it is possible to define the electrolysis progress as a function of the chemical oxygen demand (COD), a global parameter which does not depend on the nature of the organics and intermediates.

$$C_xH_yO_z + \left(x + \frac{y}{4} - \frac{Z}{2}\right)O_2 \rightarrow xCO_2 + \frac{y}{2}H_2O \tag{17}$$

According to Panizza et al. [28], a kinetic model can be developed for the mineralization of organics using a BDD anode by making the following further assumptions: (i) adsorption of organics on the electrode surface is negligible, and (ii) the reaction of hydroxyl radicals on organics is fast and controlled by mass transport of organics to the anode surface. Under these conditions, the limiting current density, i_{lim}, can be defined by Eq. (18).

$$i_{lim} = 4Fk_m[COD] \tag{18}$$

where k_m is the average mass transfer coefficient of the reactor ($m\,s^{-1}$).

Under galvanostatic conditions, two different operating regimes are defined: (i) under charge transfer and (ii) under mass transfer. Table 5 presents the equations of the model that describe the variation of COD and the instantaneous current efficiency (ICE) as a function of time during electrolysis at the BDD anode of a solution containing organics.

This model is very useful because it allows the prediction of the variation of COD during the electrolysis, whatever the organic content of the solution.

Another advantage is the irreversibility of the oxidation via hydroxyl radicals. Consequently, it doesn't require any separator in the electrochemical cell. The setup used at the lab-scale constitutes very often with

TABLE 5 Equations That Describe the Chemical Oxygen Demand (COD) ($mol\,O_2\,m^{-3}$) and Instantaneous Current Efficiency (ICE) During 2-Naphthol Incineration at the BDD Electrode

	Instantaneous current efficiency (ICE)	Chemical oxygen demand (COD) ($mol\,O_2\,m^{-3}$)
$i_{appl.} < i_{lim}$ under current-limit control	ICE = 1	$COD(t) = COD^{\circ}\left(1 - \frac{\alpha A k_m}{V_R} t\right)$
$i_{appl.} > i_{lim}$ under mass transport control	$ICE = \exp\left(-\frac{A k_m}{V_R} t + \frac{1-\alpha}{\alpha}\right)$	$COD(t) = \alpha COD^{\circ} \exp\left(-\frac{A k_m}{V_R} t + \frac{1-\alpha}{\alpha}\right)$

A, area of electrode (m^2); V_R, the reservoir volume (m^3); $\alpha = i_{appl.}/i^{\circ}_{lim}$.
Reproduced with permission from M. Panizza, G. Cerisola, Application of diamond electrodes to electrochemical processes, Electrochim. Acta 51 (2005) 191–199. Copyright (2005) Elsevier.

or without a reservoir, and a typical electrochemical cell consists of planar monopolar electrodes in batch mode or recycling. This simple design allows the control of hydrodynamic conditions and enables the calculation of the mass transfer coefficient. Fig. 4 presents a typical setup in discontinuous process (A) and batch reactor (B).

Various numerical models proposed in the literature highlight the role of hydroxyl radicals during the electroxidation of organics on a BDD anode. It has been shown that the region of chemical activity of hydroxyl radicals is very close to the BDD surface (<20 nm) [29,30]. Only organic species that reach the anodic surface can be oxidized by electrogenerated

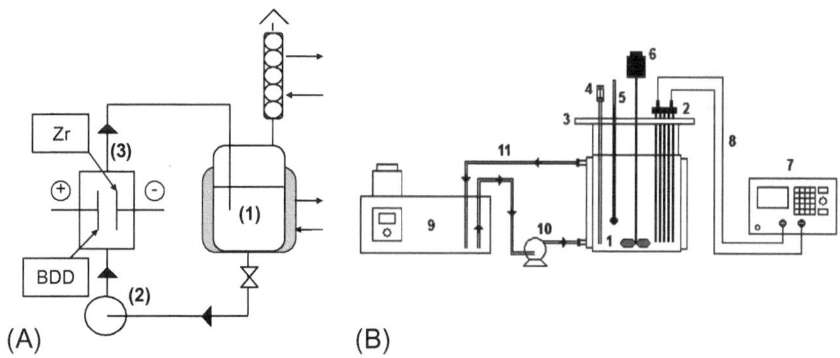

FIG. 4 (A) Discontinuous process with a single-compartment electrochemical reactor, (1) 0.5 L tank, (2) centrifugal pump, and (3) electrochemical cell. (B) (1) Batch reactor system, (2) BDD electrodes, (3) lid, (4) sampling cell, (5) thermometer, (6) mechanical mixer, (7) DC power supply, (8) electrical connections, (9) thermostat, (10) pump. *Reproduced with permission from Y. Lan, C. Coetsier, C. Causserand, K. Groenen Serrano, On the role of salts for the treatment of wastewaters containing pharmaceuticals by electrochemical oxidation using a boron doped diamond anode, Electrochim. Acta 231 (2017) 309–318. Copyright 2017. Elsevier; B.K. Körbahti, S. Taşyürek, Electrochemical oxidation of ampicillin antibiotic at boron-doped diamond electrodes and process optimization using response surface methodology, Environ. Sci. Pollut. Res. 22(5) (2015) 3265–3278. Copyright 2015. Springer.*

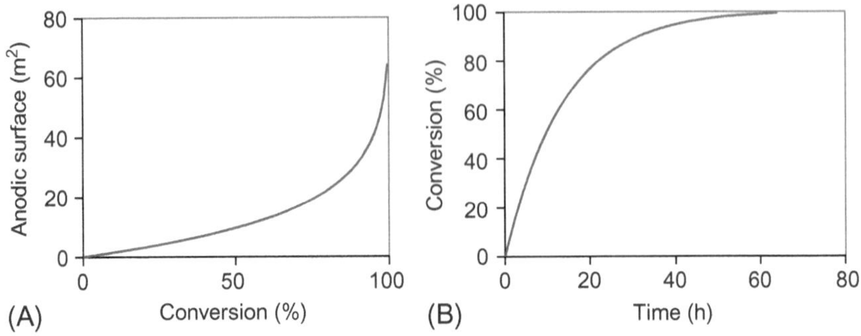

FIG. 5 (A) Surface needed, A, as a function of conversion ((COD°-COD(t))/COD°). (B) Conversion value as a function of time. Operating conditions $V=1\,m^3$, $i=i^{\circ}_{lim}=241\,A\,m^{-2}$, $A=1\,m^2$. *Reproduced with permission from A. Savall, Electrochemical Reactors Design for the Oxidation of Organic Pollutants, in: G. Kreysa, K. Ota, R.F. Savinell (Eds.), Encyclopedia of Applied Electrochemistry, Springer-Verlag, New York, 2014. Copyright 2014. Springer.*

hydroxyl radicals. This is the reason why electrochemical oxidation via hydroxyl radicals is considered as direct oxidation. This heterogeneous process is subject to mass transfer limitations. By using the model presented in Table 5, the required ratio A/V_R to obtain a desired COD conversion can be calculated for 1 h of electrolysis of $1\,m^3$ of a polluted solution under a constant current density equal to the limiting value calculated at $t=0$. Fig. 5A shows that very high surface areas of anode must be used when the desired COD conversion tends to 100%. Fig. 5B shows how the electrolysis time varies in order to reach a specific conversion value.

To render the process more efficient, the applied current density can be modulated in order to operate at limiting current density to minimize the side reaction of oxygen evolution. Another way could be to use the electrode surface to generate a strong oxidant that is sufficiently stable to diffuse through the solution and chemically react with the refractory compound. In addition, the electrosynthesis of powerful oxidants and thus selective oxidations may be exploited to render the products biodegradable. These oxidation mediators can be metallic redox couples such as Ag (II), Ce (IV), or strong oxidizing chemicals generated from salts such as active chlorine or peroxocompounds.

ACTIVE CHLORINE

The presence or the deliberate addition of Cl^- in a solution to be treated by electrochemical oxidation can accelerate the process of organic matter degradation. Indeed, the organic matter can be oxidized at the electrode and also at the center of the solution by chemical reaction with the active chlorine.

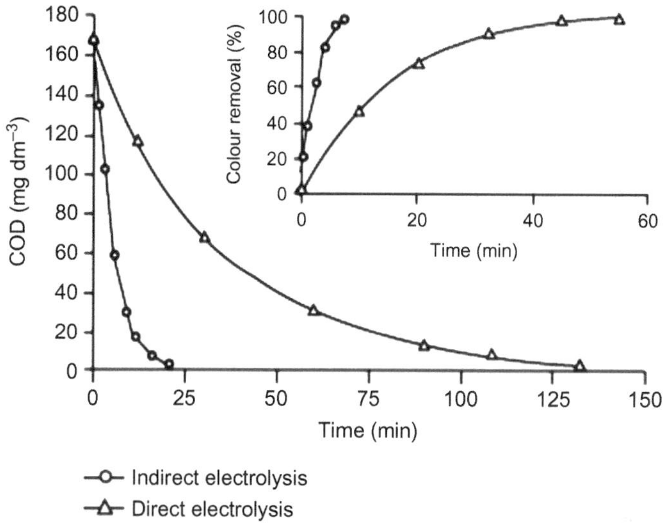

FIG. 6 Comparison of the COD variation and color removal (inset) during direct and mediated electrolysis ($[Cl^-]=1.2\,g\,L^{-1}$). *Reproduced with permission from M. Panizza, A. Barbucci, R. Ricotti, G. Cerisola, Electrochemical degradation of methylene blue, Sep. Purif. Tech. 54 (3) (2007) 382–387. Copyright 2007. Elsevier.*

For example, Panizza et al. [31] compared the COD abatement and color removal for direct and indirect electrolysis of a methylene blue solution (Fig. 6). It appeared that indirect electrolysis facilitates faster mineralization and discoloration of the solution. This is due to the strong bleaching properties of active chlorine, and the indirect electrolysis allows the mass transfer limitations to be overcome.

The electrochemical chlorination can be used for water disinfection, treatment of wastewater containing refractory organics and chloride ions, landfill leachates, pharmaceuticals, pesticides, dyes, etc. Active chlorine is generated by the oxidation of chloride ions:

$$2\,Cl^- \rightarrow Cl_{2\,(aq)} + 2e \quad (19)$$

$$Cl_2 + H_2O \rightarrow HClO + H^+ + Cl^- \quad (20)$$

$$HClO = H^+ + ClO^- \quad (21)$$

It may be in the form of Cl_2, ClO^-, or HClO, depending on the pH, as shown in Fig. 7.

At 25°C for a chloride concentration of 5×10^{-3} M, Cl_2 hydrolysis is almost complete at pH > 4 and hypochlorous acid and hypochlorite are the main chlorine species in the pH range 6–9.

Due to its oxidizing power and its chemical structure characterized by the Cl—O bond polarization, the reactions of $HClO/ClO^-$ with organics

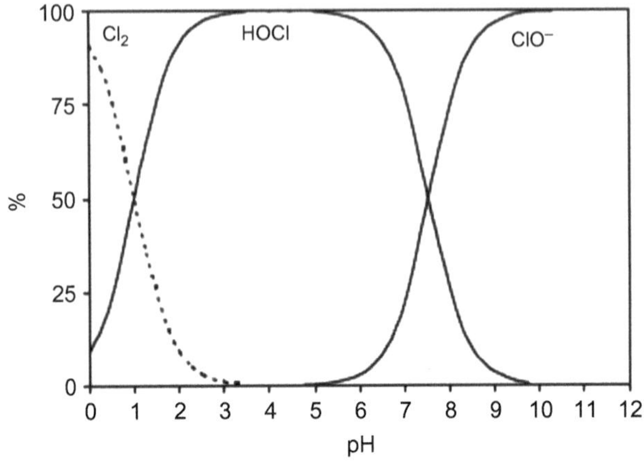

FIG. 7 Relative distribution of the main aqueous chlorine species as a function of pH at 25° C and for a chloride concentration of 5×10^{-3} M. *Reproduced with permission from M. Deborde, U. von Gunten, Reactions of chlorine with inorganic and organic compounds during water treatment-kinetics and mechanisms: a critical review, Water Res. 42 (2008) 13–51. Copyright 2008. Elsevier.*

can be described as (i) oxidation reactions, (ii) addition reactions, and (iii) electrophilic substitution reactions at nucleophilic sites [32]. In contrast to hydroxyl radicals, the reaction between organics and $HClO/OCl^-$ is selective.

Various studies using DSA electrodes show that the use of sodium chloride as an electrolyte improves the efficiency or enables the anodic oxidation of many organics such as phenol, glucose, dyestuff, pesticides, and endocrine disruptors. This is in comparison to other salts like sodium sulfate, sodium carbonate, and nitrates. As an example, the oxidation of methomyl (a pesticide) using a Ti/RuO_2 anode does not occur in the absence of chlorides, while it takes place in the presence of NaCl [33].

Among all parameters that influence the production of active chlorine, the anode material is the key parameter. The DSA active anodes such as RuO_2 and IrO_2 are the ideal anodes to produce active chlorine. Kraft et al. compared the production efficiency of hypochlorite for the dimensionally-stable anode using active coatings of IrO_2 and IrO_2/RuO_2, platinum, and BDD electrodes [34]. They highlighted that DSA clearly outperforms BDD and Pt. Indeed, in the case of inactive electrodes such as BDD, the formation of active chlorine competes with the water discharge. Nevertheless, the comparative studies on IrO_2 and BDD electrodes for the electrochemical oxidation of artificial human urine containing NaCl and NH_4Cl were investigated by Li et al. [35]. It was found that the final products of inorganic ions, including ClO_4^- and NO_3^-, were formed on BDD, and for the IrO_2 electrode the final highly toxic substances, including N_2H_4 and NO_2^-, were left in the solution. This incomplete oxidation was attributed to the low oxidizing activity of the generated hydroxyl radicals that were strongly chemisorbed on the surface of active IrO_2 electrodes.

The pH value affects the efficiency of the formation of active chlorine: a lower pH favors the chlorine evolution reaction. In contrast, a high pH is expected to favor the oxidation of water over chloride.

Moreover, in the case of an undivided cell using an active anode, an increasing flow rate has a negative impact on active chlorine production by promoting the removal of chlorine at the cathode and by homogenizing and increasing the pH.

In addition, besides direct electron transfer reactions, the electroxidation pathway of Cl^- also includes the chemical oxidation reactions with hydroxyl radicals. These lead to the formation of undesired by-products during the electrochemical oxidation of chloride anions, these by-products being chlorites, chlorates, and perchlorates. The possible pathway is exhibited in the reaction scheme of Fig. 8, with the reactions listed in Table 6.

The effects of ClO_3^- and ClO_2^- in mammals arise from oxidative damage to red blood cells resulting in hemolytic anemia and methaemoglobin formation [36]. The World Health Organization has proposed a guideline value for drinking water of $0.2\,mg\,L^{-1}$ for ClO_2^- and $0.7\,mg\,L^{-1}$ for ClO_3^- [37]. Perchlorate is suspected of disrupting the thyroid gland and is a carcinogenic potential. The WHO recommends that drinking water contains a maximum of 0.07 and $0.015\,mg\,L^{-1}$ for the Environmental Protection Agency [38]. Furthermore, the oxidizing ability of chlorate and perchlorate is lower than hypochloric acid or hypochlorite anions. In addition, the presence of chloride species during the electrochemical oxidation of organics can also form halogenated-organic compounds (HOC) or chloroamines. Various studies have shown that the formation of such compounds is attributed to addition and substitution reactions between organics and the in situ-generated chlorine species and radicals. For example, the treatment by electroxidation using BDD of landfill leachate containing $2587\,mg\,L^{-1}$ of Cl^- has led to the formation of chloroform (55% of the total amount of HOC), trihalomethanes, haloacetonitriles, haloketones, and 1,2-dichloroethane. Such by-products are highly toxic, often

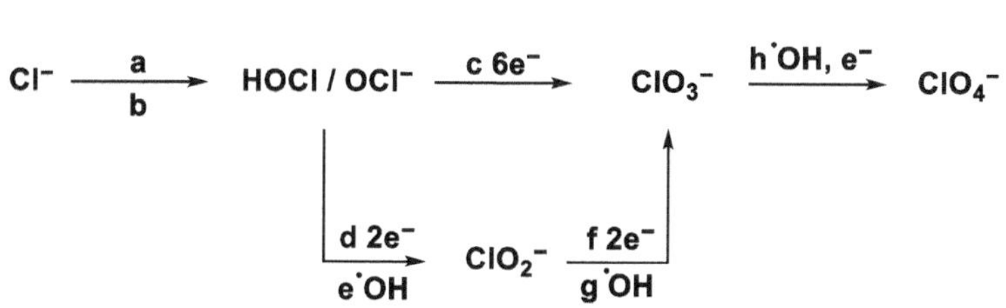

FIG. 8 Different pathways for the electrochemical oxidation of chloride ions by electron transfer or oxygen atom transfer via hydroxyl radicals. The equations are given in Table 6.

TABLE 6 Reactions of Chlorine Species During Electrolysis

	Reactions	
a	$Cl^- \rightarrow Cl^\bullet + e^-$ $Cl^\bullet + Cl^\bullet \rightarrow Cl_2$ $Cl_2 + H_2O \rightarrow HOCl + Cl^- + H_2O$	$E° = 1.36\,V$
b	$Cl^- + {}^\bullet OH \leftrightarrow ClOH^{-\bullet}$	$k = 4.3 \times 10^9\,M^{-1}\,s^{-1}$
c	$6HOCl^- + 3H_2O \rightarrow 2ClO_3^- + 4Cl^- + 12H^+ + 3/2O_2 + 6e^-$ $6OCl^- + 3H_2O \rightarrow 2ClO_3^- + 4Cl^- + 6H^+ + 3/2O_2 + 6e^-$	$E° = 0.46\,V$
d	$OCl^- + 2OH^- \rightarrow ClO_2^- + H_2O + 2e^-$	(Alkaline solution)
e	${}^\bullet OH + OCl^- \rightarrow {}^\bullet ClO + OH^-$ ${}^\bullet OH + OCl^- \rightarrow ClO_2^- + H^+$	$k = 9 \times 10^9\,M^{-1}\,s^{-1}$ $k > 1 \times 10^9\,M^{-1}\,s^{-1}$
f	$ClO_2^- + 2\,OH^- \rightarrow ClO_3^- + H_2O + 2e^-$	(Alkaline solution)
g	${}^\bullet OH + ClO_2^- \rightarrow {}^\bullet ClO_2 + OH^-$ ${}^\bullet OH + {}^\bullet ClO_2 \rightarrow ClO_3^- + H^+$	$k = 6 \times 10^9\,M^{-1}\,s^{-1}$ $k = 4 \times 10^9\,M^{-1}\,s^{-1}$
h	$ClO_3^- \rightarrow {}^\bullet ClO_3 + e^-$ ${}^\bullet ClO_3 + {}^\bullet OH \rightarrow HClO_4$	

Reproduced with permission from Y. Lan, C. Coetsier, C. Causserand, K. Groenen Serrano, On the role of salts for the treatment of wastewaters containing pharmaceuticals by electrochemical oxidation using a boron doped diamond anode, Electrochim. Acta 231 (2017) 309–318. Copyright 2017. Elsevier.

with higher rates of toxicity than the initial organics [39]. Degradation is possible for some of them by extending the electrolysis duration as Costa et al. showed, using a BDD anode for the treatment of acid black solution containing NaCl [40].

However, the formation of such by-products must be minimized or prevented during the electrochemical generation of active chlorine.

Over the last few years, numerous reviews [41–44] have collected the studies performed on the electrochemical oxidation of synthetic solutions or wastewater containing organics in the presence or absence of sodium chloride. It appears that various operating conditions may affect the efficiency of organic degradation and encourage the formation of undesired products.

To provide the optimal conditions of electrolysis is not easy. The system is complex with potential interactions between parameters. However, these results provide evidence of general trends: among all parameters, one can select the anode material and its morphology, the applied current intensity and the concentration of organics, and the geometry of the cell and its hydrodynamics.

Due to the high overpotential of oxygen evolution on the BDD anode, the oxidation of chloride ions leads to the formation of chlorate and perchlorate. Bergmann et al. [45] showed that perchlorate production on the

BDD anode is approximately 50–100 times higher than on the platinum and the DSA anode.

Moreover, the porous structure of the DSA anode enhances the generation of active chlorine by maintaining a low pH, which locally-favors a hypochlorite evolution. By contrast, if the electrochemical cell is undivided, increasing the flow rate (or the stirring rate) leads to the reduction of the hypochlorite at the cathode and hinders the indirect oxidation process [46].

In the case of BDD, to avoid or to limit the formation of undesired by-products such as chlorates, perchlorates, or HOC, it is important to manage the production of hydroxyl radicals. The choice of the applied current intensity depends on the amount of organic matter:

- When the process is limited by charge transfer and when the kinetic of the reaction between organics and hydroxyl radicals is fast, there are sufficient organic materials at the anode surface to scavenge the $^{\bullet}OH$ at the anode surface and therefore inhibit the formation of perchlorate [47].
- While under mass transfer control, many hydroxyl radicals are produced; consequently, many oxygen atoms can be transferred to the chlorine species and organics to form perchlorates and HOC.

For example, the electrolysis of a solution containing an antibiotic (ciprofloxacin) and 2.2 mM of chloride using a BDD anode has shown that the TOC removal at $i \approx i^{\circ}_{\mathrm{lim}}$ reached 90%. This is dramatically decreased to 40% when $\frac{i}{i^{\circ}_{\mathrm{lim}}} > 30$ with a significant quantity of HOC (measured as adsorbable organic halogens, AOX). Moreover, the formation of ClO_4^- was only observed at high current densities $\left(\frac{i}{i^{\circ}_{\mathrm{lim}}} > 30\right)$ and appeared when organic matter was nearly completely degraded [48].

In the case of the formation of chlorinated organic compounds, which are persistent during the electrochemical process, a second step to remove them, such as adsorption on activated carbon, is required.

PEROXO-COMPOUNDS

Peroxodisulfate and Sulfate Radicals

In the case of the organic removal by electrochemical processes, sulfate ions, as well as chlorides, can be present in effluents or natural water, or added to the solution as an electrolyte to render the solution conductive. Unlike chlorides, sulfates are not considered to be pollutants; the United States Environmental Protection Agency has listed them under the secondary drinking water standards with a maximum concentration of 1.43 mM.

Peroxodisulfate, $S_2O_8{}^{2-}$ can be electrochemically formed from the oxidation of sulfate species at very high potentials (>2 V/SHE), as shown in Eqs. (22), (23):

$$2HSO_4{}^{-} \rightarrow S_2O_8{}^{2-} + 2H^+ + 2e \quad E^\circ = 2.123\,V \tag{22}$$

$$2SO_4{}^{2-} \rightarrow S_2O_8{}^{2-} + 2e \quad E^\circ = 2.01\,V \tag{23}$$

The electrochemical production of peroxodisulfates strongly depends on the electrode material selected for the oxidation of sulfuric acid. The main criterion is the high overpotential value for the oxygen evolution reaction. Before the year 2000, platinum was the first choice of anode for the electrosynthesis of peroxodisulfates. Elbs and Schönherr [49], using ammonium cations, obtained a current efficiency of 85%. They ascribed the effect of various cations on the current efficiency to an unspecified catalytic effect. According to Smit and Hoogland [50], the coverage of the surface of platinum by oxygen was increasingly replaced by patches of adsorbed $SO_4{}^{2-}$, on which $S_2O_8{}^{2-}$ formation occurs instead of oxygen evolution. Another explanation is given by Kasatkin and Rakov [51]: these additives block up the active centers and favor the recombination of atomic oxygen, which reacts with sulfate species. Using a BDD anode, it has been shown that the formation of peroxodisulfuric acid coincides with the formation of oxygen [52]. Provent et al. [53] and Khamis et al. [54] have evidenced the formation of peroxodisulfate on the BDD surface using microelectrodes and scanning electrochemical microscopes, respectively.

In sulfuric acid solutions, sulfate anion, hydrogenosulfate anion, and undissociated sulfuric acid coexist [55,56]. Only $HSO_4{}^{-}$ and molecular H_2SO_4 react with $OH^\bullet$ radicals to form sulfate radicals, following reactions (24) and (25) [57–59]:

$$HSO_4{}^{-} + OH^\bullet \xrightarrow{k_1} SO_4{}^{-\bullet} + H_2O \tag{24}$$

$$H_2SO_4 + OH^\bullet \xrightarrow{k_2} SO_4{}^{-\bullet} + H_3O^+ \tag{25}$$

$$SO_4{}^{-\bullet} + SO_4{}^{-\bullet} \xrightarrow{k_3} S_2O_8{}^{2-} \tag{26}$$

The rate constants k_1, k_2, and k_3 equal $4.7 \times 10^5\,L\,mol^{-1}\,s^{-1}$, $1.4 \times 10^7 L\,mol^{-1}\,s^{-1}$, and $7.6 \times 10^8\,L\,mol^{-1}\,s^{-1}$, respectively [57].

To produce peroxodisulfate electrochemically with a high current efficiency, it is necessary to use a divided cell, such as in Fig. 9, to avoid the reduction of the peroxodisulfate at the cathode. Serrano et al. [52] showed that high current efficiency (95%) for electrosynthesis of peroxodisulfuric acid was achieved in concentrated H_2SO_4 (>2 M) for moderate temperatures (8–10°C). This current efficiency is higher than that obtained in industrial electrosynthesis on platinum electrodes (about 70%–75%) [60,61].

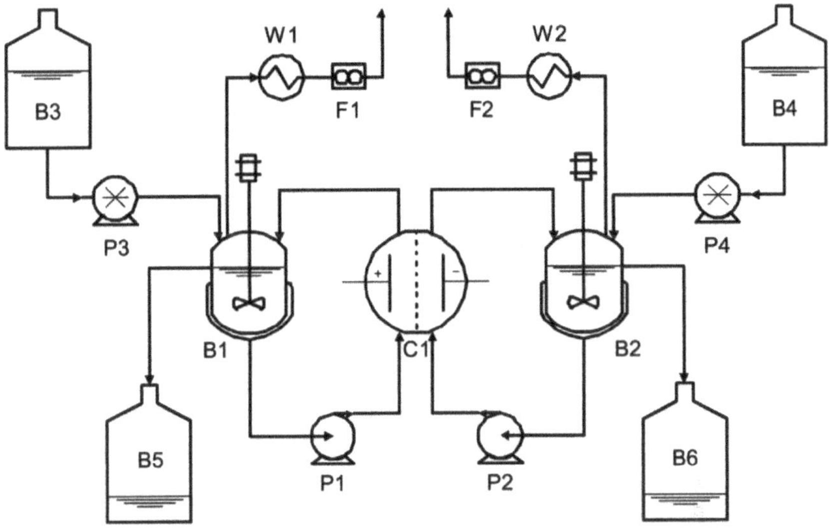

FIG. 9 Setup used for electrochemical oxidation of sulfuric acid on Si/Diamond anodes. C1: electrochemical cell; B1 and B2: anodic and cathodic (resp.) thermoregulated tank (250mL); B3 and B4 feed tank anodic and cathodic (10L); B5 and B6: purge tank; P1 and P2: centrifugal pump feeding the anodic and cathodic compartment. *Reproduced with permission from K. Serrano, P.A. Michaud, C. Comninellis, A. Savall, Electrochemical preparation of peroxodisulfuric acid using boron doped diamond thin film electrodes, Electrochim. Acta 48 (4) (2002) 431–436. Copyright 2002. Elsevier.*

The reactions of peroxodisulfate with organics at room temperature are not efficient. Peroxodisulfate should be activated. This activation can be carried out by a metal reaction or by supplying energy (heat, UV, etc.), and allows the formation of sulfate radicals (Eq. 27):

$$S_2O_8{}^{2-} + \text{heat} \rightarrow 2SO_4{}^{-\bullet} \quad E^\circ\left(SO_4{}^{2-}/SO_4{}^{-\bullet}\right) = 2.4\,\text{V} \qquad (27)$$

Generally, $SO_4{}^{-\bullet}$ is more prone to electron transfer reactions than $^{\bullet}OH$. In contrast to $SO_4{}^{-\bullet}$, $^{\bullet}OH$ more readily undergoes hydrogen abstraction or addition. Thus, $SO_4^{-\bullet}$-based oxidation is presented as an alternative oxidative treatment toward the AOPs based on $^{\bullet}OH$ production. A sulfate radical has some unique features, such as being a very strong electron acceptor that enables the degradation of persistent compounds, refractory toward the hydroxyl radical. $SO_4{}^{-\bullet}$ are capable of oxidizing pollutants such as trichloroethene [62], tertbutylmethylether [63], chlorinated ethenes [64], and perfluorinated compounds [65]. Table 7 lists the kinetic constant of some organics with sulfate radicals [66]. Moreover, peroxodisulfate is relatively stable and can be produced at high concentrations, thus avoiding problems of transport limitation.

TABLE 7 Rate Constants for Reaction of $SO_4^{-\bullet}$ With Aromatic Compounds at pH 7

Aromatic compound	Rate constant ($M^{-1}\,s^{-1}$)
Anisole	4.9×10^9
Acetanilide	3.6×10^9
Benzene	3×10^9
Benzoic acid	1.2×10^9
Acetophenone	3.1×10^8
Benzamide	1.9×10^8
Trimethylanilinium ion	1.5×10^8
Benzonitrile	1.2×10^8
Nitrobenzene	$\leq 10^6$
p-Methoxybenzoic acid	3.5×10^9
p-Hydroxybenzoic acid	2.5×10^9
m-Toluic acid	2×10^9
p-Toluic acid	1.8×10^9
o-Toluic acid	1.4×10^9
Benzoic acid	1.2×10^9
p-Bromobenzoic acid	1.0×10^9
o-Bromobenzoic acid	8.7×10^8
p-Chlorobenzoic acid	3.6×10^8
p-Acetylbenzoic acid	2.0×10^8
Terephtalic acid	1.7×10^8
p-Cyanobenzoic acid	3.3×10^7
p-Nitrobenzoic acid	$\leq 10^6$

Reproduced with permission from P. Neta, V. Madhavan, H. Zemel, R.W. Fessenden, Rate constants and mechanism of reaction of sulfate radical anion with aromatic compounds, J. Am. Chem. Soc. 99 (1) (1977) 163–164. Copyright 1977. American Chemical Society.

It has been evidenced in numerous studies dedicated to electrochemical oxidation processes that the use of sodium sulfate as an electrolyte enhances the removal of organic matter. For example, Murugananthan et al. [67] observed that the effective mineralization of ketoprofen can only be achieved when the supporting electrolyte was Na_2SO_4, in comparison with $NaNO_3$ and NaCl. Rodrigo et al. underlined the higher COD removal of olive oil mill wastes and the higher decay rate for cyanides obtained using Na_2SO_4 as a electrolyte, in comparison with NaCl [68,69]. Numerous authors explained that this phenomenon was due to the generation of $S_2O_8^{2-}$. It is most probably due to the formation of $SO_4^{-\bullet}$ from hydrogenosulfate (Eq. (24)), which is more reactive than peroxodisulfate.

An important aspect in radical-based processes is related to the water matrix composition. Lutze et al. demonstrated that the presence of chloride ions and bicarbonates can affect the oxidation processes based on $SO_4^{-\bullet}$ [70]. Indeed, the reaction between sulfate radicals and chlorides leads to the formation of chloride radical ($Cl^{\bullet}$) [71].

$$SO_4^{-\bullet} + Cl^{-} \rightarrow Cl^{\bullet} + SO_4^{2-} \quad k = 2.8 \times 10^{8}\,M^{-1}s^{-1} \tag{28}$$

Chloride radicals initiate a cascade of reactions, depending on the pH. At a low pH (<5), the reaction of the undesired chlorate is favored, whereas at a higher pH (>5) chloride radicals react with water to form hydroxyl radicals.

$$Cl^{\bullet} + H_2O \rightarrow {}^{\bullet}OH + H^{+} + Cl^{-} \tag{29}$$

The conversion of sulfate radicals into hydroxyl radicals can be hindered by the presence of bicarbonates that will scavenge the $Cl^{\bullet}$.

In view to treat industrial wastewaters with low concentrations of biorefractory organic species, Vatistas et al. propose a new combined two-step process which separates the electrochemical production of peroxodisulfates from the activation by heat (conversion of $S_2O_8^{2-}$ to $SO_4^{-\bullet}$) and oxidation of organics (Fig. 10) [72]. The first step consists of electrochemically synthesizing peroxodisulfate via hydroxyl radicals using a BDD anode and a low-temperature electrolyte of high sulfate ion concentration. The peroxodisulfate, which is relatively stable, can be produced at high concentrations. The second step consists of introducing the peroxodisulfate to the preheated wastewater, where it is activated and the organic pollutants are efficiently oxidized. In this process, the temperature plays a key role. To reach an efficient production of peroxodisulfate, the temperature must be lower than 10°C, while formation of sulfate radicals for the chemical reaction organics must be performed at temperatures above 60°C.

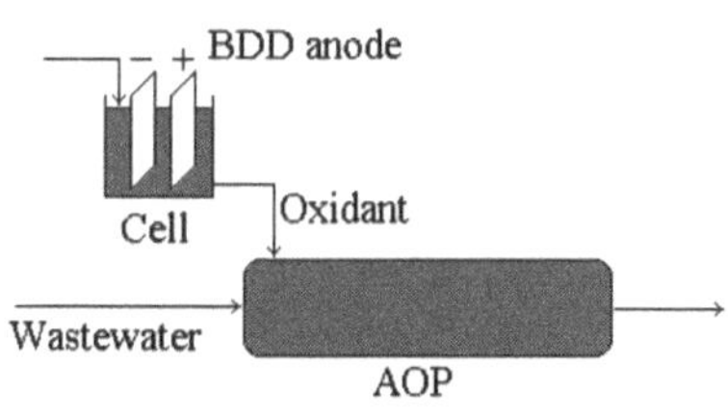

FIG. 10 Combination of oxidant production on a BDD anode and an advanced oxidation process (AOP). *Reproduced with permission from N. Vatistas, C. Comninellis, R.M. Serikawa, G. Prosperi, Oxidant production on BDD anodes and advanced oxidation processes, in: Fujishima et al. (Eds) Diamond Electrochemistry, Elsevier, Amsterdam, 2005, pp 477–501. Copyright 2005. Elsevier.*

Peroxodicarbonates

The use of anodes with a high oxygen overpotential for the electrochemical treatment of solutions containing carbonates and/or phosphates can lead to an indirect electrochemical oxidation with the mediation of peroxodicarbonate ($C_2O_6^{2-}$) or peroxodiphosphate ($P_2O_8^{2-}$), respectively.

Peroxydicarbonate ($C_2O_6^{2-}$) and percarbonate ($2Na_2CO_4.3H_2O_2$) have a wide range of uses as oxidants and bleaching agents. The main advantage is that these compounds are biodegradable, nontoxic, and environmentally-safe products. Peroxodicarbonate can be obtained by oxidation of carbonate; the overall reaction is provided in Eq. (30).

$$2CO_3^{2-} \rightarrow C_2O_6^{2-} + 2e \quad E^{\circ} = 1.8\,V/ESH \tag{30}$$

Saha et al. reported a method for the production of peroxydicarbonate on BDD in 1 M Na_2CO_3 [73]. The current efficiency of the formation of sodium peroxydicarbonate was found to significantly depend on the current density, anode material, electrolyte concentration, pH, and geometry of the cell. The maximum current efficiency for producing sodium peroxydicarbonate was 82% at a current density of $0.05\,A\,cm^{-2}$ after 30 min of electrolysis in a 1 M Na_2CO_3 solution in an undivided cell at 0°C.

Velazquez-Pena et al. [74] compared the influence of the geometry of the cell at 5°C and $pH > 10$. They found that the formation of peroxocarbonates is favored at low current densities and at high concentrations of carbonates, using a divided cell.

Another way to generate peroxydicarbonate is to remove CO_2 from the atmosphere by dissolving it in NaOH [75].

$$CO_2 + OH^- \rightarrow HCO_3- \tag{31}$$

$$2HCO_3^- + 2^{\bullet}OH \rightarrow C_2O_6^{2-} + 2H_2O \tag{32}$$

The highest current efficiency of 70% for producing peroxydicarbonate in the anode chamber was found at 5°C with a current density of $50\,mA\,cm^{-2}$ after 30 min of electrolysis.

Chardon et al. [76] performed electrolysis at low temperatures in an undivided cell using an initial concentration of 1.5 M Na_2CO_3 at $720\,mA\,cm^{-2}$ at $pH = 11$. They modelized the electrochemical generation of peroxodicarbonate, taking into account its chemical decomposition, and obtained a high peroxodicarbonate concentration at a higher current density (Fig. 11).

It is necessary to stabilize the generated peroxodicarbonate. Ruiz et al. [77] used metasilicate during the electrolysis. Chardon et al. showed that after 24 h the oxidants are completely removed at $pH = 11$, whereas 75% of the total oxidants are still present at $pH = 3.3$.

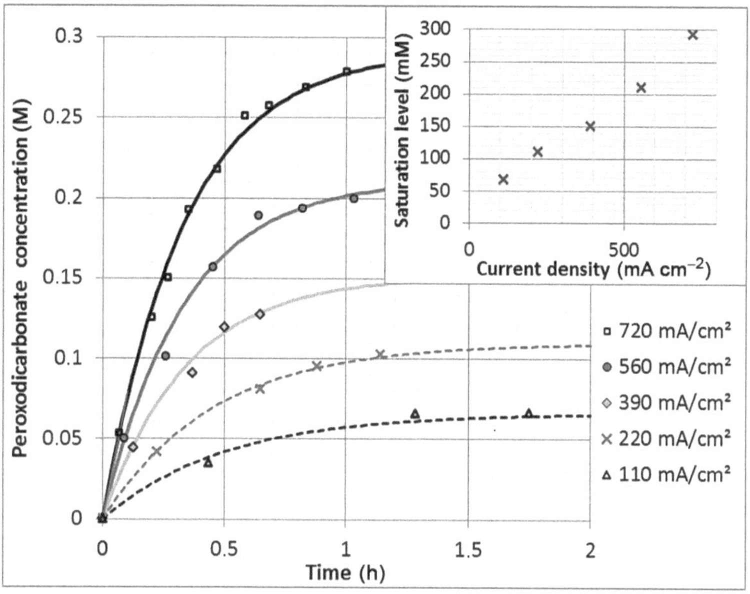

FIG. 11 Production of peroxodicarbonate over time at different current densities (symbols: experimental data; lines: model). *Reproduced with permission from C.P. Chardon, T. Matthée, R. Neuber, M. Fryda, C. Comninellis, Efficient electrochemical production of peroxodicarbonate applying Diachem® diamond electrodes, Chem. Select 2(3) (2017) 1037–1040. Copyright 2017. Wiley & Sons.*

Peroxophosphates

Peroxophosphates have a wide range of applications: among them wastewater treatment and the treatment of water polluted by petroleum. Due to its similar properties to peroxodisulfate, peroxodiphosphate can also be used in other processes as a substitute; this is more environmentally-friendly, because phosphates can be removed from aqueous wastes more easily than sulfates. Peroxodiphosphoric acid is not produced commercially; in any case, only tetrapotassium peroxophosphate was ever produced on a commercial scale. The electrosynthesis of $K_4P_2O_8$ can be carried out on a platinum anode in alkaline conditions, with a low current efficiency due to competition from water discharge. The addition of certain reagents (fluoride, thiocyanate, or nitrate) blocks the anode sites of the oxygen evolution in the process. Consequently, direct oxidation of phosphate to peroxodiphosphate is favored over the process of water oxidation; as a result, higher current efficiencies (70%–75%) are obtained.

$$\mathrm{PO4}^{3-} \rightarrow \mathrm{PO4}^{2-\bullet} + e^- \quad E^\circ = 2.26\,\mathrm{V/ESH} \tag{33}$$

$$\left(\mathrm{PO4}^{2-}\right)^\bullet + \left(\mathrm{PO4}^{2-}\right)^\bullet \rightarrow \mathrm{P_2O_8}^{4-} \tag{34}$$

However, some of the reagents can be toxic or highly corrosive to platinum. Thus, in order to obtain impurity-free peroxodiphosphate, extensive purification is required. This greatly increases manufacturing costs, and therefore is not an option for the treatment of wastewater.

In this context, Canizares et al. [78] demonstrated that the production of peroxodiphosphate by electrolysis with a conductive diamond in K_3PO_4 in alkaline conditions (pH = 12.5, $i = 1250\ \mathrm{A\ m^{-2}}$, $T = 25°\mathrm{C}$) leads to the synthesis of a high-quality product with good efficiency and a conversion yield (over 70%) with high stability. However, they observed that electrolysis at high K_3PO_4 concentrations damaged the surface of the diamond electrode and small corrosion circles appeared. They recommended the use of 1 M K_3PO_4.

In addition, the electrolysis of phosphate solutions at acidic pHs leads to the generation of peroxomonophosphate [79]. At an acidic pH, the dihydrogenophosphate is oxidized at the anode and forms the associated radical (Eq. 35). The phosphate radicals, once formed, react with hydroxyl radicals generated from water oxidation to form peroxomonophosphoric acid:

$$H_2PO_4^- \rightarrow (H_2PO_4)^\bullet + e^- \quad (35)$$

$$(H_2PO_4)^\bullet + {}^\bullet OH \rightarrow H_3PO_5 \quad k = 4 \times 10^9\ \mathrm{L\,mol^{-1}s^{-1}} \quad (36)$$

Nevertheless, the process is less efficient than the generation of peroxodiphosphate in alkaline solutions, due to the lower chemical stability of the peroxomono compound.

CONCLUSION

If wastewater is treated at the source, electrochemistry can be considered as a good option for the removal of persistent organic pollutants. This process can be used alone, or combined with other processes. Direct electrochemical oxidation consists of the transfer of electrons between electroactive species and the anode. This process has revealed the following limitations:

- Not all pollutants are electroactive.
- During the anodic oxidation, fouling can occur at the anode surface, which can affect or block the transfer of electrons.
- The rate of organic removal is limited by mass transfer, due to the depletion of organic matter in the solution. This limitation by mass transfer leads to a decrease in the current efficiency during electrolysis, which greatly enhances the energetic cost.

Indirect electrochemical oxidation allows all of these limitations to be overcome. The mediators are strong oxidants that are electrogenerated at the electrode from water (hydroxyl radicals), oxygen (ozone and hydrogen peroxide), or from salts (active chlorine or peroxocompounds). Salts may already be present in the effluent, or they can be added to render the solution conductive. The electrogenerated mediators can be classified into one of the following two groups:

(i) Very reactive species with strong oxidizing power. For this case, the chemical reaction with organics is not selective and complete mineralization can be reached. Oxidation is irreversible. The use of a divided electrochemical cell is not required, but the technological implementation is more simple. Moreover, reusing treated wastewater might be considered. The hydroxyl radicals, produced by direct water oxidation, belong to this group. Their half-life is very low (10^{-9} s in water), and their action takes place exclusively close to the anode. Unlike direct oxidation, consequently, this process has mass transfer limitations.

(ii) The other oxidants presented in this chapter belong to the selective oxidants group.

Under specific conditions, they are sufficiently stable to be able to perform homogeneous chemical reactions with organic matter. Their main advantage is their ability to treat dilute solutions, or even micropollutants. Complete mineralization of organics is not feasible, or is difficult to achieve. In this context, the coupling of processes may be a good option. For example, indirect electrochemical oxidation allows the solution to be rendered biodegradable before being treated biologically.

The operating conditions, such as pH, the hydrodynamics of the cell, or applied current density, come into play in the performances of the process, not only for the conversion rate but also for the stability of the electrogenerated oxidant. However, the anode material is the key parameter. Because all oxidants are formed at high potentials, the competitive reaction is the formation of oxygen. An anode material with a high oxygen overpotential is required. In this context, the development of the boron doped diamond (BDD) electrode had a significant impact on the electrosynthesis of the following strong oxidants: ozone, hydroxyl radicals, peroxodisulfate (whose activated form is sulfate radical), peroxodiphosphate, and peroxodicarbonate. Both high current efficiency and good corrosion resistance were obtained.

However, in the presence of chlorides the use of a BDD electrode should be carefully managed. For instance, if the applied current intensity is much higher than the limiting current intensity, the formation of toxic by-products such as chlorates, perchlorates, or halogenated organic compounds may occur. A specific study is therefore needed to seek the appropriate operating conditions to limit the formation of undesired, toxic species. Preference will be given to DSA IrO_2/RuO_2, the most efficient anodes for the active chlorine generation. These anodes have been shown to provide good performances in the chlor-alkali industry and in the disinfection of private swimming pools.

Acknowledgments

I would like to thank all my collaborators, in particular André Savall, and PhD students that I have been working with in the field of electrochemical processes for wastewater treatment over the past 20 years.
I would also like to thank Sophie Chambers for proofreading the manuscript.

References

[1] D.W. Kirk, H. Sharifian, F.R. Foulkes, Anodic oxidation of aniline for wastewater treatment, J. Appl. Electrochem. 15 (1985) 285–292.
[2] S. Stucki, R. Kötz, B. Carcer, W. Suter, Electrochemical waste water treatment using high overvoltage anodes part II: anode performance and applications, J. Appl. Electrochem. 21 (1991) 99–104.
[3] S. Trasatti, Electrocatalysis by oxides—Attempt at a unifying approach, J. Electroanal. Chem. 111 (1980) 125–131.
[4] Y.V. Pleskov, A.Y. Sakharova, M.D. Krotova, L.L. Bouilov, P.V. Spitsyn, Photoelectrochemical properties of semiconductor diamond, J. Electroanal. Chem. 228 (1987) 19–27.
[5] J.J. Carey, C.S. Christ, S.N. Lowery, Method of Electrolysis Employing a Doped Diamond Anode to Oxidize Solutes in Wastewater, (1995). Patent US 5399247, https://www.google.com/patents/US5399247.
[6] T.A. Egerton, P.A. Christensen, Photoelectrocatalysis Processes, in: S. Parson (Ed.), Advanced Oxidation Processes for Water and Wastewater Treatment, IWA, London, 2004, pp. 167–180.
[7] C. Martinez-Huittle, E. Brillas, Electrochemical alternatives for drinking water disinfection, Angew. Chem. Int. Ed. 47 (2008) 1998–2005.
[8] G.R.P. Malpass, D.W. Miwa, S.A.S. Machado, P. Olivi, A.J. Motheo, Oxidation of the pesticide atrazine at DSA® electrodes, J. Hazard. Mater. 137 (2006) 565–572.
[9] M. Panizza, G. Cerisola, Application of diamond electrodes to electrochemical processes, Electrochim. Acta 51 (2005) 191–199.
[10] U. von Gunten, Ozonation of drinking water: Part I. Oxidation kinetics and product formation, Water Res. 37 (2003) 1443–1467.
[11] J. Hoigné, H. Bader, Rate constants of reactions of ozone with organic and inorganic compounds in water-II dissociating organic compounds, Water Res. 17 (1983) 185–194.
[12] B.P. Chaplin, Critical review of electrochemical advanced oxidation processes for water treatment applications, Environ. Sci.: Processes Impacts 16 (2014) 1182–1203.
[13] L.M. Da Silva, L.A. de Faria, J.F.C. Boodts, Electrochemical ozone production: influence of the supporting electrolyte on kinetics and current efficiency, Electrochim. Acta 48 (2003) 699–709.
[14] P.C. Foller, C.W. Tobias, The anodic evolution of ozone, J. Electrochem. Soc. 129 (3) (1982) 506–515.
[15] S. Stucki, G. Theis, R. Kötz, H. Devantay, H.J. Christen, In situ production of ozone in water using a membrel electrolyzer, J. Electrochem. Soc. 132 (1985) 367–371.
[16] A. Kraft, M. Stadelmann, M. Wünsche, M. Blaschke, Electrochemical ozone production using diamond anodes and a solid polymer electrolyte, Electrochem. Commun. 8 (2006) 883–886.

[17] K. Arihara, C. Terashima, A. Fujishima, Electrochemical production of high-concentration ozone-water using freestanding perforated diamond electrodes, J. Electrochem. Soc. 154 (4) (2007) E71–E75.
[18] M.H.P. Santana, L.M. Da Silva, A.C. Freitas, J.F.C. Boodts, K.C. Fernandes, L.A. De Faria, Application of electrochemically generated ozone to the discoloration and degradation of solutions containing the dye reactive Orange 122, J. Hazard. Mater. 164 (2009) 10–17.
[19] N. Oturan, I. Sires, M.A. Oturan, E. Brillas, Degradation of pesticides in aqueous medium by electro-Fenton and related methods. A review, J. Environ. Eng. Landsc. Manag. 19 (2009) 235–255.
[20] G.V. Buxton, C.L. Greenstock, W.P. Helman, A.B. Ross, Critical review of rate constants for reactions of hydrated electrons, hydrogen atoms and hydroxyl radicals (•OH/•O-) in aqueous solution, J. Phys. Chem. Ref. Data 17 (2) (1988) 513–886.
[21] G. Merga, B.S.M. Rao, H. Mohan, J.P. Mittal, Reactions of OH and SO_4•- with some Halobenzenes and Halotoluenes: a radiation chemical study, J. Phys. Chem. 98 (1994) 9158–9164.
[22] S. Sudhakaran, G.L. Amy, QSAR models for oxidation of organic micropollutants in water based on ozone and hydroxyl radical rate constants and their chemical classification, Water Res. 47 (2013) 1111–1122.
[23] J.E. Vitt, D.C. Johnson, The importance of anodic discharge of H_2O in anodic oxygen-transfer reactions, J. Electrochem. Soc. 139 (3) (1992) 774–778.
[24] M. Pourbaix, Atlas of Electrochemical Equilibria in Aqueous Solutions, second ed., Nace International Cebelcor, Houston, 1974.
[25] I.H. Yeo, S. Kim, R. Jacobson, D.C. Johnson, Electrocatalysis of anodic oxygen transfer reactions: comparison of structural data with electrocatalytic phenomena for bismuth-doped lead dioxide, J. Electrochem. Soc. 136 (5) (1989) 1395–1401.
[26] C. Comninellis, Electrocatalysis in the electrochemical conversion/combustion of organic pollutants for waste water treatment, Electrochim. Acta 39 (11 – 12) (1994) 1857–1862.
[27] A. Wieckowski, The classification of adsorption processes of organic compounds on platinum electrode. The role of water molecules chemisorbed on platinum, Electrochim. Acta 26 (8) (1981) 1121–1124.
[28] M. Panizza, P.A. Michaud, G. Cerisola, C. Comninellis, Anodic oxidation of 2-naphthol at boron doped diamond electrodes, J. Electroanal. Chem. 507 (2001) 206–214.
[29] B. Marselli, J. García-Gómez, P.A. Michaud, M.A. Rodrigo, C. Comninellis, Electrogeneration of hydroxyl radicals on boron-doped diamond electrodes, J. Electrochem. Soc. 150 (3) (2003) D79–D83.
[30] K. Groenen-Serrano, E. Weiss-Hortala, A. Savall, P. Spiteri, Role of hydroxyl radicals during the competitive electrooxidation of organic compounds on a boron-doped diamond anode, Electrocatalysis 4 (2013) 346–352.
[31] M. Panizza, A. Barbucci, R. Ricotti, G. Cerisola, Electrochemical degradation of methylene blue, Sep. Purif. Technol. 54 (3) (2007) 382–387.
[32] M. Deborde, U. von Gunten, Reactions of chlorine with inorganic and organic compounds during water treatment-kinetics and mechanisms : a critical review, Water Res. 42 (2008) 13–51.
[33] B.N. Grgur, D.Ž. Mijin, A kinetic study of the Methomyl electrochemical degradation in the chloride containing solutions, Appl. Catal. B 147 (2014) 429–438.
[34] A. Kraft, Electrochemical water disinfection: a short review, Platin. Met. Rev. 52 (3) (2008) 177–185.
[35] H. Li, Q. Yu, B. Yang, Z. Li, L. Lei, Electro-catalytic oxidation of artificial human urine by using BDD and IrO_2 electrodes, J. Electroanal. Chem. 738 (2015) 14–19.
[36] M.S. Siddiqui, Chlorine-ozone interactions: formation of chlorate, Water Res. 30 (9) (1996) 2160–2170.

[37] Chlorite and Chlorate in Drinking-Water, Background Document for Development of WHO Guidelines for Drinking-Water Quality WHO/SDE/WSH/05.08/86 (2005). http://www.who.int/water_sanitation_health/dwq/chemicals/chlorateandchlorite 0505.pdf.

[38] Interim Drinking Water Health Advisory for Perchlorate EPA 822-R-08-025, U.S. EPA, Washington DC (2008). https://nepis.epa.gov/.

[39] A.D. McIntyre, C.F. Mills (Eds.), Ecological toxicology research: effects of heavy metal and organohalogen compounds, Proc. Nato Science Committee Conference on Eco-Toxicology Plenum Press, 1974.

[40] C.R. Costa, F. Montilla, E. Morallón, P. Olivi, Electrochemical oxidation of acid black 210 dye on the boron-doped diamond electrode in the presence of phosphate ions: effect of current density, pH, and chloride ions, Electrochim. Acta 54 (2009) 7048–7055.

[41] A. Kraft, Doped diamond: a compact review on a new, versatile electrode material, Int. J. Electrochem. Sci. 2 (2007) 355–385.

[42] C.A. Martinez-Huitle, S. Ferro, Electrochemical oxidation of organic pollutants for the wastewater treatment: direct and indirect processes, Chem. Soc. Rev. 35 (2006) 1324–1340.

[43] J.M. Peralta-Hernandez, M. Mendez-Tovar, R. Guerra-Sanchez, C.A. Martınez-Huitle, J. L. Nava, A brief review on environmental application of boron doped diamond electrodes as a new way for electrochemical incineration of synthetic dyes, Int. J. Electrochem. 2012 (2011) 1–18.

[44] C.A. Martínez-Huitle, M.A. Rodrigo, I. Sirés, O. Scialdone, Single and coupled electrochemical processes and reactors for the abatement of organic water pollutants: a critical review, Chem. Rev. 115 (2015) 13362–13407.

[45] M.E.H. Bergmann, J. Rollin, T. Iourtchouk, The occurrence of perchlorate during drinking water electrolysis using BDD anodes, Electrochim. Acta 54 (7) (2009) 2102–2107.

[46] O. Scialdone, S. Randazzo, A. Galia, G. Silvestri, Electrochemical oxidation of organics in water: role of operative parameters in the absence and in the presence of NaCl, Water Res. 43 (8) (2009) 2260–2272.

[47] A. Donaghue, B.P. Chaplin, Effect of select organic compounds on perchlorate formation at boron-doped diamond film anodes, Environ. Sci. Technol. 47 (21) (2013) 12391–12399.

[48] Y. Lan, C. Coetsier, C. Causserand, K. Groenen Serrano, On the role of salts for the treatment of wastewaters containing pharmaceuticals by electrochemical oxidation using a boron doped diamond anode, Electrochim. Acta 231 (2017) 309–318.

[49] K. Elbs, O. Schönherr, Z. Elektrochem. 2 (1895) 245.

[50] W. Smit, J.G. Hoogland, The mechanism of the anodic formation of the peroxodisulphate ion on platinum—I. Establishment of the participating anion, Electrochim. Acta 16 (1) (1971) 1–18.

[51] E.V. Kasatkin, A.A. Rakov, Kinetics and mechanism of low-temperature electrochemical oxidation at high anode potentials, Electrochim. Acta 10 (1965) 131–140.

[52] K. Serrano, P.A. Michaud, C. Comninellis, A. Savall, Electrochemical preparation of peroxodisulfuric acid using boron doped diamond thin film electrodes, Electrochim. Acta 48 (4) (2002) 431–436.

[53] C. Provent, W. Haenni, E. Santoli, P. Rychen, Boron-doped diamond electrodes and microelectrode-arrays for the measurement of sulfate and peroxodisulfate, Electrochim. Acta 49 (2004) 3737–3744.

[54] D. Khamis, E. Mahé, F. Dardoize, D. Devilliers, Peroxodisulfate generation on boron-doped diamond microelectrodes array and detection by scanning electrochemical microscopy, J. Appl. Electrochem. 40 (10) (2010) 1829–1838.

[55] N.B. Librovich, V.D. Maiorov, Ionic-molecular composition of aqueous sulfuric acid solutions at 25 °C, Russ. Chem. Bull. 26 (3) (1977) 621–623.

[56] P.Y. Jiang, Y. Katsumura, R. Nagaishi, M. Domae, K. Ishikawa, K. Ishigure, Pulse radiolysis study of concentrated sulfuric acid solutions. formation mechanism, yield and reactivity of sulfate radicals, J. Chem. Soc. Faraday Trans. 88 (1992) 1653–1658.
[57] H. Caro, Zur Kenntnis des Kalkstickstoffs, Z. Angew. Chem. 11 (1898) 845.
[58] H. Palme, Studien über die Zersetzung der Überschwefelsäure, Z. Anorg. Chem. 112 (1920) 97.
[59] I.M. Kolthoff, I.K. Miller, The chemistry of persulfate. I. the kinetics and mechanism of the decomposition of the persulfate ion in aqueous medium, J. Am. Chem. Soc. 73 (1951) 3055–3059.
[60] N. Ibl, H. Vogt, J. O'M Bockris, B.E. Conway, E. Yeager, R.E. White (Eds.), Comprehensive Treatise of Electrochemistry, 2, Plenum Press, New York, 1981, p. 224.
[61] J. Balej, M. Kaderavek, Influence of sulphuric acid concentration on current yield of peroxodisulphate, Collect. Czechoslov. Chem. Commun. 44 (1979) 1510–1520.
[62] C. Liang, I.L. Lee, I.Y. Hsu, C.P. Liang, Y.L. Lin, Persulfate oxidation of trichloroethylene with and without iron activation in porous media, Chemosphere 70 (3) (2008) 426–435.
[63] C. George, H. El Rassy, J.M. Chovelon, Reactivity of selected volatile organic compounds (VOCs) toward the sulfate radical (SO_4•—), In. J. Chem. Kinet. 33 (9) (2001) 539–547.
[64] N.S. Shah, X. He, H.M. Khan, J.A. Khan, K.E. O'Shea, D.L. Boccelli, D.D. Dionysiou, Efficient removal of endosulfan from aqueous solution by UV-C/peroxides: a comparative study, J. Hazard. Mater. 263 (2013) 584–592.
[65] H. Hori, E. Hayakawa, H. Einaga, S. Kutsuna, K. Koike, T. Ibusuki, H. Kiatagawa, R. Arakawa, Decomposition of environmentally persistent perfluorooctanoic acid in water by photochemical approaches, Environ. Sci. Technol. 38 (22) (2004) 6118–6124.
[66] P. Neta, V. Madhavan, H. Zemel, R.W. Fessenden, Rate constants and mechanism of reaction of sulfate radical anion with aromatic compounds, J. Am. Chem. Soc. 99 (1) (1977) 163–164.
[67] M. Murugananthan, S.S. Latha, G. Bhaskar Raju, S. Yoshihara, Anodic oxidation of ketoprofen—an anti-inflammatory drug using boron doped diamond and platinum electrodes, J. Hazard. Mater. 180 (2010) 753–758.
[68] P. Canizares, L. Martinez, R. Paz, C. Saez, J. Lobato, M.A. Rodrigo, Treatment of Fenton-refractory olive oil mill wastes by electrochemical oxidation with boron-doped diamond anodes, J. Chem. Technol. Biotechnol. 81 (2006) 1331–1337.
[69] P. Canizares, M. Diaz, J.A. Dominguez, J. Lobato, M.A. Rodrigo, Electrochemical treatment of diluted cyanide aqueous wastes, J. Chem. Technol. Biotechnol. 80 (2005) 565–573.
[70] H.V. Lutze, N. Kerlin, T.C. Schmidt, Sulfate radical-based water treatment in presence of chloride : formation of chlorate, inter-conversion of sulfate radicals into hydroxyl radicals and influence of bicarbonate, Water Res. 53 (2014) 370–377.
[71] W.J. McElroy, A laser photolysis study of the reaction of SO_4- with Cl- and the subsequent decay of Cl_2 in aqueous solution, J. Phys. Chem. 94 (6) (1990) 2435–2441.
[72] N. Vatistas, C. Comninellis, R.M. Serikawa, G. Prosperi, Oxidant production on BDD anodes and advanced oxidation processes, in: Fujishima et al., (Ed.), Diamond Electrochemistry, Elsevier, Amsterdam, 2005, pp. 477–501.
[73] M.S. Saha, T. Furuta, Y. Nishiki, Electrochemical synthesis of sodium peroxycarbonate at boron-doped diamond electrodes, Electrochem. Solid-State Lett. 6 (7) (2003) D5–D7.
[74] S. Velazquez-Pena, C. Saez, P. Canizares, I. Linares-Hernandez, V. Martinez-Miranda, C. Barrera-Diaz, M.A. Rodrigo, Production of oxidants via electrolysis of carbonate solutions with conductive-diamond anodes, Chem. Eng. J. 230 (2013) 272–278.
[75] M.S. Saha, T. Furuta, Y. Nishiki, Conversion of carbon dioxide to peroxycarbonate at boron-doped diamond electrode, Electrochem. Commun. 6 (2004) 201–204.
[76] C.P. Chardon, T. Matthée, R. Neuber, M. Fryda, C. Comninellis, Efficient electrochemical production of peroxodicarbonate applying Diachem® diamond electrodes, Chem. Select 2 (3) (2017) 1037–1040.

[77] E.J. Ruiz, Y. Meas, R. Ortega, J.L. Jurado, Electrochemical synthesis of sodium percarbonate in aqueous solution, ECS Trans. 3 (18) (2007) 29–35.
[78] P. Canizares, F. Larrondo, J. Lobato, M.A. Rodrigo, C. Saez, Electrochemical synthesis of peroxodiphosphate using boron-doped diamond anodes, J. Electrochem. Soc. 152 (11) (2005) D191–196.
[79] E. Weiss, C. Saez, K. Groenen Serrano, P. Canizares, A. Savall, M.A. Rodrigo, Electrochemical synthesis of peroxomonophosphate using boron-doped diamond anodes, J. Appl. Electrochem. 38 (1) (2008) 93–100.

Further Reading

[1] M. Panizza, G. Cerisola, Direct and mediated anodic oxidation of organic pollutants, Chem. Rev. 109 (2009) 6541–6569.
[2] E. Weiss, K. Groenen Serrano, A. Savall, A comparison study of electrochemical degradation of phenol on boron doped diamond and lead dioxide, J. Appl. Electrochem. 38 (2008) 329–337.
[3] B.K. Körbahti, S. Taşyürek, Electrochemical oxidation of ampicillin antibiotic at boron-doped diamond electrodes and process optimization using response surface methodology, Environ. Sci. Pollut. Res. 22 (5) (2015) 3265–3278.
[4] A. Savall, Electrochemical reactors design for the oxidation of organic pollutants, in: G. Kreysa, K. Ota, R.F. Savinell (Eds.), Encyclopedia of Applied Electrochemistry, Springer-Verlag, New York, 2014.

CHAPTER

7

Indirect Electrochemical Oxidation by Using Ozone, Hydrogen Peroxide, and Ferrate

Cristina Sáez, Manuel Andrés Rodrigo*, Ana S. Fajardo†, Carlos A. Martínez-Huitle†*

*University of Castilla La Mancha, Ciudad Real, Spain

†Federal University of Rio Grande do Norte, Natal, Brazil

OUTLINE

Electrochemical Water and Wastewater Treatment
https://doi.org/10.1016/B978-0-12-813160-2.00007-9

PRODUCTION OF OXIDANTS IN THE ANODE DURING THE ELECTROLYSIS OF WATER

The most important processes during the electrolysis of wastewater are expected to be produced on the surface of the anode, although they do not necessarily imply the direct oxidation of the organics but the formation of oxidants. In fact, direct oxidation of organics is typically inefficient and sometimes leads to the formation of polymeric materials rather than the mineralization of the waste. At this point, mediated oxidation does not get the attention required in the literature, despite it being the most relevant mechanism to explain the removal of organics in most of the references found in the literature [1].

During the electrolysis of wastewater, water molecules are oxidized; this oxidation leads to the formation of hydroxyl radicals in a first stage (Eq. 1). The average lifetime of these radicals is rather short, meaning that their action is limited to the very nearness of the anode surface. For this reason, it is nearly impossible to distinguish between the direct oxidation processes and the processes mediated by these powerful oxidants; both fit well kinetically to mass transfer controlled processes.

$$H_2O \rightarrow {}^{\bullet}OH + H^+ + e^- \qquad (1)$$

According to literature [2,3], in many cases these hydroxyl radicals can interact chemically with the surface of the electrodes (M), leading to a transient higher oxidation state of the metals (or metal oxides) contained in these electrodes (Eq. 2) which, in turn, become the final responsible of the oxidation of the organics (Eq. 3). Hydroxyl radicals are then no longer available to oxidize other chemical species contained in the water. Therefore these types of anodic oxidation processes cannot be properly considered as advanced oxidation processes (AOPs). In the literature, these electrodes are so-called active electrodes, with the platinum, Ru-MMO, and Ir-MMO materials the ones that clearly exhibit this behavior. The soft chemical oxidation carried out by this transient higher oxidation state is related to the low efficiency of this type of electrolysis, which sometimes ends with the formation of polymers. Formation of oxygen as a final product of water discharge is also typical for this type of electrode (Eq. 4).

$$M + {}^{\bullet}OH \rightarrow MO + H^+ + e^- \qquad (2)$$

$$MO + R \rightarrow RO + M \qquad (3)$$

$$MO \rightarrow M + \tfrac{1}{2}O_2 \qquad (4)$$

Conversely, there are anodes for which the chemical interaction between the hydroxyl radicals and the elements contained on the surface of the electrodes is not possible and, hence, hydroxyl radicals are available to oxidize other species. Consequently, as the components of the electrode are not susceptible to be oxidized, these radicals can only attack species contained in the electrolyte (wastewater) or they can combine with each other to form oxygen, hydrogen peroxide, or ozone, all of them through well-known mechanisms (Eqs. 5–7).

$$2^{\bullet}OH \rightarrow H_2O + 1/2\,O_2 \tag{5}$$

$$2^{\bullet}OH \rightarrow H_2O_2 \tag{6}$$

$$O_2 + 2^{\bullet}OH \rightarrow O_3 + H_2O \tag{7}$$

All these electrodes are known as nonactive electrodes and because of the active role of the hydroxyl radicals, these electrochemical processes clearly fall down into the definition of AOPs [4]. One important point in these later processes is the possibility of the interaction of the produced hydroxyl radicals with the different species contained in the wastewater (electrolyte). The most important species is water itself, with a concentration near 55.55 M. Below this species, the most important are typically the salts that are contained in concentrations that are several folds below. For instance, for wastewater containing 400 ppm of NaCl, which is a very high concentration in urban wastewater, the molar concentration of chloride is only 6.8 mM, >4 log units below the concentration of water. A very high concentration of sodium sulfate (like the 5000 ppm used in many works found in the literature to simulate industrial wastewater) results in only 35.2 mM of sulfate, again several log units below the concentration of water. Anyway, these concentrations are rather high as compared to the concentration of organics. Thus, a concentration of 100 ppm of phenol corresponds to only 1.2 mM of organic. This is graphically illustrated in Fig. 1.

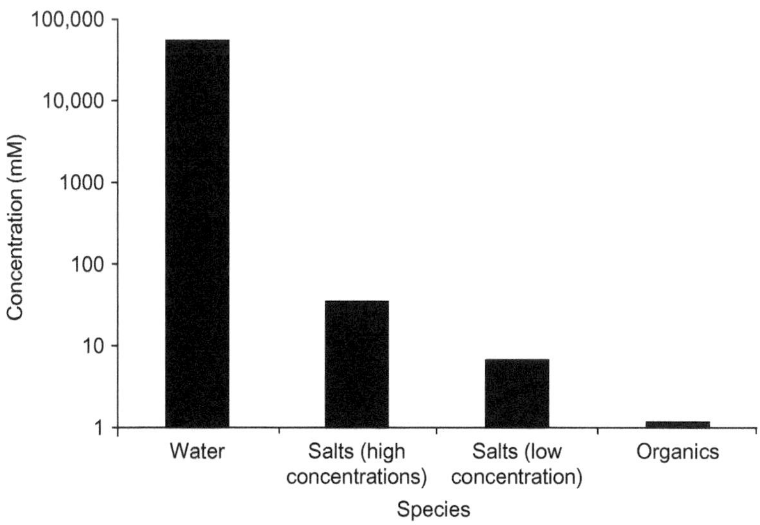

FIG. 1 Typical concentrations of chemical species surrounding hydroxyl radicals formed during electrolysis.

This means that the possibilities of the interaction of hydroxyl radicals with the different species contained in wastewater typically decrease in the sequence water > anions of inorganic salts > organics. Hence, oxygen, ozone, and hydrogen peroxide, which are oxidants directly generated from water oxidation, are key expected elements in the electrolysis.

Despite the significant differences observed in the concentration of oxidants' precursors, the formation of oxidants from the anions of the salts contained in wastewater is known to be very important. It is associated with the formation of radicals from the interaction of the salt anion and the hydroxyl radicals (Eqs. 5–11) [5].

$$Cl^- + {}^{\bullet}OH \rightarrow ClO^- + H^+ + e^- \tag{8}$$

$$HSO_4^- + {}^{\bullet}OH \rightarrow (SO_4^-)^{\bullet} + H_2O \tag{9}$$

$$HPO_4^{2-} + {}^{\bullet}OH \rightarrow \left(PO_4^{2-}\right)^{\bullet} + H_2O \tag{10}$$

$$HCO_3^- + {}^{\bullet}OH \rightarrow (CO_3^-)^{\bullet} + H_2O \tag{11}$$

Then, the recombination of these radicals allows the formation of more stable species (Eqs. 12–17), which can extend the effects of the oxidation to the bulk.

$$(SO_4^-)^{\bullet} + {}^{\bullet}OH \rightarrow HSO_5^- \tag{12}$$

$$(H_2PO_4)^{\bullet} + {}^{\bullet}OH \rightarrow H_3PO_5 \tag{13}$$

$$CO_3^{-\bullet} + {}^{\bullet}OH \rightarrow HCO_4^- \tag{14}$$

$$(SO_4^-)^{\bullet} + (SO_4^-)^{\bullet} \rightarrow S_2O_8^{2-} \tag{15}$$

$$\left(PO_4^{2-}\right)^{\bullet} + \left(PO_4^{2-}\right)^{\bullet} \rightarrow P_2O_8^{4-} \tag{16}$$

$$(CO_3^-)^{\bullet} + (CO_3^-)^{\bullet} \rightarrow C_2O_6^{2-} \tag{17}$$

The formation of oxidants on the surface of the anodes also occurs with active electrodes. However, in active electrodes this formation is not related to the hydroxyl radicals and because of that, there are important differences in the speciation obtained. A typical example is the oxidation of chloride ions [6]. With Ru-MMO electrodes, this oxidation is very efficient; important concentrations of chlorine and hypochlorite are obtained and no perchlorate is formed. However, with a nonactive electrode, perchlorate is the final product in the oxidation of chloride and this can only be explained in terms of the effect of hydroxyl radicals in the electrolysis carried out with these electrodes (Eqs. 18–21). For this reason the removal of organics in chloride media is more efficient with active electrodes for which the chlorine promotes the mediated oxidation than with nonactive electrodes, in which perchlorate has no effect on the mediated oxidation (because its action is kinetically limited at room temperature) [7].

$$Cl^- + {}^{\bullet}OH \rightarrow ClO^- + H^+ + e^- \quad (18)$$

$$ClO^- + {}^{\bullet}OH \rightarrow ClO_2^- + H^+ + e^- \quad (19)$$

$$ClO_2^- + {}^{\bullet}OH \rightarrow ClO_3^- + H^+ + e^- \quad (20)$$

$$ClO_3^- + {}^{\bullet}OH \rightarrow ClO_4^- + H^+ + e^- \quad (21)$$

Another important point that should be noted is the large differences observed between the behavior of the nonactive electrodes during electrolysis at low and high current densities, which were explained in terms of the production of hydroxyl radicals; we again point out the importance of the hydroxyl radicals with active electrodes [8]. Hence, mediated electrolysis in active electrodes involves only oxidants that do not require the production of hydroxyl radicals, with chlorine being the most typical. Conversely, mediated oxidation with nonactive electrodes leads to a very important cocktail of oxidants, including ozone and hydrogen peroxide. Interaction between these oxidants can also be very important; it is known how hydrogen peroxide and ozone can combine to form hydroxyl radicals [9]. Likewise, peroxosalts are also activated by ozone and hydrogen peroxide, leading to the formation of radicals of the precursor peroxosalts and/or to hydroxyl radicals (Fig. 2). This means that, very frequently,

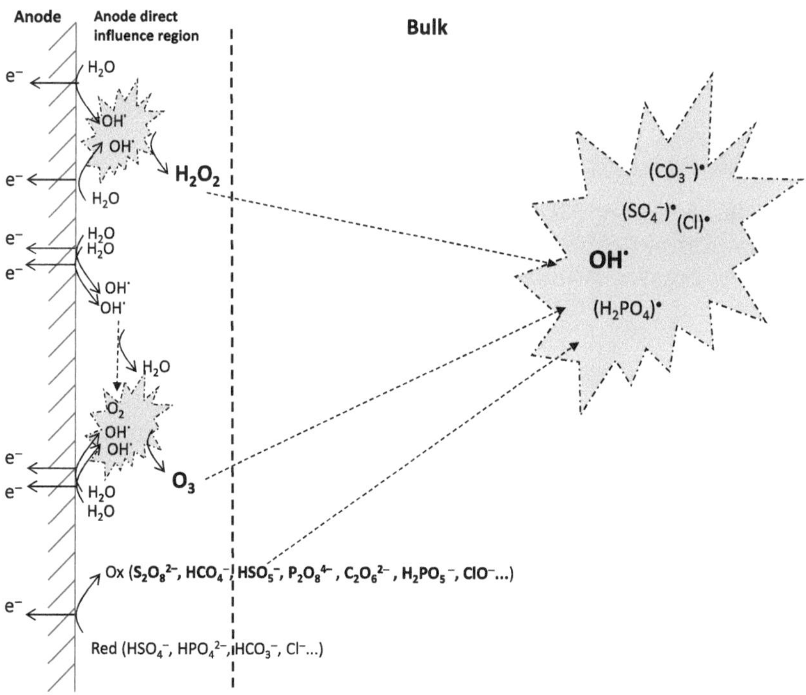

FIG. 2 Scheme of the main routes associated with the anodic formation of oxidants.

oxidants produced during electrolysis cannot be measured because they transform into other species very rapidly. This is particularly important in the case of the nonactive electrode because of the large amount of oxidants formed. The effect of these oxidants is observed in the enhanced efficiency of these processes. Thus, from the viewpoint of the mediated oxidation, it can be said that the best oxidant produced electrochemically is the one that cannot be measured, meaning that the oxidant is rapidly used for oxidizing species once it is formed.

IMPROVEMENT OF TREATMENT EFFICIENCY WITH CATHODIC PROCESSES

During the electrolysis of aqueous wastes, not only anodic reactions but also cathodic reduction processes have to be considered. Typically, water reduction to form hydrogen (Eq. 22) is the most important cathodic reaction. This reaction has no impact on the remediation of wastewater and, for this reason, materials with low overpotencials are typically used in order to try to attain low cell voltages [10] and, at least, to not increase the operation cost of the wastewater treatment unnecessarily. Besides this side reaction, the reduction of organics (Eq. 23) and of dissolved oxygen (Eq. 24) can also take place.

$$H_2O + e^- \rightarrow 1/2\,H_2 + OH^- \tag{22}$$

$$R{-}X + 2e^- + 2H^+ \rightarrow R{-}H + HX \tag{23}$$

$$O_2 + 4e^- + 4H^+ \rightarrow 2H_2O \tag{24}$$

The first reaction (23) does not have an important impact on the treatment performance because the reversibility of most of the processes related to organic pollutants is rather low. Likewise, the extension of the second reaction (24) in typical cathodes is low because of the low solubility of oxygen, which, despite being produced in the anode, is stripped during electrolysis. However, operating in this way practically means surrendering the search for more efficient electrochemical processes and neglecting the fact that the cathodic reaction can be used to improve the treatability of waste. In fact, there are two potentially important ways to use the cathode during the electrolysis of wastewater: the reductive dehalogenation and the production of hydrogen peroxide.

Reductive dehalogenation of organohalogenated compounds appears as one of the most studied and interesting reductive processes, with the catalytic properties toward the reductive dehalogenation of C—X bonds of the cathodic materials being a very relevant variable [11–14]. The electrochemical reduction allows the production of less toxic and more biodegradable products, which, in addition, are susceptible to be more easily electrochemically oxidized on the anode surface or by the oxidants electrogenerated on the anode.

However, this is not the most important cathodic process. In fact, this merit can be assigned to the cathodic generation of hydrogen peroxide by the two-electron reduction of oxygen (directly injected as pure gas or bubbled air) at the cathode surface in acidic/neutral media (Eq. 25).

$$O_2 + 2e^- + 2H^+ \rightarrow H_2O_2 \quad E^0 = 0.682\,\text{vs NHE} \tag{25}$$

Hydrogen peroxide is kinetically more active than oxygen at room temperature. This means that the cathodic reaction is run in order to produce oxidants, allowing the oxidation capacity of the whole system to increase theoretically up to double as compared with a conventional electrochemical cell (Fig. 3). Then, the combination of the anodic and cathodic processes can lead to the formation of a very important cocktail of oxidants with complex interactions among them; these are responsible for the very high efficiencies found during the electrolysis of wastewater (Fig. 4).

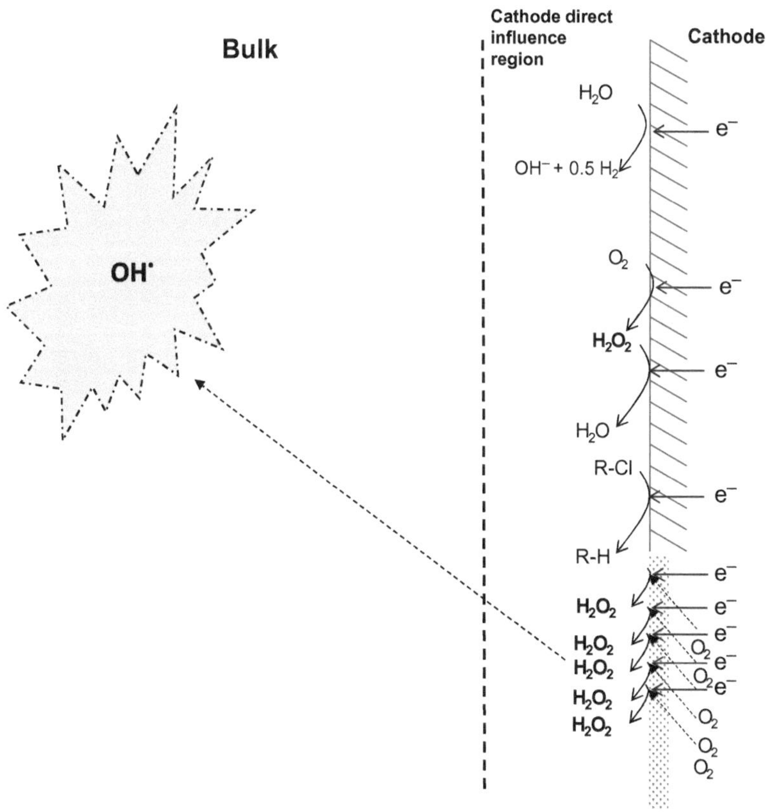

FIG. 3 Scheme of the main routes associated with the cathodic formation of oxidants.

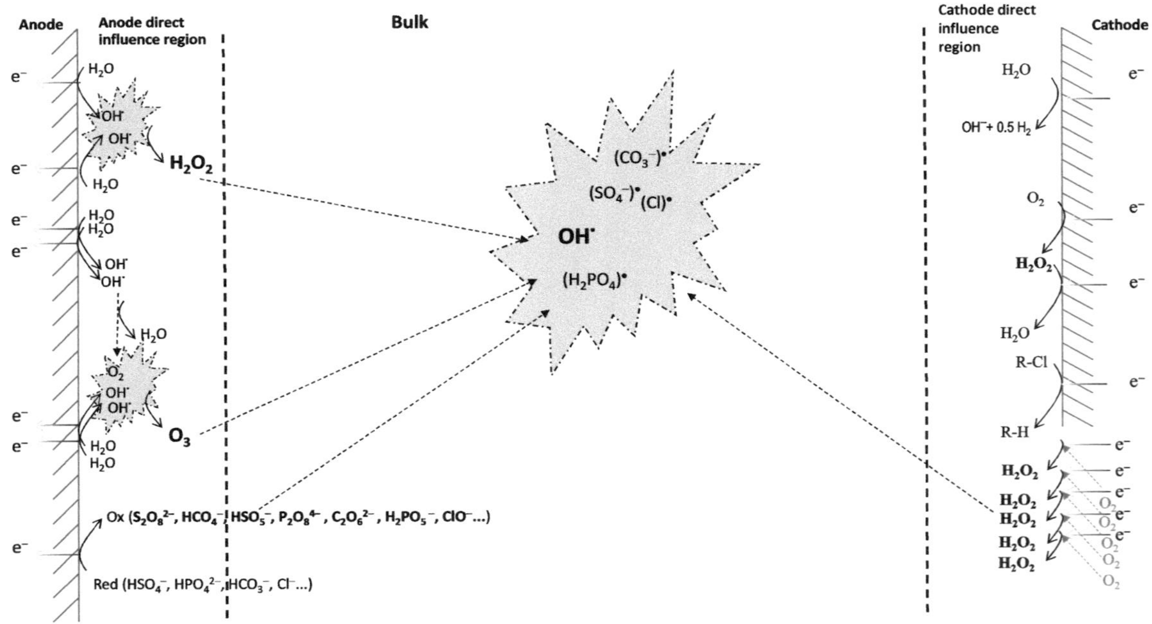

FIG. 4 Production of oxidants in the electrochemical cell.

Depending on the cell configuration, cathode properties, and operational conditions, the further reduction of H_2O_2 can occur, leading to the formation of water as a stable final product (Eq. 26). Likewise, in an undivided cell, H_2O_2 can also be oxidized to oxygen at the anode, according to Eqs. (27), (28), producing the weak oxidant hydroperoxyl radical ($HO_2^{\bullet}$) as an intermediate.

$$H_2O_2 + 2H^+ + 2e^- \rightarrow 2H_2O \quad (26)$$

$$H_2O_2 \rightarrow HO_2^{\bullet} + H^+ + e^- \quad (27)$$

$$HO_2^{\bullet} \rightarrow O_2 + H^+ + e^- \quad (28)$$

One of the major limitations of the production rate and current efficiency of the reduction of oxygen to produce H_2O_2 is the low solubility of oxygen in water ($40\,mg\,dm^{-3}$ in an oxygen atmosphere or $8\,mg\,dm^{-3}$ in an air atmosphere at 1 atm, 25°C, and deionized water) [15–17]. Thus, in a conventional system the oxygen concentration is so low that its cathodic reduction is not favored. Because of this, the way in which the oxygen is supplied to the system is critical in the development of H_2O_2 electrolyzers.

The conventional aeration system consists of bubbling the gas reactant directly into the electrolyte at atmospheric pressure, but the process is often kinetically controlled by the mass transfer of oxygen to the cathodic surface. This problem has been partially overcome by the use of gas diffusion electrodes (GDEs), in which air or oxygen is directly supplied to the cathode without the need for it to be dissolved in the electrolyte [18–20] (Fig. 5A). Those electrodes work as triple phase contactors in which the gas reactant is directly supplied to the cathode-electrolyte interface. This allows the minimization of the mass transfer resistance, therefore favoring the generation of H_2O_2. However, the generous air flow needed to prevent percolation of water through the electrode surface utilization efficiencies leads to very low oxygen utilization efficiency (<1%), with the subsequent waste of energy and oversizing of compressors in the event of potential industrial application. Moreover, the utilization of such electrodes enhances the complexity of the cell, making development of the process difficult on a prototype scale. Recently, some authors [16,21] proposed the use of pressurized reactors to enhance the oxygen solubility in water, speeding up its mass transfer toward the cathode surface. Another alternative proposed in the literature [21,22] is the use of an aeration device based on the Venturi effect without the requirement of an external compressor to supply air to the system. This jet aerator creates suction that produces the entrance of air into the water flow. A fraction of the air sucked in the system becomes dissolved, saturating water in oxygen while the rest of it remains undissolved in the form of air bubbles, supersaturating the water.

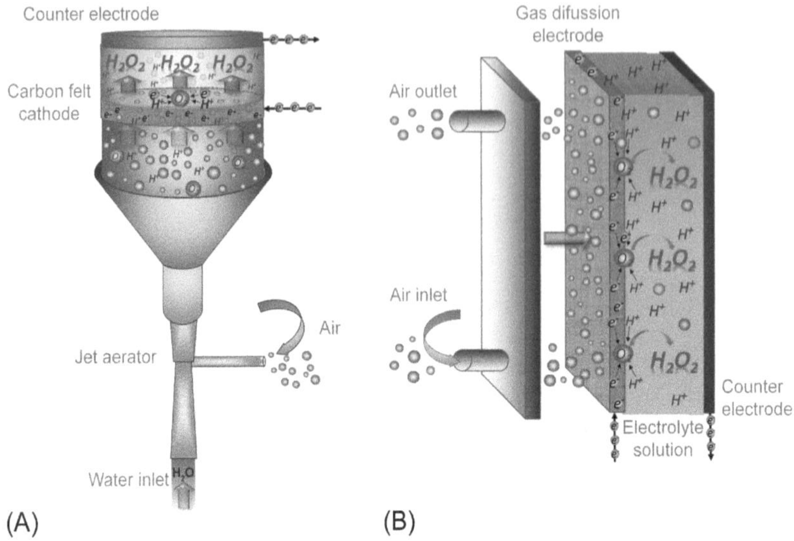

FIG. 5 (A) Jet cell; (B) GDE cell. *From J.F. Perez, J. Llanos, C. Saez, C. Lopez, P. Canizares, M.A. Rodrigo, Electrochemical jetcell for the in-situ generation of hydrogen peroxide, Electrochem. Commun. 71 (2016) 65–68. With permission from Elsevier Ltd.*

That is, with the use of a jet aeration system, water momentarily contains a concentration of oxygen higher than that predicted by Henry's law.

Besides the aeration system, the choice of an adequate cathodic material to selectively promote the two-electron pathway is a key point in the development of H_2O_2 electrolyzers. Although different metals and amalgams (Ag, Pt—Hg, or Pd—Hg) have been investigated, carbonaceous materials show the most favorable characteristics (high selectivity and working current density, commercial availability, cost, and environmental compatibility) for use as cathodic materials in the electrogeneration of hydrogen peroxide. In recent years, many different types of carbonaceous materials have been tested, including graphite, carbon felt, carbon sponge, reticulated vitreous carbon, and carbon cloth [19,20,23–26].

In this way, different authors have recently proposed the use of three-dimensional (including meshes, foams, or packed) electrodes in different flow-through reactor concepts to increase the mass transport rate, achieving interesting results (Fig. 5B).

A wide range of approaches has been reported for the enhancement of the H_2O_2 production such as deposition of metals or the addition of polytetrafluoroethylene (PTFE). In this point, it is worth mentioning that cathodes fabricated by the combination of different carbon powders and PTFE or by the deposition of carbon and PTFE mixtures on carbonaceous supports have achieved quite interesting results. This includes considerably increasing the working current density with respect to unmodified materials and, therefore, the rate of production [16,22]. The improved electroactivity for hydrogen peroxide of the modified electrodes is attributed to the formation of a microporous- and partially hydrophobic-layer in which the formation of triple-phase (oxygen gas-electrolyte-electrode) contact points is promoted.

OZONE OXIDATION

Production of ozone during the electrolysis of water or wastewater is reported in the literature for diamond and lead dioxide anodes [27]. Although concentrations obtained are significant, they are not as important as those that can be generated using other nonelectrochemical methods. One of the reasons is given in Eq. (29) in which it is shown that, by forming one molecule of ozone, oxygen has to react with hydroxyl radicals. During electrolysis, this interaction is only possible in the nearness of the anode surface, where the presence of hydrogen peroxide (formed according to reaction 5) can promote the decomposition of both species, leading to the formation of hydroxyl radicals.

$$H_2O_2 + 2O_3 + \rightarrow 2^{\bullet}OH + 3O_2 \quad (29)$$

In addition, the dissolved oxygen concentration is very limited because of its low solubility in water, restraining in turn the maximum concentration of ozone. Likewise, as pointed out before, the synergistic interaction reported with hydrogen peroxide is also exhibited with other oxidants, suggesting that there are many potential mechanisms that promote its decomposition during the electrolysis of wastewater.

These concentrations cannot be raised by coupling electrolysis with other nonelectrochemical technologies. Thus, the potential decomposition of hydrogen peroxide and ozone by ultraviolet (UV) (shown in Eqs. 30, 31, respectively) or by high-frequency ultrasound (US) makes the trials of a combination of electrolysis with other AOP technologies in order to increase the global efficiency of the process not very useful.

$$H_2O_2 + h\nu \rightarrow 2^{\bullet}OH \quad (30)$$

$$H_2O + O_3 + h\nu \rightarrow 2^{\bullet}OH + O_2 \quad (31)$$

The electrochemical generation of ozone was reviewed by Christensen et al. [28] in 2013, discussing the important issues that must be considered for producing ozone electrochemically. These include the types of electrochemical cells, the effects of temperature, the anode type and composition, the current density and electrolyte composition, and the pH.

The types of cells employed in the electrochemical generation of O_3 reflect those generally employed in electrolysis. In the simplest cell configuration, the electrodes are immersed directly in the electrolyte [29] and the gases evolved at the anode and cathode are mixed. Divided cells employ an inert separator such as glass [29–31], Teflon [31,32], or Nafion [33,34] and the anode and cathode gases are produced separately. Zero gap, filter-press, and membrane electrolyte assembly (MEA)-based cells employ a solid polymer electrolyte (SPE) membrane (typically Nafion) as the electrolyte, with the anode and cathode being pressed tightly against the membrane [28,35], allowing the transport of protons to maintain the conductivity [36]. By varying the zero gap, several cells have been used, for example, feeding humidified O_2 to a gas diffusion cathode to generate H_2O_2 [37], using an air-breathing cathode/Nafion "MEA" separated from the anode by acidic electrolyte [32], and a spiral wound MEA-based cell [38]. Unfortunately, there are a number of challenges to be addressed when employing zero gap cells with air-breathing cathodes [28].

Some problems are associated with the electrochemical production of ozone [28], such as flooding (which is observed with air breathing cathodes), calcification, low conductivity and solubility of ozone in water (it depends on the pH), membrane resistance (at neutral pH conditions), and lower ozone yields. Generally speaking, electrochemical ozone generation is carried out either in a single pass/flow operation [39] or in batch recycle mode [40] in which the anolyte, which was put in contact with ozone, is returned to the electrochemical cell. In the former case, ozone is generated in both gas and liquid phases continuously; in the latter, once saturation of the anode solution has taken place, all the ozone produced is released into the gas phase.

On the other hand, other operating parameters can influence the efficacy of ozone production, such as temperature, current density, supporting electrolytes, and electrodes. Among these, the nature of the electrocatalytic material as anode seems to be the determining factor for the production of higher ozone yields. For this reason, we report some results based on the existing literature about the use of different anodes for producing ozone, electrochemically.

In the 1980s [28], the most used materials were Pt and PbO_2 [41] by using aqueous acid electrolytes. After that, a few studies were performed with Pt [42]; however, poor current efficiencies were achieved with respect to the results by using PbO_2. Regarding the PbO_2 electrodes, the most active between the α and β forms of PbO_2, the use of β-PbO_2 is recommended [43–45]; for this reason, it continues to be investigated. During the production of ozone in the absence of added F^- or fluoride-containing electrolytes, the current efficiencies observed at PbO_2 are ca. 3%–10% at current densities of ca. 1.0 A cm^{-2} in aqueous H_2SO_4 or $HClO_4$ [41,46,47]. Meanwhile, when NaF is added, an increase of the current efficiency to 21% [47] or 10% [30] was achieved.

Among the investigated anode materials, the following can be mentioned: glassy carbon [32,48], Ni/Sb-SnO_2 [49], IrO_2-Nb_2O_5 [50,51], tantalum oxide [52–54], TiO_2 [55–57], and boron doped diamond (BDD) [34,38,58–61]. In the case of IrO_2-Nb_2O_5, it obtained a lower current efficiency of ozone at 0°C in 3.0M H_2SO_4 (1%), increasing up to 12% when 0.03M KPF_6 [50] was employed as a supporting electrolyte by applying 800mA cm^{-2}. Using a layered anode of TiO_2 in 0.01M $HClO_4$ solution, 9% of the efficiency was attained at 50mA cm^{-2} at 15°C [57]. Tantalum oxide-based anodes have shown a current efficiency of approximately 12% by applying 200mA cm^{-2} using model tapwater at neutral pH.

BDD anodes show high anodic stability and a wide potential window for water discharge. Also, these anodes are inert and with poor adsorptive qualities [62]. A few studies have reported the electrochemical O_3 generation by BDD electrolysis ([34,38,58–61]. Michaud and coworkers report that the main product of water electrolysis in H_2SO_4 is peroxydisulfate ($S_2O_8^{2-}$), and in $HClO_4$ the reactive oxygen species, such as hydrogen peroxide, hydroxyl radicals, and ozone [34,63]. Both Michaud et al. [34] and Katsuki et al. [60] report lower ozone current efficiencies with BDD anodes in aqueous acidic solutions in divided cells. Conversely, researchers investigated the electrolysis of deionized water (<1 μS cm^{-1}) using zero gap cells with Nafion membranes, achieving current efficiencies of 24% and 47% for ozone production, respectively [58,59,62].

Meanwhile, in 2006, Kraft and coworkers employed BDD-coated niobium expanded metal electrodes (29mm × 45mm) as anode and cathode with a Nafion membrane positioned in a pipe flow-through reactor (40 $dm^3 h^{-1}$) to electrolyze a solution at 115mA cm^{-2}, obtaining 24% efficiency [64]. By increasing the conductivity of the anolyte, it was found that it caused a drop in ozone current efficiency; for example, at a flow rate of 10 $dm^3 h^{-1}$ and 77mA cm^{-2}, the current efficiency dropped to ca. 2% when the conductivity was increased from 1 to 2000 μS cm^{-1}.

Arihara and coworkers have reported the generation of ozone at 50mm × 15mm × 2mm with freestanding mesh BDD plates [58,59]. A Nafion membrane was used between both electrode materials (BDD as anode and Pt mesh as cathode) and deionized water (<1 μS cm^{-1}) was employed as the catholyte and anolyte, achieving 47% of efficiency with a flow rate of 2 $dm^3 min^{-1}$ at 12°C by applying 530mA cm^{-2}. Nevertheless, the energy requirements were higher, achieving 175 kWh kg^{-1} O_3 at 530mA cm^{-2}, which is not comparable with the cold corona discharge (13 and 29 kWh kg^{-1}) [65]. At higher current densities, the efficiency was found to decrease because the autolysis ($H_2O_2 + O_3 \rightarrow {}^{\bullet}OH + HO_2^{\bullet} + 2O_2$) [28,66] is achieved, due to the increased solution concentration and an increase in side reactions such as the production of hydrogen peroxide.

Nishiki and colleagues employed a BDD/Nafion water interface as a flow cell [38]. The ozone is generated directly below the wire counter electrode, then it diffuses through the Nafion to the Nafion/water interface by using tapwater (total hardness 70 mg L^{-1}, chloride concentration 10 mg L^{-1}), tapwater with added Ca^{2+} (as $CaCl_2$, to assess the effect of hardness) or Cl^- (as NaCl), and pure water. In tapwater, at a constant current of 0.8A and cell voltage of 16 V, a current efficiency of 2.5% was attained; 8% was achieved at 10 V for pure water. However, electrochemical reactions in parallel (possible passivation of BDD and production of active chlorine) were favored, decreasing the ozone production. In the case of water containing Ca^{2+}, a membrane calcification effect was observed. Durability studies over 250 h, electrolysis under constant and interrupted current conditions supported the benefits of on/off operation, with the ozone current efficiency remaining constant at around 3% under current interrupt operation whereas it fell steadily to <1% at constant current. Real application was confirmed and disinfection of water inoculated with Pseudomonas aeruginosa, *Escherichia coli*, and *Staphylococcus epidermidis* was achieved, generating ozone at current efficiencies >20% at 0°C in water, or water with a nonfluoride electrolyte by using BDD or Ni/Sb-SnO_2 electrodes. In 2004, Cheng and Chan reported the electrochemical generation of ozone by using an anode of Sb-SnO_2 in 0.1 M $HClO_4$ with a maximum current efficiency of 15% at cell voltages <3 V [67]. Meanwhile, Wang et al. [49] reported that the activity of Sb-SnO_2 anodes was due to Ni doping. Then, Ni/Sb-SnO_2 electrodes produced 35%, but it is higher when compared with the results obtained by Foller and Tobias [41] employing Sb-SnO_2/Ti mesh with a maximum current efficiency of ca. 4% in 5.0 M H_2SO_4 at 0°C. For this reason, Sb-SnO_2 electrodes are generally considered poor materials for the electrogeneration of ozone. Christensen et al. [68] reported current efficiencies up to 50% (equivalent to 18 kWh kg^{-1} O_3) when employing Ni/Sb-SnO_2 anodes with platinized Ti mesh cathodes in a divided glass cell utilizing a Nafion membrane separator and 0.5 M H_2SO_4 as the electrolyte. Several studies on Ni/Sb-SnO_2 electrodes have reported the mechanism of ozone evolution at these anodes [28]. In 2006, the electrochemical generation of ozone in deionized water employing an MEA-based cell with a static anolyte and Pt/porous carbon air-breathing cathode [69] was studied, achieving a current efficiency of 15% with 2.0 V and requiring 48 kWh kg^{-1} by applying 17 mA cm^{-2}.

Ni/Sb-SnO_2 anodes were used for a cell and stack, operated in both passive and forced airflow (10 dm^3 min^{-1}) [39]. Deionized water (18.2 MΩ cm) was fed to the anode(s) at flow rates from 0.2 to 4.0 dm^3 min^{-1} by applying 60 mA cm^{-2}, with cell voltages between 3 and 6 V for the single cell. The four-cell stack gave a maximum current efficiency of ca. 22% at an anolyte flow rate of 5.9 dm^3 min^{-1}, 29.8 mA cm^{-2}, and 3.0 V, consuming 42 kWh kg^{-1} O_3 at 24.3 mA cm^{-2} and 5.4 dm^3 min^{-1}. Meanwhile, the PbO_2 anode achieved 19% of efficiency with an energy requirement of 40–50 kWh kg^{-1} O_3 (pure oxygen feed) or 60–80 kWh kg^{-1} O_3 (hydrogen evolved) [70]. Stucki and coworkers using the commercial Membrel electrolyzer achieved 20% of efficiency with a PbO_2 anode [71] requiring 65 kWh kg^{-1} O_3. Da Silva and coworkers [45] reported an energy requirement of 70 kWh kg^{-1} with 13% of efficiency by using a PbO_2 anode (H_2 evolving cathode) [28]. Conversely, Arihara and coworkers obtained 47% of ozone efficiency and an energy cost of 140 kWh kg^{-1} O_3 using BDD anodes [59]. It is important to note that the use of membrane electrochemical cells in combination with diamond electrodes has allowed obtaining concentrations as high as 160 mg h^{-1} with current densities of 200 mA cm^{-2}, reaching efficiencies as high as 23% [61]. These values are better than the 10% current efficiency [72] obtained with a lead dioxide anode doped with Fe^{3+} (OFM-Fe-PbO_2). This points out the drawbacks noted before but, at the same time, it also notes the promise of this technology.

HYDROGEN PEROXIDE

As explained before, hydrogen peroxide can be obtained on both the anode and cathode of an electrochemical cell during the electrolysis of wastewater. However, by a proper design of the cell with good cathode materials, its production can be increased by several times. In addition, nonactive anodes also promote its formation, although the cocktail of oxidants formed acts against the stability of this species and produces its decomposition to radicals, which produces an increase in the efficiency of the mineralization of the waste.

Hydrogen peroxide has an interesting chemistry because of its ability to function as an oxidant as well as a reductant in both acid and alkaline solutions. On the whole, at room temperature and atmospheric pressure, hydrogen peroxide is a very powerful oxidizing agent and a poor reducing agent.

Generally, H_2O_2 is employed as a reagent in chemical synthesis and a bleaching agent in the paper or textile industry and environmental applications. In this last case, H_2O_2 is used as a green oxidant for the destruction of organic and inorganic pollutants in watercourses because its decomposition only leads to water and oxygen. However, it is a selective oxidant whose activation is generally required. In fact, its further activation to hydroxyl radicals (a nonselective oxidant) gives rise to the so-called H_2O_2-based AOPs, which show great performance for the degradation of most organics and organometallic pollutants until their complete mineralization into CO_2, water, and inorganic ions. This activation can be done via transition metals (Fe, Cu, Ti, (Eq. 32)), ozone (Eq. 29), UV light (Eq. 30), or by other oxidants present in the solutions [73]. Recent studies

[19,24,73,74] have shown that the application of electrolysis in an electrochemical reactor equipped with a BDD anode and a GDE cathode gives rise to electrooxidation with an electrogenerated H_2O_2 process. This means the organics are destroyed mainly by hydroxyl radical mediated oxidation and, to a much lesser extent, by reactive oxygen species such as H_2O_2 and hydroperoxyl radicals. The oxidation ability of this procedure is strongly enhanced by the additional production of OH· in the bulk, using Fenton-based electrochemical advanced oxidation processes.

$$H_2O + Fe^{2+} \rightarrow {}^{\bullet}OH + OH^- + Fe^{3+} \quad (32)$$

Note that H_2O_2 is a compound that has very low oxidation power and can only attack reduced sulfur compounds, cyanides, chlorine, and certain organics such as aldehydes, formic acid, and some nitro-organic and sulfo-organic compounds [75]. It has been well known since 1882 that H_2O_2 can be continuously supplied to a solution contained in an electrolytic cell from the two-electron reduction of dissolved O_2 gas at a carbonaceous cathode with a high surface area [10]. In an acidic solution, this reduction process takes place according to reaction (33) with standard potential $E° = 0.68\,V/SHE$:

$$O_{2(g)} + 2H^+ + 2e^- \rightarrow H_2O_2 \quad (33)$$

This reaction is energetically easier than the four-electron reduction of O_2 to water with $E° = 1.23\,V/SHE$. It has been found that H_2O_2 production and stability depend on factors such as cell configuration, cathode properties, and operation conditions. For example, its electrochemical reduction at the cathode by reaction (34) and, to a much lesser extent, its disproportion in the bulk by reaction (35) are general parasitic reactions that result in the loss of oxidant with a drop in current efficiency [75]:

$$H_2O_2 + 2e^- \rightarrow 2OH^- \quad (34)$$

$$2H_2O_2 \rightarrow O_{2(g)} + 2H_2O \quad (35)$$

When an undivided cell is used, H_2O_2 is also oxidized to oxygen at the anode by reactions (36), (37)with the formation of hydroperoxyl radical ($HO_2^{\bullet}$) as intermediate [75,76]:

$$H_2O_2 \rightarrow HO_2^{\bullet} + H^+ + e^- \quad (36)$$

$$HO_2^{\bullet} \rightarrow O_{2(g)} + H^+ + e^- \quad (37)$$

The reaction (36) competes with the anodic oxidation of other products, generating other reactive oxygen species (ROS) that can be used to destroy the POPs contained in the electrolyzed solution. The strongest ROS is the radical ${}^{\bullet}OH$, which is formed at a high O_2-overvoltage anode (M) from water oxidation by reaction (38) [75,76]:

$$M + H_2O \rightarrow M(^{\bullet}OH) + H^+ + e^- \quad (38)$$

The mineralization action of this radical is largely ineffective for classical electrodes such as Pt, but it is much more efficient when a BDD thin layer is used as anode. Operating at the same high current within the water discharge region, a much higher quantity of reactive BDD($^{\bullet}$OH) than Pt($^{\bullet}$OH) is produced so that aromatics and unsaturated compounds such as carboxylic acids can be completely converted into CO_2 [73]. The low adsorption capability of $^{\bullet}$OH on BDD favors its dimerization to H_2O_2 by reaction (39) whereas the high oxidation power of this anode facilitates the generation of other weaker oxidants such as ozone by reaction (40) and $S_2O_8^{2-}$ from oxidation of SO_4^{2-} and/or HSO_4^- present in the electrolyte by reactions (41), (42), respectively, if sulfuric acid is used to set the solution pH [75].

$$2\,BDD(^{\bullet}OH) \rightarrow BDD + H_2O_2 \quad (39)$$

$$3\,H_2O \rightarrow O_{3(g)} + 6\,H^+ + 6\,e^- \quad (40)$$

$$2\,SO_4^{2-} \rightarrow S_2O_8^{2-} + 2\,e^- \quad (41)$$

$$2\,HSO_4^- \rightarrow S_2O_8^{2-} + 2\,H^+ + 2\,e^- \quad (42)$$

The removal of POPs by anodic oxidation (AO) with a BDD anode is then based on their direct reaction at the anode and/or their mediated oxidation with ROS such as BDD($^{\bullet}$OH) and in less extension O_3 as well as with other weaker oxidizing species formed from the anion of the electrolyte, such as $S_2O_8^{2-}$. In an undivided BDD/O_2 cell, POPs can be additionally oxidized by other ROS such as the weak oxidants H_2O_2 and $HO_2^{\bullet}$ generated from reactions (33), (36), and (39), leading to the anodic oxidation with electrogenerated H_2O_2 (AO-H_2O_2). This indirect electrooxidation method has been used to test the superiority of EAOPs based on Fenton's chemistry, such as EF and PEF with Pt/O_2 or BDD/O_2 cells, as discussed below [75]. The following major advantages for this indirect electrooxidation method compared with the chemical Fenton process have been claimed [76]:

- **(i)** The on-site production of H_2O_2 that avoids the risks related to its transport, storage, and handling.
- **(ii)** The possibility of the control of the degradation kinetics to allow mechanistic studies.
- **(iii)** The higher degradation rate of organic pollutants because of the continuous regeneration of Fe^{2+} at the cathode, which also minimizes sludge production.
- **(iv)** The feasibility of overall mineralization at a relatively low cost if the operation parameters are optimized.

A related EAOP is PEF in which the solution treated under EF conditions is illuminated with artificial UVA light to enhance the degradation process of POPs. A disadvantage for the industrial application of PEF is the high electrical cost of lamps supplying UVA light. An interesting alternative possibility is the use of sunlight as an inexpensive and renewable energy source of wavelength >300 nm [75,76]. This method, called solar photoelectro-Fenton (SPEF), enhances photolytic processes that are extended in the visible region, as in the case of reactions occurring from 300 to 480 nm.

FERRATES

Although these species are not typically produced during the electrolysis of wastewater, the authors consider it to be worth writing some paragraphs about it in this chapter. In fact, their presence is believed to occur in most electrochemical systems in which iron species are present. In some papers, it has been proposed that this occurrence may have a very relevant role to explain the performance of the electrolysis.

Ferrates have been very promising oxidants for decades with a potential great applicability in water treatment and water reclamation because they can perform as efficient disinfectants. Also, their reduction product (iron (III) species) behaves as coagulant reagents that may help to simultaneously reduce the turbidity of the water. It also has applications in the weapons industry and in super-iron batteries. Unfortunately, there is not an efficient process for their production and this fact opens chances for the evaluation of their electrochemical production.

Electrochemical synthesis of ferrate using iron-based electrodes in alkaline solutions has shown promising results, but it has also revealed important problems such as the formation of passivation layers on the surface of the electrodes. It is important to take into account that ferrates are not stable under the typical range of pHs found in wastewater and to produce them, strongly alkaline media (even up to 14 M NaOH) are required. One interesting process to produce ferrates is to use diamond anodes [77,78] because of the production of hydroxyl radicals and the capability of these radicals to oxidize iron to its highest oxidation state. However, this process has to face two important drawbacks: the first is the lack of iron precursors reaching the anode surface where the hydroxyl radicals are produced. The second is related to the service life of the electrodes: at these conditions the surface of the anode is damaged and hence the process becomes nonfeasible from an economical point of view.

NEW APPLICATIONS OF OXIDANTS ELECTROCHEMICALLY GENERATED FOR TREATING WASTEWATERS

Recently, the production of a few oxidants has been used in combination with other AOPs to improve the degradation efficiency of organic and inorganic substances. This is the case of hydrogen peroxide and ozone, referred to as the peroxone process, and has a synergetic effect on the degradation efficiency of pollutants. In the conventional peroxone process, while O_3 and H_2O_2 oxidize the target pollutants, the reaction between O_3 and H_2O_2 also takes place and forms hydroxyl radicals according to [79]:

$$O_3 + H_2O_2 \rightarrow OH^{\bullet} + O_2^{-\bullet} + O_2 \tag{43}$$

The formed hydroxyl radicals are known to be very powerful oxidants and can oxidize most recalcitrant pollutants at very high reaction rates. For that reason, high mineralization efficiencies of target pollutants can be achieved by the peroxone process. Furthermore, O_3 and H_2O_2 do not cause the formation of toxic byproducts as they leave only H_2O and O_2. Therefore, the peroxone process is considered an environmentally friendly and effective AOP in wastewater treatment. The most important limitation of this process is the addition of external H_2O_2. For this reason, in situ generation of H_2O_2 by electrochemical means seems to be the solution. In this case, the process is a combination of conventional electrolysis and ozonation (electroperoxone (*E*-peroxone)) [79]. The synergism of the electrocatalytic ozonation process allows achieving higher degradation efficiencies than that achieved for each individual process alone. In the E-peroxone process, ozone is produced the same way as in the conventional peroxone process, but the use of a carbon-based cathode favors the production of H_2O_2 from oxygen in the sparged gas mixture (O_3 and O_2). This provides in situ generation of H_2O_2 at controllable rates without wasting O_2. The *E*-peroxone process is convenient, cost-effective, and safe compared to the conventional peroxone process.

In this process, carbon-based cathodes are used to provide H_2O_2 electrochemically because these materials are very efficient in terms of H_2O_2 and $^{\bullet}OH$ production. Bakheet and coworkers [80] studied a carbon-polytetrafluoroethylene electrode for the *E*-peroxone approach, observing that the concentration of H_2O_2 increased linearly throughout an hour by only O_2 sparging into the system; also, no $^{\bullet}OH$ was produced. Subsequently, several cathode materials were studied to understand the nature of the electrode between carbon-polytetrafluoroethylene, carbon felt, and reticulated vitreous carbon (RVC) for an *E*-peroxone system [81]. High amounts of H_2O_2 were produced in the *E*-peroxone system; 252.9 mg L^{-1} was achieved for the carbon-polytetrafluoroethylene cathode, followed by RVC then carbon felt. On the other hand, the carbon-polytetrafluoroethylene cathode provided 86%–80% as current efficiency, but the same material can approach current efficiencies up to 100% while providing H_2O_2 generation [5,82,83] Wu et al. [84] evaluated carbon nanotube (CNT)-polytetrafluoroethylene, carbon black (CB)-polytetrafluoroethylene, and active carbon (AC)-polytetrafluoroethylene as three different gas diffusion cathodes in an *E*-peroxone system considering $^{\bullet}OH$ generation. The most efficient cathode was CNT-polytetrafluoroethylene, producing 4.7×10^{-13} mol L^{-1} of $^{\bullet}OH$. Concluding that, the use of a carbon-polytetrafluoroethylene electrode as the cathode material provides the most efficient production of H_2O_2 and $^{\bullet}OH$. However, the use of porous materials can be an important property of the carbon materials to improve the generation of H_2O_2, such as AC and CNT, with which ozone can react easily on the surface. In the *E*-peroxone process, applied current is another important parameter for effective electrogeneration of H_2O_2. Wang et al. [82] evaluated the effect of applied current using carbon-polytetrafluoroethylene as the cathode and Pt as the anode with currents ranging from 100 to 500 mA. A linear increase in the concentration of H_2O_2 was observed, achieving current efficiencies of about 86.9%–95.9%. This indicates that electrogeneration of H_2O_2 can be provided efficiently with high current efficiencies and that the production rate of H_2O_2 can be controlled by the applied current.

As the novel electrochemical/chemical approach has demonstrated significant efficiency during the treatment of different organic pollutants, a literature review of the reported *E*-peroxone processes, with different carbonaceous cathodes as well as their operating conditions, is summarized in Table 1. Concerning the degradation and mineralization of recalcitrant organic pollutants in water and wastewater, the combination of ozone and hydrogen peroxide with carbon-based cathodes under optimal operating conditions will lead to higher degradation, removal rates, and efficiencies compared to using ozonation and anodic electrolysis, individually. The synergetic effects of the *E*-peroxone are primarily attributed to the intensive production of highly oxidative and nonselective reagents such as hydroxyl radicals in the oxidation medium. The high degradation and mineralization rates and the simplicity of the process make it an excellent alternative to other AOPs.

TABLE 1 Examples of Complete Removal Efficiency of Organic Pollutants by the *E*-peroxone Process With Different Cathodes

Cathode compound	C_0 ($mg\,dm^{-3}$)	Experimental conditions	Anode	Time (h)	O_3 flow rate ($L\,min^{-1}$)	Current (mA)	Ref.
Carbon-PTFE							
Ibuprofen + NOM	20+0.2	300 cm^3, no supporting electrolyte used, [O_3] 10–40 $mg\,L^{-1}$, pH: 3–10, NOM addition: 5–30 $mg\,L^{-1}$	Pt	2	0.25	50–400	[85]
Venlafaxine	20	300 cm^3, 0.05 M Na_2SO_4, [O_3] 6.5–42 $mg\,L^{-1}$, pH: 3.5–10.5	Pt	2	0.25	50–450	[86]
Diclofenac	0.4	0.05 M Na_2SO_4, [O_3] 2–10 $mg\,L^{-1}$, pH8	Pt	0.4	0.25	40–120	[87]
Gemfibrozil Bezafibrate Ibuprofen Clofibric acid *p*-chlorobenzoic acid	0.4	No supporting electrolyte used, [O_3] 2–10 $mg\,L^{-1}$, pH8	Pt	0.4	0.25	40–120	[87]
Clofibric acid	100	400 cm^3, 0.01 M Na_2SO_4	BDD Pt	1	0.25	40–320	[88]
Oxalic acid	180	400 cm^3, 0.05 M Na_2SO_4, [O_3] 0–100 $mg\,L^{-1}$, pH: 3–11	BDD Pt	2	0.4	100–500	[82]
1,4-Dioxane	200	400 cm^3, 0.05 M Na_2SO_4, [O_3] 10–40 $mg\,L^{-1}$, pH: 3–11	Pt $TiRuO_2^-$ IrO_2 TiBDD	2	0.3	400	[89]

Continued

TABLE 1 Examples of Complete Removal Efficiency of Organic Pollutants by the *E*-peroxone Process With Different Cathodes—cont'd

Cathode compound	C_0 ($mg\,dm^{-3}$)	Experimental conditions	Anode	Time (h)	O_3 flow rate ($L\,min^{-1}$)	Current (mA)	Ref.
Diethyl phthalate	20	400 cm^3, 0.05 M Na_2SO_4, [O_3] 118 $mg\,L^{-1}$	Pt	1	0.4	400	[81]
Methylene blue	180	400 cm^3, 0.05 M Na_2SO_4, [O_3] 33–75 $mg\,L^{-1}$	Pt	2	0.4	100–500	[90]
Orange II	200	400 cm^3, 0.01 and 0.05 M Na_2SO_4, 0.01 and 0.2 NaCl, [O_3] 42–157 $mg\,L^{-1}$, pH3–7	Pt	1.5	0.4	50–500	[80]
Leachate	6635 COD 1650 TOC	200 cm^3, 0.01 and 0.05 M Na_2SO_4, 0.01 and 0.2 NaCl, [O_3] 35–118 $mg\,L^{-1}$, pH3–7	Pt	6	0.3	50–500	[91]
Synthetic and real surface water	3 and 6 of NOM	Continuous feed, [O_3] 5.2 $mg\,L^{-1}$	RuO_2/IrO_2	3	80	20–100	[92]
BDD							
Orange II	200	400 cm^3, 0.01 and 0.05 M Na_2SO_4, 0.01 and 0.2 NaCl, [O_3] 42–157 $mg\,L^{-1}$, pH3–7	Pt	1.5	0.4	50–500	[80]
Activated carbon fiber							
Amoxicillin	100	1000 cm^3, 0.05 M Na_2SO_4, [O3] 10–40 $mg\,L^{-1}$, pH3–11	Pt	1	0.4	100–400	[93]
Nitrobenzene	12	500 cm^3, Na_2SO_4, [O_3] 8.33 $mg\,L^{-1}$, pH7	Ti	–[a]	–[a]	50–100	[94]

[a] *Not determined.*

CONCLUDING REMARKS

The results reported in the existing literature evidence the potential use of oxidants, such as ozone, hydrogen peroxide and ferrate for degrading of different organic pollutants. Ozone electrochemically generated is an efficient oxidant but more studies are necessary regarding the electrodes used. Meanwhile, the higher oxidant character of the homogenous •OH radicals produced by Electroperoxone seems to be a suitable approach to be applied for treating wastewaters in terms of efficiency and energy consumption as well as the use of cheaper electrode materials.

Acknowledgments

Carlos A. Martínez-Huitle acknowledges the funding provided by the Alexander von Humboldt Foundation (Germany) and Coordenação de Aperfeiçoamento de Pessoal de Nível Superior (Brazil) as a Humboldt fellowship for Experienced Researcher (88881.136108/2017-01) at the Johannes Gutenberg-Universität Mainz, Germany.

References

[1] C.A. Martínez-Huitle, M.A. Rodrigo, I. Sirés, O. Scialdone, Single and coupled electrochemical processes and reactors for the abatement of organic water pollutants: a critical review, Chem. Rev. 115 (2015) 13362–13407.
[2] G. Foti, D. Gandini, C. Comninellis, A. Perret, W. Haenni, Oxidation of organics by intermediates of water discharge on IrO2 and synthetic diamond anodes, Electrochem. Solid-State Lett. 2 (1999) 228–230.
[3] C. Comninellis, A. Kapalka, S. Malato, S.A. Parsons, I. Poulios, D. Mantzavinos, Advanced oxidation processes for water treatment: advances and trends for R&D, J. Chem. Technol. Biotechnol. 83 (2008) 769–776.
[4] B. Marselli, J. Garcia-Gomez, P.A. Michaud, M.A. Rodrigo, C. Comninellis, Electrogeneration of hydroxyl radicals on boron-doped diamond electrodes, J. Electrochem. Soc. 150 (2003) D79–D83.
[5] I. Sires, E. Brillas, M.A. Oturan, M.A. Rodrigo, M. Panizza, Electrochemical advanced oxidation processes: today and tomorrow. A review, Environ. Sci. Pollut. Res. 21 (2014) 8336–8367.
[6] C. Comninellis, A. Nerini, Anodic-oxidation of phenol in the presence of nacl for wastewater treatment, J. Appl. Electrochem. 25 (1995) 23–28.
[7] A. Sanchez-Carretero, C. Saez, P. Canizares, M.A. Rodrigo, Electrochemical production of perchlorates using conductive diamond electrolyses, Chem. Eng. J. 166 (2011) 710–714.
[8] P. Canizares, C. Saez, J. Lobato, R. Paz, M.A. Rodrigo, Effect of the operating conditions on the oxidation mechanisms in conductive-diamond electrolyses, J. Electrochem. Soc. 154 (2007) E37–E44.
[9] C. Barrera-Diaz, P. Canizares, F.J. Fernandez, R. Natividad, M.A. Rodrigo, Electrochemical advanced oxidation processes: an overview of the current applications to actual industrial effluents, J. Mex. Chem. Soc. 58 (2014) 256–275.
[10] C.A. Martinez-Huitle, S. Ferro, Electrochemical oxidation of organic pollutants for the wastewater treatment: direct and indirect processes, Chem. Soc. Rev. 35 (2006) 1324–1340.

[11] R.W. Presley, M.C. Helvenston, B. Zhao, Electro-reductive dehalogenation on palladized graphite electrodes, Abstracts of Papers of the American Chemical Society 213; 102-ENVR, 1997.
[12] J. Radjenovic, M.J. Farre, Y. Mu, W. Gernjak, J. Keller, Reductive electrochemical remediation of emerging and regulated disinfection byproducts, Water Res. 46 (2012) 1705–1714.
[13] Y. Yang, Y. Li, S. Walse, W. Mitch, Destruction of methyl bromide Sorbed to activated carbon by thiosulfate or electrolysis, Environ. Sci. Technol. 49 (2015) 4515–4521.
[14] Y. Li, J.M. Kemper, G. Datuin, A. Akey, W.A. Mitch, R.G. Luthy, Reductive dehalogenation of disinfection byproducts by an activated carbon-based electrode system, Water Res. 98 (2016) 354–362.
[15] M. Geng, Z. Duan, Prediction of oxygen solubility in pure water and brines up to high temperatures and pressures, Geochim. Cosmochim. Acta 74 (2010) 5631–5640.
[16] O. Scialdone, A. Galia, C. Gattuso, S. Sabatino, B. Schiavo, Effect of air pressure on the electro-generation of H2O2 and the abatement of organic pollutants in water by electro-Fenton process, Electrochim. Acta 182 (2015) 775–780.
[17] E. Petrucci, A. Da Pozzo, L. Di Palma, On the ability to electrogenerate hydrogen peroxide and to regenerate ferrous ions of three selected carbon-based cathodes for electro-Fenton processes, Chem. Eng. J. 283 (2016) 750–758.
[18] R.M. Reis, A.A.G.F. Beati, R.S. Rocha, M.H.M.T. Assumpcao, M.C. Santos, R. Bertazzoli, M.R.V. Lanza, Use of gas diffusion electrode for the in situ generation of hydrogen peroxide in an electrochemical flow-by reactor, Ind. Eng. Chem. Res. 51 (2012) 649–654.
[19] W.R.P. Barros, R.M. Reis, R.S. Rocha, M.R.V. Lanza, Electrogeneration of hydrogen peroxide in acidic medium using gas diffusion electrodes modified with cobalt (II) phthalocyanine, Electrochim. Acta 104 (2013) 12–18.
[20] J.F. Carneiro, R.S. Rocha, P. Hammer, R. Bertazzoli, M.R.V. Lanza, Hydrogen peroxide electrogeneration in gas diffusion electrode nanostructured with Ta2O5, Appl. Catal. A Gen. 517 (2016) 161–167.
[21] J.F. Perez, J. Llanos, C. Saez, C. Lopez, P. Canizares, M.A. Rodrigo, The jet aerator as oxygen supplier for the electrochemical generation of H2O2, Electrochim. Acta 246 (2017) 466–474.
[22] J.F. Perez, J. Llanos, C. Saez, C. Lopez, P. Canizares, M.A. Rodrigo, Electrochemical jet-cell for the in-situ generation of hydrogen peroxide, Electrochem. Commun. 71 (2016) 65–68.
[23] M. Panizza, G. Cerisola, Electrochemical generation of H2O2 in low ionic strength media on gas diffusion cathode fed with air, Electrochim. Acta 54 (2008) 876–878.
[24] R.B. Valim, R.M. Reis, P.S. Castro, A.S. Lima, R.S. Rocha, M. Bertotti, M.R.V. Lanza, Electrogeneration of hydrogen peroxide in gas diffusion electrodes modified with tert-butyl-anthraquinone on carbon black support, Carbon 61 (2013) 236–244.
[25] A. Verdaguer-Casadevall, D. Deiana, M. Karamad, S. Siahrostami, P. Malacrida, T. W. Hansen, J. Rossmeisl, I. Chorkendorff, I.E.L. Stephens, Trends in the electrochemical synthesis of H2O2: enhancing activity and selectivity by electrocatalytic site engineering, Nano Lett. 14 (2014) 1603–1608.
[26] J.F. Perez, A. Galia, M.A. Rodrigo, J. Llanos, S. Sabatino, C. Saez, B. Schiavo, O. Scialdone, Effect of pressure on the electrochemical generation of hydrogen peroxide in undivided cells on carbon felt electrodes, Electrochim. Acta 248 (2017) 169–177.
[27] M. Fryda, T. Matthee, Disinfecting system for low conductivity liquids, especially water, comprises electrochemical cell with electrodes for generating oxidants, with unit for thoroughly mixing the oxidants and liquid. Condias Gmbh. Patent number: DE202005003720-U1; WO2006092125-A1; US2008156642-A1.
[28] P.A. Christensen, T. Yonar, K. Zakaria, The electrochemical generation of ozone: a review, Ozone Sci. Eng. 35 (2013) 149–167.

[29] G.L. Putnam, R.W. Moulton, W.W. Fillmore, L.H. Clark, Electrolytic ozone, J. Electroanal. Chem. 93 (1948) 211–221.
[30] L.M. Da Silva, L.A. De Faria, J.F.C. Boodts, Eletrochemical ozone production: influence of the supporting electrolyte on kinetics and current efficiency, Electrochim. Acta 48 (2003) 699–709.
[31] J.D. Seader, C.W. Tobias, Ozone by electrolysis of sulfuric acid, Ind. Eng. Chem. 44 (1952) 2207–2211.
[32] P.C. Foller, G.H. Kelsall, Ozone generation via the electrolysis of fluoboric acid using glassy-carbon anodes and air depolarized cathodes, J. Appl. Electrochem. 23 (1993) 996–1010.
[33] J. Kim, G.V. Korshin, Examination of in situ generation of hydroxyl radicals and ozone in a flow-through electrochemical reactor, Ozone Sci. Eng. 30 (2008) 113–119.
[34] P.A. Michaud, M. Panizza, L. Ouattara, T. Diaco, G. Foti, C. Comninellis, Electrochemical oxidation of water on synthetic boron-doped diamond thin film anodes, J. Appl. Electrochem. 33 (2003) 151–154.
[35] S.D. Han, J.D. Kim, K.C. Singh, R.S. Chaudhary, Electrochemical generation of ozone using solid polymer electrolyte—state of the art, Indian J. Chem. 43A (2004) 1599–1614.
[36] J. Larminie, A. Dicks, Fuel Cell Systems Explained, Wiley, Chichester, 2003, pp. 69–72.
[37] P. Tatapudi, J.M. Fenton, Simultaneous synthesis of ozone and hydrogen-peroxide in a proton-exchange-membrane electrochemical reactor, J. Electrochem. Soc. 141 (1994) 1174–1178.
[38] Y. Nishiki, N. Kitaori, K. Nakamuro, Performances of small-sized generator of ozone-dissolved water using boron-doped diamond electrodes, Ozone Sci. Eng. 33 (2011) 114–120.
[39] Y. Cui, Y. Wang, B. Wang, H. Zhou, K.-Y. Chan, X.-Y. Li, Electrochemical generation of ozone in a membrane electrode assembly cell with convective flow, J. Electrochem. Soc. 156 (2009) E75–E80.
[40] S. Stucki, G. Theis, R. Kötz, H. Devantay, H.J. Christen, In situ production of ozone in water using a membrel electrolyzer, J. Electrochem. Soc. 132 (1985) 367–371.
[41] P.C. Foller, C.W. Tobias, The mechanism of the disintegration of lead dioxide anodes under conditions of ozone evolution in strong acid electrolyte, J. Electrochem. Soc. 129 (1982) 567–570.
[42] F. Okada, K. Naya, Highly efficient and long-lifetime ozone water production system realized using a felt separator, J. Electrochem. Soc. 156 (2009) E125–E131.
[43] L.M. Da Silva, L.A. De Faria, J.F.C. Boodts, Green processes for environmental application. Electrochemical ozone production, Pure Appl. Chem. 73 (2001) 1871–1884.
[44] M.I. Awad, M.M. Saleh, Electrochemical generation of ozone at pbo2-loaded platinum screens, J. Solid State Electrochem. 14 (2010) 1877–1883.
[45] L.M. Da Silva, D.V. Franco, L.G. Sousa, I.C. Goncalves, Characterization of an electrochemical reactor for the ozone production in electrolyte-free water, J. Appl. Electrochem. 40 (2010) 855–864.
[46] L.M. Da Silva, M.H.P. Santana, Electrochemistry and green chemical processes: electrochemical ozone production, Quim Nova 26 (2003) 880–888.
[47] E.R. Kötz, S. Stucki, Ozone and oxygen evolution on pbo2 electrodes acid—solution, J. Electroanal. Chem. 228 (1987) 407–415.
[48] G.F. Potapova, O.V. Shestakova, Defining specifications for concentrated ozone production on type SU-1 modules, Russ. J. Electrochem. 31 (1995) 730–733.
[49] Y.-H. Wang, S.A. Cheng, K.-Y. Chan, X.Y. Li, Electrolytic generation of ozone on antimony- and nickel-doped tin oxide electrode, J. Electrochem. Soc. 152 (2005) D197–D200.
[50] M.H.P. Santana, L.A. De Faria, J.F.C. Boodts, Investigation of the properties of Ti/[IrO_2-Nb_2O_5] electrodes for simultaneous oxygen evolution and electrochemical ozone production, EOP, Electrochim. Acta 49 (2004) 1925–1935.

[51] M.H.P. Santana, L.A. De Faria, J.F.C. Boodts, Effect of preparation procedure of IrO_2-Nb_2O_5 anodes on surface and Electrocatalytic properties, J. Appl. Electrochem. 35 (2005) 915–924.

[52] L.M. Da Silva, D.V. Franco, L.A. De Faria, J.F.C. Boodts, Surface, kinetics and electrocatalytic properties of Ti/ (IrO_2 + Ta_2O_5) electrodes, prepared using controlled cooling rate, for ozone production, Electrochim. Acta 49 (2004) 3977–3988.

[53] K. Kaneda, M. Ikematsu, K. Kitsuka, M. Iseki, H. Matsuura, T. Higuchi, T. Hattori, T. Tsukamoto, M. Yasuda, Si/TiOx/Pt/TaOx electrodes fabricated by sputtering for electrochemical ozone generation, J. Appl. Phys. Pt. 1-Reg. Papers Brief Commun. Rev. Papers 45 (2006) 6417–6419.

[54] K. Kaneda, M. Ikematsu, Y. Koizumi, H. Minoshima, T. Rakuma, D. Takaoka, M. Yasuda, Ozone generation by a TaOx and Pt composite insulator-coated Ti electrode, Electrochem. Solid-State Lett. 8 (2005) J13–J16.

[55] J.E. Graves, D. Pletcher, R.L. Clarke, F.C. Walsh, The electrochemistry of magneli phase titanium—oxide ceramic electrodes. 2. Ozone generation at ebonex and ebonex lead dioxide anodes, J. Appl. Electrochem. 22 (1992) 200–203.

[56] K. Kitsuka, K. Kaneda, M. Ikematsu, M. Iseki, K. Mushiake, T. Ohsaka, Ex situ and in situ characterization studies of spin-coated TiO2 film electrodes for the electrochemical ozone production process, Electrochim. Acta 55 (2009) 31–36.

[57] K. Kitsuka, K. Kaneda, M. Ikematsu, M. Iseki, K. Mushiake, T. Ohsaka, n-Type TiO_2 thin films for electrochemical ozone production, J. Electrochem. Soc. 157 (2010) F30–F34.

[58] K. Arihara, C. Terashima, A. Fujishima, Application of freestanding perforated diamond electrodes for efficient ozone-water production, Electrochem. Solid-State Lett. 9 (2006) D17–D20.

[59] K. Arihara, C. Terashima, A. Fujishima, Electrochemical production of high-concentration ozone-water using freestanding perforated diamond electrodes, J. Electrochem. Soc. 154 (2007) E71–E75.

[60] N. Katsuki, E. Takahashi, M. Toyoda, T. Kurosu, M. Iida, S. Wakita, Y. Nishiki, T. Shimamune, Water electrolysis using diamond thin-film electrodes, J. Electrochem. Soc. 145 (1998) 2358–2362.

[61] A. Kraft, M. Stadelmann, M. Wunsche, M. Blaschke, Electrochemical ozone production anodes and a solid polymer using diamond electrolyte, Electrochem. Commun. 8 (2006) 883–886.

[62] M. Panizza, G. Cerisola, Application of diamond electrodes to electrochemical processes, Electrochim. Acta 51 (2005) 191–199.

[63] P.A. Michaud, E. Mahe, W. Haenni, A. Perret, C. Comninellis, Preparation of peroxodisulfuric acid using boron-doped diamond thin film electrodes, Electrochem. Solid-State Lett. 3 (2000) 77–79.

[64] H. Bader, J. Hoigné, Determination of ozone in water by indigo method, Water Res. 15 (1981) 449–456.

[65] A. Seidel, Kirk-Othmer Encyclopedia of Chemical Technology, John Wiley & Sons, Inc., Hoboken, NJ, 2006, pp. 768–822.

[66] K. Sehested, H. Corfitzen, J. Holcman, E.J. Hart, On the mechanism of the decomposition of acidic O_3 solutions, thermally or H_2O_2-initiated, J. Phys. Chem. A 102 (1998) 2667–2672.

[67] S.A. Cheng, K.Y. Chan, Electrolytic generation of ozone on an antimony-doped tin dioxide coated electrode, Electrochem. Solid-State Lett. 7 (2004) D4–D6.

[68] P.A. Christensen, W.F. Lin, H. Christensen, A. Imkum, J.M. Jin, G. Li, C.M. Dyson, Room temperature electrochemical generation of ozone with 50% current efficiency in 0.5M sulfuric acid at cell voltages $< 3V$, Ozone Sci. Eng. 31 (2009) 287–293.

[69] Y.-H. Wang, S.A. Cheng, K.-Y. Chan, Synthesis of Ozone from air via a polymer-electrolyte-membrane cell with a doped tin oxide anode, Green Chem. 8 (2006) 568–572.

[70] M. Katoh, Y. Nishiki, S. Nakamatsu, Polymer electrolyte—type electrochemical ozone generator with an oxygen cathode, J. Appl. Electrochem. 24 (1994) 489–494.
[71] S. Stucki, H. Baumann, H.J. Christen, R. Kötz, Performance of a pressurized electrochemical ozone generator, J. Appl. Electrochem. 17 (1987) 773–778.
[72] L.G. De Sousa, D.V. Franco, L.M. Da Silva, Electrochemical ozone production using electrolyte-free water for environmental applications, J. Environ. Chem. Eng. 4 (2016) 418–427.
[73] I. Sires, J.A. Garrido, R.M. Rodriguez, P.I. Cabot, F. Centellas, C. Arias, E. Brillas, Electrochemical degradation of paracetamol from water by catalytic action of Fe2+, Cu2+, and UVA light on electrogenerated hydrogen peroxide, J. Electrochem. Soc. 153 (2006) D1–D9.
[74] O. Scialdone, A. Galia, S. Sabatino, Electro-generation of H_2O_2 and abatement of organic pollutant in water by an electro-Fenton process in a microfluidic reactor, Electrochem. Commun. 26 (2013) 45–47.
[75] E. Brillas, I. Sirés, M.A. Oturan, Electro-fenton process and related electrochemical technologies based on fenton's reaction chemistry, Chem. Rev. 109 (2009) 6570–6631.
[76] C.A. Martinez-Huitle, E. Brillas, Decontamination of wastewaters containing synthetic organic dyes by electrochemical methods: A general review, Appl. Catal. B Environ. 87 (2009) 105–145.
[77] P. Canizares, M. Arcis, C. Saez, M.A. Rodrigo, Electrochemical synthesis of ferrate using boron doped diamond anodes, Electrochem. Commun. 9 (2007) 2286–2290.
[78] A. Sanchez-Carretero, M.A. Rodrigo, P. Canizares, C. Saez, Electrochemical synthesis of ferrate in presence of ultrasound using boron doped diamond anodes, Electrochem. Commun. 12 (2010) 644–646.
[79] O. Turkay, Z. Girgin Ersoy, S. Barısçi, Review—the application of an electro-peroxone process in water and wastewater treatment, J. Electrochem. Soc. 164 (2017) E94–E102.
[80] B. Bakheet, S. Yuan, Z. Li, H. Wang, J. Zuo, S. Komarneni, Y. Wang, Electro-peroxone treatment of Orange II dye wastewater, Water Res. 47 (2013) 6234–6243.
[81] M. Hou, Y. Chu, X. Li, H. Wang, W. Yao, G. Yu, S. Murayama, Y. Wang, Electro-peroxone degradation of diethyl phthalate: Cathode selection, operational parameters, and degradation mechanisms, J. Hazard. Mater. 319 (2015) 61–68.
[82] H. Wang, S. Yuan, J. Zhan, Y. Wang, G. Yu, S. Deng, J. Huang, B. Wang, Mechanisms of enhanced total organic carbon elimination from oxalic acid solutions by electro-peroxone process, Water Res. 80 (2015) 20–29.
[83] M. Zhou, Q. Tan, Q. Wang, Y. Jiao, N. Oturan, M.A. Oturan, Degradation of organics in reverse osmosis concentrate by electro-Fenton process, J. Hazard. Mater. 215 (2012) 287–293.
[84] D. Wu, G. Lu, R. Zhang, Q. Lin, Z. Yan, J. Liu, Y. Li, Enhanced hydroxyl radical generation in the combined ozonation and electrolysis process using carbon nanotubes containing gas diffusion cathode, Environ. Sci. Pollut. Res. 22 (2015) 15812–15820.
[85] X. Li, Y. Wang, S. Yuan, Z. Li, B. Wang, J. Huang, S. Deng, G. Yu, Degradation of the anti-inflammatory drug ibuprofen by electro-peroxone process, Water Res. 63 (2014) 81–93.
[86] X. Li, Y. Wang, J. Zhao, H. Wang, B. Wang, J. Huang, S. Deng, G. Yu, Electro-peroxone treatment of the antidepressant venlafaxine: Operational parameters and mechanism, J. Hazard. Mater. 300 (2015) 298–306.
[87] W. Yao, X. Wang, H. Yang, G. Yu, S. Deng, J. Huang, B. Wang, Y. Wang, Removal of pharmaceuticals from secondary effluents by an electro-peroxone process, Water Res. 88 (2016) 826–835.
[88] Z. Lin, W. Yao, Y. Wang, G. Yu, S. Deng, J. Huang, B. Wang, Perchlorate formation during the electro-peroxone treatment of chloride-containing water: effects of operational parameters and control strategies, Water Res. 88 (2016) 691–702.

[89] B. Bakheet, C. Qiu, S. Yuan, Y. Wang, G. Yu, S. Deng, J. Huang, B. Wang, Inhibition of polymer formation in electrochemical degradation of p-. nitrophenol by combining electrolysis with ozonation, Chem. Eng. J. 252 (2014) 17–21.
[90] S. Yuan, Z. Li, Y. Wang, Effective degradation of methylene blue by a novel electrochemically driven process, Electrochem. Commun. 29 (2013) 48–51.
[91] Z. Li, S. Yuan, C. Qiu, Y. Wang, X. Pan, J. Wang, C. Wang, J. Zuo, Effective degradation of refractory organic pollutants in landfill leachate by electro-peroxone treatment, Electrochim. Acta 102 (2013) 174–182.
[92] Y. Li, W. Shen, S. Fu, H. Yang, G. Yu, Y. Wang, Inhibition of bromate formation during drinking water treatment by adapting ozonation to electro-peroxone process, Chem. Eng. J. 264 (2015) 322–328.
[93] W. Guo, Q.-L. Wu, X.-J. Zhou, H.-O. Cao, J.-S. Du, R.-L. Yin, N.-Q. Ren, Enhanced amoxicillin treatment using the electro-peroxone process: key factors and degradation mechanism, RSC Adv. 5 (2015) 52695–52702.
[94] X. Zhang, Y. Zhou, C. Zhao, Z. Sun, Z. Zhang, Z.A. Mirza, G. Saylor, J. Zhai, H. Zheng, Electric field induced activated carbon fiber (ACF) cathode transition from an initiator/a promoter into an electrocatalyst in ozonation process, Chem. Eng. J. 304 (2016) 129–133.

CHAPTER

8

Electro-Fenton Process: Background, New Developments, and Applications

Nihal Oturan, Mehmet A. Oturan

Université Paris-Est, Laboratoire Géomatériaux et Environnement (LGE), Marne-la-Vallée, France

OUTLINE

INTRODUCTION

The continuous release of different types of organic micropollutants arising from different types of wastewater (industrial chemicals, dyes, pesticides, pharmaceuticals and personal care products (PPCPs), and

Electrochemical Water and Wastewater Treatment
https://doi.org/10.1016/B978-0-12-813160-2.00008-0

other human activities) into bodies of water constitutes a growing concern as these micropollutants are, in general, persistent organics and therefore accumulate in the aquatic environment. The occurrence of the persistent organic pollutants (POPs) is well documented and concerns all countries worldwide [1–5]. POPs being persistent and harmful substances, they constitute a common concern due to their high potential toxicological impact and risk for human health [6,7]. These pollutants are not eliminated efficiently in conventional wastewater treatment plants (WWTPs) because they have not been designed for removing these types of pollutants [8]. Therefore, WWTPs constitute one of the most important sources for releasing POPs into the natural water system.

Due to the potential adverse effects (carcinogenic and mutagenic as well as damaging to the human hormonal, reproductive, and immune systems) on human health and ecological systems of most of POPs, there is great concern to prevent their release into bodies of water [9,10]. For this, the development of simple, efficient, and cost-effective processes is a major challenge for researchers and industries [7,8,11]. The removal of toxic/persistent micropollutants from effluents in a source before discharge into the body of water is of prime importance [12–14] because this is the most efficient and economical way to protect the environment. Among available technologies, the so-called advanced oxidation processes (AOPs), defined first by Glaze et al. [15], constitute promising technologies. These processes are based on the in situ production of strong oxidants, mainly hydroxyl radicals ($^{\bullet}OH$), the second most powerful oxidizing species ($E° = 2.80$ V/SHE), being able to oxidize nonselectively any organic pollutant up to mineralization (transformation to CO_2 and water) (reactions 1a and 1b). $^{\bullet}OH$ can be produced by different activation modes such as chemistry (Fenton process), photochemistry (UV photolysis, photo-Fenton, and heterogeneous photocatalysis), sonoloysis, and electrochemistry [14,16,17].

$$^{\bullet}OH + \text{Organic pollutant} \rightarrow \text{Primary intermediates} \quad (1a)$$

$$\text{Primary intermediates} + {}^{\bullet}OH \rightarrow\rightarrow\rightarrow CO_2 + H_2O + \text{inorganic ions} \quad (1b)$$

Electrochemical advanced oxidation processes (EAOPs) are the most recent AOPs. They have been developed as a clean, efficient, and economical alternative to destroy POPs in water [13,18,19]. The most popular EAOPs are electro-oxidation (or anodic oxidation) and electro-Fenton (EF) processes [12,14,20]. The electro-oxidation process being the subject of the other chapters, we focus in this chapter mainly on the electro-Fenton process.

PRINCIPLES AND FUNDAMENTALS OF THE EF PROCESS

The chemical generation of $^{\bullet}OH$ happens conventionally by the classical Fenton reaction (2) through a mixture of H_2O_2 and a soluble ferrous iron salt, the so-called "Fenton's reagent" [14,16]:

$$H_2O_2 + Fe^{2+} \rightarrow Fe^{3+} + OH^- + {}^{\bullet}OH \tag{2}$$

Indeed, H.J.H. Fenton observed at the end of the 19th century an enhancement of the oxidation power of H_2O_2 in the presence of ferrous iron ions [21]. Four decades later, Haber and Wise [22] evidenced the kinetics and mechanism of this reaction, evidencing the reason of the enhancement of oxidation efficiency by generation of $^{\bullet}OH$ from Fenton's reagent through reactions (1a) and (1b).

The Fenton reaction provides a simple system to produce $^{\bullet}OH$. However, the process needs high amounts of reagent to produce enough $^{\bullet}OH$. This implies an economical aspect as well as the formation of ferric sludge requiring supplementary processing. More importantly, high reagent concentrations favor the rate of wasting reactions (2) and (3), resulting in low efficiency:

$$^{\bullet}OH + H_2O_2 \rightarrow H_2O + HO_2^{\bullet} \tag{3}$$

$$^{\bullet}OH + Fe^{2+} \rightarrow Fe^{3+} + OH^- \tag{4}$$

These reactions are doubly harmful as they waste $^{\bullet}OH$ and consume reagents. Therefore, the drop in mineralization efficiency conducts to the formation of oxidation reaction intermediates being potentially toxic [23]. To avoid these reactions, the concentration of Fenton's reagent should be controlled.

On the other hand, electrochemistry offers an excellent way to generate and control the concentration of Fenton's reagent, thus allowing the development of the electro-Fenton (EF) process. This process has been developed to overcome the drawbacks of the Fenton process through a controlled synthesis of H_2O_2 in the solution to be treated and catalyzing the Fenton reaction by electrochemical regeneration of Fe^{2+}. H_2O_2 is generated by a two-electron reduction of O_2 from compressed air (reaction 5) in an electrochemical cell (Fig. 1) at an acidic medium of around pH 3 [24,25]. The electrochemical supply of H_2O_2 is of great interest because the in situ generation avoids the risks linked to its transport and storage, in addition to the reagent cost savings.

$$O_2 + 2H^+ + 2e^- \rightarrow H_2O_2 \tag{5}$$

Thereby the Fenton reaction can proceed if a catalytic amount of about 10^{-4} M of ferrous (or ferric) iron is externally added to the solution.

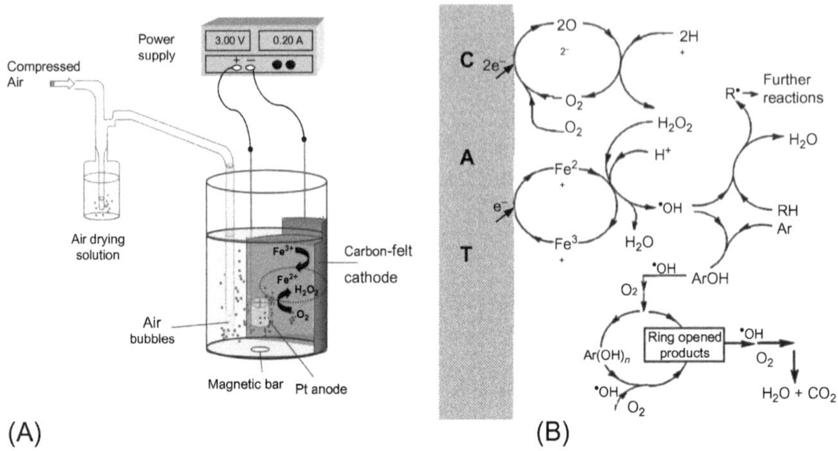

FIG. 1 (A) Schematic representation of an open and undivided lab-scale, two-electrode tank reactor with a 60 cm^2 carbon-felt cathode fed with compressed air to produce $^{\bullet}OH$, and (B) schematic representation of the main reactions involved in the EF process. RH denotes an unsaturated organic species leading to the H atom abstraction reaction and Ar denotes an unsaturated organic (mainly aromatic) compound involving an additional reaction. *Modified with permission from E. Brillas, I. Sirés, M.A. Oturan, Electro-Fenton process and related electrochemical technologies based on Fenton's reaction chemistry, Chem. Rev. 109 (2009) 6570–6631. Copyright 2009 ACS Publications.*

The Fenton reaction then generates ferric iron that can be reduced at the cathode (reaction 6), at the same potential serving H_2O_2 generation to regenerate Fe^{2+} ions [26,27]:

$$Fe^{3+} + e^- \rightarrow Fe^{2+} \quad (6)$$

Thus, in the EF process, Fenton's reagent is generated in a continuous and controlled manner in an electrochemical reactor. This leads to the continuous formation of $^{\bullet}OH$ without accumulation of Fenton's reagent, thus avoiding wasting reactions [28,29]. The Fenton's reagent production rate (reactions 5 and 6) can be controlled mainly by the application of a suitable current between the anode and cathode of a two-electrode cell or by controlling the potential of a suitable cathode (three-electrode cells) [12,30,31]. Though some works are carried out in a divided cell [32–35], the majority of published reports were performed in nondivided and open electrochemical reactors due to their simplicity of design and practical advantages.

Therefore, the EF process presents several major advantages relative to the chemical Fenton system: (i) on site production of H_2O_2 avoiding the risks related to its transport, storage, and handling; (ii) the possibility of controlling the degradative oxidation kinetics and performing mechanistic studies; (iii) a higher removal rate of organic pollutants due to the catalysis of Fenton's reaction by continuous regeneration of Fe^{2+} at low and constant concentration (avoiding wasting reaction 4); (iv) preventing the use of chemical reagents and sludge formation; and (v) the potential to reach overall mineralization at a relatively low cost by optimization of the operating parameters [12,14,18,36].

The in situ generation of H_2O_2 is primordial in the EF process. Therefore, this process uses suitable cathode materials. The first cathode material employed for the in situ formation of H_2O_2 was a mercury pool [37,38]. Due to its toxicity, this cathode was later replaced by different carbonaceous materials such as the carbon-PTFE-O_2 gas diffusion cathode [24], reticulated vitreous carbon [39], and three-dimensional cathodes such as carbon felt [30] and carbon sponge [40]. Later, other carbon materials such as activated carbon fiber [41], graphite [42,43], and carbon nanotubes [44] were used. Carbon is the cathode material of interest because it constitutes a nontoxic, well-conductive, and stable material exhibiting high overpotential for H_2 evolution and low catalytic activity for H_2O_2 reduction. Recently, other materials such as a boron doped diamond (BDD) [45] and Ni foam [46] were also tested as a cathode material.

Three-dimensional carbon cathodes, particularly carbon felt and carbon sponge, provide high efficiency in wastewater treatment by the EF process because they present a high surface-to-volume ratio, a large specific area, and large mass transfer coefficients of dissolved O_2, which increases the rates of generation of H_2O_2 (reaction 5) and consequently the formation of $^{\bullet}OH$ through Fenton's reaction (2). Moreover, these cathodes have a very low cost and allow easy handling [47–49]. The most employed carbon felt is RVG 2000 (from Carbon-Lorraine or Mersen, France) with the following characteristics: 99.9% carbon content, 0.088 g cm^{-3} density, 0.7 $m^2\ g^{-1}$ specific surface area, and 0.5 cm depth.

The nature of anode material (M) is also of great importance in the EF process because heterogeneous hydroxyl radicals $M(^{\bullet}OH)$ can also be generated at the anode surface from water oxidation, according to reaction (7):

$$M + H_2O \rightarrow M(^{\bullet}OH) + H^+ + e^- \quad (7)$$

The amount of $M(^{\bullet}OH)$ and its contribution to electro-oxidation of organic matter depend strongly on the type of anode material [20,50]. In fact, $M(^{\bullet}OH)$ is produced at the range of O_2 evolution overvoltage. Active anodes such as Pt and DSA have low overvoltage of water oxidation. The amount of $M(^{\bullet}OH)$ generated is relatively low compared to nonactive anodes such as BDD or PbO_2 that have large O_2 evolution overvoltage. Besides $M(^{\bullet}OH)$ are chemisorbed in the former case (and less available) while they are physisorbed in the latter case and thus more easily available for oxidation of organic micropollutants [51–53]. In this context, BDD constitutes an excellent anode material providing high removal efficiency. The use of BDD in the EF process allows high mineralization efficiency. In the latter case, the EF process takes advantage of the oxidation reactions arising from the simultaneous production of homogeneous ($^{\bullet}OH$) and heterogeneous (BDD($^{\bullet}OH$)) hydroxyl radicals on the anode surface and bulk solution, respectively [54,55]. This anode is particularly efficient for mineralization of carboxylic acids, which are more recalcitrant to classical anodes such as Pt. Oturan et al. [55] studied the effect of different anodes on the oxidation efficiency and mineralization power of the process through the treatment of synthetic solutions of the antibiotic tetracycline (Fig. 2). It was shown that the BDD anode exhibits significantly better mineralization efficiency in electro-oxidation of the drug than Pt or DSA (Ti/RuO_2-IrO_2) anodes. The mineralization power became extraordinary when used as an anode in the EF process: the solution TOC 98% was reduced by 98% with the BDD anode versus 74% TOC reduction with the Pt anode after 4 h treatment under the same operating conditions.

However, the material acquisition cost and relatively higher energy consumption motivated the researchers to search for alternative materials presenting lower operational costs but also higher efficiency. Therefore, ceramic electrodes based on substoichiometric titanium oxides with

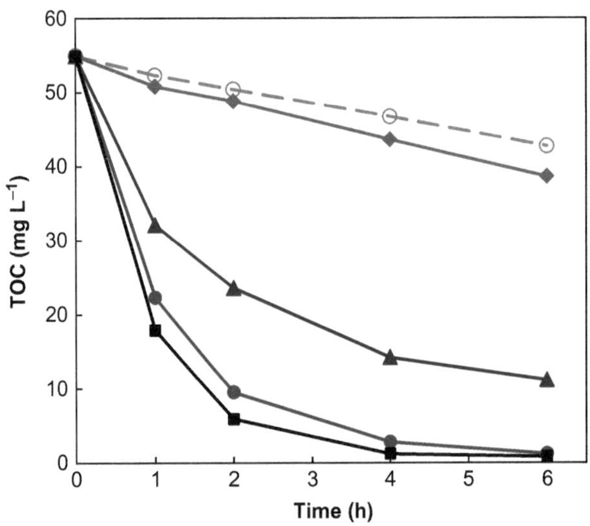

FIG. 2 Comparative mineralization efficiency (in terms of TOC removal) of three different anodes during mineralization of 100 mg L^{-1} tetracycline in 0.05 M Na_2SO_4 aqueous solution by electro-oxidation (without catalyst) with DSA/stainless steel (○), Pt/carbon-felt (◆), and BDD/carbon-felt (●) cells and by the EF process (with 0.2 mM Fe^{2+}) using Pt/carbon-felt (▲) and BDD/carbon-felt (■) cells at 500 mA constant current and room temperature. *Reproduced with permission from N. Oturan, E. Brillas, M.A. Oturan, Unprecedented total mineralization of atrazine and cyanuric acid by anodic oxidation and electro-Fenton with a boron-doped diamond anode, Environ. Chem. Lett. 10 (2012) 165–170. Copyright 2012 Elsevier.*

general formula Ti_nO_{2n-1} were recently developed [56,57] as low cost alternative anodes because they are produced from TiO_2, one of most abundant feedstocks available on the planet [58]. Among these oxides, the composition belonging to the Magnéli phase (with $3 < n < 5$) and, in particular, Ti_4O_7 and Ti_5O_9 exhibit high electrical conductivity, good corrosion resistance, and chemical stability even in extremely corrosive media [58]. Recent studies have also shown that Ti_4O_7 ceramic electrodes behave as a nonactive anode for hydroxyl radical formation from water oxidation [59]. Recently, the Ti_4O_7 anode has been tested in electro-oxidation kinetics and the mineralization effectiveness of the antibiotic amoxicillin, showing performances approaching those of the BDD anode [60]. More recently, this anode was utilized in the EF process for oxidation/mineralization of amoxicillin aqueous solutions, providing quite identical results compared to the BDD anode [61]. Thus, this highlights its great potential for replacing the BDD anode in EF for low-cost processing.

EFFECT OF OPERATING PARAMETERS

Apart from electrode materials, the oxidation/mineralization rate of organics in water/wastewater and the process efficiency in the EF process depends on operational parameters such as the solution pH, the supporting electrolyte composition, the stirring (or liquid flow) rate, air (or O_2) feeding, temperature, pollutant concentration, the catalyst nature, and concentration and applied current (or potential).

Some of these parameters do not necessitate an optimization: The value of solution pH (which is one of the most influent parameters) is now well established: it is optimal at pH 2.8 [62] but the process can be performed properly at the range $2.5 < pH < 3.5$ [31,63]. Higher pH values provoke catalyst loss by the precipitation of ferric iron while low pH values lead to the formation of peroxonium ion ($H_3O_2^+$) from protonation of H_2O_2. This species is electrophilic and is less reactive toward Fe^{2+} [64].

Experiments have been generally carried out in the presence of 0.05 M Na_2SO_4 as a supporting electrolyte. Perchlorate and nitrate salts have also been tested [63]. The use of chloride salt leads to the formation of active chlorine species and can affect the oxidation/mineralization efficiency of the process. This topic is extensively covered in the chapter on "mediated oxidation." The activation energy of Fenton's reaction being low, mostly ambient temperatures are employed. An increase to 35–40° C was found to slightly increase the mineralization efficiency [31] while higher temperatures are detrimental because of the decrease in O_2 solubility and decomposition of H_2O_2 [65]. The stirring rate depends on the cell design and volume; it should be set to ensure mass transport to the electrodes. The air (or O_2) feeding flow rate should be set in order to saturate the solution in dissolved O_2 (carbon-felt cathode cells) or to feed suitable gas diffusion electrode (GDE) cells.

Apart from pH, the most important operating parameters of the EF process are the applied potential (three-electrode cells) or current (two-electrode cells) and catalyst nature and concentration. In the first case, the potential of the cathode is set to a value on the plateau corresponding to a two-electron reduction wave of O_2 to form H_2O_2 in an optimal way. This value is dependant on the nature of the cathode used, that is, to H_2 evolution overpotential and H_2O_2 reduction potential. The optimized potential values of −0.5 V/SCE for carbon-felt/graphite-felt cathodes [29,66], −1.6 V/SCE for reticulated vitreous carbon (RVC) [67], and −0.9 V/SCE for PTFE [42] were reported in the literature.

On the other hand, the majority of experiments have been performed under galvanostatic conditions, owing to an easy optimization of current density or intensity to generate $^{\bullet}OH$ by minimizing side reactions. This parameter is of great importance for EAOPs and particularly for the EF process because it governs the formation rate of H_2O_2 (reaction 5) and the regeneration rate of the catalyst (ferrous iron) (reaction 6) and consequently the generation rate of $^{\bullet}OH$ via the Fenton reaction (2) as well as the formation rate of $M(^{\bullet}OH)$ on the anode surface (reaction 7). In general, the degradation rate or mineralization efficiency increases with current up to an optimal value for which parasitic side reactions such as four-electron reduction of O_2 can take place (reaction 8) in detriment to H_2O_2 formation (reaction 5), H_2 evolution (reaction 9) at the cathode, and oxidative/reductive decay of H_2O_2 at the anode (reaction 10) and at the cathode (reaction 11) begin to harm the oxidation process [49,68]. High currents beyond the optimal value can also lead to enhancement of the rate of $^{\bullet}OH/M(^{\bullet}OH)$ wasting reactions (12) and (13) with high energy consumption, thus lowering the specific current efficiency or mineralization current efficiency (MCE) [18].

$$2H_2O \rightarrow O_2(g) + 4H^+ + 4e^- \quad (8)$$

$$2H^+ + 2e^- \rightarrow H_2(g) \quad (9)$$

$$H_2O_2 + 2H^+ + 2e^- \rightarrow 2H_2O \quad (10)$$

$$H_2O_2 \rightarrow O_2(g) + 2H^+ + 2e^- \quad (11)$$

$$2M(^{\bullet}OH) \rightarrow 2M + O_2(g) + 2H^+ + 2e^- \quad (12)$$

$$H_2O_2 + {}^{\bullet}OH \rightarrow HO_2^{\bullet} + H_2O \quad (13)$$

A typical example of this behavior is given in Fig. 3 [50]. Fig. 3A shows the effect of current density on the concentration decay of 0.2 mM

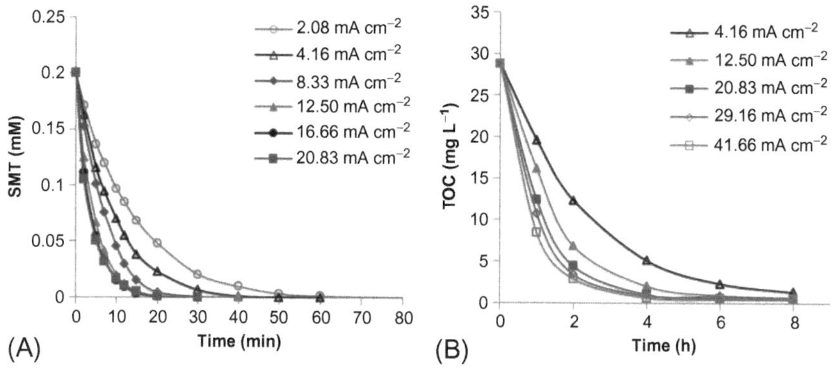

FIG. 3 Effect of current applied current density on (A) the decay kinetics and (B) TOC removal efficiency during EF treatment of 0.2 mM (57.5 ppm) of 300 mL SMT solution in 0.05 M Na_2SO_4 solution at pH3 and room temperature using the BDD anode/graphite-felt cathode. *Reproduced with permission from F. Sopaj, N. Oturan, J. Pinson, F. Podvorica, M.A. Oturan, Effect of the anode materials on the efficiency of the electro-Fenton process for the mineralization of the antibiotic sulfamethazine, Appl. Catal. B: Environ. 199 (2016) 331–341. Copyright 2016 Elsevier.*

(57.5 ppm) antibiotic sulfamethazine (SMT), determined by HPLC using 0.2 mM ferrous iron as a catalyst and a BDD cathode. The decay kinetic of the SMT concentration is enhanced by increasing the current from 2.08 to 12.5 mA cm^{-2} for which SMT completely disappeared at 20 min. Further increases in current density do not lead to any significant improvement, highlighting that the system has attained its maximum oxidation power at 12.5 mA cm^{-2} current. The effect of applied current on the mineralization efficiency (in terms of TOC removal rate) of this same solution (corresponding to 28.8 mg L^{-1} initial carbon content) during EF treatment is depicted in Fig. 3B. Sae trend is observed; mineralization power attained its optimal value for 20.83 mA cm^{-2} current density for which an outstanding mineralization rate of >97% was already reached after 4 h electrolysis. This trend linked to the kinetic limitation is generally observed during oxidative degradation and/or mineralization of POPs by the EF process.

The third important operating parameter in the EF process is the nature and concentration of the catalyst used. Production of OH through an electrochemically supported Fenton's reaction needs the use of a soluble redox couple M^n/Mn^{n-1}. Among different redox couple investigated (Co^{3+}/Co^{2+}, Cu^{2+}/Cu^+, Ag^{2+}/Ag^+, Mn^{3+}/Mn^{2+}...) to catalyze Fenton's reaction, the couple Fe^{3+}/Fe^{2+} has been found to be the best catalyst because of its nontoxic character and efficiency at low concentrations [47,69–71]. The catalyst concentration for an optimal efficiency depends on the cathode used. In the case of a carbon-felt cathode, the amount of Fe^{2+} (or Fe^{3+}) in the range 0.05–0.2 mM was found to be optimal depending on the structure of the initial pollutant whereas higher concentrations of 0.5–1.0 mM were reported for carbon-PTFE GDE cathodes [12,31,40] due to the faster electrogeneration of Fe^{2+} in the former case. High catalyst concentrations harm process efficiency due to the improvement of the rate of the $^{\bullet}OH$ wasting reaction (4). Fe^{2+} concentration should be lower than the value making the reaction (4) competitive with the oxidation of organic pollutants or their intermediates (particularly for oxidation of carboxylic acids to reach high mineralization rates). Fig. 4 shows that the optimal value for catalyst concentration was found as 0.1 and 0.2 mM for oxidation of antimicrobial ciprofloxacin [72] and insecticide imidaclopride [73], respectively. In both cases, lower iron concentrations are not enough to efficiently catalyze Fenton's reaction (2) whereas higher concentrations than optimal value improve the rate of wasting reaction (4) and decrease oxidation ability of the process. It is worth noting that the oxidation kinetic is very fast in both cases and ciprofloxacin is oxidized efficiently even at a low applied current of 100 mA.

In short, the influent parameters of these process can be optimized to achieve the best oxidation efficiency, MCE, and lowest energy cost. In general, a pH near 3.0, optimized current and catalyst concentration, a suitable anode and cathode material and their geometry, a high air (or O_2) flow rate (for high H_2O_2 formation rate), a high stirring rate

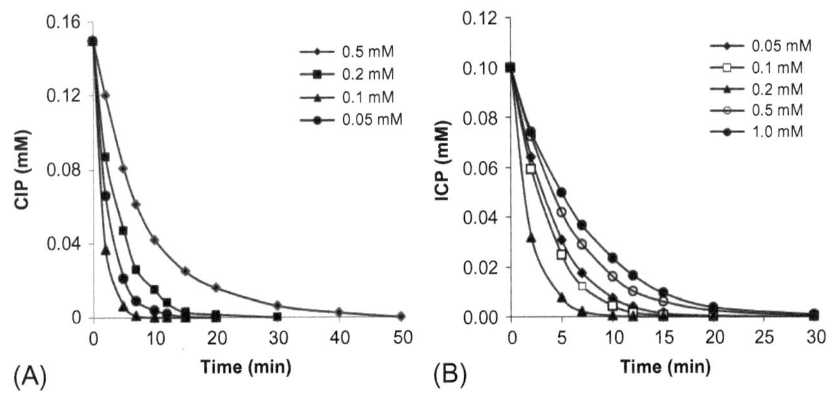

FIG. 4 Effect of catalyst concentration on concentration decay of (A) ciprofloxacin (antimicrobial) and (B) imidaclopride (insecticide) in 0.05 M Na_2SO_4 solution during electro-Fenton treatment using the Pt anode/graphite-felt (70 cm^2) cathode at pH 3 and room temperature. Other experimental conditions: (A) $[CIP]_0 = 0.15$ mM, catalyst: Fe^{2+}, $I = 400$ mA, $V = 200$ mL and (B) $[ICP]_0 = 0.1$ mM, catalyst: Fe^{3+}, $I = 100$ mA, $V = 200$ mL. *Reproduced with permission from M.S. Yahya, N. Oturan, K. El Kacemi, M. El Karbane, C.T. Aravindakumar, M.A. Oturan, Oxidative degradation study on antimicrobial agent ciprofloxacin by electro-Fenton process: kinetics and oxidation products, Chemosphere 117 (2014) 447–454 (copyright 2014 Elsevier) and M. Turabik, N. Oturan, B. Gözmen, M.A. Oturan, Efficient removal of insecticide imidaclopride from water by electrochemical advanced oxidation processes, Environ. Sci. Pollut. Res. Int. 21 (2014) 8387–8397 (copyright 2014 Springer Science + Business Media).*

(tank reactors), or liquid flow rate (flow reactors) (to obtain fast homogenization), and temperatures up to 35°C are required for optimal process efficiency. For instance, there are very limited studies for modeling the EF process due to the complexity of the mineralization procedure involving several oxidation steps. However, an interesting attempt for kinetic modeling of phenol was made by Mousset et al. [74], taking into account formation of different intermediates and their successive oxidation steps until full mineralization.

APPLICATION OF EF PROCESS TO REMOVAL OF ORGANIC MICROPOLLUTANTS

Because of its first applications to the treatment of water polluted by chlorophenoxy acid herbicides [28,29,32,75], the EF process has been successfully applied to the treatment of water polluted by toxic/persistent organic pollutants such as industrial pollutants, synthetic dyes, PPCPs, etc. These pioneering works have shown that the EF process constitutes an efficient way to quickly destroy POPs and mineralize their solution, as demonstrated in the case of the herbicide 2,4-D (2,4-dichlorophenoxyacetic acid) with a mineralization rate higher than 95% in terms of TOC removal and identification of oxidation intermediates formed during the treatment. It was highlighted that the attack of $^{\bullet}OH$ on the aromatic ring of 2,4-D leads to the formation of its hydroxylated derivatives, along with other processes involving dechlorination and dehydrogenation. These primary intermediates have then oxidized to polyhydroxylated products and quinones before undergoing an oxidative ring opening reaction to generate short-chain carboxylic acids. This behavior was also observed during the treatment of polychlorinated pesticides such as pentachlorophenol in which the ipso-attack of $^{\bullet}OH$ on chlorine positions leads to the formation of hydroxylated intermediates, which further oxidized to tetrachloro-*o*-benzoquinone and tetrachloro-p-benzoquinone [76]. These intermediates, and particularly the quinones, being toxic, the complete destruction of the aromatic/cyclic derivatives is of great importance for the detoxification of treated solutions through almost mineralization. Afterward, the high oxidation power of the process has been demonstrated for different families of pesticides such as herbicides of organophosphorus [63], imidazolines [77], phenylurea [78], phosphonates [71], triazines [79,80], and triketones [81] families as well as insecticides [73], fungicides [82], and a mixture of pesticides [83]. In all cases, an almost complete mineralization of treated solutions was reached.

The EF process has been also successfully applied by several research groups to remove organic pollutants from different types of wastewater and/or effluents such as industrial pollutants [84,85], olive oil mill wastewater [86], landfill leachates [87,88], and reverse osmosis concentrates [89]. On the other hand, a very large number of works has been focused on the effective removal of dyes with full color removal and almost complete mineralization rates [41,48,90–95]. These studies concerned the decolorization of dyes, the determination of decay kinetics and rate constants of the reaction with $^{\bullet}OH$, mineralization yields, and oxidative degradation pathways.

Recently, a number of studies have been focused on the removal of emerging pollutants, mainly PPCPs, due to their occurrence in natural water streams and their potential toxic effects on aquatic species [3,13]. These compounds have been detected, even at low concentrations, in surface water, ground water, and top water [4]. The groups of Brillas and Oturan have carried out detailed studies for the destruction of PPCPs in water media, starting with a pioneering work on the mineralization of paracetamol [96], followed by several other works; some examples being antimicrobials [72,97], antibiotics [98,99], nonsteroidal antiinflammatory drugs [100,101], beta-blockers [102,103], diuretics, antihistaminic [104], and so on. These studies focused mainly on oxidative degradation and mineralization efficiency, including the determination of oxidation kinetic, mechanistic assessments, and degradation pathways. A typical example for an oxidative degradation mechanism is given in Fig. 5, which highlights different paths involved in the mineralization of the aqueous solution of the antibiotic sulfachloropyridazine during its EF processing [105]. Most of these studies have also included an acute toxicity assessment using the Microtox bioluminescence method based on the inhibition of marine bacteria *Vibrio fischeri* showing the effective removal of toxicity when the treatment was prolonged toward almost mineralization rates [23,105].

Due to the constant production rate of $^{\bullet}OH/M(^{\bullet}OH)$ at given experimental conditions, the oxidative degradation of organic pollutants follows pseudo first-order reaction kinetics. This allows determining the apparent rate constants (k_{app}) of oxidation reactions using kinetic analysis based on the plot $Ln(C_0/C) = f(t)$ [14]. The k_{app} values thus obtained can be then utilized for the determination of absolute (second-order) rate constants by applying the competition kinetic method. The method is based on the use of a standard competitor) with a well-known second-order rate constant, such as *p*-hydroxybezoic acid ($k_{2(S)} = 2.19 \times 109\ M^{-1}\ s^{-1}$ [106]. The EF process is then performed with a solution containing equal concentrations of both organic pollutants (P) and standard competitors (S). The concentration decay was simultaneously measured by HPLC analysis to determine the pseudo first-order rate constants $k_{app(P)}$ and $k_{app(S)}$. The value of the absolute rate constant $k_{2(P)}$ is then calculated from Eq. (14):

$$k_{2(P)} = \frac{k_{app\ (P)}}{k_{app(S)}} \times k_{2(S)} \tag{14}$$

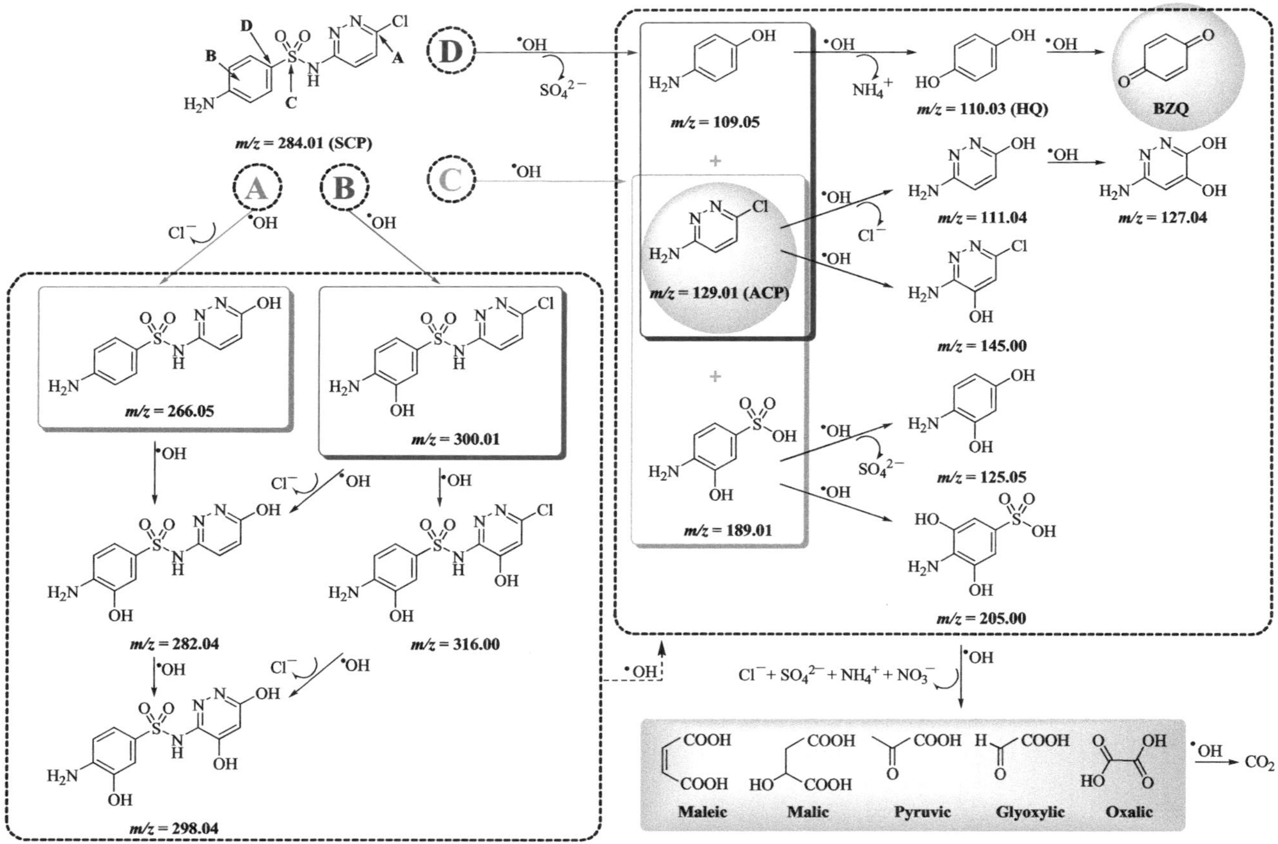

FIG. 5 Reaction pathways for the mineralization of sulfachloropyridazine by $^{\bullet}OH$ generated in the EF process at 300 mA constant current using Pt/carbon-felt cell. *Reproduced with permission from A. Dirany, I. Sirés, N. Oturan, A. Özcan, M.A. Oturan, Electrochemical treatment of Sulfachloropyridazine: kinetics, reaction pathways, and toxicity evolution, Environ. Sci. Technol. 46 (2012) 4074–4082. Copyright 2012 ACS publications.*

Very high $k_{2(P)}$ values (10^9–10^{10} M^{-1} s^{-1}) were obtained by this method for oxidation by $^{\bullet}OH$ of organic pollutants from different families. These large $k_{(2,P)}$ values are expected for hydroxylation of aromatic compounds due to their high reactivity for $^{\bullet}OH$ [107].

The determination of $k_{2(P)}$ values can give interesting information about the effect of substituents on the reactivity of aromatics toward $^{\bullet}OH$. For instance, a study of Oturan et al. [108] highlighted that the affinity of chlorophenols for $^{\bullet}OH$ depended on the number of aromatic chlorine substituents; an increasing number of chlorine substituents decreased the rate constant values (Table 1).

It was also shown more recently that the EF process can be applied to soil remediation through the treatment of soil washing (SW)/soil flushing (SF) solutions to destroy the organic pollutants from polluted soil. The main criteria affecting the combined process of EF/SW have been identified as: (i) applied current intensity (or density) and electrode material, (ii) nature and dose of extracting agent leading to a less or more complex matrix, and (iii) the environmental impact (acute ecotoxicity and biodegradability of the SW solution and the impact on soil microbial activity). The influence of these parameters on the SW/EF and SF/EF integrated processes has been recently reviewed. These studies have been mainly focused on hydrophobic organic contaminants (HOCs) such as petroleum hydrocarbons [109], polycyclic aromatic hydrocarbons including phenanthrene (PHE) [110,111], or pesticides [110].

As an example, interesting work has been done by Mousset et al. [111] on the recycling ability of the extracting agent in the application of the EF process to SW solutions. This parameter is of importance for the treatment of SW/SF solutions aimed at reducing both the operating cost of reagents for the washing/flushing operation and energy requirements during the

TABLE 1 Absolute Rate Constants (k_2) for the Oxidation of Chlorophenols (CPs) by $^{\bullet}OH$ Determined Under EF Conditions By Using the Competition Kinetic Method With Benzoic Acid (BA) as a Standard Competitor

Chlorophenol	k_2 (10^9) M^{-1} s^{-1}
4-Chlorophenol	7.75 ± 0.07
2,6-Dichlorophenol	6.13 ± 0.05
2,4,5-Trichlorophenol	5.72 ± 0.05
2,3,5,6-Tetrachlorophenol	4.95 ± 0.07
Pentachlorophenol	3.56 ± 0.06

Operation conditions: $[CP]_0 = [BA]_0 = 0.124$ mM, $[Fe^{3+}] = 0.2$ mM, pH = 3, $I = 50$ mA.

Reproduced from N. Oturan, M. Panizza, M.A. Oturan, Cold incineration of chlorophenols in aqueous solution by electro-Fenton process. Effect of number and position of chlorine atoms on the degradation kinetics, J. Phys. Chem. A 113 (2009) 10988–10993, with permission. Copyright 2009, ACS publications.

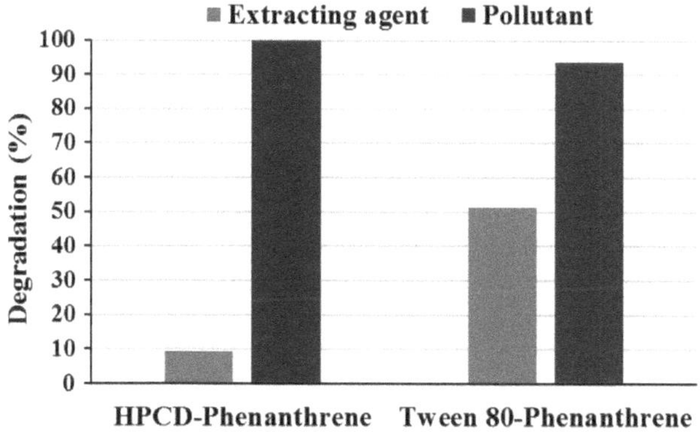

FIG. 6 Influence of nature of extracting agent (HPCD (10 g L^{-1}) or Tween 80 (0.75 g L^{-1})) on PHE (17 mg L^{-1}) degradation. Operating conditions: current density: 13.3 mA cm^{-2}, catalyst concentration: $[Fe^{2+}]=0.05$ mM, anode: Pt, cathode: carbon felt. *Reproduced with permission from E. Mousset, N. Oturan, E.D. van Hullebusch, G. Guibaud, G. Esposito, M.A. Oturan, Influence of solubilizing agents (cyclodextrin or surfactant) on phenanthrene degradation by electro-Fenton process—study of soil washing recycling possibilities and environmental impact, Water Res. 48 (2014) 306–316. Copyright 2014 Elsevier.*

EF treatment of the SW solution. To clarify this parameter, synthetic solutions of Tween 80 (0.75 g L^{-1}) and cyclodextrin HPCD (10 g L^{-1}) containing PHE at the same initial concentration (17 mg L^{-1}) have been treated by EF in a cell equipped with a carbon-felt cathode and a Pt anode (Fig. 6). In the case of Tween 80, 95% of PHE and 50% of the surfactant were removed after 4 h of treatment. Conversely, in the case of HPCD, PHE was completely removed after 4 h while HPCD was barely degraded at a 10% yield. The higher degradation rate of PHE in the presence of HPCD could be explained by the formation of a ternary complex promoting its reaction with $^{\bullet}OH$. On the other hand, the amount of HPCD required to solubilize the same amount of PHE is 13 times higher compared to Tween 80. Therefore, considering the 90% removal of HPE, the amount of HPCD destroyed is more important than that of Tween 80. Therefore, Tween 80 has better recycling abilities compared to HPCD because of the less solubilization power (concentration of washing agent to solubilize pollutant PHE) of the latter one.

RECENT DEVELOPMENTS IN EF PROCESS

Recently, there have been new improvements aiming to rectify some inconveniences of the process that delay its implementation at an industrial level for large-scale applications: use of heterogeneous catalysts (EF-Pyrite) and coupling with microbiology (bio-EF) or solar energy (solar photo-EF, which is the subject of another chapter) in addition to the developments gained in electrode materials (discussed above).

HETEROGENEOUS EF

The efficiency of various heterogeneous catalysts such as iron oxides, iron supported on zeolite, alginate beads, etc., was tested by different researchers in the classical Fenton process and the EF process [112,113]. A new and promising alternative has been recently explored for catalyzing the EF process by the utilization of natural pyrite (FeS_2) mineral as a catalyst. The use of FeS_2 as a heterogeneous catalyst constitutes a natural source of Fe^{2+} and H^+ ions, according to the reactions (15) and (16) [114], bringing several advantages to the EF process and overcoming its drawbacks such as self-regulation of the solution pH around of 3 (optimal pH value for EF process) and self-regulation of the Fe^{2+} concentration in the solution to be treated. Indeed, FeS_2 is oxidized by dissolved oxygen in aqueous solutions (reaction 8), providing Fe^{2+} and H^+ ions. On the other hand, FeS_2 also reacts with H_2O_2 generated at the cathode (reaction 5), producing ferric iron through reaction (16), which is then reduced to Fe^{2+} by reaction (17). Thus, both the continuous generation of H_2O_2 and Fenton's reaction result in a self-regulated process for Fe^{2+} production [115].

$$2FeS_2(s) + 7O_2(g) + 2H_2O \rightarrow 2Fe^{2+} + 4SO_4^{2-} + 4H^+ \tag{15}$$

$$2FeS_2(s) + 15H_2O_2 \rightarrow 2Fe^{3+} + 14H_2O + 4SO_4^{2-} + 2H^+ \tag{16}$$

$$2FeS_2(s) + 14Fe^{3+} + 8H_2O \rightarrow 15Fe^{2+} + 2SO_4^{2-} + 16H^+ \tag{17}$$

The generation of H^+ and Fe^{2+} from reactions (15)–(17) is noteworthy because it promotes an acidic medium and a suitable catalyst concentration to the EF process, ensuring a favorable H_2O_2/Fe^{2+} molar ratio for the Fenton reaction (2) to generate $^{\bullet}OH$ and avoid wasting reactions.

The use of pyrite as a heterogeneous catalyst for the EF process allowed the description of a new FF process, the so-called EF-Pyrite process [116]. It was found that pyrite can release and regulate the appropriate concentration of Fe^{2+} ions and pH required by the Fenton reaction (2). Moreover, the continuous electrochemical generation of H_2O_2 accelerates reaction (16) and regeneration of the Fe^{2+} (reaction 6), contributing to the Fe^{2+}/Fe^{3+} regulation cycle, enhancing the efficiency of the process. The main operating parameters influencing process efficiency such as applied current pyrite dosage, anode materials, and initial pollutant concentration have been investigated [115] for the EF-pyrite process. For example, Fig. 7 shows the evolution of iron ions and solution pH. Fe^{2+} ions released to the solution by pyrite are quickly oxidized to Fe^{3+} through the Fenton reaction while the reduction of ferric iron through reactions (6) and (17)

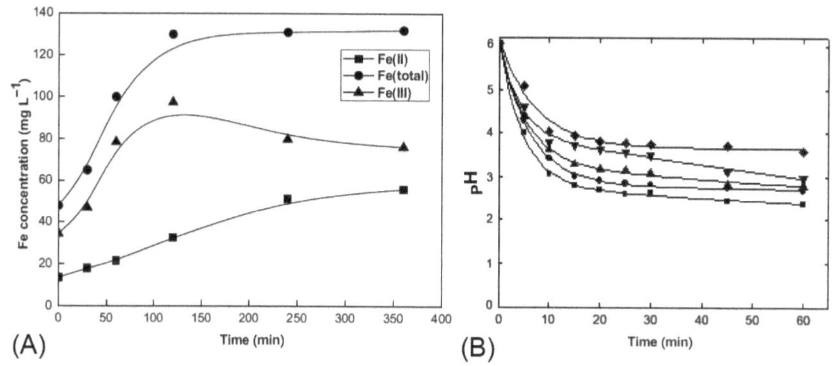

FIG. 7 (A) Concentration profile of aqueous ferrous, ferric, and total iron during the degradation of AHPS textile dye solution containing $2\,g\,L^{-1}$ pyrite by EF-Pyrite process ($[AHPS]_0 = 175\,mg\,L^{-1}$, $I = 300$ mA, $[Na_2SO_4] = 50$ mM, and (B) variation of solution pH with amount of pyrite powder in suspension by bubbling compressed air at $1\,L\,min^{-1}$: (♦) $0.5\,g\,L^{-1}$, (▼) $2.0\,g\,L^{-1}$, (▲) $0.4.0\,g\,L^{-1}$, (●) $5.0\,g\,L^{-1}$, and (■) $6.0\,g\,L^{-1}$. *Modified from S. Ammar, M.A. Oturan, L. Labiadh, A. Guersalli, R. Abdelhedi, N. Oturan, E. Brillas, Degradation of tyrosol by a novel electro-Fenton process using pyrite as heterogeneous source of iron catalyst, Water Res. 74 (2015) 77–87 (copyright 2015) and L. Labiadh, M.A. Oturan, M. Panizza, N. Ben Hamadi, S. Ammar, Complete removal of AHPS synthetic dye from water using new electro-fenton oxidation catalyzed by natural pyrite as heterogeneous catalyst, J. Hazard. Mater. 297 (2015) 34–41 (copyright 2015) with permission from Elsevier.*

occurs at a slower rate (Fig. 7A). It was also demonstrated that a suitable pH value around 3 is quickly reached during first 10–15 min (depending on initial FeS_2 amount) of electrolysis and remained without significant change throughout the experiment (Fig. 7B).

The amount of pyrite added to the solution strongly affects the oxidation kinetics of organic pollutants because it regulates the concentration of free Fe^{2+} for the Fenton reaction (2). Therefore the optimal pyrite dosage was investigated for oxidative degradation of azo dye AHPS during its EF-pyrite treatment (Fig. 8). Degradation kinetics of AHPS are enhanced by raising the pyrite dosage from 0.5 to $2\,g\,L^{-1}$ (Fig. 8A). A further increase in pyrite dosage resulted in a detrimental effect because an excess of Fe^{2+} ions released from the catalyst provokes wasting $^{\bullet}OH$, according to the reaction (4), decreasing oxidation efficiency. This dosage of $2\,g\,L^{-1}$ was then utilized to mineralize the antibiotic sulfamethazine aqueous solution (Fig. 8B). This figure shows well the effect of the anode material in the EF-pyrite process, highlighting the higher mineralization performance of the process with the BDD (95% TOC removal) anode than the Pt (86%) anode and exemplifying the superiority of EF-pyrite over the classical EF process, allowing better %MCE values [117].

Table 2 gives an overview of the studies reported on the performances in mineralization efficiency of the EF-pyrite process and a comparison with the classical EF process using soluble iron (II) salts.

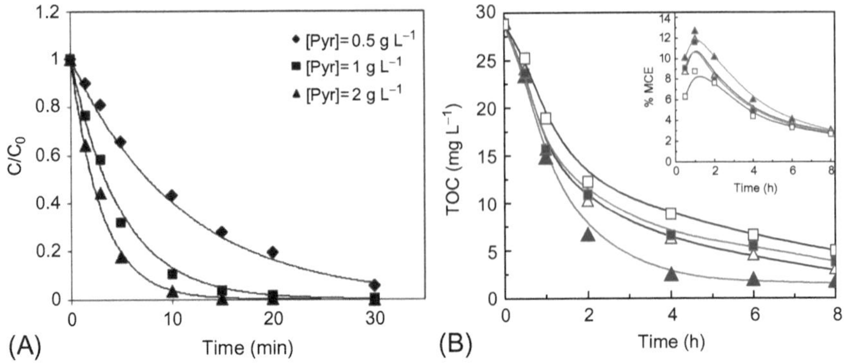

FIG. 8 (A) Effect of pyrite dosage on the oxidative degradation rate 30 mL of 175 mg L^{-1} AHPS solution in EF-pyrite process and (B) TOC removal kinetics for the mineralization of 200 mL of 0.2 mM sulfamethazine solution using an (▲, △) BDD/carbon-felt and (■, □) Pt/carbon-felt cells. The inset panel presents the corresponding MCE curves for (▲, ■) EF-Pyrite with 2.0 g L^{-1} pyrite and (△, □) classical EF with 0.2 mM Fe^{2+}. Experimental conditions: pH = 3, I = 300 mA, $[Na_2SO_4]$ = 50 mM. *Modified from (A) L. Labiadh, M.A. Oturan, M. Panizza, N. Ben Hamadi, S. Ammar, Complete removal of AHPS synthetic dye from water using new electro-fenton oxidation catalyzed by natural pyrite as heterogeneous catalyst, J. Hazard. Mater. 297 (2015) 34–41 and (B) N. Barhoumi, N. Oturan, H. Olvera-Vargas, E. Brillas, A. Gadri, S. Ammar, M.A. Oturan, Pyrite as a sustainable catalyst in electro-Fenton process for improving oxidation of sulfamethazine. Kinetics, mechanism and toxicity assessment, Water Res. 94 (2016) 52–61 with permission from Elsevier.*

TABLE 2 TOC Removal Efficiencies Obtained During the EF-Pyrite Treatment of Different Contaminants and Comparison With Classic (Homogeneous) EF Process

Contaminant	Cell configuration	% TOC removal (EF-pyrite)	%TOC removal (Classic EF)
Levofloxacin (Barhoumi et al. [99])	BDD-carbon felt undivided cell (300 mA) 1 g L^{-1} of pyrite	95 in 8 h treatment	–
Tyrosol (Ammar et al. [116])	BDD-carbon felt undivided cell (300 mA) 1 g L^{-1} of pyrite	89 in 6 h treatment	88 in 6 h treatment
Synthetic dye (AHPS) (Labiadh et al. [115])	BDD-carbon felt undivided cell (450 mA) 2 g L^{-1} of pyrite	>90 in 5 h treatment	70 in 5 h treatment
Sulfamethazine (Barhoumi et al. [117])	BDD-carbon felt undivided cell (300 mA) 2 g L^{-1} of pyrite	95 in 8 h treatment	90 in 8 h treatment

Overall, it was demonstrated that the use of pyrite as a heterogeneous catalyst can constitute a potential alternative for performing the EF process at natural pH with several advantages:

- Self-regulation of the catalyst (Fe^{2+}) without the addition of soluble iron salt with the possibility of recycling solid FeS_2.
- Self-regulation of the solution pH to around 3 (optimal value for the EF process), avoiding an external addition of acids.
- Slightly superior performance than classic homogeneous EF, which is due to the self-regulation system of the catalyst and solution pH provided by the mineral.

BIOELECTRO-FENTON (BIO-EF)

The treatment of persistent organic pollutants with AOPs (including EAOPs) leads to the formation of oxidation intermediates that are often more toxic than the mother pollutant [23,105]. Therefore, a suitable treatment requires an efficient destruction of the aromatic/cyclic intermediates responsible for the toxicity. For this, it is necessary to reach relatively high mineralization rates during treatment. On the other hand, longer treatment times to reach almost complete mineralization degrees lead to the consumption of high amounts of electrical energy that is one of the drawbacks of the EF process for industrial-level applications. A typical example is given in Fig. 9, showing the increase of the energy consumption per unit TOC mass (EC_{TOC} in kWh (g TOC)$^{-1}$) (Fig. 9B) and the MCE (Fig. 9C) with TOC removal degree (Fig. 9A) during the EF treatment of the pharmaceuticals ranitidine and furosemide [104]. As can be seen in this figure, the TOC removal is fast in the beginning of electrolysis but the mineralization rate becomes slower over longer treatment times due to the formation of hardly oxidizable end-products such as short-chain carboxylic acids, which are the common ultimate end-products of AOPs. On the other hand, carboxylic acids are not toxic and are biodegradable. In this context a right treatment strategy consists in electrolysis by the EF process up to reaching a suitable biodegradability level and then giving way to biological treatment. Indeed, it was demonstrated that AOPs enhance the biodegradability of initially recalcitrant effluents. For instance the biodegradability of landfill leachate, measured by BDO_5/DCO ratio, was increased from 0.2 to 0.5 [118], from 0.13 to 0.40 [119], and from 0.1 to 0.3 by the Fenton and photo-Fenton processes, respectively. The enhancement of biodegradability was also demonstrated by EAOPs [120,121]. More recently, our group showed that the EF process can be used as a pre- or posttreatment stage prior to or after biological treatment [104,122–124].

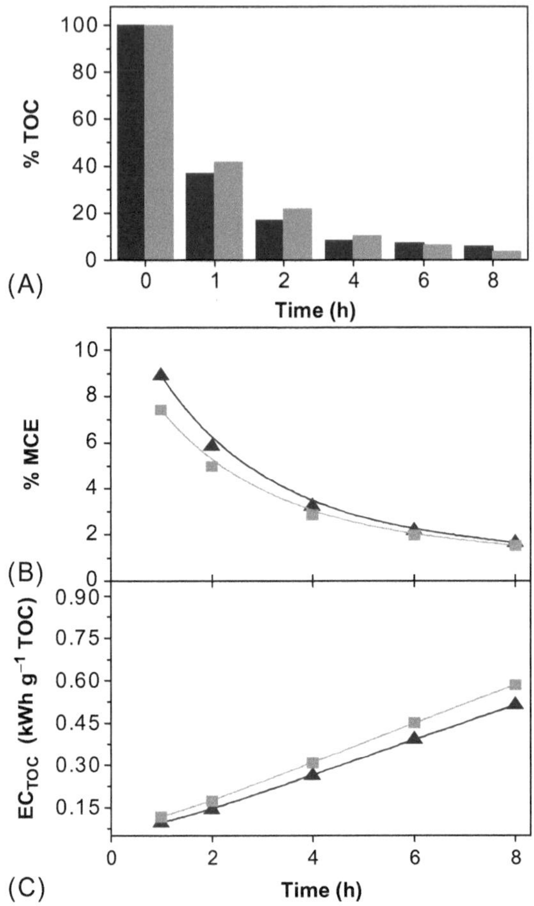

FIG. 9 Evolution of TOC removal (A), MCE (B), and EC_{TOC} (C) during mineralization of 230 mL of 0.1 mM FRSM (■) and RNTD (■) aqueous solutions (both in 0.05 M Na_2SO_4 medium, at pH 3.0, room temperature and 500 mA of current) using a BDD/carbon-felt cell with $[Fe^{2+}]=0.1$ mM. The equations utilized to determine MCE and EC_{TOC} (in (g TOC)$^{-1}$) are given in Ref. [12]. *Reproduced from H. Olvera-Vargas, N. Oturan, D. Buisson, M. A. Oturan, A coupled Bio-EF process for mineralization of the pharmaceuticals Furosemide and Ranitidine: feasibility assessment, Chemosphere 155 (2016) 606–613, with permission. Copyright 2016 Elsevier.*

Therefore coupling EF with biological treatment (Bio-EF) can allow the removal of POPs from water in a cost-effective way. A short EF processing of polluted solutions increases the biodegradability with a suitable mineralization rate while almost complete mineralization can be completed by means of cheaper biological processes [125].

An outstanding work using the EF process as a pretreatment stage in a hybrid Bio-EF process was reported by Olvera-Vargas et al. [122] that highlighted the ability of the EF process to increase the biodegradability index of the recalcitrant drug metoprolol (β-blocker) solution from

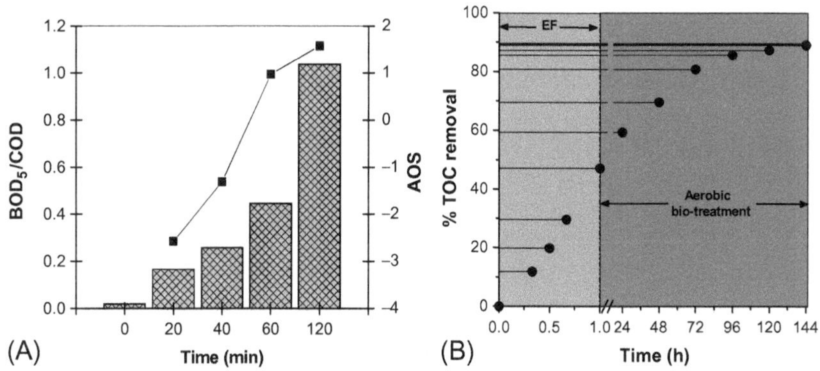

FIG. 10 Treatment of b-blockers metoprolol by Bio-EF process: (A) Evolution of the BOD_5/COD (☒) ratio and the AOS (-■-) during the EF-processing of 220 mL of 0.1 mM of MTPL solution in 0.05 M Na_2SO_4 and 0.1 mM Fe^{2+} at pH3.0 using an EF-BDD cell at 300 mA and room temperature, and (B) Time course of the overall TOC removal 26.74 mg L^{-1} (corresponding to 0.1 mM) solution. Biological phase was conducted at aerobic conditions using a mixture 12 pure cultures of microorganisms. *Reproduced with permission from H. Olvera-Vargas, T. Cocerva, N. Oturan, D. Buisson, M. A. Oturan, Bioelectro-Fenton: a sustainable integrated process for removal of organic pollutants from water: application to mineralization of metoprolol, J Hazard Mater 319 (2016) 13–23. Copyright 2016 Elsevier.*

0.012 to 0.44 after a short electrolysis time of 1 h at 300 mA (Fig. 10). The evolution of the biodegradability (in terms of BOD_5/COD ratio) is presented in Fig. 10A. This figure shows also the rise in the average oxidation state (AOS), another indicator of biocompatibility, which reaches a value of 1.0 after 1 h treatment. This value of AOS shows that the remaining organic carbon in the solution is mainly in oxidized form and available for biodegradation. Fig. 10B indicates that the initial TOC (26.74 mg L^{-1}) is reduced by 46% after 1 h EF pretreatment and then is efficiently reduced by 90% after 4 days incubation of pretreated solution under aerobic conditions. The toxicity tests of the Microtox method using *V. fischeri* bacteria showed the efficient destruction of harmful intermediates during the EF treatment stage [122].

In short, this study showed that the energy required for an almost complete mineralization of 90% with EF alone was reduced by 85% with Bio-EF because the electrolysis time was reduced from 6 to 1 h. This result has been reached by enhancing biodegradability during a short treatment time with the EF process with partial mineralization of the solution by means of transforming recalcitrant organics to biodegradable products. The EF can also be used as a post-treatment stage [124]. In this case, biodegradable organics are degraded in the first stage (biological degradation) with significant reduction of solution TOC. The remaining persistent organic pollutants are then destroyed during a short EF post-treatment stage. This strategy can be interesting for the effluents with a high biodegradability index.

CONCLUDING REMARKS AND PERSPECTIVES

The EF process is currently well established in bench scale with developments during the last decade concerning electrode materials, catalyst nature and concentration, and oxidation and mineralization kinetics as well as other operating parameters such as pH, air sparging rate, temperature, etc. Therefore this technology has now reached a state of development in which its industrial application for the degradation of persistent/toxic effluents is strongly encouraged. Nonetheless, its implementation for integration to WWTPs as an advanced and sustainable step needs further development to make it more sustainable and cost effective.

In this context the use of a solid catalyst offers the opportunity of operating at neutral or near neutral pHs and constitutes a new way to make EF a most sustainable and cost-effective process. For instance, the use of pyrite avoids the external addition of acids (to set pH about 3) and allows recycling the catalyst, thus reducing extra operational costs. This is also important for the implementation of the bio-EF process because biological steps need near-neutral pH values.

On the other hand, the integration of the EF process to multistage treatment systems can be more advantageous compared to individual applications because of synergetic effects and performances capitalized on during the coupling of processes. In this sense, the upgrading and scaling up of the bio-EF process is notably important in order to take advantage of both processes: EF and biodegradation. EF can be used as pre- or posttreatment in a coupling with biological treatment. This action can reduce remarkably the electrical energy consumption because, as shown above, a short electrolysis time at a suitable current can transform recalcitrant organics to biodegradable molecules and consequently lead to an important drop in the overall operating costs. Short treatment times are also beneficial for increasing the lifetime of the electrode materials.

It is also worthy to note that the use of the substoichiometric anodes such as Ti_4O_7 as an alternative to expensive anode materials such as BDD or Pt reduce strongly the investment cost for large-scale implementation. The modification of the surface of the carbon-felt cathode is another way to enhance the H_2O_2 generation rate and consequently the process efficiency.

Finally upgrading the bench level results to pilot level is primordial for future development. Our team is currently working to reach this. Further research is needed on scaling up and optimization for industrial application. The design of an electrochemical reactor for integrated systems with easy control of the operating parameters affecting the performance of the EF unit and that of the whole system is an essential point needing unceasing efforts of the scientific community. We are currently working to reach this perspective.

References

[1] S.K. Khetan, T.J. Collins, Human pharmaceuticals in the aquatic environment: a challenge to green chemistry, Chem. Rev. 107 (2007) 2319–2364.

[2] T. Ternes, The occurrence of micropollutants in the aquatic environment: a new challenge for water management, Water Sci. Technol. 55 (2007) 327–332.

[3] K. Kümmerer, The presence of pharmaceuticals in the environment due to human use—present knowledge and future challenges, J. Environ. Manag. 90 (2009) 2354–2366.

[4] S. Mompelat, B. Le Bot, O. Thomas, Occurrence and fate of pharmaceutical products and by-products, from resource to drinking water, Environ. Int. 35 (2009) 803–814.

[5] D.J. Lapworth, N. Baran, M.E. Stuart, R.S. Ward, Emerging organic contaminants in groundwater: a review of sources, fate and occurrence, Environ. Pollut. 163 (2012) 287–303.

[6] J.P. Bound, N. Voulvoulis, Pharmaceuticals in the aquatic environment—a comparison of risk assessment strategies, Chemosphere 56 (2004) 1143–1155.

[7] M.O. Barbosa, N.F.F. Moreira, A.R. Ribeiro, M.F.R. Pereira, A.M.T. Silva, Occurrence and removal of organic micropollutants: an overview of the watch list of EU decision 2015/495, Water Res. 94 (2016) 257–279.

[8] I.O. Uribe, A. Mosquera-Corral, J.L. Rodicio, S. Esplugas, Advanced technologies for water treatment and reuse, AICHE J. 61 (2015) 3146–3158.

[9] C. Bouissou-Schurtz, P. Houeto, M. Guerbet, M. Bachelot, C. Casellas, A.-C. Mauclaire, P. Panetier, C. Delval, D. Masset, Ecological risk assessment of the presence of pharmaceutical residues in a French national water survey, Regul. Toxicol. Pharmcol. 69 (2014) 296–303.

[10] C. Vecoli, L. Montano, M.G. Andreassi, Environmental pollutants: genetic damage and epigenetic changes in male germ cells, Environ. Sci. Pollut. Res. Int. 23 (2016) 23339–23348.

[11] P.R. Gogate, A.B. Pandit, A review of imperative technologies for wastewater treatment I: oxidation technologies at ambient conditions, Adv. Environ. Res. 8 (2004) 501–551.

[12] E. Brillas, I. Sirés, M.A. Oturan, Electro-Fenton process and related electrochemical technologies based on Fenton's reaction chemistry, Chem. Rev. 109 (2009) 6570–6631.

[13] I. Sirés, E. Brillas, Remediation of water pollution caused by pharmaceutical residues based on electrochemical separation and degradation technologies: a review, Environ. Int. 40 (2012) 212–229.

[14] M.A. Oturan, J.J. Aaron, Advanced oxidation processes in water/wastewater treatment: principles and applications. A review, Crit. Rev. Environ. Sci. Technol. 44 (2014) 2577–2641.

[15] W.H. Glaze, J.W. Kang, D.H. Chapin, The chemistry of water-treatment processes involving ozone, hydrogen-peroxide and ultraviolet-radiation, Ozone-Sci. Eng. 9 (1987) 335–352.

[16] J.J. Pignatello, E. Oliveros, A. MacKay, Advanced oxidation processes for organic contaminant destruction based on the Fenton reaction and related chemistry, Crit. Rev. Environ. Sci. Technol. 36 (2006) 1–84.

[17] C.A. Martínez-Huitle, E. Brillas, Electrochemical alternatives for drinking water disinfection, Angew. Chem. Int. Ed. Eng. 47 (2008) 1998–2005.

[18] I. Sirés, E. Brillas, M.A. Oturan, M.A. Rodrigo, M. Panizza, Electrochemical advanced oxidation processes: today and tomorrow. A review, Environ. Sci. Pollut. Res. Int. 21 (2014) 8336–8367.

[19] C.A. Martínez-Huitle, M.A. Rodrigo, I. Sirés, O. Scialdone, Single and coupled electrochemical processes and reactors for the abatement of organic water pollutants: a critical review, Chem. Rev. 115 (2015) 13362–13407.

[20] M. Panizza, G. Cerisola, Direct and mediated anodic oxidation of organic pollutants, Chem. Rev. 109 (2009) 6541–6569.
[21] H.J.H. Fenton, Oxidation of tartaric acid in presence of iron, J. Chem. Soc. 65 (1894) 899–910.
[22] F. Haber, J. Weiss, The catalytic decomposition of hydrogen peroxide by iron salts, Proc. Roy. Soc. A 147 (1934) 332–351.
[23] N. Oturan, S. Trajkovska, M.A. Oturan, M. Couderchet, J.J. Aaron, Study of the toxicity of diuron and its metabolites formed in aqueous medium during application of the electrochemical advanced oxidation process "electro-Fenton", Chemosphere 73 (2008) 1550–1556.
[24] E. Brillas, R.M. Bastida, E. Llosa, J. Casado, Electrochemical destruction of aniline and 4-chloroaniline for waste-water treatment using a carbon-PTFE O_2-fed cathode, J. Electrochem. Soc. 142 (1995) 1733–1741.
[25] M.A. Oturan, J. Pinson, Hydroxylation by electrochemically generated OH radicals. Mono- and polyhydroxylation of benzoic acid: products and isomers distribution, J. Phys. Chem. 99 (1995) 13948–13954.
[26] M.A. Oturan, J. Pinson, D. Deprez, B. Terlain, Polyhydroxylation of salicylic acid by electrochemically generated OH radicals, New J. Chem. 16 (1992) 705–710.
[27] E. Brillas, E. Mur, J. Casado, Iron(II) catalysis of the mineralization of aniline using a carbon-PTFE O_2-fed cathode, J. Electrochem. Soc. 143 (1996) L49–L53.
[28] E. Brillas, J.C. Calpe, J. Casado, Mineralization of 2,4-D by advanced electrochemical oxidation processes, Water Res. 34 (2000) 2253–2262.
[29] M.A. Oturan, An ecologically effective water treatment technique using electrochemically generated hydroxyl radicals for in situ destruction of organic pollutants. Application to herbicide 2,4-D, J. Appl. Electrochem. 30 (2000) 475–482.
[30] M.A. Oturan, J.L. Peiroten, P. Chartrin, A.J. Acher, Complete destruction of p-nitrophenol in aqueous medium by electro-Fenton method, Environ. Sci. Technol. 34 (2000) 3474–3479.
[31] B. Boye, M.M. Dieng, E. Brillas, Degradation of herbicide 4-chlorophenoxyacetic acid by advanced electrochemical oxidation methods, Environ. Sci. Technol. 36 (2002) 3030–3035.
[32] M.A. Oturan, J.J. Aaron, N. Oturan, J. Pinson, Degradation of chlorophenoxyacid herbicides in aqueous media, using a novel electrochemical method, Pestic. Sci. 55 (1999) 558–562.
[33] F. Ferrag-Siagh, F. Fourcade, I. Soutrel, H. Ait-Amar, H. Djelal, A. Amrane, Electro-Fenton pretreatment for the improvement of tylosin biodegradability, Environ. Sci. Pollut. Res. Int. 21 (2014) 8534–8542.
[34] H. Lin, H. Zhang, X. Wang, L. Wang, J. Wu, Electro-Fenton removal of Orange II in a divided cell: reaction mechanism, degradation pathway and toxicity evolution, Sep. Purif. Technol. 122 (2014) 533–540.
[35] A. Thiam, E. Brillas, F. Centellas, P.L. Cabot, I. Sirés, Electrochemical reactivity of Ponceau 4R (food additive E124) in different electrolytes and batch cells, Electrochim. Acta 173 (2015) 523–533.
[36] P.V. Nidheesh, R. Gandhimathi, Trends in electro-Fenton process for water and wastewater treatment: an overview, Desalination 299 (2014) 1–15.
[37] T. Tzedakis, A. Savall, M.J. Clifton, The electrochemical regeneration of fenton reagent in the hydroxylation of aromatic substrates—batch and continuous-processes, J. Appl. Electrochem. 19 (1989) 911–921.
[38] M.A. Oturan, J. Pinson, J. Bizot, D. Deprez, B. Terlain, Reaction of inflammation inhibitors with chemically and electrochemically generated hydroxyl radicals, J. Electroanal. Chem. 334 (1992) 103–109.
[39] A. Alvarez-Gallegos, D. Pletcher, The removal of low level organics via hydrogen peroxide formed in a reticulated vitreous carbon cathode cell, part 1. The electrosynthesis of hydrogen peroxide in aqueous acidic solutions, Electrochim. Acta 44 (1998) 853–861.

[40] A. Özcan, Y. Şahin, A.S. Koparal, M.A. Oturan, Carbon sponge as a new cathode material for the electro-Fenton process. Comparison with carbon felt cathode and application to degradation of synthetic dye basic blue 3 in aqueous medium, J. Electroanal. Chem. 616 (2008) 71–78.
[41] A.M. Wang, J.H. Qu, J. Ru, H.J. Liu, J.T. Ge, Mineralization of an azo dye acid red 14 by electro-Fenton's reagent using an activated carbon fiber cathode, Dyes Pigments 65 (2005) 227–233.
[42] A. Da Pozzo, L. Di Palma, C. Merli, E. Petrucci, An experimental comparison of a graphite electrode and a gas diffusion electrode for the cathodic production of hydrogen peroxide, J. Appl. Electrochem. 35 (2005) 413–419.
[43] P.V. Nidheesh, R. Gandhimathi, N.S. Sanjini, $NaHCO_3$ enhanced rhodamine B removal from aqueous solution by graphite-graphite electro Fenton system, Sep. Purif. Technol. 132 (2014) 568–576.
[44] J. Fu, X. Zhang, L. Lei, Fe-modified multi-walled carbon nanotube electrode for production of hydrogen peroxide, Acta Phys. -Chim. Sin. 23 (2007) 1157–1162.
[45] K. Cruz-Gonzalez, O. Torres-Lopez, A. Garcia-Leon, J.L. Guzman-Mar, L.H. Reyes, A. Hernandez-Ramirez, J.M. Peralta-Hernandez, Determination of optimum operating parameters for acid yellow 36 decolorization by electro-Fenton process using BDD cathode, Chem. Eng. J. 160 (2010) 199–206.
[46] E. Bocos, O. Iglesias, M. Pazos, M.A. Sanroman, Nickel foam a suitable alternative to increase the generation of Fenton's reagents, Proc. Saf. Environ. 101 (2016) 34–44.
[47] N. Oturan, M. Zhou, M.A. Oturan, Metomyl degradation by electro-Fenton and electro-Fenton like processes: a kinetics study of the effect of the nature and concentration of some transition metal ions as catalyst, J. Phys. Chem. A 114 (2010) 10605–10611.
[48] M. Panizza, M.A. Oturan, Degradation of alizarin red by electro-Fenton process using a carbon-felt cathode, Electrochim. Acta 56 (2011) 7084–7087.
[49] N. Oturan, J. Wu, H. Zhang, V.K. Sharma, M.A. Oturan, Electrocatalytic destruction of the antibiotic tetracycline in aqueous medium by electrochemical advanced oxidation processes: effect of electrode materials, Appl. Catal. B Environ. 140–141 (2013) 92–97.
[50] F. Sopaj, N. Oturan, J. Pinson, F. Podvorica, M.A. Oturan, Effect of the anode materials on the efficiency of the electro-Fenton process for the mineralization of the antibiotic sulfamethazine, Appl. Catal. B Environ. 199 (2016) 331–341.
[51] M.A. Rodrigo, P.A. Michaud, I. Duo, M. Panizza, G. Cerisola, C. Comninellis, Oxidation of 4-chlorophenol at boron-doped diamond electrode for wastewater treatment, J. Electrochem. Soc. 148 (2001) D60–D64.
[52] M. Panizza, P.A. Michaud, G. Cerisola, C. Comninellis, Anodic oxidation of 2-naphthol at boron-doped diamond electrodes, J. Electroanal. Chem. 507 (2001) 206–214.
[53] B. Marselli, J. Garcia-Gomez, P.-A. Michaud, M.A. Rodrigo, C. Comninellis, Electrogeneration of hydroxyl radicals on boron-doped diamond electrodes, J. Electrochem. Soc. 150 (2003) D79–D83.
[54] E. Brillas, M.A. Banos, M. Skoumal, P.L. Cabot, J.A. Garrido, R.M. Rodriguez, Degradation of the herbicide 2,4-DP by anodic oxidation, electro-Fenton and photoelectro-Fenton using platinum and boron-doped diamond anodes, Chemosphere 68 (2007) 199–209.
[55] N. Oturan, E. Brillas, M.A. Oturan, Unprecedented total mineralization of atrazine and cyanuric acid by anodic oxidation and electro-Fenton with a boron-doped diamond anode, Environ. Chem. Lett. 10 (2012) 165–170.
[56] J.R. Smith, F.C. Walsh, R.L. Clarke, Electrodes based on Magneli phase titanium doxides: the properties and applications of Ebonex(R) materials, J. Appl. Electrochem. 28 (1998) 1021–1033.
[57] L. Guo, Y. Jing, B.P. Chaplin, Development and characterization of ultrafiltration TiO2 Magneli phase reactive electrochemical membranes, Environ. Sci. Technol. 50 (2016) 1428–1436.

[58] F.C. Walsh, R.G.A. Wills, The continuing development of Magnéli phase titanium suboxides and Ebonex® electrodes, Electrochim. Acta 55 (2010) 6342–6351.

[59] D. Bejan, J.D. Malcolm, L. Morrison, N.J. Bunce, Mechanistic investigation of the conductive ceramic Ebonex® as an anode material, Electrochim. Acta 54 (2009) 5548–5556.

[60] S.O. Ganiyu, N. Oturan, S. Raffy, M. Cretin, R. Esmilaire, E. van Hullebusch, G. Esposito, M.A. Oturan, Sub-stoichiometric titanium oxide (Ti_4O_7) as a suitable ceramic anode for electrooxidation of organic pollutants: a case study of kinetics, mineralization and toxicity assessment of amoxicillin, Water Res. 106 (2016) 171–182.

[61] N. Oturan, S.O. Ganiyu, S. Raffy, M.A. Oturan, Sub-stoichiometric titanium oxide as a new anode material for electro-Fenton process: Application to electrocatalytic destruction of antibiotic amoxicillin, Appl. Catal. B: Environ. 217 (2017) 214–223.

[62] Y. Sun, J.J. Pignatello, Photochemical reactions involved in the total mineralization of 2,4-D by iron(3+)/hydrogen peroxide/UV, Environ. Sci. Technol. 27 (1993) 304–310.

[63] M. Diagne, N. Oturan, M.A. Oturan, Removal of methyl parathion from water by electrochemically generated Fenton's reagent, Chemosphere 66 (2007) 841–848.

[64] J. Feng, X. Hu, P.L. Yue, H.Y. Zhu, G.Q. Lu, degradation of azo-dye Orange II by a photoassisted Fenton reaction, Ind. Eng. Chem. Res. 42 (2003) 2058–2066.

[65] G.R. Agladze, G.S. Tsurtsumia, B.-I. Jung, J.-S. Kim, G. Gorelishvili, Comparative study of hydrogen peroxide electro-generation on gas-diffusion electrodes in undivided and membrane cells, J. Appl. Electrochem. 37 (2007) 375–383.

[66] N. Daneshvar, S. Aber, V. Vatanpour, M.H. Rasoulifard, Electro-Fenton treatment of dye solution containing Orange II: influence of operational parameters, J. Electroanal. Chem. 615 (2008) 165–174.

[67] C. Badellino, C.A. Rodrigues, R. Bertazzoli, Oxidation of herbicides by in situ synthesized hydrogen peroxide and Fenton's reagent in an electrochemical flow reactor: study of the degradation of 2,4-dichlorophenoxyacetic acid, J. Appl. Electrochem. 37 (2007) 451–459.

[68] H. Zazou, N. Oturan, H. Zhang, M. Hamdani, M.A. Oturan, Comparative study of electrochemical oxidation of herbicide 2,4,5-T: kinetics, parametric optimization and mineralization pathway, Sust Environ. Res. 27 (2017) 15–23.

[69] N. Oturan, M.A. Oturan, Degradation of three pesticides used in viticulture by electrogenerated Fenton's reagent, Agron. Sustain. Dev. 25 (2005) 267–270.

[70] M. Pimentel, N. Oturan, M. Dezotti, M.A. Oturan, Phenol degradation by advanced electrochemical oxidation process electro-Fenton using a carbon felt cathode, Appl. Catal. B Environ. 83 (2008) 140–149.

[71] B. Balci, M.A. Oturan, N. Oturan, I. Sirés, Decontamination of aqueous glyphosate, (aminomethyl) phosphonic acid and glufosinate solutions by electro-Fenton-like process with Mn^{2+} as the catalyst, J. Agric. Food Chem. 57 (2009) 4888–4894.

[72] M.S. Yahya, N. Oturan, K. El Kacemi, M. El Karbane, C.T. Aravindakumar, M.A. Oturan, Oxidative degradation study on antimicrobial agent ciprofloxacin by electro-Fenton process: kinetics and oxidation products, Chemosphere 117 (2014) 447–454.

[73] M. Turabik, N. Oturan, B. Gözmen, M.A. Oturan, Efficient removal of insecticide imidaclopride from water by electrochemical advanced oxidation processes, Environ. Sci. Pollut. Res. Int. 21 (2014) 8387–8397.

[74] E. Mousset, L. Frunzo, G. Esposito, E.D. van Hullebusch, N. Oturan, M.A. Oturan, A complete phenol oxidation pathway obtained during electro-Fenton treatment and validated by a kinetic model study, Appl. Catal. B Environ. 180 (2016) 189–198.

[75] E. Brillas, B. Boye, I. Sirés, J.A. Garrido, R.M. Rodriguez, C. Arias, P.L. Cabot, C. Comninellis, Electrochemical destruction of chlorophenoxy herbicides by anodic oxidation and electro-Fenton using a boron-doped diamond electrode, Electrochim. Acta 49 (2004) 4487–4496.

[76] M.A. Oturan, N. Oturan, C. Lahitte, S. Trevin, Production of hydroxyl radicals by electrochemically assisted Fenton's reagent. Application to the mineralization of an organic micropollutant, the pentachlorophenol, J. Electroanal. Chem. 507 (2001) 96–102.
[77] G. Kaichouh, N. Oturan, M.A. Oturan, K. El Kacemi, A. El Hourch, Degradation of the herbicide imazapyr by Fenton reactions, Environ. Chem. Lett. 2 (2004) 31–33.
[78] M.A. Oturan, N. Oturan, M.C. Edelahi, F.I. Podvorica, K. El Kacemi, Oxidative degradation of herbicide diuron in aqueous medium by Fenton's reaction based advanced oxidation processes, Chem. Eng. J. 171 (2011) 127–135.
[79] B. Balci, N. Oturan, R. Cherrier, M.A. Oturan, Degradation of atrazine in aqueous medium by electrocatalytically generated hydroxyl radicals. A kinetic and mechanistic study, Water Res. 43 (2009) 1924–1934.
[80] N. Borras, R. Oliver, C. Arias, E. Brillas, Degradation of atrazine by electrochemical advanced oxidation processes using a boron-doped diamond anode, J. Phys. Chem. A 114 (2010) 6613–6621.
[81] M. Murati, N. Oturan, J.J. Aaron, A. Dirany, B. Tassin, Z. Zdravkovski, M.A. Oturan, Degradation and mineralization of sulcotrione and mesotrione in aqueous medium by the electro-Fenton process: a kinetic study, Environ. Sci. Pollut. Res. 19 (2012) 1563–1573.
[82] A. Özcan, Y. Şahin, A.S. Koparal, M.A. Oturan, Propham mineralization in aqueous medium by anodic oxidation using boron-doped diamond anode. Experimental parameters' influence on degradation kinetics and mineralization efficiency, Water Res. 42 (2008) 2889–2898.
[83] A. Kesraoui-Abdessalem, N. Bellakhal, N. Oturan, M. Dachraoui, M.A. Oturan, Treatment of a mixture of three pesticides by photo- and electro-Fenton processes, Desalination 250 (2010) 450–455.
[84] E. Brillas, E. Mur, R. Sauleda, L. Sanchez, J. Peral, X. Domènech, J. Casado, Aniline mineralization by AOP's: anodic oxidation, photocatalysis, electro-Fenton and photoelectro-Fenton processes, Appl. Catal. B: Environ. 16 (1998) 31–42.
[85] B. Gözmen, M.A. Oturan, N. Oturan, O. Erbatur, Indirect electrochemical treatment of bisphenol a in water via electrochemically generated Fenton's reagent, Environ. Sci. Technol. 37 (2003) 3716–3723.
[86] F. Kaplan, A. Hesenov, B. Gozmen, O. Erbatur, Degradations of model compounds representing some phenolics in olive mill wastewater via electro-Fenton and photoelectro-Fenton treatments, Environ. Technol. 32 (2011) 685–692.
[87] H. Zhang, D. Zhang, J. Zhou, Removal of COD from landfill leachate by electro-Fenton method, J. Hazard. Mater. 135 (2006) 106–111.
[88] N. Oturan, E.D. van Hullebusch, H. Zhang, L. Mazeas, H. Budzinski, K. Le Menach, M.A. Oturan, Occurrence and of removal organic micropollutants in landfill leachates treated by electrochemical advanced oxidation processes, Environ. Sci. Technol. 49 (2015) 12187–12196.
[89] M.H. Zhou, Q.Q. Tan, Q. Wang, Y.L. Jiao, N. Oturan, M.A. Oturan, Degradation of organics in reverse osmosis concentrates by electro-Fenton process, J. Hazard. Mater. 215-216 (2012) 287–293.
[90] E. Guivarch, M.A. Oturan, Le problème de la contamination des eaux par les colorants synthétiques: comment les détruire? L'Act. Chim. 277–278 (2004) 65–69.
[91] S. Ammar, R. Abdelhedi, C. Flox, C. Arias, E. Brillas, Electrochemical degradation of the dye indigo carmine at boron-doped diamond anode for wastewaters remediation, Environ. Chem. Lett. 4 (2006) 229–233.
[92] A.R. Khataee, V. Vatanpour, A.R.A. Ghadim, Decolorization of CI acid blue 9 solution by UV/Nano-TiO_2, Fenton, Fenton-like, electro-Fenton and electrocoagulation processes: a comparative study, J. Hazard. Mater. 161 (2009) 1225–1233.
[93] M. Panizza, G. Cerisola, Electro-Fenton degradation of synthetic dyes, Water Res. 43 (2009) 339–344.

[94] S. Garcia-Segura, F. Centellas, C. Arias, J.A. Garrido, R.M. Rodríguez, P.L. Cabot, E. Brillas, Comparative decolorization of monoazo, diazo and triazo dyes by electro-Fenton process, Electrochim. Acta 58 (2011) 303–311.
[95] M. Diagne, V.K. Sharma, N. Oturan, M.A. Oturan, Depollution of indigo dye by anodic oxidation and electro-Fenton processes using boron-doped diamond anode, Environ. Chem. Lett. 12 (2014) 219–224.
[96] I. Sires, C. Arias, P.L. Cabot, F. Centellas, R.M. Rodriguez, J.A. Garrido, E. Brillas, Paracetamol mineralization by advanced electrochemical oxidation processes for wastewater treatment, Environ. Chem. 1 (2004) 26–28.
[97] I. Sirés, J.A. Garrido, R.M. Rodríguez, E. Brillas, N. Oturan, M.A. Oturan, Catalytic behavior of the Fe^{3+}/Fe^{2+} system in the electro-Fenton degradation of the antimicrobial chlorophene, Appl. Catal. B-Environ. 72 (2007) 382–394.
[98] A. El-Ghenymy, R.M. Rodríguez, E. Brillas, N. Oturan, M.A. Oturan, Electro-Fenton degradation of the antibiotic sulfanilamide with Pt/carbon-felt and BDD/carbon-felt cells. Kinetics, reaction intermediates and toxicity assessment, Environ. Sci. Pollut. Res. Int. 21 (2014) 8368–8378.
[99] N. Barhoumi, L. Labiadh, M.A. Oturan, N. Oturan, A. Gadri, S. Ammar, E. Brillas, Electrochemical mineralization of the antibiotic levofloxacin by electro-Fenton-pyrite process, Chemosphere 141 (2015) 250–257.
[100] E. Guinea, C. Arias, P.L. Cabot, J.A. Garrido, R.M. Rodriguez, F. Centellas, E. Brillas, Mineralization of salicylic acid in acidic aqueous medium by electrochemical advanced oxidation processes using platinum and boron-doped diamond as anode and cathodically generated hydrogen peroxide, Water Res. 42 (2008) 499–511.
[101] L. Feng, N. Oturan, E.D. van Hullebusch, G. Esposito, M.A. Oturan, Degradation of anti-inflammatory drug ketoprofen by electro-oxidation: comparison of electro-Fenton and anodic oxidation processes, Environ. Sci. Pollut. Res. Int. 21 (2014) 8406–8416.
[102] E. Isarain-Chávez, R.M. Rodriguez, P.L. Cabot, F. Centellas, C. Arias, J.A. Garrido, E. Brillas, Degradation of pharmaceutical beta-blockers by electrochemical advanced oxidation processes using a flow plant with a solar compound parabolic collector, Water Res. 45 (2011) 4119–4130.
[103] I. Sirés, N. Oturan, M.A. Oturan, Electrochemical degradation of beta-blockers. Studies on single and multicomponent aqueous solutions, Water Res. 44 (2010) 3109–3120.
[104] H. Olvera-Vargas, N. Oturan, D. Buisson, M.A. Oturan, A coupled bio-EF process for mineralization of the pharmaceuticals furosemide and ranitidine: feasibility assessment, Chemosphere 155 (2016) 606–613.
[105] A. Dirany, I. Sirés, N. Oturan, A. Özcan, M.A. Oturan, Electrochemical treatment of Sulfachloropyridazine: kinetics, reaction pathways, and toxicity evolution, Environ. Sci. Technol. 46 (2012) 4074–4082.
[106] J.B. De Heredia, J. Torregrosa, J.R. Dominguez, J.A. Peres, Kinetic model for phenolic compound oxidation by Fenton's reagent, Chemosphere 45 (2001) 85–90.
[107] G.V. Buxton, C.L. Greenstock, W.P. Helman, A.B. Ross, Critical review of rate constants for reactions of hydrated electrons, hydrogen atoms and hydroxyl radicals ($^{\cdot}OH/O^{-}$) in aqueous solutions, J. Phys. Chem. Ref. Data 17 (1988) 513–886.
[108] N. Oturan, M. Panizza, M.A. Oturan, Cold incineration of chlorophenols in aqueous solution by electro-Fenton process. Effect of number and position of chlorine atoms on the degradation kinetics, J. Phys. Chem. A 113 (2009) 10988–10993.
[109] D. Huguenot, M. Emmanuel, E.D. van Hullebusch, M.A. Oturan, Combination of surfactant enhanced soil washing and Electro-Fenton processes for the treatment of soils contaminated by petroleum hydrocarbons, J. Environ. Manage. 153 (2015) 40–47.
[110] E. Rosales, M. Pazos, M.A. Longo, M.A. Sanroman, Influence of operational parameters on electro-Fenton degradation of organic pollutants from soil, J. Environ. Sci. Health A 44 (2009) 1104–1110.

[111] E. Mousset, N. Oturan, E.D. van Hullebusch, G. Guibaud, G. Esposito, M.A. Oturan, Influence of solubilizing agents (cyclodextrin or surfactant) on phenanthrene degradation by electro-Fenton process—study of soil washing recycling possibilities and environmental impact, Water Res. 48 (2014) 306–316.
[112] O. Iglesias, J. Gómez, M. Pazos, M.A. Sanromán, Electro-Fenton oxidation of imidacloprid by Fe alginate gel beads, Appl. Catal. B Environ. 144 (2014) 416–424.
[113] P.V. Nidheesh, Heterogeneous Fenton catalysts for the abatement of organic pollutants from aqueous solution: a review, RSC Adv. 5 (2015) 40552–40577.
[114] P. Bonnissel-Gissinger, M. Alnot, J.J. Ehrhardt, P. Behra, Surface oxidation of pyrite as a function of pH, Environ. Sci. Technol. 32 (1998) 2839–2845.
[115] L. Labiadh, M.A. Oturan, M. Panizza, N. Ben Hamadi, S. Ammar, Complete removal of AHPS synthetic dye from water using new electro-fenton oxidation catalyzed by natural pyrite as heterogeneous catalyst, J. Hazard. Mater. 297 (2015) 34–41.
[116] S. Ammar, M.A. Oturan, L. Labiadh, A. Guersalli, R. Abdelhedi, N. Oturan, E. Brillas, Degradation of tyrosol by a novel electro-Fenton process using pyrite as heterogeneous source of iron catalyst, Water Res. 74 (2015) 77–87.
[117] N. Barhoumi, N. Oturan, H. Olvera-Vargas, E. Brillas, A. Gadri, S. Ammar, M.A. Oturan, Pyrite as a sustainable catalyst in electro-Fenton process for improving oxidation of sulfamethazine. Kinetics, mechanism and toxicity assessment, Water Res. 94 (2016) 52–61.
[118] A. Lopez, M. Pagano, A. Volpe, A.C. Di Pinto, Fenton's pre-treatment of mature landfill leachate, Chemosphere 54 (2004) 1005–1010.
[119] J.L. De Morais, P.P. Zamora, Use of advanced oxidation processes to improve the biodegradability of mature landfill leachates, J. Hazard. Mater. 123 (2005) 181–186.
[120] I. Oller, S. Malato, J.A. Sanchez-Perez, Combination of advanced oxidation processes and biological treatments for wastewater decontamination-A review, Sci. Total. Environ. 409 (2011) 4141–4166.
[121] F. Ferrag-Siagh, F. Fourcade, I. Soutrel, H. Ait-Amar, H. Djelal, A. Amrane, Electro-Fenton pretreatment for the improvement of tylosin biodegradability, Environ. Sci. Pollut. Res. 21 (2014) 8534–8542.
[122] H. Olvera-Vargas, T. Cocerva, N. Oturan, D. Buisson, M.A. Oturan, Bioelectro-Fenton: a sustainable integrated process for removal of organic pollutants from water: application to mineralization of metoprolol, J. Hazard. Mater. 319 (2016) 13–23.
[123] C. Trellu, O. Ganzenko, S. Papirio, Y. Pechaud, N. Oturan, D. Huguenot, E.D. van Hullebusch, G. Esposito, M.A. Oturan, Combination of anodic oxidation and biological treatment for the removal of phenanthrene and Tween 80 from soil washing solution, Chem. Eng. J. 306 (2016) 588–596.
[124] O. Ganzenko, C. Trellu, S. Papirio, N. Oturan, D. Huguenot, E.D. van Hullebusch, G. Esposito, M.A. Oturan, Bio-electro-Fenton: evaluation of a combined biological-advanced oxidation treatment for pharmaceutical wastewater, Environ. Sci. Pollut. Res. (2018). in press, https://doi.org/10.1007/s11356-017-8450-6.
[125] O. Ganzenko, D. Huguenot, E.D. van Hullebusch, G. Esposito, M.A. Oturan, Electrochemical advanced oxidation and biological processes for wastewater treatment: A review of the combined approaches, Environ. Sci. Pollut. Res. 21 (2014) 8493–8524.

CHAPTER

9

Photo and Solar Fenton Processes for Wastewater Treatment

Juan M. Peralta-Hernández, Shraddha Vijay, Oscar Rodríguez-Narváez, Martin A. Pacheco-Álvarez

University of Guanajuato, México

OUTLINE

INTRODUCTION

Water is an essential natural resource for life. It is a primary need for health, and it should, therefore, be considered as one of the basic human rights. In today's societies, water has become a very precious commodity. Due to its scarcity, it is a necessity of life, and economic development is also dependent on the availability of water. The natural water cycle has a great capacity for purification. But, this ease of regeneration and its apparent abundance make the usual waste dump: pesticides, chemical waste, heavy metals, radioactive waste, etc. The degradation of the waters comes from ancient times, and it has been just in this century that this problem has spread to rivers and seas all over the world. In this context, so-called

https://doi.org/10.1016/B978-0-12-813160-2.00009-2

advanced oxidation processes (AOPs) have shown their importance for both potable and residual wastewater treatment when compared with conventional treatments [1–3]. In the literature, it is possible to find different studies on the use of AOPs for the removal of pollutants from wastewater. These publications include those related to the Fenton oxidation process due to its effectiveness in the decontamination of wastewater from petrochemical, textile, pharmaceutical, paper, and other industries [1,4–6].

The Fenton oxidation reaction was defined for the first time in 1894 by Henry J. Fenton when he discovered that hydrogen peroxide (H_2O_2) could be activated with ferrous ions to oxidize organics [7]. The process consists mainly of the addition of iron salts in the presence of H_2O_2 solution, preferably at acid medium, for the generation of $^{\bullet}OH$ free radicals. Thus, the combination of H_2O_2 solution and iron salts is called the reactive Fenton reaction (Eqs. 1, 2) [5,8].

$$Fe^{2+} + H_2O_2 \rightarrow Fe^{3+} + {}^{\bullet}OH + OH^- \quad (1)$$

$$Fe^{3+} + H_2O_2 \rightarrow Fe^{2+} + H^+ + {}^{\bullet}OH_2 \quad (2)$$

Perhydroxyl radicals ($^{\bullet}OH_2$) can be also generated via the Fenton reaction, together with free $^{\bullet}OH$ radicals formed, which start a chain oxidation reaction to remove oxidizable matter. However, $^{\bullet}OH_2$ radicals have lower oxidation power than $^{\bullet}OH$ [9].

Furthermore, during the Fenton reaction, the consumption of free $^{\bullet}OH$ radicals can take place [7,9]. Some excess of H_2O_2 and/or iron ions at the beginning of the process, with respect to the amount of pollutant to be treated, limits the efficiency of the process (Eqs. 3, 4; Fig. 1) [5].

$$H_2O_2 + {}^{\bullet}OH \rightarrow {}^{\bullet}OH_2 + H_2O \quad (3)$$

$$Fe^{2+} + {}^{\bullet}OH \rightarrow Fe^{3+} + OH^- \quad (4)$$

Similarly, the recombination of free $^{\bullet}OH$ radicals and/or the reaction of these with $^{\bullet}OH_2$ radicals can result in the reduction of the Fenton reaction yield (Eqs. 6, 7).

$$^{\bullet}OH + {}^{\bullet}OH \rightarrow H_2O_2 \quad (6)$$

$$^{\bullet}OH + {}^{\bullet}OH_2 \rightarrow H_2O + O_2 \quad (7)$$

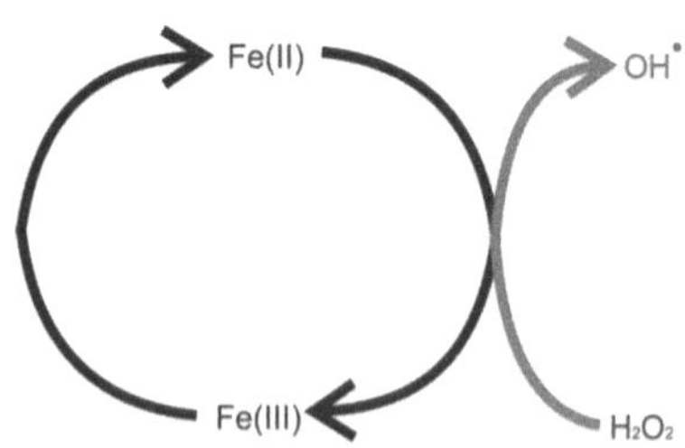

FIG. 1 Schematic representation of the Fenton process, illustrating the reaction between iron ions (Fe^{2+}/Fe^{3+}) and hydrogen peroxide (H_2O_2) to generate free hydroxyl radicals ($^{\bullet}OH$).

Iron can be added in its ferrous (Fe^{2+}) or ferric (Fe^{3+}) form. Ferrous iron salts are used in minor proportion as a consequence of the slow generation of $^{\bullet}OH$ radicals in the initial stages of the process, because this would start with Eq. (2).

However, the regeneration of Fe^{2+} is favored (Eq. 8) when a combination of hydrogen peroxide and ultraviolet (UV) or solar radiation, commonly called photo-Fenton UV or photo-Fenton solar processes, respectively, is attained [2], and at the same time, a great amount of $^{\bullet}OH$ radicals is produced [7]. This fact can be explained by the formation of aqueous Fe^{3+} and ($Fe(OH)^{2+}$) oxalate complexes, which are capable of absorbing UV and visible radiation [10], suffering photo reduction, and generating $^{\bullet}OH$ [7,11].

$$Fe(OH)^{2+} + h\upsilon \rightarrow Fe^{2+} + {}^{\bullet}OH \quad (8)$$

Additionally, it is possible to carry out the UV/H_2O_2 reaction where there is cleavage of the molecule into hydroxyl radicals with a quantum yield of two $^{\bullet}OH$ radicals formed per quantum of radiation absorbed (Eq. 9). Meanwhile, H_2O_2 is known that can be reduced as indicated below (Eq. 10).

$$H_2O_2 + h\upsilon \rightarrow 2^{\bullet}OH \quad (9)$$

$$H_2O_2 + HO_2^{-} \rightarrow O_2 + H_2O + OH^{-} \quad (10)$$

Iron oxalate complexes are also photoactive; in fact, they give a greater quantum yield of iron Fe^{2+} ions when compared to the Fe^{3+} hydroxyl complexes. Also, the absorption level is prolonged to 570nm, leading to an improvement of the absorption coefficient. Possible reactions can be given as:

$$Fe(C_2O_4)_3{}^{3-} + h\upsilon \rightarrow Fe^{2+} + 2C_2O_4{}^{2-} + C_2O_4{}^{\bullet -} \quad (11)$$

$$Fe(C_2O_4)_3{}^{3-} + C_2O_4{}^{\bullet -} \rightarrow Fe^{2+} + 3C_2O_4{}^{2-} + 2CO_2 \quad (12)$$

Also, by the action of the reaction of intermediate oxalate radicals with the Fe^{3+} oxalate complexes, additional Fe^{2+} ions are produced. Thus, the generated Fe^{2+} ions react with H_2O_2 leading to free $^{\bullet}OH$ in accordance with Eq. 1. The Fe^{3+} ions generated in the process react again with H_2O_2, which leads to the regeneration of the Fe^{2+} ions (Eq. 2; Fig. 2).

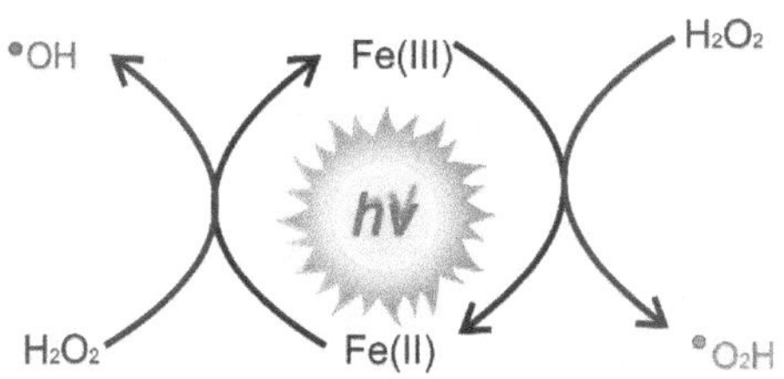

FIG. 2 Reactions that occur when the Fenton process is irradiated with UV.

The free hydroxyl radicals formed in the system are highly reactive oxidants and promote the oxidative degradation of compounds (RH) in water, which may lead to a total mineralization of the organic compounds.

$$^{\bullet}OH + RH \rightarrow R^{\bullet} + H_2O \quad (13)$$

$$R^{\bullet} + O_2 \rightarrow RO_2^{\bullet} \quad (14)$$

PRINCIPLES AND FUNDAMENTALS OF THE PHOTOELECTRO-FENTON PROCESS

In spite of the high oxidative efficacy of the Fenton reagent, it also presents several disadvantages for its application such as (a) storage and (b) transport of concentrated solutions of H_2O_2, and (c) the generation of Fe^{3+} sludge.

In this context, the electric current produced by the redox reactions onto the electrode surface resulted in the production of reactive intermediates that could degrade the organic pollutants. With this procedure, it is possible to carry out the treatment of liquids, gases, and solids. The major advantage is its compatibility with the environment because the main reactant, the electron, is a clean reagent that does not generate residue; it is also possible to control the production [12]. The in situ electrochemical production of H_2O_2 in acidic media and the regeneration of Fe^{2+} by the simultaneous reduction of oxygen and Fe^{3+} on the cathode can solve the problems regarding the application of the Fenton reagent discussed above. In this sense, electro-Fenton process is an alternative when the Fe^{2+} or Fe^{3+} ions are added to the solution and the hydrogen peroxide is electrogenerated the O_2 reduction at the cathode of an undivided electrolytic cell [13].

$$O_2 + 2e^- + 2H^+ \rightarrow H_2O_2 \quad (15)$$

In this way, the electrogenerated H_2O_2 can be catalytically transformed to $^{\bullet}OH$ radicals by means of the catalytic addition of ferrous ions, as shown schematically in Eq. (1) [14–16].

In order to carry out adequate H_2O_2 production, it is necessary to maintain the use of optimum operating conditions, such as: pH control (2.8–3), temperature (25–35°C), and the utilization of an appropriate cathode material, often based on carbon, such as reticulated vitreous carbon (RCV), carbon cloths, and recently BDD [17].

Recently, the interest to use the processes that involve the application of artificial (UV) or solar light to carry out the treatment of different wastewaters, has increased. The most commonly energy sources types used, in these processes, are UVA ($\lambda = 315$–400 nm), UVB ($\lambda = 285$–315 nm), and UVC ($\lambda < 285$ nm). In most cases, the energies supplied by UV lamps are commonly used as sources (Fig. 3) [21,22].

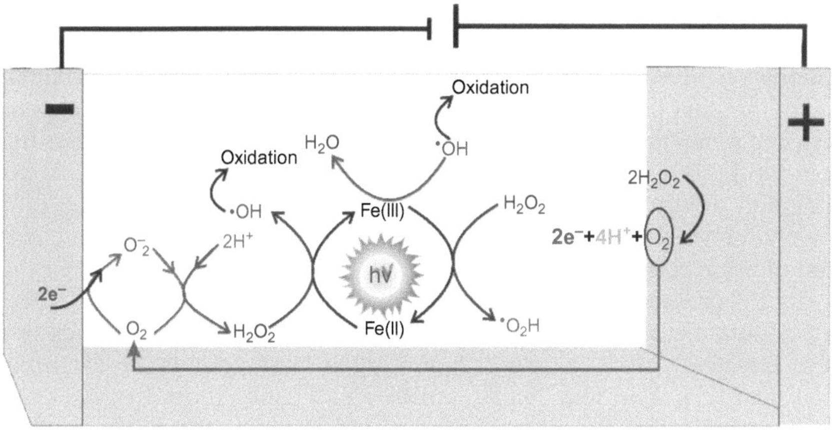

FIG. 3 Simplified schematic representation of the photoelectron-Fenton process.

According to the information discussed above, it is possible to consider the development of an interesting new family of photoelectrochemical processes, such as: photoelectro-Fenton (PEF), solar photoelectro-Fenton (SPEF), photoelectrocatalysis (PEC) as well as hybrid systems [23].

The photoelectro-Fenton process using artificial UVA light and the solar photoelectro-Fenton application have been visualized and widely developed by different researchers in the world, mainly by the Brillas' group. These processes consist of the treatment of the pollutant solution under EF conditions along with the simultaneous irradiation with UVA or solar light to accelerate the degradation rate of pollutants. Oxidizing $^{\bullet}OH$ are produced from Fenton's reaction (Eq. 8) while the undesired accumulation of refractory Fe^{3+} ions that decelerates the treatment is avoided by the reductive photolysis of $[Fe(OH)]^{2+}$, (Eq. 16), thereby regenerating Fe^{2+} (i.e., the catalyst in Fenton's reaction) and producing more radicals [24]:

$$Fe^{3+} + e \rightarrow Fe^{2+} \tag{16}$$

As an example, reaction (17) shows the photolytic process for Fe(III)-oxalate complexes ($Fe(C_2O_4)^+$, $Fe(C_2O_4)_2^-$ and $Fe(C_2O_4)_3^{3-}$), formed as the ultimate products of aromatics.

$$Fe(OH)^{2+} + h\upsilon \rightarrow Fe^{2+} + {}^{\bullet}OH \tag{17}$$

$$Fe(OOCR)^{2+} + h\upsilon \rightarrow Fe^{2+} + CO_2 + R^{\bullet} \tag{18}$$

$$2Fe(C_2O_4)_n^{(3-2n)} + h\upsilon \rightarrow 2Fe^{2+} + (2n-1)C_2O_4^{2-} + 2CO_2 \tag{19}$$

It is significant that the high electrical cost of the UV source used in PEF processes is the main disadvantage of this technique for commercial applications. To solve this matter, the Brillas group has proposed the alternative use of the solar photoelectro-Fenton system, where sunlight—an inexpensive and renewable energy source with $\lambda > 300$ nm—directly irradiates the solution [25,26]. The higher intensity of the UV light provided by solar radiation along with the additional absorption at $\lambda > 400$ nm (e.g., for the photolysis of Fe(III)-carboxylate complexes) improved the mineralization of SPEF compared with PEF.

Table 1 presents some illustrative research for the degradation of different pollutants under the photoelectro-Fenton process. This approach has been effectively used to treat different organic compounds, for example, Acid Red [27], Acid Red 29 [28], tartrazine [29] and Allura Red AC [30].

The application of solar photo-assisted Fenton reactors is another alternative that has recently been studied to treat wastewaters. For example, a solar reactor with a capacity of 2.5 L was used to degrade Acid Yellow 36 dye and Congo Red di-azo dye [33,38]. Significant efforts to develop higher-capacity reactors, in terms of volume, have been spent, for example, the reactor used to treat Direct Yellow 4 azo dye [39] which has a capacity to depollute 10 L of effluent.

TABLE 1 Typical Reactor Conditions for Using the Photoelectro-Fenton Process for Wastewater Treatment

Reference	Year	Pollutant	Experimental condition	% COD removal
[27]	2014	Acid Red 1	Reactor capacity = 0.15 L, 0.1 L of 236 mg/L pollutant, 0.5 mM Fe^{+2}, 0.05 M Na_2SO_4, pH = 3.0, 100 mA/cm^2, 35°C, flow rate 300 cm^3/min, flow reactor with 3 cm^2 electrodes, 360 min in batch	100
[28]	2012	Acid Red 29	Reactor capacity = 150 mL, 0.1 L of 236 mg/L pollutant, 0.5 mM Fe^{+2}, 0.05 M Na_2SO_4, pH = 3.0, 16.6–100 mA/cm^2, 35°C, flow rate 300 mL/min, flow reactor with 3 cm^2 electrodes, 360 min in batch	92
[29]	2014	Tartrazine	Reactor capacity = 150 mL, 0.13 L of 278 mg/L pollutant, 0.5 mM Fe^{+2}, 0.05 M Na_2SO_4, pH = 3.0, 200 mA, 25°C, flow rate 1 L/min, flow reactor with 3 cm^2 electrodes, 240 min in batch	100

TABLE 1 Typical Reactor Conditions for Using the Photoelectro-Fenton Process for Wastewater Treatment—cont'd

Reference	Year	Pollutant	Experimental condition	% COD removal
[30]	2015	Allura Red AC	Reactor capacity = 130 mL, 0.13 L of 230 mg/L pollutant, 0.5 mM Fe^{+2}, 0.05 M Na_2SO_4, pH = 3.0, 33.3–150 mA/cm^2, 25°C, flow rate 1 L/min, flow reactor with 3 cm^2 electrodes, 360–480 min in batch	95–98
[31]	2010	Basic Red 46	Reactor capacity = 1 L, 0.8 L of 2–20 mg/L pollutant, 0.3 mM Fe^{+3}, 0.05 M Na_2SO_4, pH = 3.0, 100 mA, 25°C, flow rate 140 mL/min, flow reactor with 5.4 and 1 cm^2 electrodes (cathode and anode) respectively, 27 min in batch	NA
[32]	2011	Basic Yellow 28	Reactor capacity = 2.5 L, 2 L of 20 mg/L pollutant, 0.1 mM Fe^{+3}, 0.05 M Na_2SO_4, pH = 2.0–6.0, 100 mA/cm^2, room temp., flow rate 140 mL/min, flow reactor with 5.4 and 11.5 cm^2 electrodes (cathode and anode) respectively, 6 h in batch	94.7
[33]	2015	Congo Red	Reactor capacity = 100 mL, 0.1 L of 0.260 mM pollutant, 0.5 mM Fe^{+2}, 0.05 M Na_2SO_4, pH = 3.0, 66.7 mA/cm^2, 35°C, flow rate 300 mL/min, flow reactor with 3 cm^2 electrodes (cathode and anode) respectively, 360 min in batch	97
[34]	2015	E122 (Azorubine), E124(Ponceau 4R), E129(Allura Red AC)	Reactor capacity = 130 mL, 0.13 L of 0.260 mM pollutant, 0.5 mM Fe^{+2}, 0.05 M Na_2SO_4, pH = 3.0, 66.7 mA/cm^2, 35°C, flow rate 300 mL/min, flow reactor with 3 cm^2 electrodes (cathode and anode) respectively, 360 min in batch	100

Continued

TABLE 1 Typical Reactor Conditions for Using the Photoelectro-Fenton Process for Wastewater Treatment—cont'd

Reference	Year	Pollutant	Experimental condition	% COD removal
[34]	2015	Evans Blue	Reactor capacity = 100 mL, 0.1 L of 0.245 mM pollutant, 0.5 mM Fe^{+2}, 0.05 M Na_2SO_4, pH = 3.0, 16.7–100 mA/cm^2, 35°C, flow rate 300 mL/min, flow reactor with 3 cm^2 electrodes (cathode and anode) respectively, 300–360 min in batch	>95%
[35]	2006	Indigo Carmine	Reactor capacity = 100 mL, 0.1 L of 0.9 g/L pollutant, 0.25 mM Cu^{+2} and 1.0 mM Fe^{+2}, 0.05 M Na_2SO_4, pH = 2.0–6.0, 33–150 mA/cm^2, 35°C, flow rate 12 mL/min, flow reactor with 3 cm^2 electrodes (cathode and anode) respectively, 7 h in batch	>97%
[36]	2008	Orange II	Reactor capacity = 250 mL, 0.4 L of 50 mg/L pollutant, 0.2–0.5 mM Fe^{+2}, 0.05 M Na_2SO_4, pH = 3.0, 100–300 mA/cm^2, ambient temp., flow rate 100 mL/min, flow reactor with 164 and 200 cm^2 electrodes (cathode and anode) respectively, 60 min in batch	80%
[37]	2013	Sunset Yellow FCF	Reactor capacity = 100 mL, 0.1 L of 290 mg/L pollutant, 0.5 mM Fe^{+2}, 0.05 M Na_2SO_4, pH = 3.0, 16.7–100 mA/cm^2, 35°C, flow rate 3.3 L/min, flow reactor with 3 cm^2 electrodes (cathode and anode) respectively, 240 min in batch	95%

It should be emphasized that light is almost always the limiting reagent in photochemical reactions. Photoelectro-Fenton reactions are often conducted in reactors in which the solution is recirculated through the photo reactor from an external reservoir. Table 2 summarizes some of experimental conditios used with different photo reactors to degrade different organic compounds. Despite significant advances have been performed in the last decade, the widespread application of photochemical technology for the decontamination of industrial residues and wastewaters is not

TABLE 2 Typical Solar Photoelectro-Fenton Reactor Approaches Used for Wastewater Treatment

Reference	Year	Pollutant	Experimental conditions	% removal
[38]	2011	Acid Yellow 36	Reactor capacity = 2.5 L of 108 mg/L pollutant, 0.5 mM Fe^{+2}, 0.05 M Na_2SO_4, pH = 3.0, 150 mA/cm^2, 35°C, Flow rate 3333.3 cm^3/min, flow reactor with 20 cm^2 electrodes, 90 min in batch, UV radiation 19 W/m^2, irradiated volume = 0.6 L	TOC 95 MCE 56
[33]	2015	Congo Red azo dye	Reactor capacity = 2.5 L of 362 mg/L pollutant, 0.5 mM Fe^{+2}, 0.05 M Na_2SO_4, pH = 3.0, 100 mA/cm^2, 35°C, stirred tank reactor with 20 cm^2 electrodes, 240 min in batch, UV radiation 30–32 W/m^2	DOC 97–98 MCE 49
[39]	2014	Direct Yellow 4 azo dye	Reactor capacity = 10 L of 200 mg/L pollutant, 0.5 mM Fe^{+2}, 0.05 M Na_2SO_4, pH = 3.0, 50 mA/cm^2, 35°C, Flow rate 3333.3 cm^3/min, flow reactor with 90.2 cm^2 electrodes, 240 min in batch, UV radiation 31 W/m^2, irradiated volume = 1.57 L	MCE 96–97
[34]	2015	Evans Blue azo dye	Reactor capacity = 10 L of 235 mg/L pollutant, 0.5 mM Fe^{+2}, 0.05 M Na_2SO_4, pH = 3.0, 55.4 mA/cm^2, 35°C, Flow rate 3333.3 cm^3/min, flow reactor with 90.3 cm^2 electrodes, 300 min in batch, UV radiation 30–32 W/m^2, irradiated volume = 1.57 L	MCE 88

Continued

TABLE 2 Typical Solar Photoelectro-Fenton Reactor Approaches Used for Wastewater Treatment—cont'd

Reference	Year	Pollutant	Experimental conditions	% removal
[40]	2014	Atrazine pesticide	Reactor capacity = 200 mL of 20 mg/L pollutant, 0.1 mM Fe^{+2}, 0.05 M Na_2SO_4, pH = 3.0, 13.33 mA/cm^2, 35°C, stirred tank reactor with 7.5 cm^2 electrodes, 300 min in batch, UV radiation 34 W/m^2	MCE 80
[41]	2014	Diuron pesticide	Reactor capacity = 2.5 L of 43 mg/L pollutant, 0.5 mM Fe^{+2}, 0.05 M Na_2SO_4, pH = 3.0, 50 mA/cm^2, 25°C, stirred tank reactor with 20 cm^2 electrodes, 240 min in batch, UV radiation 30–32 W/m^2	MCE 65
[26]	2007	Mecoprop pesticide	Reactor capacity = 2.5 L of 640 mg/L pollutant, 0.5 mM Fe^{+2}, 0.05 M Na_2SO_4, pH = 3.0, 50 mA/cm^2, 25°C, flow rate 3000 cm^3/min, flow reactor with 20 cm^2 electrodes, 540 min in batch	TOC 96
[42]	2017	Tebuthiuron and ametryn herbicides	Reactor capacity = 2.5 L of 0.18 Mm de Tebuthiuronm y 0.36 Mm de ametryn, 0.5 mM Fe^{+2}, 0.05 M Na_2SO_4, pH = 3.0, 50 mA/cm^2, 35°C, flow rate 3333.3 cm^3/min, flow reactor with 20 cm^2 electrodes, 240 min in batch, UV radiation 30–32 W/m^2, irradiated volume = 0.6 L	TOC 48 MCE 25.6

TABLE 2 Typical Solar Photoelectro-Fenton Reactor Approaches Used for Wastewater Treatment—cont'd

Reference	Year	Pollutant	Experimental conditions	% removal
[43]	2015	Ranitidine H2 receptor antagonist	Reactor capacity = 2.5 L of 112.6 mg/L, 0.5 mM Fe^{+2}, 0.05 M Na_2SO_4, pH = 3.0, 100 mA/cm^2, 35 °C, flow rate 3333.3 cm^3/min, flow reactor with 20 cm^2 electrodes, 360 min in batch, UV radiation 30–35 W/m^2, irradiated volume = 0.6 L	TOC 67 MCE 80
[44]	2011	Herbicide 4-chloro-2 methylphenoxyacetic acid	Reactor capacity = 10 L of 186 mg/L pollutant, 1 mM $Fe^{+2}_{,}$ pH = 3.0, 55.37 mA/cm^2, 35°C, Flow rate 3000 cm^3/min, flow reactor with 90.3 cm^2 electrodes, 120 min in batch, UV radiation 32 W/m^2, irradiated volume = 1.57	TOC 75 MCE 71
[45]	2011	Beta-blockers atenolol, metoprolol tartrate and propranolol hydrochloride	Reactor capacity = 10 L of 100 mg/L pollutant, 0.5 mM $Fe^{+2}_{,}$ 0.1 M Na_2SO_4, pH = 3.0, 55.4 mA/cm^2, 35°C, Flow rate 250 L/h, flow reactor with 90.25 cm^2 electrodes, 360 min in batch, UV radiation 19 W/m^2	TOC 94–97 MCE 95–97

TOC, Total organic carbon; DOC, Dissolver organic carbon; MCE, Mineralization current efficiency.

still a concrete reality. Engineers and chemists need to work together to solve the most important problems that limit the real industrial application of these approaches. Optimization strategies must be applied to design large-scale reactors in order to reduce economic and energy costs.

CONCLUSIONS

As summarized in this chapter, the Fenton reaction processes can be considered a promising alternative to treat wastewaters as well as a practical application. It is a fact that the most popular reactor type for use is the batch reactor type but higher removal efficiencies are usually obtained by applying coupled processes. For the reactors that use UV irradiation, it is common to place the lamp sources at the center of the reactor by using a water-cooling jacket to maintain constant the temperature. Distributed double-electrode and divided three-electrode electrochemical cells are the most common electro-Fenton reactors in which higher removal efficiencies have been achieved by employing carbon cathodes and iron anodes installed in parallel. Other electrode materials, such as platinum and boron doped diamond (BDD), are frequently used as the anode for degrading organic compounds via electro-Fenton arrangement. Nevertheless, some topics that need further study are the role of temperature, H_2O_2 dosage, the iron effect on Fenton's reagent, and solar developments as well as new strategies to improve efficiency to reduce costs. Finally, the reactor design must be improved in order to optimize conditions so that the process can be used at an efficient, effective, and profitable industrial scale.

Acknowledgment

The authors thank UG under the Project 778/2016 (Convocatoria Institucional de Apoyo a la Investigación Científica 2016-2017).

References

[1] A. Babuponnusami, K. Muthukumar, A review on Fenton and improvements to the Fenton process for wastewater treatment, J. Environ. Chem. Eng. 2 (2014) 557–572.
[2] T. Mandal, S. Maity, D. Dasgupta, S. Datta, Advanced oxidation process and biotreatment: their roles in combined industrial wastewater treatment, Desalination 250 (2010) 87–94.
[3] I. Grčić, D. Vujević, J. Šepčić, N. Koprivanac, Minimization of organic content in simulated industrial wastewater by Fenton type processes: a case study, J. Hazard. Mater. 170 (2009) 954–961.
[4] B. Bianco, I. De Michelis, F. Vegliò, Fenton treatment of complex industrial wastewater: optimization of process conditions by surface response method, J. Hazard. Mater. 186 (2011) 1733–1738.
[5] P. Ghosh, A.N. Samanta, S. Ray, COD reduction of petrochemical industry wastewater using Fenton's oxidation, Can. J. Chem. Eng. 88 (2010) 1021–1026.
[6] A. Coelho, A.V. Castro, M. Dezotti, G.L.J. Sant'Anna, Treatment of petroleum refinery sourwater by advanced oxidation processes, J. Hazard. Mater. 137 (2006) 178–184.
[7] J.J. Pignatello, E. Oliveros, A. MacKay, Advanced oxidation processes for organic contaminant destruction based on the Fenton reaction and related chemistry, Crit. Rev. Environ. Sci. Technol. 36 (2006) 1–84.
[8] E. Neyens, J. Baeyens, A review of classic Fenton's peroxidation as an advanced oxidation technique, J. Hazard. Mater. 98 (2003) 33–50.
[9] X. Domenech, F. Wilson, W. Jardim, M. Litter, (2004). Procesos avanzados de oxidación para la eliminación de contaminantes. In: M.A. Blesa and B. Sánchez (editores), Eliminación de contaminantes por fotocatálisis heterogénea, Colección Documentos Ciemat, (Capítulo 1).

[10] V. Kavitha, K. Palanivelu, The role of ferrous ion in Fenton and photo-Fenton processes for the degradation of phenol, Chemosphere 55 (2004) 1235–1243.
[11] A.P.S. Batista, R.F. Pupo, Parameters affecting sulfonamide photo-Fenton degradation-iron complexation and substituent group, J. Photochem. Photobiol. A Chem. 232 (2012) 8–13.
[12] V. Vatanpour, N. Daneshvar, M.H. Rasoulifard, Electro-Fenton degradation of synthetic dye mixture: influence of intermediates, J. Environ. Eng. Manage 19 (2009) 277–282.
[13] E. Brillas, E. Mur, R. Sauleda, L. Sanchez, J. Peral, X. Domènech, J. Casado, Aniline mineralization by AOPs: anodic oxidation, photocatalysis, electro-Fenton and photoelectro-Fenton processes, Appl. Catal. B Environ. 16 (1998) 31–42.
[14] S.H. Bossmann, E. Oliveros, S. Göb, S. Siegwart, E.P. Dahlen, L. Payawan, M. Straub, M. Wörner, A.M. Braun, New evidence against hydroxyl radicals as reactive intermediates in the thermal and photochemically enhanced Fenton reactions, J. Phys. Chem. 102 (1998) 5542–5550.
[15] E. Oliveros, O. Legrini, M. Hohl, T. Müller, A.M. Braun, Industrial waste water treatment: large scale development of a light-enhanced Fenton reaction, Chem. Eng. Process. 36 (1997) 397–405.
[16] C. Walling, Fenton's reagent revisited, Acc. Chem. Res. 8 (1975) 125–131.
[17] K. Cruz-González, O. Torres-López, A. García-León, J.L. Guzmán-Mar, L.H. Reyes, A. Hernández-Ramírez, J.M. Peralta-Hernández, Determination of optimum operating parameters for acid yellow 36 decolorization by electro-Fenton process using BDD cathode, Chem. Eng. J. 160 (2010) 199–206.
[18] K. Cruz-González, O. Torres-López, A. García-León, E. Brillas, A. Hernández-Ramírez, J.M. Peralta-Hernández, Optimization of electro-Fenton/BDD process for decolorization of a model azo dye wastewater by means of response surface methodology, Desalination 286 (2012) 63–68.
[19] O. García, E. Isarain-Chávez, S. García-Segura, E. Brillas, J.M. Peralta-Hernández, Degradation of 2,4-dichlorophenoxyacetic acid by electro-oxidation and electro-Fenton/BDD processes using a pre-pilot plant, Electrocatalysis 4 (2013) 224–234.
[20] O. García, E. Isarain-Chávez, A. El-Ghenymy, E. Brillas, J.M. Peralta-Hernández, Degradation of 2,4-D herbicide in a recirculation flow plant with a Pt/air-diffusion and a BDD/BDD cell by electrochemical oxidation and electro-Fenton process, J. Electroanal. Chem. 728 (2014) 1–9.
[21] C.A. Martínez-Huitle, E. Brillas, Decontamination of wastewaters containing synthetic organic dyes by electrochemical methods: a general review, Appl. Catal. B Environ. 87 (2009) 105–145.
[22] A. Dirany, S. Efremova-Aaron, N. Oturan, I. Sires, M.A. Oturan, J.J. Aaron, Study of the toxicity of sulfamethoxazole and its degradation products in water by a bioluminescence method during application of the electro-Fenton treatment, Anal. Bioanal. Chem. 400 (2011) 353–360.
[23] I. Sirés, E. Brillas, M.A. Oturan, A. Mehmet A, M.A. Rodrigo, M. Panizza, Electrochemical advanced oxidation processes: today and tomorrow. a review. Environ. Sci. Pollut. Res. 21 (2014) 216.
[24] Y. Luo, L. Xu, M. Rysz, Y. Wang, H. Zhang, P.J.J. Alvarez, Occurrence and transport of tetracycline, sulfonamide, quinolone, and macrolide antibiotics in the Haihe River basin, China, Environ. Sci. Technol. 45 (2011) 1827–1833.
[25] C. Flox, P.L. Cabot, F. Centellas, J.A. Garrido, R.M. Rodríguez, C. Arias, E. Brillas, Solar photoelectro-Fenton degradation of cresols using a flow reactor with a boron-doped diamond anode, Appl. Catal. B Environ. 75 (2007) 17–28.
[26] C. Flox, J.A. Garrido, R.M. Rodríguez, P.L. Cabot, F. Centellas, C. Arias, E. Brillas, Mineralization of herbicide mecoprop by photoelectro-Fenton with UVA and solar light, Catal. Today 129 (2007) 29–36.

[27] X. Florenzaa, A. Solano, F. Centellas, C. Martínez-Huitle, E. Brillas, S. Garcia-Segura, Degradation of the azo dye acid red 1 by anodic oxidation and indirect electrochemical processes based on Fenton's reaction chemistry. Relationship between decolorization, mineralization and products, Electrochim. Acta 142 (2014) 276–288.
[28] L. Almeida, S. García-Segura, C. Arias, N. Bocchi, E. Brillas, E. Electrochemical, Mineralization of the azo dye acid red 29 (Chromotrope 2R) by photoelectro-Fenton process, Chemosphere 89 (2012) 751–758.
[29] A. Thiam, M. Zhou, E. Brillas, I. Sirés, Two-step mineralization of Tartrazine solutions: study of parameters and by-products during the coupling of electrocoagulation with electrochemical advanced oxidation processes, Appl. Catal. B Environ. 150 (2014) 116–125.
[30] A. Thiam, I. Sirés, J. Garrido, R. Rodríguez, E. Brillas, Decolorization and mineralization of Allura red AC aqueous solutions by electrochemical advanced oxidation processes, J. Hazard. Mater. 290 (2015) 34–42.
[31] A. Khataee, M. Zarei, M. Moradkhannejhad, Application of response surface methodology for optimization of azo dye removal by oxalate catalyzed photoelectro-Fenton process using carbon nanotube-PTFE cathode, Desalination 258 (2010) 112–119.
[32] M. Iranifam, M. Zarei, A. Khataee, Decolorization of C.I. Basic yellow 28 solution using supported ZnO nanoparticles coupled with photoelectro-Fenton process, J. Electroanal. Chem. 659 (2011) 107–112.
[33] A. Solano, S. Garcia-Segura, C.A. Martínez-Huitle, E. Brillas, Degradation of acidic aqueous solutions of the diazo dye Congo red by photo-assisted electrochemical processes based on Fenton's reaction chemistry, Appl. Catal. B Environ. 168–169 (2015) 559–571.
[34] V. Antonin, S. García-Segura, M. Santos, E. Brillas, Degradation of Evans blue diazo dye by electrochemical processes based on Fenton's reaction chemistry, J. Electroanal. Chem. 747 (2015) 1–11.
[35] C. Flox, S. Ammar, C. Arias, E. Brillas, A. Vargas-Zavala, R. Abdelhedi, Electro-Fenton and photoelectro-Fenton degradation of indigo carmine in acidic aqueous medium, Appl. Catal. B Environ. (2006) 93–104.
[36] J. Peralta-Hernández, Y. Meas-Vong, F. Rodríguez, T. Chapman, M. Maldonado, L. Godínez, Comparison of hydrogen peroxide-based processes for treating dye-containing wastewater: decolorization and destruction of Orange II azo dye in dilute solution, Dyes Pigm. 76 (2008) 656–662.
[37] F. Moreira, S. García-Segura, V. Vilar, R. Boaventura, E. Brillas, Decolorization and mineralization of sunset yellow FCF azo dye by anodic oxidation, electro-Fenton, UVA photoelectro-Fenton and solar photoelectro-Fenton processes, Appl. Catal. B Environ. 142–143 (2013) 877–890.
[38] E.J. Ruiz, C. Arias, E. Brillas, A. Hernández-Ramírez, J.M. Peralta-Hernández, Mineralization of acid yellow 36 azo dye by electro-Fenton and solar photoelectro-Fenton processes with a boron-doped diamond anode, Chemosphere 82 (2011) 385–392.
[39] S. García-Segura, E. Brillas, Advances in solar photoelectro-Fenton: decolorization and mineralization of the direct yellow 4 diazo dye using an autonomous solar pre-pilot plant, Electrochim. Acta 140 (2014) 384–395.
[40] B.R. Garza-Campos, J.L. Guzmán-Mar, L.H. Reyes, E. Brillas, A. Hernández-Ramírez, E. J. Ruiz-Ruiz, Coupling of solar photoelectro-Fenton with a BDD anode and solar heterogeneous photocatalysis for the mineralization of the herbicide atrazine, Chemosphere 97 (2014) 26–33.
[41] A.R.F. Pipi, I. Sirés, A.R. De Andrade, E. Brillas, Application of electrochemical advanced oxidation processes to the mineralization of the herbicide diuron, Chemosphere 109 (2014) 49–55.

[42] F. Gozzi, I. Sirés, A. Thiam, S.C. de Oliveira, A. Machulek Junior, E. Brillas, Treatment of single and mixed pesticide formulations by solar photoelectro-Fenton using a flow plant, Chem. Eng. J. 310 (2017) 503–513.
[43] H. Olvera-Vargas, N. Oturan, M.A. Oturan, E. Brillas, Electro-Fenton and solar photoelectro-Fenton treatments of the pharmaceutical ranitidine in pre-pilot flow plant scale, Sep. Purif. Technol. 146 (2015) 127–135.
[44] S. García-Segura, L. Cesar Almeida, N. Bocchi, E. Brillas, Solar photoelectro-Fenton degradation of the herbicide 4-chloro-2-methylphenoxyacetic acid optimized by response surface methodology, J. Hazard. Mater. 194 (2011) 109–118.
[45] E. Isarain-Chávez, R. María Rodríguez, P. Lluís Cabot, F. Centellas, C. Arias, J. Antonio Garrido, E. Brillas, Degradation of pharmaceutical beta-blockers by electrochemical advanced oxidation processes using a flow plant with a solar compound parabolic collector, Water Res. 45 (2011) 4119–4130.

Further Reading

[1] A. Thiam, I. Sirés, E. Brillas, Treatment of a mixture of food color additives (E122, E124 and E129) in different water matrices by UVA and solar photoelectro-Fenton, Water Res. 81 (2015) 178–187.

CHAPTER

10

Photo-Electrochemical Technologies for Removing Organic Compounds in Wastewater

Elisama V. dos Santos, Onofrio Scialdone*[†]

*Federal University of Rio Grande do Norte, Natal, Brazil

[†]University of Palermo, Palermo, Italy

OUTLINE

https://doi.org/10.1016/B978-0-12-813160-2.00010-9

INTRODUCTION

Water contamination is associated with industrial, agricultural, and urban human activities. The contamination of water by organic chemicals is considered a serious problem because it affects the environment, living organisms, and human health, but also because of economic activities related to the use of water. It is a critical issue that recent water framework directives are trying to address to ensure good water quality and healthy ecosystems [1,2]. For instance, many industries, including textiles, cosmetics, paper, leather, agricultural, pharmaceuticals, and food, produce large volumes of wastewater polluted with high concentrations of dyes and other components. As an example, about 280,000 tons of textile dyes are currently discharged in effluents every year and introduced into the aquatic environment [3]. As a result, they have been detected in rivers, lakes, oceans, and even drinking water all over the world. Currently, major attention is given to persistent and emerging pollutants, especially if they are likely to present endocrine-disrupting activity. Therefore, wastewater is a serious problem that requires a rapid solution in order to prevent more environmental damages.

During recent decades, scientists developed innovative technologies for wastewater remediation. Those technologies use physical, chemical, and biological principles to remove and/or eliminate contaminants from the water. Thus, these techniques require a high equipment cost and usually present low effectiveness, particularly for the remediation of many organic compounds. Biological treatments are certainly the cheapest processes, but the presence of toxic or biorefractory molecules may hinder the removal [4–6]. Over the past three decades, advanced oxidation processes (AOPs) have been developed as more effective methods of removing recalcitrant and persistent organic compounds from wastewater [7]. AOPs are based on the in situ generation of powerful oxidizing agents, such as hydroxyl radicals ($^{\bullet}OH$), with an adequate concentration to achieve the effective removal of organic pollutants. The hydroxyl radical is the second strongest oxidant known after fluorine, displaying a high standard reduction potential of $E°$ ($^{\bullet}OH/H_2O$) = 2.80 V (SHE) and rate constants for reaction with several contaminants in the order of 10^6–10^{10} M^{-1} s^{-1} [8].

In recent years, new AOPs based on electrochemical technology, the so-called electrochemical advanced oxidation processes (EAOP), have been developed as more effective technologies to remove persistent organic compounds from wastewater [9]. EAOPs include anodic oxidation (AO), in which $^{\bullet}OH$ are generated at the anode surface, as well as the electro-Fenton (EF) and photo electro-Fenton (PEF) processes. Another process studied for the treatment of wastewater is photocatalysis (PC). This process uses a semiconductor, usually TiO_2 known as the photocatalyst, under light illumination (UV or solar) to generate electron-hole pairs with the ability to degrade many organic compounds by producing the strong oxidant $^{\bullet}OH$ at its surface [10]. During the 21st century, the combination of electrochemistry and PC in the so-called photoelectrocatalysis (PEC) was widely studied. The aim of this chapter is to review the studies devoted in recent years to the utilization of photoelectrocatalysis for the treatment of wastewater contaminated by organic pollutants.

FUNDAMENTALS OF PHOTOELECTROCATALYSIS

PEC is based in the injection of an electron (e_{CB}^-) from the valence band (VB) of a semiconductor, which is occupied, to the conductive band (CB), which is completely empty, generating a positively charged vacancy or hole (h_{VB}^+). Traditional photocatalysis has been extensively developed using the nanocrystalline anatase of TiO_2 for light-induced oxidation of organic pollutants in waters [11–13]. In 1964, Kato et al. [14] published their work on the photocatalytic oxidation of tetralin (1,2,3,4-tetrahydronaphthalene) by a TiO_2 suspension, followed by the study of McLintock et al. [15] that investigated the photocatalytic oxidation of propylene. In the 1970s, Fujushima and Honda [16] showed the possibility of using the photo-excited semiconductor titanium dioxide (TiO_2) for splitting water into hydrogen and oxygen in a photoelectrochemical solar cell with a process that was analyzed from the electrochemical point of view by Brockis et al. [17,18] in the early 1980s. TiO_2 is a semiconductor material with very attractive proprieties such as low cost, low toxicity, and a wide band gap of 3.2 eV, which results in good stability and prevents photocorrosion [19,20]. Fig. 1 illustrates the mechanism of the processes taking place in an n-type semiconductor such as TiO_2 [11].

Photocatalysis involves the irradiation of anatase TiO_2 nanoparticles in a colloidal suspension or deposited as a thin film on Ti by UV photons of sufficient energy ($\lambda < 380\,nm$) to promote an electron from the valence band to the conduction band (e_{CB}), generating a positively charged vacancy or hole (h_{VB}^+) as follows [21–23]:

$$TiO_2 + h\nu \rightarrow e_{CB} + h_{VB}{}^{+} \quad (1)$$

Organic compounds can be oxidized by the photo-generated h_{VB}^+, that is, a strong oxidizing species, whereas the promoted e_{CB} is a potential

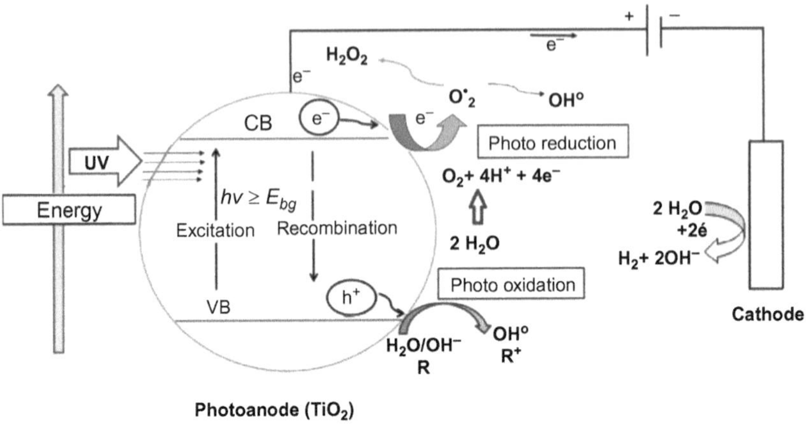

FIG. 1 Mechanism of photoelectrocatalytic process using TiO_2 photocatalyst and the reactions that occur at the surface. *Reproduced with permission from R. Daghrir, P. Drogui, D. Robert, Photoelectrocatalytic technologies for environmental applications, J. Photochem. Photobiol. A Chem. 238 (2012) 41–52. Copyright 2012, Elsevier.*

reductor and, in particular, organic compounds by heterogeneous hydroxyl radicals formed from the following reaction between the photo-generated vacancy and adsorbed water:

$$h_{VB}{}^{+} + H_2O \rightarrow {}^{\bullet}OH + H^{+} \tag{2}$$

Organic pollutants can be oxidized up to complete mineralization by the strong oxidant $^{\bullet}OH$ (although there is no definitive evidence for the formation of free hydroxyl radicals from h_{vb}^{+}). The e_{CB}^{-} can react with adsorbed O_2 to form the superoxide radical $O_2^{\bullet -}$ from the reaction (3).

$$e_{CB} + O_2 \rightarrow O_2{}^{\bullet -} \tag{3}$$

Moreover, other reactive oxygen species (ROS) such as H_2O_2 and hydroperoxyl radical $HO_2{}^{\bullet}$ can be produced via reactions (4), (5), and (6):

$$O_2{}^{\bullet -} + H^{+} \rightarrow HO_2{}^{\bullet} \tag{4}$$

$$2HO_2{}^{\bullet} \rightarrow H_2O_2 + O_2 \tag{5}$$

$$H_2O_2 + O_2{}^{\bullet -} \rightarrow {}^{\bullet}OH + {}^{-}OH + O_2 \tag{6}$$

The major loss in efficiency of photocatalysis is due to the recombination of electrons promoted to the valence band, either with unreacted holes or with adsorbed hydroxyl radicals [24–26]:

$$e_{CB} + h_{vb}{}^{+} \rightarrow TiO_2 + heat \tag{7}$$

$$e_{CB} + {}^{\bullet}OH \rightarrow {}^{-}OH \tag{8}$$

The reaction (7) represents the main disadvantage for the efficient use of classical photocatalysis. Electrochemical technology can provide a much higher efficiency for wastewater remediation using photoelectrocatalysis. This photo-assisted method consists of the application of either a constant current density j or a constant anodic potential (E_{anod}) to a TiO_2-based thin film subjected to UV illumination. In this situation, the photo-induced electrons are continuously extracted from the anode by an external electrical circuit. This causes the inhibition of reactions (3), (7), (8), and favors the production of a higher amount of holes from reaction (1) and heterogeneous $^{\bullet}OH$ from reaction (2), thus largely enhancing organic oxidation in comparison to photocatalysis [17,27,28].

PHOTO-ANODE TYPE

The PEC applicability and efficiency for organic pollutant remediation is directly associated with the nature of photo-anodes by their intrinsic photocatalytical properties. Moreover, the efficiency of a semiconductor photocatalyst depends on electron and whole recombination; therefore, the presence of oxygen or other substances is important as a good scavenger in the PEC system. In general, in a heterogeneous photocatalysis system, photo-induced molecular transformations or reactions take at the surface of a catalyst. The research on the synthesis of semiconductors for the removal of organic compounds has attracted increasing interest over the past two decades [29–31]. Currently, different materials are used in PEC for the removal of organic compounds such as dyes, pharmaceuticals, pesticides, organochlorines, and real wastewater [32–34].

There are a few articles that have reviewed either the role of every constituent (photo-anode and electrolyte) or the kinetics of photo-excited electrons for removing organic compounds. However, a detailed discussion on the role of the band gap, morphology, the composition of metal oxides, and the thickness of the metal oxide layer to develop efficient photo-anodes with suitable evidence has been overlooked in most of these articles [35].

A large number of applications have also generated a renewed scientific interest in PEC and indeed on photo-assisted reactions on semiconducting metal oxides in general. The most typical photo-anodes are based on TiO_2 because it offers important advantages as a result of non-toxic, optical, low-cost, and biocompatibility properties. However, there are other different materials used with photo-anode such as, WO_3, and ZnO_2 [10].

TiO_2 Photo-Anodes

In 1958, Kennedy et al. [36] at the University of Edinburgh studied the photo-adsorption of O_2 on TiO_2 in order to try to more fully understand the photocatalytic process for this material, They concluded that electrons were transferred to O_2 as a result of photoexcitation, and the resulting reduced form of O_2 adsorbed on the TiO_2 surface. These authors found a correlation between the ability of the TiO_2 sample to photocatalytically decompose chlorazol sky blue and to photo-adsorb O_2. There are many studies with TiO_2 because it has a sufficiently positive valence band edge to oxidize water to oxygen; it is an extremely stable material in the presence of aqueous electrolyte solutions; and it shows a prolonged lifetime for the photo-generated charge carries (e^-_{CB} and $h_{vb}{}^+$) and the spatial charge separation promoted by the anatase crystalline structure. The number of works on TiO_2 increased significantly after the publication of the 1972 paper in Nature on water photolysis [16].

Frequently, three main types of TiO_2 structures are described: rutile, anatase, and brookite. Rutile is the most stable phase for particles above 35 nm in size, as described by Fujishima et al. [37]. Anatase is the most stable phase for nanoparticles below 11 nm. On the other hand, brookite has been found to be the most stable for nanoparticles in the 11–35 nm range, although it was found that anatase is the only phase obtained for their nanocrystalline samples [38]. These three types of structures have different activities for photocatalytic reactions, but most of the practical work has been carried out with anatase. Fig. 2A and C show two low energy surfaces, (101) and (001), which are common for natural crystals. The most predominant face for anatase nanocrystals is (101), which is corrugated with alternating rows of 5-coordinate Ti atoms and bridging oxygen, which are at the edges of the corrugations [12]. The (100) surface (Fig. 2B) is less common on typical nanocrystals, but it was observed on rod-like anatase grown hydrothermally under basic conditions [12]. Moreover, this surface has double rows of 5-coordinate Ti atoms alternating with double rows of bridging oxygen [39,40].

WO_3 Photo-Anodes

The WO_3 for the decomposition of organic compounds both in liquid and gas phases has become an excellent alternative material, owing largely to the fact that it presents smaller band gap energy around 2.5–2.7 eV when it absorbs visible light up to 470 nm and generates $^{\bullet}OH$ radicals via valence band injection, but it was often studied in the framework of PEC for solar water splitting. The main drawback of tungsten oxides is an inadequate electron transfer to O_2 by single-electron reduction ($O_2 + e^- \rightarrow O_2^-$, $-0.284\ V_{NHE}$; $O_2 + H^+ + e^- \rightarrow HO_2$, $-0.046\ V_{NHE}$) [41,42]),

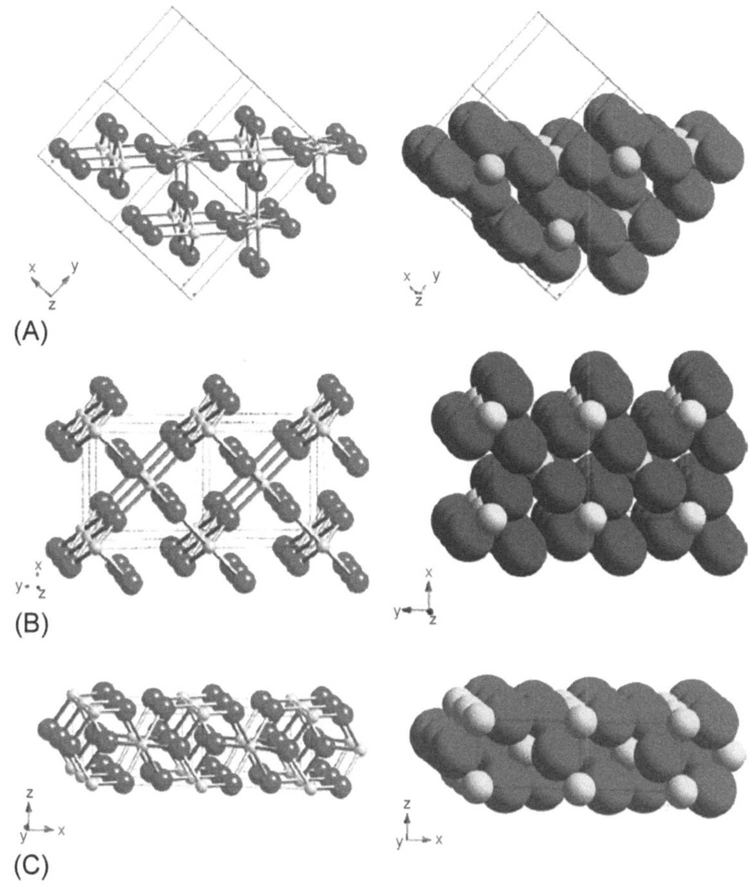

FIG. 2 Schematic representations of selected low-index faces of rutile: (A) (110); (B) (100); and (C) (001). *Reproduced with permission from A. Fujishima, X. Zhang, D.A. Tryk, TiO_2 photocatalysis and related surface phenomena, Surf. Sci. Rep. 63 (2008) 515–582. Copyright 2008, Elsevier.*

because the conduction band-edge potential is higher than the reduction potential of the oxygen couple, ($E^{\circ}(O_2/O_2^{\bullet-}) = -0.33$ VSHE). Currently, Abe et al. described that Pt loading was an attractive solution to enhance the photocatalytic properties of WO_3 because of the contribution for reduction of dioxygen ($O_2 + 2H^+ + 2e \rightarrow H_2O_2$, +0.682 V_{NHE}; $O_2 + 4H^+ + 4e \rightarrow 2H_2O$, +1.23 V_{NHE} [43]). Hence, the photo-generated electron/hole couple readily recombines, leading to low photonic efficiencies in many applications such as wastewater treatment [44]. Therefore, the utilization of both electrochemical and photocatalytic techniques is particularly useful to facilitate efficient charge separation. Indeed, applying an external electric field favors the separation of these photo-generated pairs and subsequently formed hydroxyl radicals, which contributes to the oxidation of organic compounds and oxidation products, resulting in a faster treatment of various organic compounds such as methyl orange [44] than in photocatalysis or electrocatalysis.

ZnO_2 Photo-Anodes

One of the most widely studied semiconductor materials is ZnO because of its attractive properties, such as a large optical absorption coefficient and high chemical and thermal stability. It has also been demonstrated that ZnO is an efficient photocatalyst for the degradation of pollutants [45]. The more relevant advantage of ZnO in comparison with TiO_2 is that it absorbs over a larger fraction of the UV spectrum and the corresponding threshold of ZnO is 425 nm. Furthermore, the ZnO photodegradation mechanism is likely to be similar to that of TiO_2. The mechanism of heterogeneous photocatalysis is presented below [45]:

$$ZnO + h\upsilon \rightarrow h^{+} + e^{-} \quad (9)$$

$$h^{+} + e^{-} \rightarrow \text{heat} \quad (10)$$

$$H_2O + h^{+} \rightarrow H^{+} + OH^{\bullet} \quad (11)$$

$$h^{+} + OH^{-} \rightarrow OH^{\bullet} \quad (12)$$

$$e^{-} + O_2 \rightarrow O^{\bullet}{}_2{}^{-} \quad (13)$$

$$O^{\bullet}{}_2{}^{-} + HO^{\bullet}{}_2 + H^{+} \rightarrow H_2O_2 + O_2 \quad (14)$$

$$O^{\bullet}{}_2{}^{-} + \text{organics species} \rightarrow \text{organics species} - OO^{\bullet} \quad (15)$$

$$^{\bullet}OH + \text{organics species} \rightarrow \text{intermediate} \rightarrow P \quad (16)$$

When light is irradiated over ZnO, valence band electrons are promoted to the conduction band, generating a hole (Eq. 9) or interacting separately with other molecules. The holes at the ZnO valence band can oxidize the adsorbed water molecules or hydroxide ions to produce hydroxyl radicals (Eqs. 10, 11) while electrons in the conduction band on the catalyst surface can reduce O_2 to superoxide anion (Eq. 13). This radical can give rise to organic peroxides or hydrogen peroxide in the presence of organic scavengers (Eqs. 14, 15). The hydroxyl radical attacks the organic compounds and intermediates formed, allowing the wastewater treatment (Eqs. 15, 16) [46,47].

Modified TiO_2 Materials

For metal oxides, doping with suitable cation/anion modifies the band gap, which may modulate their electrical properties. Capasso envisaged the application of band gap engineering in semiconductor devices in the late 1980s [48]. He proposed that band gaps of semiconductors can be modified for specific applications by altering the growth method of the materials. The band gap of nanocrystals can be tuned by altering the size and shape of the materials through alloying, doping, strain tuning, and band gap edge-warping techniques.

PHOTO-ASSISTED ELECTROCHEMICAL TREATMENTS

The application of PEC to wastewater remediation has been preferentially centered on the treatment of synthetic aqueous solutions. The photo-assisted treatments are based on the irradiation of a contaminated solution or a photoactive electrode with UV or solar light. These are UVA ($\lambda = 315$–400 nm), UVB ($\lambda = 285$–315 nm), and UVC ($\lambda < 285$ nm) light supplied by UV lamps as energy sources. Different emerging photo-assisted electrochemical approaches have been proposed [11,30,49–51]. In particular, photo-Fenton (H_2O_2/Fe^{2+}/UV system) and heterogeneous TiO_2 photocatalysis (TiO_2/UV system) are considered the most promising photo-assisted AOPs. A brief description of such methods is given below.

(I) The photo-Electro Fenton method, in which the contaminated solution treated under electro-Fenton conditions is exposed to UV illumination favoring the generation of homogenous $^{\bullet}OH$ and the photodegradation of complexes of Fe(III) with organics.

(II) The photoelectrocatalytic method, where a TiO_2-based thin film anode is irradiated with a UV light to produce electron excitation and positively charge holes, which can oxidize organics and water giving heterogeneous $^{\bullet}OH$. The constant current density or bias potential applied to the semiconductor electrode promotes the efficient separation of electron-hole pairs from electron transfer by an external circuit and accelerates the production of holes and heterogeneous $^{\bullet}OH$ as oxidizing species.

In recent years, different works have evaluated the utilization of PEF to remove dyes, pesticides, and organochlorides [19,52,53]. In this context, it is worth mentioning that the commercialization of TiO_2-based photocatalytic products commenced in the mid-1900s in Japan. Table 1 collects the relevant data for the treatment of water polluted by different contaminants by PEC under selected conditions using undivided cells with both TiO_2 and other semiconductors. In the following, some selected examples are reported (see Table 2).

A recent paper of Frontistis et al. [54] reported on photoelectrocatalytic oxidation over immobilized Ti/TiO_2 films (consisting of 75:25 anatase: rutile,) in the presence of simulated solar light for the degradation of bisphenol-A in water. These researchers evaluated the degradation of bisphenol-A by photocatalysis and electrochemical oxidation using a

TABLE 1 Treatment of Several Polluted Effluents by Coupled PEC Treatments

Effluent	Experimental conditions	Removal	Ref.
Chlortetracycline hydrochloride (CTC)	1 L 230 μg L^{-1}, Na_2SO_4 = 0.0 5 M, Light intensity = 6900 μW cm^{-2}, pH = 2.5–8.5, j = 3, 54 mA cm^{-2} H_2O_2-PEC	92.3–97.3 CTC 51.6–67.9 TOC	[49]
Bisphenol-A	0.06 L 300 μg L^{-1} Bisphenol-A 8.4 mg L^{-1} of DOC, pH = 8, 5 mM $HClO_4$ as the supporting electrolyte, j = 0.02–0.32 mA cm^{-2} Photocatalysis + Electrochemical Oxidation	82% BPA 53% DOC	[54]
Tetracycline (TC)	0.02 L 10 mg L^{-1}, pH = 5.5, 0.02 M Na_2SO_4, time = 3 h, TiO_2 nanotube, 2.5 mW cm^{-2}	80% removal of TC	[55]
Acid Orange 7 (AO7)	0.1 L of 15 mg L^{-1} AO7, pH = 3.0 and 11, j = 2.0 mA cm^{-2}, 0.05 M Na_2SO_4, 30.3 and 31.2 W m^{-2} Solar photocatalysis Anodic oxidation-H_2O_2 Solar photoelectrocatalysis (SPEC)	40% TOC removal-AO-H_2O_2 100% TOC removal-SPEC	[56]
Remazol Brilliant Orange 3R (RBO)	0.03 L, 5×10^{-5} M, RBO 0.5 M in 0.25 M NaCl, $NaClO_4$, KNO_3 and Na_2SO_4, pH = 2–11, 50 mW cm^{-2} PEC	Complete discoloration of RBO and mineralization of the dye (70%)	[57]
Acid Yellow 1	1 L, 100 mg L^{-1} acid yellow 1, 0.01 M Na_2SO_4, pH = 6.5		[58]
	Photolysis	7.8% TOC	
	Photoelectrocatalytic oxidation	18.7% TOC	
	Ozonation	38.7% TOC	
	Ozonation + UV	42.1% TOC	
	Ozonation + Photocatalysis	47.2% TOC	
	Ozonation + photoelectrocatalysis	99.4% TOC	

TABLE 1 Treatment of Several Polluted Effluents by Coupled PEC Treatments—cont'd

Effluent	Experimental conditions	Removal	Ref.
Procion Blue dye	0.130 L, 50 mg L^{-1} procion blue, 0.34 M NaCl and 0.014 M Na_2CO_3, pH = 2,0, j = 20 mA cm^{-2}		[59]
	Photolysis	62% COD	
	Anodic Oxidation	84% COD	
	Photolysis + Anodic Oxidation	96% COD	
Benzotriazole (BTA)	0.1 L, 2×10^{-4} M BTA, 0.05–0.5 M Na_2SO_4, pH = 6.2 15 W UV lamp (253.7 nm) Potential applied = 0–0.8 V; treatment, time = 180 min Electrochemical oxidation Direct photolysis photocatalysis photoelectrocatalysis	EO (0% removal) DP (65% removal) PO (70% removal) PEO (89.8%–90% removal)	[60]
4-chlorophenol	20 mg L^{-1}, 0.01 M Na_2SO_4, pH = 3.2–8.0, UV light intensity 2.5 mW cm^2, treatment time = 2 h; pH 3.25–8.00; C	EO (0% removal) DP (44% removal); PO (62% removal) PEO(82%–86% removal)	[61]

TABLE 2 Production of Oxidants by Coupled PEC Treatments

Pollutant	Experimental conditions	Removal	Ref.
Diclofenac sodium	0.120 L, 10 mg L^{-1}, 0.05 M Na_2SO_4, pH = 5.62–10.0, j = 5 μA cm^{-2}, 1.0–10.0 mM $S_2O_8^{2-}$	19.4 removal PEC 86.3% removal 10 mM $S_2O_8^{2-}$ PEC	[64]
C.I. Basic Blue 3 (BB3)	0.5 L, 20 mg L^{-1} in 0.1–2.0 mM $S_2O_8^{2-}$, pH = 7.0, light intensity = 21.9 W m^2	100% removal	[69]
Phenol, nalidixic acid, mixture of pesticides	21 mg L^{-1} of oxamyl, 5.1 mg L^{-1} of methomyl, 5 mg L^{-1} of imidacloprid, 8 mg L^{-1} of dimethoate, and 13.5 mg L^{-1} of pyrimethanil. TOC = 50 mg L^{-1}, 0.7–6.2 mM H_2O_2, PVF-TiO_2-Fe oxide		[34]
Remazol Brilliant Orange 3R	0.03 L, 0.05 mM solution in 0.50 M NaCl, pH = 6.0, the electrode (TiO_2) to UV illumination and Pt with counter electrode (9 cm^{-2}), 5 W	100% color, 30 min 70% TOC, 240 min	[65]
Acid Orange 7 azo	0.1 L of 15 mg L^{-1} AO7 solution in 0.05 M Na_2SO_4 at pH 7.0 and 35°C and anodic current density (j_{anod}) of 1.0 mA cm^{-2}	100% azo dye 100% color 40% TOC	[56]
Oxyfluorfen (OXY) and Sodium dodecyl sulfate (SDS)	0.8 L, 100 mg L^{-1}, SDS = 100–5000 mg L^{-1}, 500 mg L^{-1} $NaHCO_3$, pH = 6.8, intensity of 930 W cm^{-2}, j = 30 mA cm^{-2} Photolytic (PO) Photo-electrolytic (PEC)	PO 8.5% TOC 16% OXY PEC 100% TOC 100% OXY	[66]
Naphthol Blue Black (NBB)	8.8×10^{-5} M NBB in 0.5 M NaCl and 8.8×10^{-5} M NBB in 0.5 M Na_2SO_4, pH = 4.5–12.0, E = 1.10–1.12 V	100% removal	[71]

TABLE 2 Production of Oxidants by Coupled PEC Treatments—cont'd

Pollutant	Experimental conditions	Removal	Ref.
Ponceau 6R	0.2 L of 0.1 mM dye, 2.0 mM of ammonium persulfate, pH = 2, graphite electrodes		[94]
Petroleum	0.8 L, 359–1298 mg L^{-1}, SDS = 100–5000 mg L^{-1}, 500 mg L^{-1} $NaHCO_3$, pH = 7.4, intensity of 930 W cm^{-2}, j = 30 mA cm^{-2}, 24 kHz and maximum ultrasonic power 200 W BDD-electrolysis Photolysis US US/BDD-electrolysis Photo/BDD-electrolysis	100% TOC	[67]
Rhodamine B	2.09 × 10–4 M in 7.04 × 10^{-3} M NaCl, 2.09 × 10^{-4} M in 7.04 × 10^{-3} $HClO_4$, j = 0–90 mA cm^{-2} No UV irradiation UV irradiation	No UV irradiation 20% COD UV irradiation 100% COD	[36]
Caffeine	600 mL of 100 mg L^{-1} caffeine in industrial effluent (13 mg L^{-1} TOC)., BDD/SS cell (78 cm^2 area), liquid flow rate = 21.4 L h^{-1}, j = 30 mA cm^{-2}		[95]
	EO, 750 min	92% TOC	
	UVC (4 W, interelectrodes) + EO, 750 min	93% TOC	

stirred reactor directly illuminated with a solar simulator for the solar PEC. The experiments were made changing the current density in the range 0.02–0.32 mA cm^{-2}, the TiO_2 loading, the organic concentration (120–820 μg L^{-1}), and the pH using a galvanostatic-potentiostat with the TiO_2 film operating as the anode and zirconium as the cathode. The degradation of bisphenol-A was favored using current densities up to 0.04 mA cm^{-2} and a lower organic concentration. The PEC process was compared with pure photocatalysis or electrochemical oxidation alone. It was shown that the application of the potential and photocatalysis resulted in a synergistic effect, with an increase of the photocatalytic efficiency between 24% and 97%. The effective promotion of photoelectrocatalysis was supposed to be associated with the efficient separation of photo-generated holes from electrons, thus maximizing their utilization as primary oxidants or a source to generate secondary oxidants such as hydroxyl radicals [54].

Furthermore, the efficient degradation of several dyes by PEC with a TiO_2 photo-anode has been well proven. Bessegato et al. [62] used TiO_2 nanotube electrodes prepared by the sol-gel method to treat 100 mg L^{-1} of Acid Yellow 1 in 0.01 Na_2SO_4 under a UV-B 36 W lamp and $E_{anod} = +2.0$ V on TiO_2NT anode and graphite cathodes, using irradiation from a UV-B lamp of 36 W concomitant to an ozone flow of 1.25×10^{-4} mol min^{-1}. After 30 min, the color removal was 100% for O_3 + PEC treatment. However, a similar behavior was obtained for the ozonation process. Nevertheless, when the TOC removal was compared, the mineralization was four times higher compared to the O_3 + PEC system after 60 min of treatment. The authors reported that this integrated technique is an efficient way to increase the degradation rate of Acid Yellow 1, due to the enhanced $^{\bullet}OH$ generation formed by PEC and ozonation that accelerates the destruction of pollutants.

Y. Ding et al. [60] described a strong improvement of the TOC removal for a water solution (100 mL) of 2×10^{-4} M Benzotriazole and 0.5 M Na_2SO_4 at pH 6.2 treated by photoelectrocatalysis in comparison to results achieved by electrochemical oxidation and photolysis, using TiO_2 as the working electrode and Pt as the counter electrode with a central 15 W UVA lamp in batch operation mode. The authors observed, after 165 min, a negligible removal of Benzotriazole for electrochemical oxidation, removal of about 65% for photolysis, and removal of about 90% for the photoelectrochemical process. An interesting study from Liu et al. [55] focused on tetracycline degradation in 0.02 M Na_2SO_4. The authors compared the efficiency removal achieved via PC, EO, and PEC using a TiO_2 nanopore arrays (TNPs) electrode. PEC offered a synergetic effect for tetracycline degradation on the TNPs electrode; in particular, the rate constant for the PEC process of the TNPs electrode was ~6–7 times as high as for the PC process. The removal rate of tetracycline was accelerated in the presence of 2.5 mW cm^{-2} light intensity for PEC. Sergi-Garcia et al. [56] demonstrated that a novel synthesized TiO_2 photo-anode was

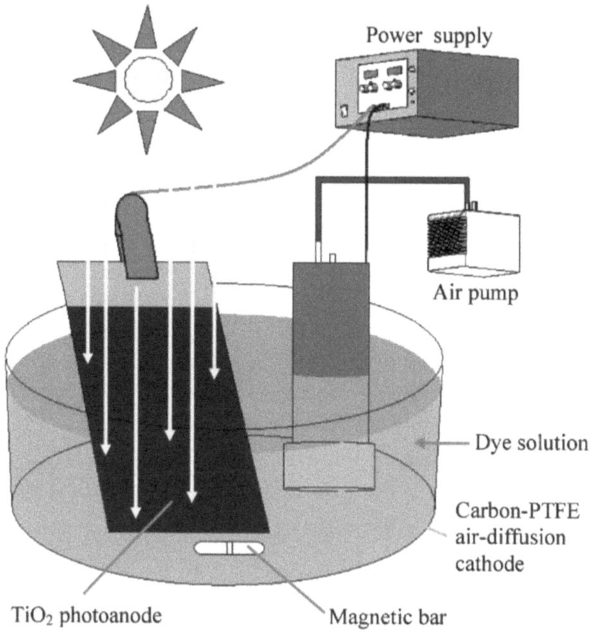

FIG. 3 Schematic representations of the photoelectrochemical reactor used for the treatment of the water solution of tetracycline. *Reproduced with permission from S. Garcia-Segura, S. Dosta, J.M. Guilemany, Enric Brillas. Solar photoelectrocatalytic degradation of Acid Orange 7 azo dye using a highly stable TiO_2 photoanode synthesized by atmospheric plasma spray. Appl. Catal. B Environ. 132–133 (2013) 142–150. Copyright 2013, Elsevier.*

particularly effective in removing and decolorizing AO7 azo dye using SPEC under direct sunlight irradiation in a photoelectrochemical cell with an air-diffusion cathode that generates H_2O_2. A $5\,cm^2$ photo-anode and a $3\,cm^2$ carbon-polytetrafluoethylene (PTFE) air-diffusion cathode were used as electrodes. They found a pH = 7.0 as optimal for 100% decolorization of a water solution (100 mL) of Acid Orange 7 ($15\,mg\,L^{-1}$) and 0.05 M Na_2SO_4 using the cell of Fig. 3 at $j_{anod} = 1.0\,mA\,cm^{-2}$ and 35°C after 120 min of electrolysis under direct sunlight. The SPEC process gave a synergistic effect concerning individual SPC and AO-H_2O_2 processes under comparable conditions as a result of the larger production of $^{\bullet}OH$ generated by the higher amounts of photo-generated holes that can be separated from photo-generated electrons.

PRODUCTION OF OXIDANTS BY PEC

One of the key points to explain the high efficiencies reached by PEC and other electrochemical technologies in the removal of organic pollutants is the understanding of the role of mediated oxidation processes in the overall oxidation carried out during the treatment. Several works have described the best performance of coupled photo-assisted treatments. For example, Zanoni et al. [57] studied the treatment of a water solution of 0.05 mM Remazol Brilliant Orange with different supporting electrolytes at pH 6.0 in the anodic compartment of a divided cell operating with a titanium dioxide thin-film photoelectrode illuminated by UV light with $E_{anod} = +1.0$ V (SCE) upon 450 W Xe-Hg arc illumination for 30 min. The authors observed that: (1) the photoelectrocatalytic treatment gave drastically higher abatements than the photocatalytic and electrochemical ones; (2) for the photoelectrocatalytic process, the utilization of NaCl as a supporting electrolyte instead of $NaClO_4$ and Na_2SO_4 resulted in a dramatic enhancement of the organic removal. This result can be explained focusing on the ability of the holes formed at the photo-anode from reaction (1) to oxidize Cl^- to active chlorine species (Cl_2, HClO, and ClO^-) and probably other radicals such as $Cl^\bullet$ and $Cl_2^\bullet$ by reactions (17)–(19) (which oxidize organics more quickly than heterogeneous $^\bullet OH$) [57]:

$$2Cl^- \rightarrow Cl_2 + 2e^- \tag{17}$$

$$Cl_2 + H_2O \rightarrow HClO + Cl^- + H^+ \tag{18}$$

$$HClO + h\upsilon \rightarrow Cl^\bullet + {}^\bullet OH \tag{19}$$

This behavior can be explained because chloride ions might be oxidized at a less positive potential under photoelectrocatalytic conditions, as denoted by Eq. (20).

$$TiO_2 - h_{vb}{}^+ + Cl^-{}_s \rightarrow TiO_2 - Cl_s{}^\bullet \tag{20}$$

However, the most referenced example of mediated electrochemical oxidation arises from the effect of sodium sulfate on the oxidation of organics. Sulfate is commonly contained in most of the wastewater flow of the textile industry and can be easily oxidized to persulfate by many types of anode materials. For this reason, light irradiation activation can promote the formation of high persulfate by UV light irradiation and PEC. In this case, irradiation can be applied naturally (solar driven) or artificially (using UV lamps) [63]. Peroxydisulfate ($S_2O_8{}^{2-}$) is a strong oxidant ($E° = 2.05$ V) and has been reported to be effective for degradation of organic compounds in wastewater in acidic or basic media through direct chemical oxidation where peroxydisulfate is used as a sacrificial material.

Liu et al. [64] introduced persulfate ($S_2O_8{}^{2-}$) into the PEC system, which consisted of a Y-Bi_2MoO_6 photo-anode and a copper foil cathode (Bi_2MoO_6/Cu/PEC/PS system). They used the system for the removal of diclofenac.

In the presence of an applied potential of 1.5 V vs SCE and of an initial pH of 5.6, the removal of diclofenac (initial concentration of 10 mg L^{-1})

was increased from 19.4% to 86.3% by the addition of 10 mM persulfate. The effect of persulfate can be due to the generation of active free radicals involving $SO_4^{-\bullet}$ (Eq. 21) and $HO^\bullet$, which was confirmed by electron spin resonance analysis. Photochemical-activated decomposition of $S_2O_8^{2-}$ ion SO_4^- radical has been proposed as a method of accelerating the process [65–69], as summarized in Eq. (21) and in the following reactions. The decomposition can also be explained because the electrochemical oxidation ROS such as heterogeneous $^\bullet OH$ by reaction (33) are generated, although physisorbed is the strongest oxidant of pollutants.

$$S_2O_8^{2-} + h\upsilon \rightarrow 2(SO_4^-)^\bullet \tag{21}$$

$$SO_4^{-\bullet} + RH_2 \rightarrow SO_4^{2-} + H^+ + RH \tag{22}$$

$$RH + S_2O_8^{2-} \rightarrow R + SO_4^{2-} + H^+ + SO_4^{-\bullet} \tag{23}$$

$$SO_4^{-\bullet} + RH \rightarrow R^\bullet + SO_4^{2-} + H^+ \tag{24}$$

$$2R^\bullet \rightarrow RR\,(\text{dimer}) \tag{25}$$

$$SO_4^{-\bullet} + H_2O \rightarrow HSO_4^- + OH^\bullet \tag{26}$$

$$HSO_4^- \rightarrow H^+ + SO_4^{2-} \tag{27}$$

$$OH^\bullet + S_2O_8^{2-} \rightarrow HSO_4^- + SO_4^{-\bullet} + ½O_2 \tag{28}$$

$$SO_4^{-\bullet} + OH^\bullet \rightarrow HSO_4^- + ½O_2 \tag{29}$$

$$2OH^\bullet \rightarrow H_2O_2 \tag{30}$$

$$OH^\bullet + H_2O_2 \rightarrow H_2O + HO_2^\bullet \tag{31}$$

$$S_2O_8^{2-} + H_2O_2 \rightarrow H^+ + 2SO_4^{2-} + O_2 \tag{32}$$

Li et al. [70] studied the decolorization of a water solution (50 mL) of 0.015 mM rhodamine B and 0.10 M Na_2SO_4 at Ti/TiO_2, $E_{anod} = +0.50$ V (SCE), exposed to 30 W UVA, attaining the best color removal of 95% after 70 min of electrolysis increasing the pH.

Xie et al. [51] studied the azo dye orange degradation in an undivided cell using an air-diffusion reticulated vitreous carbon as the cathode for the H_2O_2 electrogeneration and TiO_2-Ti mesh as the photo-anode for the photoelectrocatalysis under UV light irradiation. They observed that the utilization of electro-Fenton-PEC enhanced the degradation of Orange-G dye. The synergistic removal achieved by EF-PEC was due to the contribution of Fe^{2+} by reactions (33)–(36).

$$Fe^{2+} + H_2O_2 \rightarrow Fe^{3+} + {}^\bullet OH + OH^- \tag{33}$$

$$H_2O_2 + h\upsilon \rightarrow 2{}^{\bullet}OH \tag{34}$$

$$Fe^{2+} + H_2O_2 + h\upsilon \rightarrow Fe(OH)^{2+} + {}^{\bullet}OH \tag{35}$$

$$Fe(OH)^{2+} + h\upsilon \rightarrow Fe^{2+} + {}^{\bullet}OH \tag{36}$$

and finally, the organic compounds can be degraded by reaction:

$$R + {}^{\bullet}OH \rightarrow \text{degradation products} \tag{37}$$

while Fe^{2+} regeneration also could occur by the reactions:

$$Fe^{3+} + H_2O_2 \rightarrow Fe^{2+} + H + HOO^{\bullet} \tag{38}$$

$$Fe^{3+} + HOO^{\bullet} \rightarrow Fe^{2+} + H^+ + O_2 \tag{39}$$

$$Fe(OH)^{2+} + h\upsilon \rightarrow Fe^{2+} + H + HO^{\bullet} \tag{40}$$

Hepel and Luo [71] presented an interesting study on the degradation of an 85.4 mg L^{-1} azo dye Orange G solution in 0.05 M Na_2SO_4 with 0.5 mM Fe^{2+} in an undivided cell to check the performance of nanotubes TiO_2NTs under 80 W UV illumination by several treatments. The best performance was shown by the EF-PEC process; it has been demonstrated that the decolorization and mineralization of dye solutions increased in the sequence PEC < EF < coupled PEC-PEF, with better results observed at lower pH values. By applying the same electric charge per unit volume, TOC decays of 80%, 87%, and 97% were achieved for PEC, EF, and the coupled PEC-PEF process.

EFFECT OF THE NATURE OF THE CATHODE

According to previous studies [61,72], the presence of an oxidant agent such as H_2O_2 is beneficial for the photocatalytic reaction by trapping the conduction band electron in order to generate more hydroxyl radicals. It has been known since 1882 that hydrogen peroxide can be generated in the aqueous medium by the cathodic reduction of dissolved O_2 gas at carbonaceous electrodes with high surface area such as reticulated vitreous carbon, graphite, active carbon fiber, carbon felt, and graphite-polytetrafluorethylene (PTFE) (Eq. 41) [23,73,74].

$$O_2 + 2H^+ + 2e^- \rightarrow H_2O_2 \quad E = 0.695\ \text{V/NHE} \tag{41}$$

Bañuelos et al. [75] studied a photo electron-Fenton process based on the electrogeneration of H_2O_2 from O_2 reduction at a carbon-felt (CF) cathode and the utilization of BDD anodes and UV light for malachite green degradation in 0.05 M Na_2SO_4 at pH 3. They observed that CF allowed to generate, in the absence of malachite green and of UV light, a maximum concentration of $94\,mg\,L^{-1}$ of H_2O_2 for the best adopted current density value ($21.7\,mA\,cm^{-2}$). In the presence of UV light, a higher removal of the color and TOC was achieved by photo electron-Fenton with respect EF because the production of $^{\bullet}OH$ from Fenton's reaction was enhanced by the additional photolysis of Fe(III)-hydroxy species and Fe(III) complexes with oxalate of the dye. These reactions include, for example, the oxidation of BDD($HO^{\bullet}$) to O_2 via reaction (42) and the removal of $HO^{\bullet}$ with Fe^{2+} and H_2O_2 from reactions (43) and (44) respectively.

$$BDD(HO^{\bullet}) \rightarrow 2BDD + O_2 + 2H^+ + 2e^- \tag{42}$$

$$Fe^{2+} + HO^{\bullet} \rightarrow Fe^{3+} + OH^- \tag{43}$$

$$H_2O_2 + HO^{\bullet} \rightarrow HO_2^{\bullet} + H_2O \tag{44}$$

Ding et al. [72] studied the effectiveness of the integrate PEC/EF treatment of rhodamine B in 0.05 M Na_2SO_4 at neutral pH 6.2 in an undivided and divided semicylindrical tank reactors, using Bi_2WO_6/FTO photocatalysts exposed to 300 W visible illumination. The gas diffusion cathode was composed of Fe/Fe_2O_3 core-shell nanowires supported on activated fiber. It was found that more hydrogen peroxide can be electrogenerated on the Fe/Fe_2O_3 activated fiber cathode by photo-generated electrons transferred from the Bi_2WO_6/FTO anode. On the other hand, the degradation of RhB in two separated cells was enhanced in comparison to the undivided cell.

EFFECT OF pH

According to different researchers [76–78], a factor that influences significantly various electrochemical processes for wastewater treatment is pH. Indeed, solution pH can affect the capacity of adsorption and dissociation of the organic compounds, the charge distribution to the photocatalyst surface, and the oxidation potential of the valence band. According to Daghrir et al. [11], three probable reaction mechanisms can contribute to the degradation of organic compounds in PEC processes: (i) direct attack of hydroxyl radicals, (ii) direct photo-oxidation by the positive hole in the valence band, and (iii) direct photo-reduction by the electron in the conduction band. Each of these depends on the pH value and the nature of the substrate.

Selcuk et al. [79] reported the removal of humic acid dissolved in an aqueous medium by the PEC process. They observed that the removal of humic acid was much more effective at pH 6. Furthermore, two different phenomena contributed to the color and degradation removal: the H+ production occurs on the surface and the strong effect of a positive surface charge applied on the photoelectrode improves adsorption of the HA on the semiconductor surface. Wang et al. [61] studied the effect of pH of the solution on the 4-chlorophenol degradation at the TiO_2 nanotube with a constant bias of potential of 0.6 V and a light intensity of $2.5\,mW\,cm^{-2}$. The degradation was enhanced in both the alkaline medium and the acid medium. Hepel et al. [71] studied photoelectrochemical processes for the removal of Naphthol Blue Black (NBB) dye in aqueous solutions using a nanocrystalline WO_3 film on a Pt mesh electrode as the photo-anode. The authors observed a different behavior for low and high pH. In the presence of chlorides, at low pH (1.0–4.5), photooxidation of the medium favors the formation of Cl and Cl_2^- radicals, which are powerful oxidants for the NBB dye. In alkaline solutions, the formation of OH radicals was predominant over Cl and Cl_2^- radicals, but the kinetics of NBB degradation were much slower than in acidic media.

APPLICATION OF PHOTOELECTROCATALYTIC PROCESSES

Inactivation of Bacteria

Bessegato et al. [80] investigated the use of W/WO_3 electrodes prepared by electrochemical anodization applied in the photoelectrocatalytic disinfection of *Candida parapsilosis* using ultraviolet-visible (UV-Vis) and visible irradiation. The results showed that, under optimized experimental conditions of $0.1\,mol\,L^{-1}$ NaCl, pH = 1.5 V and UV-Vis irradiation, it was possible to promote total inactivation and around 70%–80% of mineralization of organic matter. On the other hand, the results obtained through the microbial counts showed that after 1 min of treatment, 99.99% inactivation of $1.6 \times 10^6\,CFU\,mL^{-1}$ of *C. parapsilosis* was obtained using both visible and ultraviolet radiation.

Lei et al. [81a] evidenced the applicability of PEC to disinfect microorganisms from aqueous environments. They applied to inactivate Candida parapsilosis sensu stricto in planktonic and biofilm calls, and PEC efficiency was compared to that of PC and PEC treatment. They indicated that the PEC oxidation performed at the Ti/TiO_2 and Ti/TiO_2-Ag nanotube photo-anodes is a powerful strategy to inactivate C. parapsilosis sensu stricto.

PEC technology using TiO_2 as a photo-anode has been applied by Liu et al. [81b], who studied the performances of the photoanode with UV and visible light for PC and PEC processes using *Escherichia coli* as the test bacteria. They determined that, under UV irradiation, 99.97% inactivation of 45 mL of $1.0 \times 10^7\,CFU\,mL^{-1}$ *E. coli* was attained within 10 min for PEC treatments while 96.40% inactivation was obtained inside 30 min for PC treatment. The TiO_2 photo-anodes were used for sunlight driven PC treatment and 1.0 mL of 1.0×10^7 CFU mL^{-1} *E. coli* cells were completely inactivated within 2 min.

Treatment of Dyes

Likewise, PEC could be a good alternative for discoloration and mineralization of dyes. Semiconductor materials such as TiO_2 [80,82], WO_3 [58,83], Bi_2WO_6 [84], BDD-$ZnWO_4$ [85], PbO_2 [86], $BiVO_4$ [87], $BiPO_4$/BiOI [88], and Fe2O3 [89] have been checked for the remediation of dye solutions by PEC. UV irradiations (254 nm) have been applied to treat textile wastewater before and after conventional treatment.

As an example, Oliveira et al. [90] studied the photoelectrocatalytic treatment of 200 mL of the solution containing 20 mg L^{-1} of a commercial azo dye, CI Disperse Blue 291 dye, using NaCl as a supporting electrolyte showing the generation of strong oxidants. They used an irradiated and biased Ti/O_2 photo-anode illuminated by a commercial lamp of 125 W cm^{-2} operating as a UV light source (315–400 nm) and platinum gauze as the counter electrode. The performances of the process were compared with that of a conventional chlorination system. After 120 min of treatment, only a maximum of TOC removal was obtained by the chlorination. The chlorination of the dye solution produced several byproducts that could also be mutagenic. However, the dye solution treated with the PEC chlorination process in 0.2 mol L^{-1} NaCl and 0.2 mol L^{-1} Na_2SO_4 for 120 min completely removed mutagenicity.

During the treatment of Naphthol Blue Black (NBB), Hepel et al. [71], showed a decrease of the NBB dye absorbance ($\lambda = 616$ nm) with time during its PEC degradation on a WO_3 film electrode in 0.5 M NaCl electrolytes containing 8.8×10^{-5} M NBB and 0.5 M Na_2SO_4 with 8.8×10^{-5} M NBB at low (4.5) and high pH (12). They observed a strong pH dependence for the NBB dye degradation rate in chloride solutions. Indeed, in an acidic medium the degradation rate was almost constant in the pH range 1.0–4.5 and decreased for pH values higher than 4.5.

Treatment of Pesticides

Most authors confirmed the oxidation ability of electrochemical technologies and, in particular, of PEC to remove pesticides. Zhao et al. [91] used Bi_2MoO_6 particles selectively deposited onto a BDD surface from a complex amorphous solution via a dip-coating method followed by calcination. They performed the degradation experiments using a cylindrical quartz reactor containing a 60 mL solution of 10 mg L^{-1} ibuprofen and

$15\,mg\,L^{-1}$ naproxen. It was shown that ibuprofen was degraded via photocatalysis and an electrooxidation process with the applied bias potential of 2.0 V. Furthermore, ibuprofen and naproxen were rapidly degraded via combined electrooxidation and photocatalysis processes under visible light irradiation.

An interesting work of Fang et al. [92] studied the PEC treatment for a $40\,mg\,L^{-1}$ dipterex solution using various electrolytes such as 0.02 M Na_2SO_4, NaCl, and $NaNO_3$ in the pH range 1.0–7.0, at $2.5\,mA\,cm^{-2}$ under $4.23\,mW\,cm^{-2}$ using TiO_2/Ni. Under these conditions, 83.5% COD decay after 2 h of treatment of $40\,mg\,L^{-1}$ dipterex was obtained. It was observed that at different reaction phases, the COD removal rate of dipterex with NaCl as the electrolyte was significantly higher than with the $NaNO_3$ and Na_2SO_4 electrolytes. The superiority of the NaCl electrolyte was related to its ability to produce active species. The TiO_2/Ni electrode under UV radiation generated electron-hole pairs by the application of current that promoted a photocatalytic reactions (2)–(8). Moreover, Cl^- is adsorbed on the surface of titanium dioxide and converted to high oxidative activity groups ($Cl^\bullet$, Cl_2, and HOCl) by electron hole pairs, enhancing the organic removal.

Rodrigo et al. [93] reported recently the treatment of soil polluted with the herbicide pendimethalin by the combination of surfactant-aided soil washing (SASW) and EAOPs. Soil-washing effluents obtained from SASW and containing the surfactant (SDS, sodium dodecyl sulfate), and pendimethalin were treated by electrolysis, photoelectrolysis, and sonoelectrolysis with diamond anodes. Photoelectrolysis was the most efficient technology for the treatment of effluents obtained at high SDS/soil ratios. Photoelectrolysis (57.59%) showed higher efficiencies for the removal of SDS followed by sonoelectrolysis (52.64%) and, finally, electrolysis (48.29%), after 8 h of treatment.

CONCLUSIONS

Efficient electrochemical methods have been developed in recent years for removing organic compounds from wastewater. Heterogeneous and homogenous PEC degradation using TiO_2 as a photocatalyst has proven a good alternative for the degradation of several persistent organic contaminants. In recent years, different materials have been investigated as semiconductors for PEC to remove organic compounds. It was also shown that the cathode could strongly help the process because it could contribute to the generation of active species ($S_2O_8^{2-}$, $Cl^\bullet$, $HO^\bullet$), depending on the operating conditions. Furthermore, it has been shown that the integration of different technologies is a good alternative for the removal of contaminants. However, many more research efforts are needed to apply these methods to an industrial level. Moreover, proper selection of pH, nature of the electrolyte, photo-anode, cathode, and reactor are very important for an application of PEC in a real scenario.

References

[1] H. Page, J. Jensen, Environmental reviews and case studies: an interdisciplinary approach to environmental planning and permitting: city of Seattle Elliott Bay seawall project case study, Environ. Pract. 17 (2015) 270–277.

[2] M. Kucukvar, B. Cansev, G. Egilmez, N.C. Onat, H. Samadi, Energy-climate-manufacturing nexus: new insights from the regional and global supply chains of manufacturing industries, Appl. Energy 184 (2016) 889–904.

[3] E. Brillas, C.A. Martínez-Huitle, Decontamination of wastewaters containing synthetic organic dyes by electrochemical methods. An updated review, Appl. Catal. B Environ. 166–167 (2015) 603–643.

[4] T. Robinson, G. McMullan, R. Marchant, P. Nigam, Remediation of dyes in textile effluent: a critical review on current treatment technologies with a proposed alternative, Bioresour. Technol. 77 (2001) 247–255.

[5] M. Solís, A. Solís, H.I. Pérez, N. Manjarrez, M. Flores, Microbial decolouration of azo dyes: a review, Process Biochem. 47 (2012) 1723–1748.

[6] A.B. dos Santos, F.J. Cervantes, J.B. van Lier, Review paper on current technologies for decolourisation of textile wastewaters: perspectives for anaerobic biotechnology, Bioresour. Technol. 98 (2007) 2369–2385.

[7] M.A. Oturan, J.-J. Aron, Advanced oxidation processes in water/wastewater treatment: principles and applications. A review, Crit. Rev. Environ. Sci. Technol. 44 (2014) 2577–2641.

[8] G.V. Buxton, C.L. Greenstock, W.P. Helman, A.B. Ross, Critical review of rateconstants for reactions of hydrated electrons, hydrogen atoms and hydroxylradicals (OH/O^-) in aqueous solution, J. Phys. Chem. Ref. Data 17 (1988) 513–886.

[9] E. Brillas, C.A. Martínez-Huitle (Eds.), Synthetic Diamond Films: Preparation, Electrochemistry, Characterization and Applications, Wiley, Weinheim, 2011.

[10] S. Garcia-Segura, E. Brillas, Applied photoelectrocatalysis on the degradation of organic pollutants in wastewaters, J. Photochem. Photobiol. C Photochem. Rev. 31 (2017) 1–35.

[11] R. Daghrir, P. Drogui, D. Robert, Photoelectrocatalytic technologies for environmental applications, J. Photochem. Photobiol. A Chem. 238 (2012) 41–52.

[12] A. Fujishima, X. Zhang, D.A. Tryk, TiO_2 photocatalysis and related surface phenomena, Surf. Sci. Rep. 63 (2008) 515–582.

[13] J. Georgieva, E. Valova, S. Armyanov, N. Philippidis, I. Poulios, S. Sotiropoulos, Bi-component semiconductor oxide photoanodes for thephotoelectrocatalytic oxidation of organic solutes and vapors: a short review with emphasis to TiO_2-WO_3 photoanodes, J. Hazard. Mater. 211–212 (2012) 30–46.

[14] S. Kato, F. Masuo, Titanium dioxide-photocatalyzed oxidation. I. Titanium dioxide-photocatalyzed liquid phase oxidation of tetralin, Kogyo Kagaku Zasshi 67 (1964) 42–50.

[15] S. McLintock, M. Ritchie, Reactions on titanium dioxide; photoadsorption and oxidation of ethylene and propylene, Trans. Faraday Soc. 61 (1965) 1007–1016.

[16] A. Fujishima, K. Honda, Electrochemical photolysis of water at a semiconductor electrode, Nature 238 (1972) 37–38.

[17] J.O.'.M. Bockris, S.U.M. Khan, O.J. Murphy, M. Szklarczyk, Technical brief on models of photoelectrocatalysis, Int. J. Hydrog. Energy 9 (1984) 243–244.

[18] A.Q. Contractor, J.O.'.M. Bockris, Photoelectrocatalysis of oxygen evolution onn-TiO_2, Electrochim. Acta 32 (1987) 121–123.

[19] J.M. Peralta-Hernández, Y. Meas-Vong, F.J. Rodríguez, T.W. Chapman, M. I. Maldonado, L.A. Godínez, In situ electrochemical and photo-electrochemical generation of the fenton reagent: a potentially important new water treatment technology, Water Res. 40 (2006) 1754–1762.
[20] M.V.B. Zanoni, J.J. Sene, M.A. Anderson, Photoelectrocatalytic degradation of Remazol brilliant Orange 3R on titanium dioxide thin-film electrodes, J. Photochem. Photobiol. A 157 (2003) 55–63.
[21] C.A. Martínez-Huitle, E. Brillas, Decontamination of wastewaters containing synthetic organic dyes by electrochemical methods: a general review, Appl. Catal. B Environ. 87 (2009) 105–145.
[22] R. Zouaghi, A. Zertal, B. David, S. Guittonneau, Photocatalytic degradation of monolinuron and linuron in an aqueous suspension of titanium dioxide under simulated solar irradiation, J. Water Sci. 20 (2007) 163–172.
[23] L. Chen, C. Zhao, D. Dionysiou, K.E. O'shea, TiO_2 photocatalytic degradation and detoxification of cylindros permopsin, J. Photochem. Photobiol. A Chem. 307–308 (2015) 115–122.
[24] P. Fernández-Ibánez, M.I. Polo-López, S. Malato, S. Wadhwa, J.W.J. Hamilton, P.S. M. Dunlop, R. D'sa, E. Magee, K. O'shea, D.D. Dionysiou, J.A. Byrne, Solarphotocatalytic disinfection of water using titanium dioxide graphenecomposites, Chem. Eng. J. 261 (2015) 36–44.
[25] M.E. Osugi, M.V.B. Zanoni, C.R. Chenthamarakshan, N.R. Tacconi, G.A. Woldemariam, S.S. Mandal, K. Rajeshwar, Toxicity assessment and degradation of di sperse azo dyes by photoelectrocatalytic oxidation on Ti/TiO_2 nanotubular array electrodes, J. Adv. Oxid. Technol. 11 (2008) 425–434.
[26] E. Scott-Emuakpor, A. Kruth, A. Raab, G. Paton, D. Macphee, Remediation of 2,4-dichlorophenol contaminated water by visible light-enhanced WO_3 photoelectrocatalysis, Appl. Catal. B Environ. 123–124 (2012) 433–439.
[27] P.Q. Li, G.H. Zhao, X. Cui, Y.G. Zhang, Y.T. Tang, Constructing stake structured TiO_2-NTs/Sb-doped SnO_2 electrode simultaneously with high electrocatalytic and photocatalytic performance for complete mineralization of refractory aromatic acid, J. Phys. Chem. C 113 (2009) 2375–2383.
[28] Y.Q. Wu, G.X. Lu, S.B. Li, The doping effect of bi on TiO_2 for photocatalytic hydrogen generation and photodecolorization of rhodamine B, J. Phys. Chem. C 113 (2009) 9950–9955.
[29] K. Rajeshwar, J.G. Ibanez, Fundamentals and Application in Pollution Abatement, Academic Press, San Diego, CA, 1997.
[30] E. Brillas, P.L. Cabot, J. Casado, M. Tarr (Ed.), Chemical Degradation Methods for Wastes and Pollutants Environmental and Industrial Applications, Marcel Dekker, New York, 2003, pp. 235–304.
[31] C.A. Martinez-Huitle, S. Ferro, Electrochemical oxidation of organic pollutants for the wastewater treatment: direct and indirect processes, Chem. Soc. Rev. 35 (2006) 1324–1340.
[32] T.A. Egerton, H. Purnama, S. Purwajanti, M. Zafar, Decolourization of dyesolutions usingphotoelectrocatalysis and photocatalysis, J. Adv. Oxid. Technol. 9 (2006) 79–85.
[33] X. Nie, J. Chen, G. Li, H. Shi, H. Zhao, P.-K. Wong, T. An, Synthesis andcharacterization of TiO_2 nanotube photoanode and its application inphotoelectrocatalytic degradation of model environmental pharmaceuticals, J. Chem. Technol. Biotechnol. 88 (2013) 1488–1497.
[34] F. Mazille, T. Schoettl, N. Klamerth, S. Malato, C. Pulgarin, Field solar degradation of pesticides and emerging water contaminants mediated by polymer films containing titanium and iron oxide with synergistic heterogeneous photocatalytic activity at neutral pH, Water Res. 44 (2010) 3029–3038.

[35] F.C. Moreira, R.A.R. Boaventura, E. Brillas, V.J.P. Vilar, Electrochemical advanced oxidation processes: a review on their application to synthetic and real wastewaters, Appl. Catal. B Environ. 202 (2017) 217–261.
[36] D.R. Kennedy, M. Ritchie, J. Mackenzie, The photosorption of oxygen and nitric oxide on titanium dioxid, Trans. Faraday Soc. 54 (1958) 119–129.
[37] A. Fujishima, T.N. Rao, D.A. Tryk, Titanium dioxide photocatalysis, J. Photochem. Photobiol. C Photochem. Rev. 1 (2000) 1–21.
[38] S.D. Burnside, V. Shklover, C. Barbe, P. Comte, F. Arendse, K. Brooks, M. Grätzel, Self-organization of TiO_2 nanoparticles in thin films, Chem. Mater. 10 (1998) 2419–2425.
[39] N. Ruzycki, G.S. Herman, L.A. Boatner, U. Diebold, Scanning tunneling microscopy study of the anatase (1 0 0) surface, Surf. Sci. 529 (2003) L239–L244.
[40] M. Mattesini, J.S. de Almeida, L. Dubrovinsky, N. Dubrovinskaia, B. Johansson, R. Ahuja, Cubic TiO_2 as a potential light absorber in solar-energy conversion, Phys. Rev. B 70 (2004) 115101–115109.
[41] S. Nishimoto, T. Mano, Y. Kameshima, M. Miyake, Photocatalytic water treatment over WO_3 under visible light irradiation combined with ozonation, Chem. Phys. Lett. 500 (2010) 86–89.
[42] T. Torimoto, N. Nakamura, S. Ikeda, B. Ohtani, Discrimination of the active crystalline phases in anatase–rutile mixed titanium(IV) oxide photocatalysts through action spectrum analyses, Phys. Chem. Chem. Phys. (2002) 5910–5914.
[43] R. Abe, H. Takami, N. Murakami, B. Ohtani, Pristine simple oxides as visible light driven photocatalysts: highly efficient decomposition of organic compounds over platinum-loaded tungsten oxide, J. Am. Chem. Soc. 130 (2008) 7780–7781.
[44] Q. Zheng, C. Lee, Visible light photoelectrocatalytic degradation of methyl orange usinganodized nanoporous WO_3, Electrochim. Acta 115 (2014) 140–145.
[45] S.S. Shinde, P.S. Shinde, C.H. Bhosale, K.Y. Rajpure, Zinc oxide mediated heterogeneous photocatalytic degradation of organic species under solar radiation, J. Photochem. Photobiol. B Biol. 104 (2011) 425–433.
[46] S. Chakrabarti, B.K. Dutta, Photocatalytic degradation of model textile dyes in wastewater using ZnO as semiconductor catalyst, J. Hazard. Mater. 112 (2004) 269–278.
[47] B. Dindar, S. Icli, Unusual photoreactivity of zinc oxide irradiated by concentrated sunlight, J. Photochem. Photobiol. A Chem. 140 (2001) 263–268.
[48] F. Capasso, Band-gap engineering: from physics and materials to new semi-conductor devices, Science 235 (1987) 172–176.
[49] R. Daghrir, P. Drogui, M.A.E. Khakani, Photoelectrocatalytic oxidation of chlortetracycline using Ti/TiO_2 photo-anode with simultaneous H_2O_2 production, Electrochim. Acta 87 (2012) 18–31.
[50] P.A. Carneiro, M.E. Osugi, C.S. Fugivara, N. Boralle, M. Furlan, M.V.B. Zanoni, Evaluation of different electrochemical methods on the oxidation and degradation of Reactive Blue 4 in aqueous solution, Chemosphere 59 (2005) 431–439.
[51] Y.B. Xie, X.Z. Li, Interactive oxidation of photoelectrocatalysis and electro-Fenton for azo dye degradation using TiO_2–Ti mesh and reticulated vitreous carbon electrodes, Mater. Chem. Phys. 95 (2006) 39–50.
[52] J.M. Peralta-Herna'ndez, Y. Meas-Vong, F.J. Rodrı'guez, T.W. Chapman, M. I. Maldonado, L.A. Godı'nez, Comparison of hydrogen peroxide-based processes for treating dye-containing wastewater: decolorization and destruction of Orange II azo dye in dilute solution, Dyes Pigm. 76 (2008) 656–662.
[53] M.H. Habibi, N. Talebian, J.-H. Choi, Characterization and photocatalytic activity of nanostructured indium tin oxide thin-film electrode for azo-dye degradation, Thin Solid Films 515 (2006) 1461–1469.
[54] Z. Frontistis, V.M. Daskalaki, A. Katsaounis, I. Poulios, D. Mantzavinos, Electrochemical enhancement of solar photocatalysis: degradation of endocrine disruptor bisphenol-a on Ti/TiO_2 films, Water Res. 45 (2011) 2996–3004.

[55] Y. Liu, X. Gan, B. Zhou, B. Xiong, J. Li, C. Dong, J. Bai, W. Cai, Photoelectrocatalytic degradation of tetracycline by highly effective TiO_2 nanopore arrays electrode, J. Hazard. Mater. 171 (2009) 678–683.

[56] S. Garcia-Segura, S. Dosta, J.M. Guilemany, E. Brillas, Solar photoelectrocatalytic degradation of acid Orange 7 azo dye using a highly stable TiO_2 photoanode synthesized by atmospheric plasma spray, Appl. Catal. B Environ. 132– 133 (2013) 142–150.

[57] M. Valnice, B. Zanoni, J.J. Sene, M.A. Anderson, Photoelectrocatalytic degradation of Remazol brilliant Orange 3R on titanium dioxide thin-film electrodes, J. Photochem. Photobiol. A Chem. 157 (2003) 55–63.

[58] Y.M. Hunge, V.S. Mohite, S.S. Kumbhar, K.Y. Rajpure, A.V. Moholkar, C.H. Bhosale, Photoelectrocatalytic degradation of methyl red using sprayed WO_3 thin films under visible light irradiation, J. Mater. Sci. Mater. Electron. 26 (2015) 8404–8412.

[59] M.G. Neelavannan, M. Revathi, C.A. Basha, Photocatalytic and electrochemical combined treatment of textile wash water, J. Hazard. Mater. 149 (2007) 371–378.

[60] Y. Ding, C. Yang, L. Zhu, J. Zhang, Photoelectrochemical activity of liquid phase deposited TiO_2 film for degradation of benzotriazole, J. Hazard. Mater. 175 (2010) 96–103.

[61] N. Wang, X. Li, Y. Wang, X. Quan, G. Chen, Evaluation of bias potential enhanced photocatalytic degradation of 4-chlorophenol with TiO_2 nanotube fabricated by anodic oxidation method, Chem. Eng. J. 146 (2009) 30–35.

[62] G.G. Bessegato, T.T. Guaraldo, J. Ferreira de Brito, M.F. Brugnera, M.V.B. Zanoni, Achievements and trends in photoelectrocatalysis: from environmental to energy applications, Electrocatalysis 6 (2015) 415–441.

[63] I. Sirés, E. Brillas, M.A. Otura, M.A. Rodrigo, M. Panizza, Electrochemical advanced oxidation processes: Today and tomorrow. A review, Environ. Sci. Pollut. Res. 21 (2014) 8336–8367.

[64] S. Liu, X. Zhao, H. Zeng, Y. Wang, M. Qiao, W. Guan, Enhancement of photoelectrocatalytic degradation of diclofenac with persulfate activated by Cu cathode, Chem. Eng. J. 320 (2017) 168–177.

[65] L.M.B. Batista, A.J. Santos, D.R. Silva, A.P.M. Alves, S. Garcia-Segura, C.A. Martínez-Huitle, Solar photocatalytic application of NbO_2OH as alternative photocatalyst for water treatment, Sci. Total Environ. 596–597 (2017) 79–86.

[66] E.V. Santos, C. Sáez, P. Cañizares, C.A. Martínez-Huitle, M.A. Rodrigo, UV assisted electrochemical technologies for the removal of oxyfluorfen from soil washing wastes, Chem. Eng. J. 318 (2017) 2–9.

[67] E.V. Santos, C. Sáez, P. Cañizares, D.R. Silva, C.A. Martínez-Huitle, M.A. Rodrigo, Treatment of ex-situ soil-washing fluids polluted with petroleum by anodic oxidation, photolysis, sonolysis and combined approaches, Chem. Eng. J. 310 (2017) 581–588.

[68] D.M. Araújo, S. Cotillas, C. Sáez, P. Cañizares, C.A. Martínez-Huitle, M.A. Rodrigo, Activation by light irradiation of oxidants electrochemically generated during rhodamine B elimination, J. Electroanal. Chem. 757 (2015) 144–149.

[69] A.R. Khatae, O. Mirzajani, UV/peroxydisulfate oxidation of C. I. Basic blue 3: modeling of key factors by artificial neural network, Desalination 251 (2010) 64–69.

[70] J. Li, L. Li, L. Zheng, Y. Xian, L. Jin, Photoelectrocatalytic degradation of rhodamine B using Ti/TiO_2 electrode prepared by laser calcination method, Electrochim. Acta 51 (2006) 4942–4949.

[71] M. Hepel, J. Luo, Photoelectrochemical mineralization of textile diazo dye pollutants using nanocrystalline WO_3 electrodes, Electrochim. Acta 47 (2001) 729–740.

[72] X. Ding, Z. Ai, L. Zhang, A dual-cell wastewater treatment system with combining anodic visible light driven photoelectro-catalytic oxidation and cathodic electro-Fenton oxidation, Sep. Purif. Technol. 125 (2014) 103–110.

[73] P.A. Carneiro, M.E. Osugi, J.J. Sene, M.A. Anderson, M.V.B. Zanoni, Evaluation of color removal and degradation of a reactive textile azo dye on nanoporous TiO_2 thin-film electrodes, Electrochim. Acta 49 (2004) 3807–3820.

[74] A. Socha, E. Sochocka, R. Podsiadły, J. Sokołowska, Electrochemical and photoelectrochemical degradation of direct dyes, Color. Technol. 122 (2006) 207–212.
[75] J.A. Bañuelos, O. García-Rodríguez, A. El-Ghenymy, F.J. Rodríguez-Valadez, L. A. Godínez, E. Brillas, Advanced oxidation treatment of malachite green dye using a low cost carbon-felt air-diffusion cathode, J. Environ. Chem. Eng. 4 (2016) 2066–2075.
[76] E. Guinea, C. Arias, P.L.'s. Cabot, J.´.A. Garrido, R.M.'ı. Rodrı'guez, F. Centellas, E. Brillas, Mineralization of salicylic acid in acidic aqueous medium by electrochemical advanced oxidation processes using platinum and boron-doped diamond as anode and cathodically generated hydrogen peroxide, Water Res. 42 (2008) 499–511.
[77] R.G. da Silva, A.R. de Andrade, Degradation of the dye reactive blue 4 by coupled Photo-assisted electrochemistry at DSA®-type electrode, J. Braz. Chem. Soc. 27 (5) (2016) 857–865.
[78] E. Guinea, F. Centellas, J.A. Garrido, R.M.'ı. Rodrı'guez, C. Arias, P.-L.'s. Cabot, E. Brillas, Solar photoassisted anodic oxidation of carboxylic acids in presence of Fe^{3+} using a boron-doped diamond electrode, Appl. Catal. B Environ. 89 (2009) 459–468.
[79] H. Selcuk, J.J. Sene, M.A. Anderson, Photoelectrocatalytic humic acid degradation kinetics and effect of pH, applied potential and inorganic ions, J. Chem. Technol. Biotechnol. 78 (2003) 979–984.
[80] G.G. Bessegato, J.C. Cardoso, B.F. da Silva, M.V.B. Zanoni, Combination of photoelectrocatalysis and ozonation: a novel and powerful approach applied in acid yellow 1 mineralization, Appl. Catal. B Environ. 180 (2016) 161–168.
[81] (a)Y. Lei, C. Chen, Y. Tu, Y. Huang, H. Zhang, Heterogeneous degradation of organic pollutants by persulfate activated by CuO-Fe_3O_4: mechanism, stability, and effects of pH and bicarbonate Ions, Environ. Sci. Technol. 49 (2015) 6838–6845; (b)X. Liu, H. Zhang, C. Liu, J. Chena, G. Lic, T. Anc, P. Wongd, H. Zhao, UV and visible light photoelectrocatalytic bactericidal performance of 100% {111} faceted rutile TiO2 photoanode, Catal. Today 224 (2014) 77–82.
[82] H. Park, W. Choi, Photoelectrochemical investigation on electron transfer mediating behaviors of polyoxometalate in UV-illuminated suspensions of TiO_2 and Pt/TiO_2, J. Phys. Chem. B 107 (2003) 3885–3890.
[83] Z. Xua, X. Li, J. Li, L. Wu, Q. Zenga, Z. Zhou, Effect of CoOOH loading on the photoelectrocatalytic performance of WO_3 nanorod array film, Appl. Surf. Sci. 284 (2013) 285–290.
[84] L. Liang, Y. Tursun, A. Nulahong, T. Dilinuer, A. Tunishaguli, G. Gao, A. Abulikemua, K. Okitsu, Preparation and sonophotocatalytic performance of hierarchical Bi_2WO_6 structures and effects of various factors on the rate of rhodamine B degradation, Ultrason. Sonochem. 39 (2017) 93–100.
[85] X. Zhao, H. Liu, J. Qu, Photoelectrocatalytic degradation of organic contaminant at hybrid BDD-$ZnWO_4$ electrode, Catal. Commun. 12 (2010) 76–79.
[86] X. Li, H. Xu, W. Yan, Effects of twelve sodium dodecyl sulfate (SDS) on electro-catalytic performance and stability of PbO_2 electrode, J. Alloys Compd. 718 (2017) 386–395.
[87] B. Zhou, X. Zhao, H. Liu, J. Qu, C.P. Huang, Visible-light sensitive cobalt-doped $BiVO_4$ (Co-$BiVO_4$) photocatalytic composites for the degradation of methylene blue dye in dilute aqueous solutions, Appl. Catal. B Environ. 99 (2010) 214–221.
[88] Y. Liu, W. Yao, D. Liu, R. Zong, M. Zhang, X. Ma, Y. Zhu, Enhancement of visible light mineralization ability and photocatalytic activity of BiPO4/BiOI, Appl. Catal. B Environ. 163 (2015) 547–553.
[89] G. Zhang, Y. Gao, Y. Zhang, Y. Guo, Fe2O3-pillared Rectorite as an efficient and stable Fenton-like heterogeneous catalyst for photodegradation of organic contaminants, Environ. Sci. Technol. 44 (16) (2010) 6384–6389.
[90] R.L. Oliveira, M.A. Andersonc, G.A. Umbuzeiro, G.J. Zocolo, M.V.B. Zanoni, Assessment of by-products of chlorination and photoelectrocatalytic chlorination of an azo dye, J. Hazard. Mater. 205–206 (2012) 1–9.

[91] X. Zhao, J. Qu, H. Liu, Z. Qiang, R. Liu, C. Hu, Photoelectrochemical degradation of anti-inflammatory pharmaceuticals at Bi_2MoO_6–boron-doped diamond hybrid electrode under visible light irradiation, Appl. Catal. B Environ. 91 (2009) 539–545.
[92] T. Fang, C. Yang, L. Liao, Photoelectrocatalytic degradation of high COD dipterex pesticide by using TiO_2/Ni photo electrode, J. Environ. Sci. 24 (6) (2012) 1149–1156.
[93] P.T. Almazán-Sánchez, S. Cotillas, C. Sáez, M.J. Solache-Ríos, V. Martínez-Miranda, P. Canizares, I. Linares-Hernández, M.A. Rodrigo, Removal of pendimethalin from soil washing effluents using electrolytic and electro-irradiated technologies based on diamond anodes, Appl. Catal. B Environ. 213 (2017) 190–197.
[94] A.A. El-Zomrawy, Kinetic studies of photoelectrocatalytic degradation of Ponceau 6R dye with ammonium persulfate, J. Saudi Chem. Soc. 17 (2013) 397–402.
[95] M.R. Ranade, A. Navrotsky, H.Z. Zhang, J.F. Banfield, S.H. Elder, A. Zaban, P.H. Borse, S.K. Kulkarni, G.S. Doran, H.J. Whitfield, Proc. Nat. Acad. Sci. 99 (2002) 6476–6481.

Further Reading

[1] A.H. Ltaïef, A. D'Angelo, S. Ammara, A. Gadri, A. Galia, O. Scialdone, Electrochemical treatment of aqueous solutions of catechol by various electrochemical advanced oxidation processes: effect of the process and of operating parameters, J. Electroanal. Chem. 796 (2017) 1–8.

CHAPTER

11

Hybrid and Sequential Chemical and Electrochemical Processes for Water Decontamination

Enric Brillas, Ignasi Sirés
University of Barcelona, Barcelona, Spain

OUTLINE

https://doi.org/10.1016/B978-0-12-813160-2.00011-0

INTRODUCTION

Environmental conservation has become a matter of major social concern owing to imposed tougher discharge regulations worldwide, even in developing countries where no attention had been paid during the massive industrialization period. Biological treatments are the most widespread methods used in municipal wastewater treatment facilities, but they require much time and need large operational areas that entail high capital expenses. They are also rather inefficient for the removal of toxic emerging pollutants such as pharmaceuticals and for the remediation of industrial effluents with high concentrations of recalcitrant xenobiotics, landfill leachates, and effluents with extreme pH values [1–6]. Conventional physicochemical processes are typically employed as pretreatment prior to biological methods because they are relatively expensive and ineffective for the destruction of contaminants in real wastewater matrices. Adsorption, coagulation, and membrane filtration are three of the most representative physical treatments, but their main drawback is secondary pollution in the form of polluted adsorbents or sludge and concentrated retentate fractions. On the other hand, traditional oxidants allow the effective degradation of certain molecules, but others are very refractory and persist in the water matrix. High energy consumption and/or a high cost of reagents are usually associated with the innovative advanced oxidation processes (AOPs), including wet oxidation, heterogeneous photocatalysis, and the Fenton process, among others. They are also unable to remove all existing organic pollutants in wastewater. In conclusion, the development of new environmentally friendly technologies with the ability to quickly mineralize nonbiodegradable and toxic organic matter from urban and industrial effluents has become an extremely urgent challenge.

Over the past two decades, great progress has been made regarding electrochemical technologies, especially for the abatement of biorefractory substances; this is well documented in several authoritative reviews [1–8]. Conventional treatments such as electrocoagulation (EC) as well as electrochemical advanced oxidation processes (EAOPs) such as electrochemical oxidation (EO) and the electro-Fenton (EF) process have been identified as promising ways to address the abovementioned concerns. Electrochemical methods present important advantages for the removal of organic contaminants from wastewater, including their high efficiency, mild operation conditions, ease of automation and transportation, versatility, and low cost, especially when they are powered with renewable energy from wind and solar sources [4–8]. The applicability of electrochemical technologies has been well proven for a plethora of synthetic wastewater types containing a diversity of target compounds and microorganisms, allowing appropriate parameters to be set for both destruction of organic matter and water disinfection. However, many drawbacks still persist, thus limiting their industrial application. These drawbacks include their structural and intrinsic problems, such as a short lifetime and/or the high cost of some electrode materials; mass transport limitations inherent to some processes; and low space-time yield and surface area to volume ratio. Also, an important challenge to be solved is the low current efficiency of the processes linked to the degree of mineralization achieved and time needed. Many research efforts have then been made for the intensification of electrochemical treatments by coupling or combining them with other chemical or electrochemical processes, aiming at a more robust performance that makes them appealing for their use on an industrial scale.

This chapter describes the application of the most interesting coupled/combined treatments involving electrochemical technology for water decontamination (i.e., strictly the degradation of organic pollutants) from synthetic and real wastewater. EC and EAOPs are considered as the most promising alternatives for these approaches. The treatments have been classified as: (i) hybrid or single-step processes in which the coupled processes operate simultaneously, and (ii) sequential or two-step processes where the combined processes are applied consecutively in a stepwise fashion. Water treatment process intensification is thus a strategy that is gaining major acceptance, especially for implementing novel or more sophisticated methods whose viability as unique treatments would be questionable from an economic standpoint. Coupling enhances the overall performance thanks to the synergy between different optimized processes, eventually reducing the treatment time and/or the operational costs. Fig. 1 summarizes the large variety of coupled/combined processes considered in the chapter.

SINGLE-STEP PROCESSES

Hybrid methods related to EC, EO, and EF as well as alternative processes such as peroxi-coagulation, UV photoelectro-Fenton (PEF), solar photoelectro-Fenton (SPEF), and photoelectrocatalysis (PEC) have been developed to enhance the removal of organic pollutants from wastewater. As can be seen in Fig. 1, the intensification can be promoted by coupling with chemical oxidants, adsorption, and membrane filtration as well as either ultrasound (US) or light irradiation. This section gives details on the particulars of these kinds of treatments.

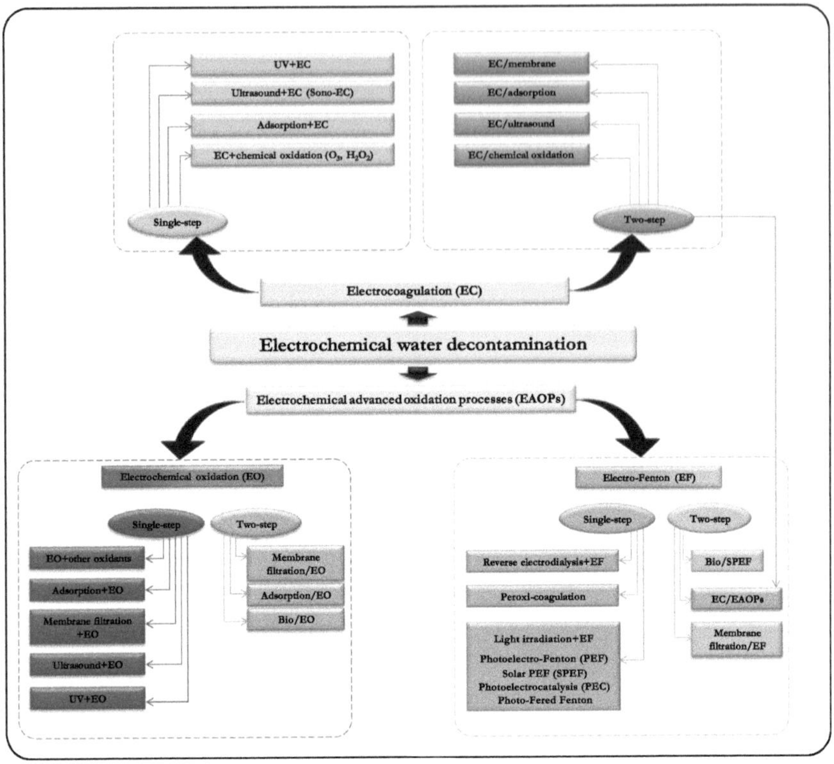

FIG. 1 Main hybrid (single-step) and sequential (two-step) processes with electrochemical technology proposed for water decontamination (i.e., removal of organics).

Electrocoagulation

The EC method involves the in situ generation of coagulants by electrolytic oxidation of a sacrificial anode (iron or aluminum) in an undivided cell, as a result of the dissolution of the electrodes immersed into the effluent. Metal ions, at an appropriate pH, can give rise to a wide variety of coagulated species and metal hydroxides that destabilize the particles and foster their aggregation or precipitate and adsorb the dissolved contaminants. Using Fe or stainless steel (SS) as the anode, Fe^{2+} is produced from reaction (1), which in the presence of dissolved O_2 gas is converted to Fe^{3+} from the global reaction (2). In contrast, an Al anode is oxidized to Al^{3+} via reaction (3). In both cases, H_2 is formed at the cathode from water reduction by reaction (4) [5,9,10].

$$Fe \rightarrow Fe^{2+} + 2e^- \quad (1)$$

$$4Fe + 10H_2O + O_2(g) \rightarrow 4Fe(OH)_3(s) + 4H_2(g) \quad (2)$$

$$Al \rightarrow Al^{3+} + 3e^- \quad (3)$$

$$2H_2O + 2e^- \rightarrow 2OH^- + H_2(g) \quad (4)$$

At pH > 5.5, $Fe(OH)_2(s)$ precipitates, remaining in equilibrium with Fe^{2+} up to pH 9.5 or with monomers such as $Fe(OH)^+$ and $Fe(OH)_3^-$ at higher pH values. In turn, $Fe(OH)_3(s)$ coagulates from pH > 1.0 and can be in equilibrium with soluble monomeric species such as Fe^{3+}, $Fe(OH)^{2+}$, $Fe(OH)_2^+$, and $Fe(OH)_4^-$. On the other hand, Al^{3+} reacts in the bulk to form soluble monomers such as $Al(OH)^{2+}$, $Al(OH)_2^+$, and $Al(OH)_3$ in acidic medium and $Al(OH)_4^-$ in an alkaline medium, precipitating as $Al(OH)_3(s)$. The insoluble flocs of $Fe(OH)_2(s)$, $Fe(OH)_3(s)$, or $Al(OH)_3(s)$ can then remove colloidal particles or soluble organic ionic species by surface complexation or electrostatic attraction, which are separated from the effluent as a sludge. The point of zero charge of these species varies between pH 5.0 for $Al(OH)_3(s)$ and pH 8.5 for $Fe(OH)_3(s)$, indicating that negative or positive particles can be adsorbed at lower or higher pH values, respectively. Reaction (4) provides the hydroxyl ions to form the insoluble species from iron or aluminum cations whereas the hydrogen gas evolved increases the turbulence, contributes to the destabilization of colloidal particles (electroflocculation process), and allows flotation of pollutants adhered onto the gas bubbles (electroflotation process). Note that apart from the phase separation, degradation of organic pollutants is also feasible in EC either by their reduction at the cathode or by the attack of oxidants such as active chlorine originated from the anodic oxidation of Cl^- present in the bulk [11].

Table 1 summarizes some relevant results obtained upon degradation of polluted effluents using different hybrid processes involving EC. Fig. 2A shows the setup of a coupled ozonation-photo-electrocoagulation system used for EC + chemical oxidation, in which the electrolytic cell is in bipolar arrangement with only the extreme electrodes connected to the DC power supply [12]. This system equipped with $100\,cm^2$ Fe electrodes was employed for treating 2.0 L of a distillery industrial wastewater with $2000\,mg\,L^{-1}$ chemical oxygen demand (COD) at pH = 6. Fig. 3A and B show that the direct use of O_3 at a mass flow rate of $2\,g\,h^{-1}$ caused a very poor decolorization and COD removal for 240 min because of the slow attack of this oxidant and the small production of the potent hydroxyl radical ($^{\bullet}OH$) by reaction (5). In contrast, these figures highlight a large loss in color and COD upon EC at current density (j) of $30\,mA\,cm^{-2}$, which was further enhanced up to 95% decolorization and 98% mineralization by applying coupled EC + O_3 (see Table 1). This improvement can be ascribed to the formation of extra amounts of $^{\bullet}OH$ via reactions (6) and (7) from injected O_3. Even better decolorization was obtained by adding $237\,mg\,L^{-1}$ H_2O_2 due to the quick $^{\bullet}OH$ generation by Fenton's reaction (8). This method, the so-called electro-peroxi-coagulation (E-PC), is a kind

TABLE 1 Treatment of Several Polluted Effluents by Selected Single-Step Processes With EC

Effluent	Experimental conditions	Removal	Ref.
EC+CHEMICAL OXIDANTS			
Distillery industrial wastewater	2 L of wastewater (2000 mg L^{-1} COD, pH = 6), system of Fig. 2A (100 cm^2 Fe), 240 min		[12]
	O_3 (2 g h^{-1})	6% color, 19% COD	
	EC (j = 30 mA cm^{-2})	90% color, 81% COD	
	EC + O_3	95% color, 98% COD	
	EC + H_2O_2 (234 mg L^{-1})	100% color, 96% COD	
Wet-spun acrylic fibers wastewater	2 L of wastewater (100 mg L^{-1} TOC, pH = 7.8), cell similar to Fig. 2A (8 Fe electrodes of 70 cm^2), 30 min		[13]
	EC (j = 35.7 mA cm^{-2})	39% TOC	
	EC + H_2O_2 (10 mM)	47% TOC	
ADSORPTION+EC			
Basic Red 2 dye	1 L of 50 mg L^{-1} dye solution, pH = 6.05, stirred Al/SS cell, 90 cm^2 electrodes, j = 11.47 mA cm^{-2}, 10.91 g L^{-1} eggshell adsorbent, 5 min	93% color	[14]
ULTRASOUND IRRADIATION+EC			
Amoxicillin	250 mL of 30 mg L^{-1} drug solution in 0.1 M KCl, cylindrical reactor (9.2 cm^2 Fe), 240 min		[15]
	EC (E_{anod} = 5 V/Hg|HgO)	72% substrate	
	US (200 W, 24 kHz) + EC	81% substrate	
Rhodamine 6G dye	4 L of 100 mg L^{-1} dye solution in 3 g L^{-1} Na_2SO_4, pH = 9.5, system of Fig. 2B (100 cm^2 Fe), liquid flow rate = 55 L h^{-1}, 300 min		[16]
	EC (j = 25 mA cm^{-2})	67% substrate	
	US (200 W) + EC	95% substrate	

TABLE 1 Treatment of Several Polluted Effluents by Selected Single-Step Processes With EC—cont'd

Effluent	Experimental conditions	Removal	Ref.
UV IRRADIATION + EC			
Urban wastewater	4L of wastewater, turbidity = 3 NTU, cylindrical flow cell (69.4 cm^2 Fe), liquid flow rate = 55 L h^{-1}, j = 1.44 mA cm^{-2}, 4 W UVC, 210 min	82% turbidity	[17]
Distillery industrial wastewater	(See above)		[12]
	EC + UVC (16 W)	91% color, 88% COD	
	EC + O_3 + UVC	97% color, 90% COD	

of PC process that will be more extensively examined below. The energy consumption per order was also assessed (Fig. 3C), revealing that the coupling of EC with H_2O_2 became less expensive than that with O_3, being more useful in practice. The better performance of E-PC compared to EC has also been reported for the degradation of wet-spun acrylic fibers wastewater [13], as shown in Table 1.

$$O_3 + H_2O \rightarrow 2^{\bullet}OH + O_2(g) \quad (5)$$

$$O_3 + Fe^{2+} \rightarrow FeO^{2+} + O_2(g) \quad (6)$$

$$FeO^{2+} + H_2O \rightarrow Fe^{3+} + {}^{\bullet}OH + OH^- \quad (7)$$

$$Fe^{2+} + H_2O_2 \rightarrow Fe^{3+} + {}^{\bullet}OH + OH^- \quad (8)$$

The phase separation of organic pollutants in EC can be enhanced by additional adsorption onto some solid present in the effluent. This approach has been investigated by Carvalho et al. [14], who added egg-shell adsorbent to a synthetic solution of 50 mg L^{-1} Basic Red 2 dye to be treated by EC in a stirred Al/SS cell at j = 11.47 mA cm^{-2}. Table 1 shows that a very short time (5 min) was needed to remove 93% color, as determined by Response Surface Methodology (RSM).

Other possible hybrid processes are based on the coupling of EC with US [15,16] and light irradiation [12,17,18]. The high energy supplied by US significantly improves the kinetics and effectiveness of the electrode processes taking place in the cell, because such radiation can: (i) improve the coagulation through vigorous mixing, (ii) destroy the

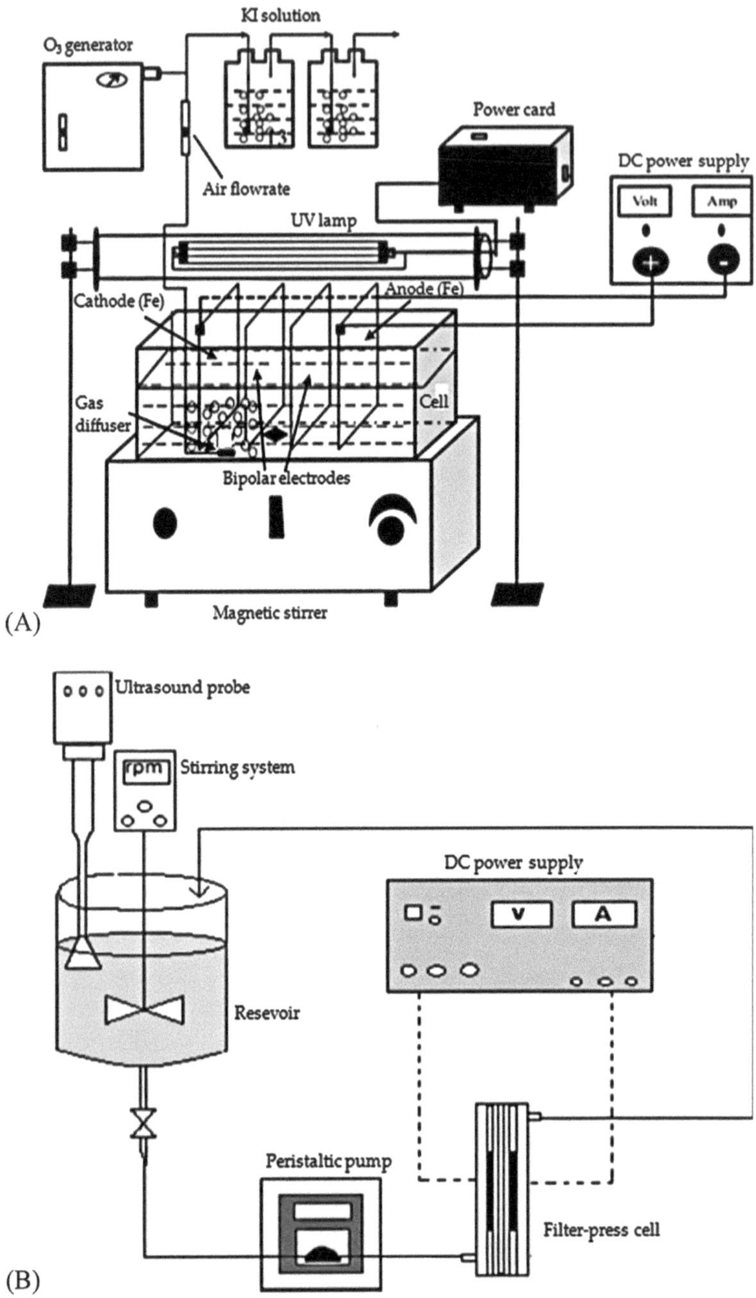

FIG. 2 (A) Experimental setup for a hybrid ozonation-photo-electrocoagulation (EC) system with an electrochemical tank reactor. (B) Sono-electrocoagulation system with a filter-press electrochemical cell. *(A) Modified from A.R.A. Aziz, P. Asaithambi, W.M.A.B.W. Daud, Combination of electrocoagulation with advanced oxidation processes for the treatment of distillery industrial effluent, Process Safety Environ. Protect. 99 (2016) 227–235. Copyright 2015, with permission from Elsevier. (B) Modified from A. Raschitor, C.M. Fernandez, I. Cretescu, M.A. Rodrigo, P. Cañizares, Sono-electrocoagulation of wastewater polluted with Rhodamine 6G, Sep. Purif. Technol. 135 (2014) 110–116. Copyright 2014. Elsevier.*

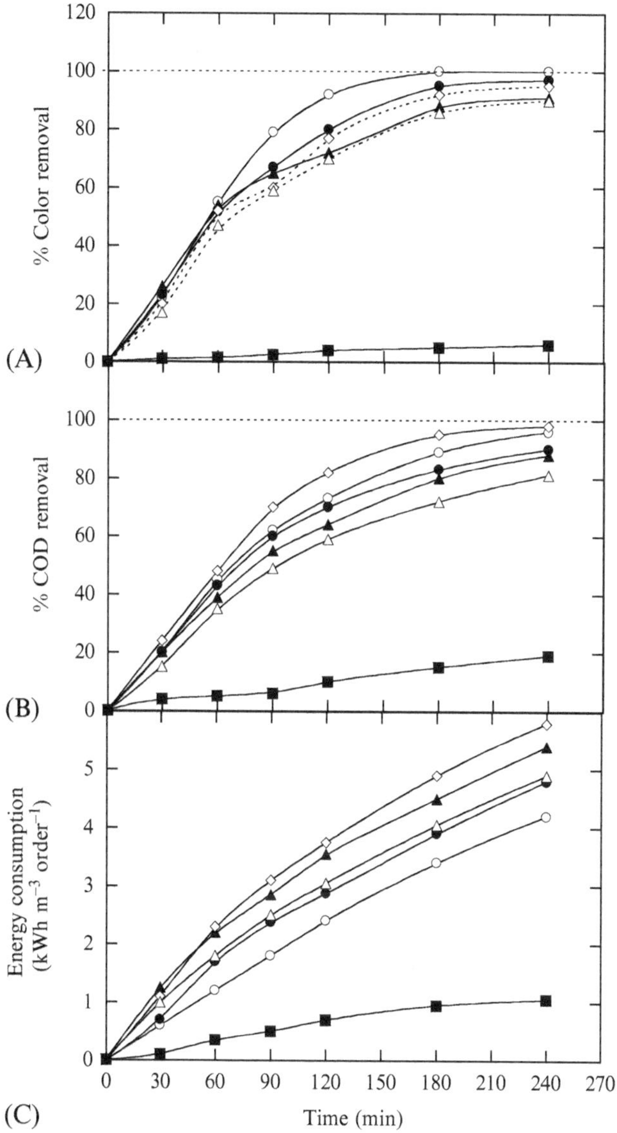

FIG. 3 Percentage of (A) color and (B) TOC removals and (C) energy consumption with time for the treatment of 2L of distillery industrial wastewater using the cell of Fig. 2A with 100 cm^2 Fe electrodes. Method: (■) ozonation, (△) EC, (◇) EC + O_3, (▲) EC + UV, (○) EC + H_2O_2, and (●) EC + UV + H_2O_2. Experimental conditions: Initial COD = 2000 mg L^{-1}, pH = 6, current density (j) = 30 mA cm^{-2}, O_3 mass flow rate = 2 g h^{-1}, H_2O_2 added = 234 mg L^{-1}, 16 W UVC lamp (λ_{max} = 254 nm). *Based on A.R.A. Aziz, P. Asaithambi, W.M.A.B.W. Daud, Combination of electrocoagulation with advanced oxidation processes for the treatment of distillery industrial effluent, Process Safety Environ. Protect. 99 (2016) 227–235. Copyright 2015. Elsevier.*

compact layer formed at the electrode surfaces by the products of electrode reactions, (iii) activate the ions and electrode surfaces through generation of defects in the crystal lattices of the electrodes, (iv) increase the temperature at the electrode surfaces as a result of friction between the liquid and the surfaces, and (v) form radicals by cavitation that can both ensure chemical polishing of the floc surface [9,16] and degrade the organic matter. Padilla et al. [15] described the degradation of 30 mg L^{-1} amoxicillin in 0.1 M KCl by EC using a cylindrical cell with an Fe anode at an anodic potential (E_{anod}) of 5 V/Hg|HgO, eventually finding larger substrate removal when a typical US of 200 W was coupled to EC (see Table 1). Fig. 2B depicts the sono-EC system with a filter-press electrochemical cell used by Raschitor et al. [16] for treating 4 L of 100 mg L^{-1} Rhodamine 6G dye with 3 g L^{-1} Na_2SO_4 by EC with Fe coupled with an US source of 200 W. A remarkable improvement in dye removal was found after 300 min of the coupled treatment compared to single EC (see Table 1), corroborating the positive influence of US. Although the high energy consumption for US generation impedes its use in real applications, the use of sustainable and free energies (solar photovoltaic or wind energy) can be envisioned.

Light irradiation coupled to EC is an alternative way to enhance the performance of EC. To this aim, an ultraviolet C (UVC) light with $\lambda < 300$ nm, usually with $\lambda_{max} = 254$ nm, illuminates the setup during effluent treatment, as shown in Fig. 2A. The irradiation can improve the removal of pollutants through (i) the photolysis of organic and/or inorganic species such as H_2O_2, which is dissociated to two $^{\bullet}OH$ from reaction (9), (ii) the photoreduction of Fe(III) species such as $Fe(OH)^{2+}$ to form Fe^{2+} and $^{\bullet}OH$ by the photo-Fenton reaction (10), and (iii) the photodecarboxylation of Fe(III)-carboxylate complexes from reaction (11) [2–6].

$$H_2O_2 + h\nu \rightarrow 2^{\bullet}OH \tag{9}$$

$$Fe(OH)^{2+} + h\nu \rightarrow Fe^{2+} + {}^{\bullet}OH \tag{10}$$

$$Fe(OOCR)^{2+} + h\nu \rightarrow Fe^{2+} + CO_2(g) + R^{\bullet} \tag{11}$$

Cotillas et al. [17] tested the effectiveness of UVC + EC, attaining a turbidity decay of up to 73% upon treatment of urban wastewater using a cylindrical flow cell with Fe anode at $j = 1.44$ mA cm^{-2} with an inner 4 W UVC lamp. A long time of 210 min was required because of the low efficiency of the process (see Table 1). Following their study on the coupling of EC with chemical oxidants, Aziz et al. [12] also applied the UVC + EC process to degrade distillery industrial wastewater, reaching higher color and COD removals as compared to EC + O_3 but lower performance if compared with EC + H_2O_2 (see Fig. 3A and B). A slight improvement was obtained when EC was simultaneously coupled to both

O_3 and UVC, but still with worse performance than $EC + H_2O_2$ (see Table 1). In a recent study on the treatment of winery wastewater by EC with SS electrodes, a superior remediation was found again by coupling it with $O_3 + H_2O_2 + UVC$ [18]. All the above results suggest that the optimum hybrid process involving EC has to be elucidated for each particular wastewater.

Electrochemical Oxidation

EO is probably the most viable EAOP because it combines two fundamental technological characteristics such as simplicity and adaptability. In this process, organic pollutants can be directly oxidized at the anode surface (M) and/or become much more rapidly destroyed by physisorbed hydroxyl radical $M(^{\bullet}OH)$ produced via reaction (12) at high applied current (or high applied anodic potential) [1]. Active anodes such as Pt, IrO_2, and RuO_2 foster the conversion to less-powerful species such as chemisorbed MO whereas high oxidation power materials such as SnO_2-Sb_2O_5, PbO_2, and boron-doped diamond (BDD) favor the electrochemical incineration of organics in the anode vicinity [1,4–6].

$$M + H_2O \rightarrow M(^{\bullet}OH) + H^+ + e^- \quad (12)$$

The oxidation ability of EO is limited by the mass transport of pollutants toward the anode, which can be enhanced, for instance, by increasing the stirring rate in tank reactors or the liquid flow rate in flow systems. The complexity of this process increases by the concomitant production of O_3 by reaction (13) and the anodic oxidation of ions contained in the effluent, such as SO_4^{2-} and Cl^- [4]. Sulfate ion is converted into the weak oxidant peroxodisulfate ($S_2O_8^{2-}$) from reaction (14) [1] and can also be oxidized to generate other strong oxidants like sulfate radicals ($SO_4^{-\bullet}$). When the treated solution contains Cl^-, this anion yields active chlorine (Cl_2, HClO, and/or ClO^-, which predominate at $pH < 3.0$, 3.0–8.0, and >8.0, respectively) in the bulk via reactions (15)–(17) [1,4–6]. In addition, these species can cause the appearance of chlorine oxyanions (chlorate and perchlorate), (oxy) chlorine radicals, chloramines, hydroxylamines, trihalomethanes, haloacetic acids, and refractory organochlorinated by-products during the electrolysis, which are undesirable toxic compounds.

$$3H_2O \rightarrow O_3 + 6H^+ + 6e^- \quad (13)$$

$$2SO_4^{2-} \rightarrow S_2O_8^{2-} + 2e^- \quad (14)$$

$$2Cl^- \rightarrow Cl_2(aq) + 2e^- \quad (15)$$

$$Cl_2(aq) + H_2O \rightarrow HClO + Cl^- + H^+ \quad (16)$$

$$HClO \leftrightharpoons ClO^- + H^+ \quad pK_a = 7.56 \quad (17)$$

Coupling between EO and O_3 has been assessed for the remediation of industrial wastewater at pH = 8.24 [19]. Experiments were performed in a cylindrical cell containing two BDD electrodes of 50 cm^2 and bubbling O_3 gas to maintain a concentration of 5 mg L^{-1}. Total COD reduction was achieved after 150 min of EO at $j = 10$ mA cm^{-2}, whereas the hybrid process accelerated the oxidation of organics to be completely removed in 115 min, thanks to the synergistic action of BDD($^{\bullet}$OH) formed from reaction (12) and $^{\bullet}$OH produced from reaction (5). Table 2 summarizes some relevant results obtained in other systems coupling EO with adsorption, membrane filtration, and US or UV irradiation.

Adsorption + EO has been tested for several carbonaceous electrodes such as reticulated vitreous carbon (RVC) [20,21], activated carbon [28], and carbon nanotubes (CNTs) [29,30], which show the ability to adsorb organics and/or ions during electrolysis. RVC electrodes of 8 cm^2 area and different porosities from 20 to 100 pores per inch (ppi) have been used to prepare three-dimensional anodes coated with nonactive PbO_2 in order to decolorize and mineralize a 0.25 mM Methyl Orange solution in 0.050 M Na_2SO_4 at pH 3.0 in a stirred tank reactor [20]. The best performance was found with a 45 ppi RVC/PbO_2 anode at a current (I) of 600 mA, owing to the stronger dye adsorption that favored its subsequent oxidation (see Table 2). Activated carbon anodes have shown great effectiveness at simultaneously removing phenol (by oxidation and adsorption) and NaCl (by adsorption) from synthetic solutions containing both components [28] (see Table 2). The use of CNTs seems more promising for this procedure because of their much larger active surface area for adsorption and oxidation. For example, Wu et al. [30] prepared a duplex-structured Ti/SnO_2-Sb/SnO_2-Sb-CNT anode of 12 cm^2 area with high efficiency to electro-oxidize 200 mL of a synthetic solution with 100 mg L^{-1} bisphenol A and 0.05 M Na_2SO_4 in a stirred tank reactor at $j = 10$ mA cm^{-2} and pH = 7.

Several authors have proposed the use of selected anodes as membrane filters to enhance the adsorption of organic pollutants in EO [22,23,31,32]. Santos et al. [22] designed a membrane made of poly(sulfone) fibers and Ti_4O_7 particles of 10 cm^2 area to be incorporated as the anode in a cross-flow cell. They found about 80% of substrate removal at $j = 10$ mA cm^{-2} after 180 min upon recirculation of 300 mL of 1 mM phenol in phosphate medium at pH = 4.5, indicating that phenol was adsorbed and oxidized at the anode when the liquid crossed through it. Fast adsorption followed by desorption and oxidation of Methylene Blue has also been reported by Vecitis et al. [23] when a 7.0 μM dye solution crossed through an electrochemically active multi-walled carbon nanotube (MWCNT) filter and a perforated SS cathode, attaining total decolorization in only 10 min of treatment by applying a cell voltage (E_{cell}) of 3.0 V (see Table 2).

TABLE 2 Degradation of Several Pollutants by Selected Single-Step Methods Based on EO

Pollutant	Experimental conditions	Removal	Ref.
ADSORPTION + EO			
Methyl Orange dye	275 mL of 0.25 mM dye solution in 0.050 M Na_2SO_4, pH = 3.0, 25°C, stirred tank reactor with 45 ppi reticulated vitreous carbon (RVC) coated with PbO_2 anode (8 cm^2 area), I = 600 mA	98% color, 45 min 55% TOC, 240 min	[20]
Phenol	100 mL of 1.1 mM phenol +17.1 mM NaCl, flow cell with 2 activated carbon electrodes (6 cm^2 area), liquid flow rate = 2.5 L h^{-1}, O_2 flow rate = 0.5 L min^{-1}, E_{cell} = 2.0 V, 300 min	90% phenol 60% TOC 20% NaCl	[21]
MEMBRANE FILTRATION + EO			
Phenol	300 mL of 1 mM phenol in phosphate medium, pH = 4.5, batch cross-flow cell upon filtration with membrane of poly(sulfone) fibers and Ti_4O_7 particles (10 cm^2), liquid flow rate = 25.8 L h^{-1}, j = 10 mA cm^{-2}, 180 min	80% phenol	[22]
Methylene Blue dye	Total adsorption of 7.0 μM dye solution, 10 mM NaCl, filter-press cell with MWNT filter anode and perforated SS cathode, liquid flow rate = 90 mL h^{-1} without current. Parallel desorption and oxidation at E_{cell} = 3.0 V, 10 min	100% color	[23]
ULTRASOUND IRRADIATION + EO			
Triclosan	600 mL of 100 mg L^{-1} drug in 0.035 M Na_2SO_4, BDD/SS cell (78 cm^2 area), liquid flow rate = 21.4 L h^{-1}, j = 30 mA cm^{-2} + US (200 W, 24 kHz), 900 min	95% drug 64% TOC	[24]
2,4-D herbicide	1.2 L of 100 mg L^{-1} herbicide solution in 3 g L^{-1} Na_2SO_4, pH = 3.5, BDD/BDD Diacell reactor (70 cm^2 area), liquid flow rate = 26.4 L h^{-1}, I = 10 A + US (200 W, 24 kHz), 145 min	100% TOC	[25]
UV IRRADIATION + EO			
Glyphosate pesticide	600 mL of 100 mg L^{-1} pesticide in 3 g L^{-1} Na_2SO_4, BDD/SS cell (78 cm^2 area), liquid flow rate = 21.4 L h^{-1}, j = 100 mA cm^{-2}		[26]
	EO, 180 min	82% TOC	
	UVC (4 W, interelectrodes) + EO, 120 min	100% TOC	
Caffeine	600 mL of 100 mg L^{-1} caffeine in industrial effluent (13 mg L^{-1} TOC)., BDD/SS cell (78 cm^2 area), liquid flow rate = 21.4 L h^{-1}, j = 30 mA cm^{-2}		[27]
	EO, 750 min	92% TOC	
	UVC (4 W, interelectrodes) + EO, 750 min	93% TOC	

More intense research has been carried out to show the advantages of coupling US with EO [24,25,33–36]. The work has been primordially developed by Rodrigo's group using flow plants with filter-press cells equipped with a BDD anode [24,25,35]. For example, they found quicker destruction of triclosan antibiotic [24] and faster total mineralization of 2,4-D herbicide [25] when a US source of 200 W and 24 kHz was coupled to the EO system (see Table 2). They interpreted this enhancement in the US + EO process as a result of two phenomena: (1) a much greater mass transport of organics toward the anode, accelerating their reaction with BDD($^{\bullet}$OH), and (2) the promotion of water decomposition to produce $^{\bullet}$OH, although this seems less significant because of the low degradation of organics when US alone was applied.

On the other hand, the combination of UV and EO has been extensively studied for textile wastewater [37] and preeminently by Rodrigo's group for urban wastewater [38], synthetic glyphosate pesticide solution [26], oxyfluorfen contained in soil [39], caffeine in industrial wastewater [27], and atrazine effluents [40] using BDD. As an example, Table 2 reveals a strong improvement in TOC removal in the EO treatment of a 100 mg L^{-1} glyphosate pesticide solution in a sulfate medium when it was irradiated with a 4 W UVC lamp. This was ascribed to the photolysis of some intermediates and concomitant production of the strong oxidant sulfate radical ($SO_4^{\bullet -}$) from photodecomposition of $S_2O_8^{2-}$ by reaction (18), an inorganic product originated from SO_4^{2-} oxidation at the BDD anode via reaction (14). In contrast, the performance of UVC + EO was not better as compared to dark EO for the degradation of caffeine in industrial wastewater (see Table 2), suggesting a poor photolytic action over its intermediates.

$$S_2O_8^{2-} + h\nu \rightarrow 2SO_4^{\bullet -} \tag{18}$$

Recently, Rodrigo's group presented some results on the treatment of progesterone [41] and phthalates [42] by combining EO with both US and UVC. However, the use of EO + US + UVC did not yield better mineralization than EO + US or EO + UVC. This suggests that the coupling of both radiation sources in EO is detrimental for degrading organics, although more data are needed to know whether this complex arrangement can be useful for some wastewater matrix.

Alternatives to Conventional Electro-Fenton

The characteristics and advantages of conventional EF for wastewater remediation have been described in the preceding chapters of this book. This subsection explores the behavior of alternative hybrid processes including PC, PEF, SPEF, and PEC. Table 3 collects the main results

TABLE 3 Degradation of Several Pollutants by Selected Single-Step Processes Based on EF

Effluent	Experimental conditions	Removal	Ref.
PC			
Chlorinated herbicides	200 mL of alachlor (175 mg L^{-1} COD), atrazine (68 mg L^{-1} COD) or chlorbromuron (47 mg L^{-1} COD) in 0.05 M Na_2SO_4, pH = 2.0, stirred tank reactor with SS anode and RVC cathode (14 cm^2), constant O_2 flow, E_{cell} = 2.5 V, 75 min	93%–96% COD	[43]
Textile wastewater	300 mL of wastewater (1310 mg L^{-1} COD, pH = 6.5), cylindrical reactor with Fe anode and graphite felt cathode (63 cm^2 area), air flow rate = 2 L min^{-1}, I = 200 mA, 120 min	59% color 71% COD	[44]
PEF			
Naproxen	2.5 L of 40 mg L^{-1} drug solution in 0.05 M $NaClO_4$, pH = 3.0, flow cell with an air-diffusion cathode (20 cm^2 area), 0.50 mM Fe^{2+}, 160 W UVA, j = 50 mA cm^{-2}, 120 min. Anode of 20 cm^2:		[45]
	BDD anode	83% TOC	
	Pt anode	80% TOC	
	IrO_2-based anode	78% TOC	
	RuO_2-based anode	76% TOC	
Methylparaben	100 mL of 158 mg L^{-1} substrate in 0.025 M Na_2SO_4 + 0.035 M NaCl, pH = 3.0, stirred tank reactor with air-diffusion cathode (3 cm^2 area), 0.50 mM Fe^{2+}.6 W UVA, j = 66.7 mA cm^{-2}, 360 min. Anode of 3 cm^2:		[46]
	BDD anode	89% TOC	
	Pt anode	43% TOC	
	IrO_2-based anode	1.9% TOC	
	RuO_2-based anode	43% TOC	
SPEF			
Dyes mixture	2.5 L of 100 mg L^{-1} DOC of food dyes under the conditions of Fig. 5, j = 100 mA cm^{-2}, 300 min. Electrolyte:		[47]
	0.05 M Na_2SO_4, 0.05 M $LiClO_4$, 0.05 M $NaNO_3$	96%–97% DOC	
	0.04 M Na_2SO_4 + 0.01 M NaCl		
	0.05 M NaCl	85% DOC	

Continued

TABLE 3 Degradation of Several Pollutants by Selected Single-Step Processes Based on EF—cont'd

Effluent	Experimental conditions	Removal	Ref.
Triclosan	10 L of 0.225 mM drug solution in 0.05 M Na_2SO_4, pH = 3.0, FM01-LC reactor with Ti/Pt anode and graphite-felt cathode (64 cm^2 area) coupled to solar CPC photoreactor, 0.50 mM Fe^{2+}, air sparging, liquid flow rate = 180 L h^{-1}, $E_{cat} = -0.35$ V/SHE, 300 min	78% TOC	[48]
PEC			
Orange G dye	30 mL of a 0.1 mM Orange G solution under the conditions of Fig. 6. $E_{anod} = 0.71$ V/SCE, 300 min		[49]
	PEC with TiO_2/Pt cell	25% dye	
	PEC with TiO_2/RVC cell	50% dye	
	PEC coupled to EF	99% dye	
Direct Yellow 52 dye	250 mL of 2 mg L^{-1} dye solution in 0.05 M Na_2SO_4, pH = 2, TiO_2/carbon-cloth cell (38 cm^2 area of anode), anode illuminated with 75 mW cm^{-2} UVA, liquid flow rate = 6 L h^{-1}		[50]
	$E_{cat} = -0.70$ V/Hg∣$HgSO_4$, 50 min	74% color	
	UVC (4 W, interelectrodes) + EO, 120 min	66% TOC	

obtained for the degradation of selected pollutants by these EAOPs, based on the electrogeneration of H_2O_2 from O_2 reduction at a carbonaceous cathode [2]:

$$O_2(g) + 2H^+ + 2e^- \rightarrow H_2O_2 \qquad (19)$$

PC consists of the continuous and simultaneous injection to the effluent of Fe^{2+} from a sacrificial Fe anode by reaction (1) and H_2O_2 produced at a suitable cathode by reaction (19). These species yield $^{\bullet}OH$ from Fenton's reaction (8), favoring the destruction of organics, which can also be coagulated by the high amounts of Fe(III) hydroxide precipitate formed, even operating at optimum pH 3.0 [2,4,5]. Usually, coagulation of organics predominates over oxidation upon increase of the applied current. The process differs from E-PC because H_2O_2 is now electrogenerated instead of being added as a chemical to the effluent. PC was proposed by Brillas' group in 1997 [2] and it has been successfully used to degrade synthetic solutions of dyes [51] and chlorinated herbicides [43] as well as of real wastewater [44,52–54]. For example, a very fast COD reduction up to 93%–96% has been reported for highly recalcitrant chlorinated herbicides such as alachlor, atrazine, or chlorbromuron in a sulfate medium using a

SS/RVC cell at $E_{cell}=2.5$ V for 75 min [43] (see Table 3). Excellent color and COD removals were also obtained for a complex textile wastewater of 1310 mg L^{-1} COD upon treatment by PC with an Fe/graphite cylindrical reactor at $I=200$ mA [44] (see Table 3).

PEF, which was proposed by Brillas' group in 1996, is based on the treatment of the solution under EF conditions but adding a UVA lamp to irradiate with UV rays of $\lambda_{max}=360$ nm in an undivided cell [2–5,55]. Organics are destroyed not only by M($^{\bullet}$OH) formed at the anode surface and $^{\bullet}$OH produced in the bulk, according to reactions (12) and (8), but also by the photolytic action of the radiation due to the production of larger amounts of $^{\bullet}$OH from the photo-Fenton reaction (10) and the photodecomposition of intermediates such as generated Fe(III)-carboxylate complexes via reaction (11). As occurs in EF, Fe^{2+} can be regenerated from cathodic reduction of Fe^{3+} by reaction (20) and electrogenerated H_2O_2 is partially lost because it is slowly oxidized at the anode to form the weak oxidant hydroperoxyl radical ($HO_2^{\bullet}$), according to reaction (21):

$$Fe^{3+} + e^- \rightarrow Fe^{2+} \tag{20}$$

$$H_2O_2 \rightarrow HO_2^{\bullet} + H^+ + e^- \tag{21}$$

The superiority of PEF over EF has been well proven for different pollutants, including pesticides, dyes, and pharmaceuticals [45,46,56–60]. Recent research has been devoted to studying the influence of the anode material and matrix composition, as can be seen in Table 3. Experiments were performed either in a 2.5 L flow plant, similar to that of Fig. 4 but with the liquid flowing through a cylindrical photoreactor with a central 160 W UVA lamp, or in a stirred tank reactor with 100 mL of solution exposed to 6 W UVA lamp. For naproxen [45] and methylparaben [46], much greater mineralization was always found using BDD instead of Pt, IrO_2-based, and RuO_2-based anodes and an inert electrolyte. This means that in both cases, the oxidative action of BDD($^{\bullet}$OH) was much higher than that of Pt($^{\bullet}$OH), IrO_2($^{\bullet}$OH), and RuO_2($^{\bullet}$OH) and even superior to that active chlorine produced from reactions (15)–(17).

The most recent version of PEF is SPEF in which the solution is directly irradiated by free and renewable sunlight as the UV source [55]. This gives rise to a much more inexpensive process because a commercial UVA lamp is no longer required. SPEF has been developed since 2007 by Brillas' group, having demonstrated its better performance compared to PEF as a result of (i) the higher energy density of UV radiation, and (ii) the advantage of operating within a λ range of 300–800 nm that can photolyze organics to a larger extent [47,48,55,61–63]. Fig. 4A and B show the schemes of the first 2.5 L solar plant constructed for SPEF with a solar planar photoreactor and the filter-press BDD/O_2-diffusion cell incorporated in the setup, respectively [61]. Fig. 5A and B depict the change of color and

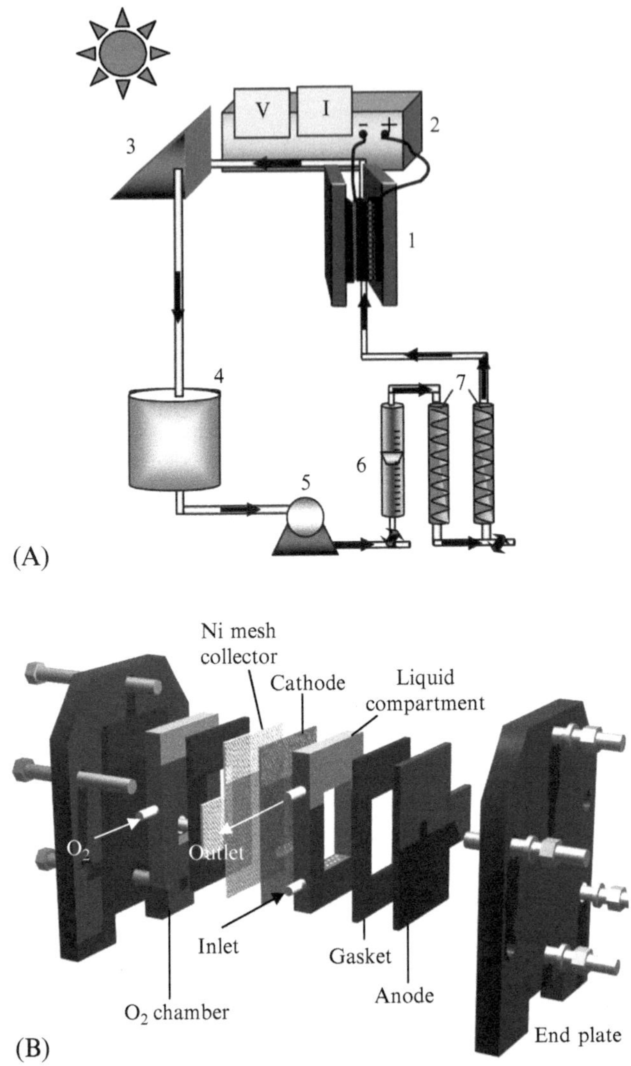

FIG. 4 Sketches of (A) the 2.5 L prepilot plant and (B) the single one-compartment filter-press electrochemical cell with a BDD anode and an O_2-diffusion cathode, both of $20\,cm^2$ exposed area, used for the solar photoelectro-Fenton (SPEF) degradation of organic pollutants in acid medium. In (A), (1) flow cell, (2) power supply, (3) solar photoreactor, (4) reservoir, (5) peristaltic pump, (6) flowmeter, and (7) heat exchangers. *Modified from C. Flox, J.A. Garrido, R.M. Rodríguez, P.L. Cabot, F. Centellas, C. Arias, E. Brillas, Mineralization of herbicide mecoprop by photoelectro-Fenton with UVA and solar light, Catal. Today 129 (2007) 29–36. Copyright 2017. Elsevier.*

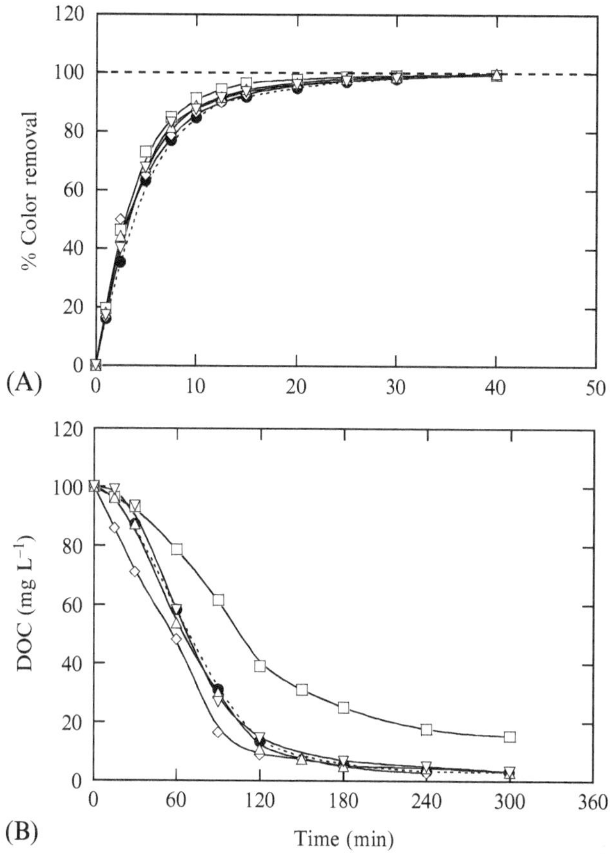

FIG. 5 Effect of the nature of the supporting electrolyte on the (A) percentage of color removal at $\lambda = 510$nm and (B) DOC decay vs. electrolysis time for the SPEF treatment of 2.5L of a mixture of three food azo dyes (33.3mgL^{-1} DOC each, equivalent to 0.139mM E122, 0.140mM E124 and 0.145mM E129) in ultrapure water with 0.05M of electrolyte and 0.50mM Fe^{2+} at pH3.0 using the flow plant of Fig. 4 at $j = 100$mAcm^{-2}, 35°C and liquid flow rate of 200Lh^{-1}. Electrolyte: (●) Na_2SO_4, (△) $LiClO_4$, (◊) $NaNO_3$, (□) NaCl, and (▽) 0.04M Na_2SO_4 +0.01M NaCl. *Based on A. Thiam, I. Sirés, E. Brillas, Treatment of a mixture of food color additives (E122, E124 and E129) in different water matrices by UVA and solar photoelectro-Fenton, Water Res. 81 (2015) 178–187. Copyright 2015. Elsevier.*

dissolved organic carbon (DOC) for the degradation of a mixture of three food azo dyes in different electrolytes at pH3.0 using the above solar plant at $j = 100$mAcm^{-2}. The effluent was completely decolorized at a quite analogous rate under all experimental conditions in about 40min of electrolysis. Conversely, DOC decayed very rapidly in 0.05M Na_2SO_4, 0.05M $LiClO_4$, 0.05M $NaNO_3$, and 0.04M Na_2SO_4 +0.01M NaCl up to attain almost total mineralization in 300min (see also Table 3) but the decrease was inhibited in 0.05M NaCl. This was explained by the formation of recalcitrant chloroderivatives upon the action of the generated active chlorine, thus demonstrating the negative influence of increasing Cl^- content in the aqueous matrix. Larger plants (up to 10 L) with more efficient solar compound parabolic collectors (CPC) have also been devised in our laboratory. In a recent paper, Pérez et al. [48] reported a 78% TOC reduction for the treatment of 10 L of an erythromycin solution in a solar plant equipped with an FM01-LC reactor with a Ti/Pt anode and a graphite-felt cathode under the application of a cathodic potential (E_{cat}) of −0.35 V/SHE for 300 min (see Table 3).

Alternative hybrid Fenton processes with light irradiation including E-PC [64–66] and Fered-Fenton [67,68] have been proposed as well. In the former case, the effluent treated by E-PC was simultaneously exposed to UV illumination. For this, Altin [66] used a stirred tank reactor equipped with two pairs of Fe electrodes of 20 cm^2 area and a 4 W UVC lamp to degrade 0.5 L of landfill leachate of 2350 $mg\,L^{-1}$ COD at pH 8.4. This author found that after adding 3 $g\,L^{-1}$ H_2O_2 and regulating the pH to 3.0, COD was reduced by 95% after 20 min of electrolysis at $I = 2.5$ A, a value much higher than the 32% determined by E-PC alone, thereby demonstrating the powerful influence of light irradiation on wastewater remediation. Fered-Fenton consists of the addition of either Fe^{2+} or Fe^{3+} to the initial wastewater along with the continuous injection of H_2O_2 during the treatment in an undivided electrolytic cell. Fe^{3+} is then cathodically reduced to Fe^{2+} by reaction (20), which reacts with added H_2O_2 to produce $^{\bullet}OH$ via Fenton's reaction (8). In addition, organics can be oxidized by M($^{\bullet}$OH) formed at the anode surface. Fig. 6A depicts a cell with double or single cathode configurations for the Fered-Fenton destruction of 3.5 L of 10 mM benzene sulfonic acid with 8 mM Fe^{2+} and 166 mM H_2O_2 at pH 2.0 [67]. The double cathode configuration was proven to be superior because it provided faster Fe^{2+} electrogeneration, accelerating the mineralization of pollutants. Fig. 6B illustrates the TOC abatement using the latter geometry for chemical Fenton and Fered-Fenton at anodic current density (j_{anod}) of 7.58 $mA\,cm^{-2}$ and cathodic current density (j_{cat}) of 7.1 $mA\,cm^{-2}$. TOC was reduced by 46% and 64% after 480 min of Fenton and Fered-Fenton, the latter showing much higher performance due to the continuous regeneration of Fe^{2+} at the surface of both cathodes.

Several authors have applied PEC combined with EF to remove organic pollutants using the semiconductor TiO_2 as a photoanode [49,50,69–71]. Xie and Li [49] proposed the use of the quartz reactor of Fig. 7A, filled with 30 mL of a 0.1 mM solution containing a dye along with 0.01 M Na_2SO_4 at pH 6.2 and an 8 W UVA lamp. In this system, H_2O_2 was electrogenerated at the RVC cathode and destroyed at the anode surface from reaction (21). The UVA irradiation of $\lambda_{max} = 365$ nm promoted the electrons from the valence to the conduction band (e^-_{CB}) with production of holes (h^+) from reaction (22), which can oxidize organics or react with H_2O to form $^{\bullet}OH$.

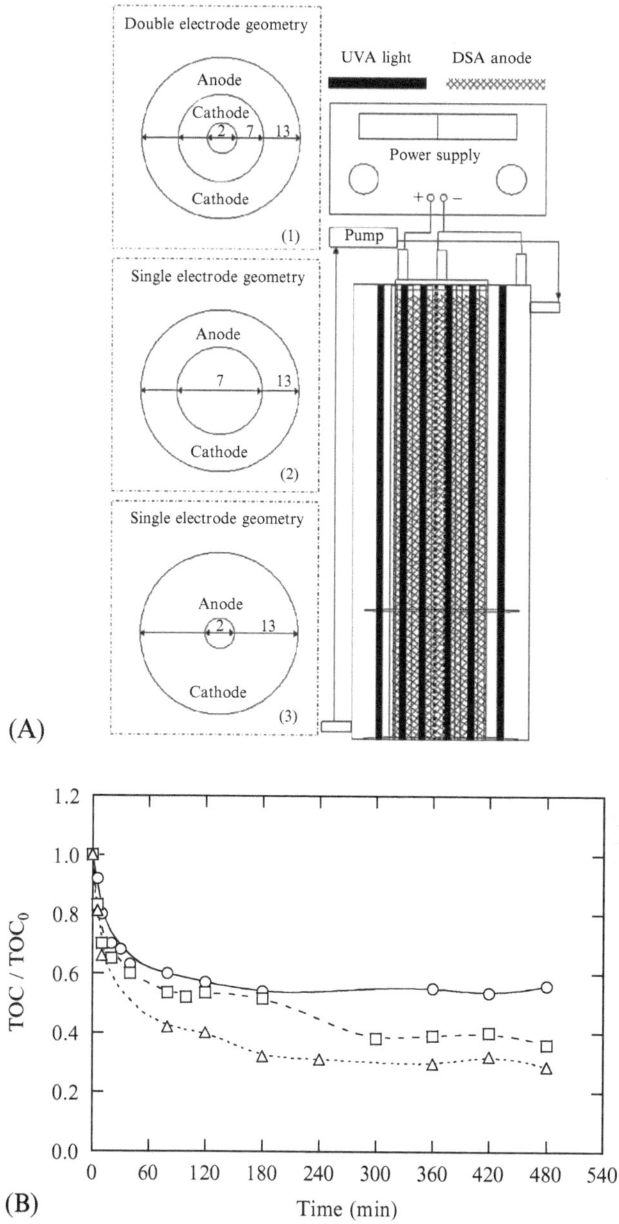

FIG. 6 (A) Experimental setup of a cylindrical tank reactor used for the Fered-Fenton treatment of organics in wastewater under galvanostatic conditions. (B) Normalized TOC decay for the treatment of 3.5 L of 10 mM benzene sulfonic acid with 8 mM Fe^{2+} and 166 mM H_2O_2 at pH 2 using the above reactor. Process: (○) Fenton process (without current), (□) Fered-Fenton, and (△) photoassisted Fered-Fenton by applying $j = 7.1\ mA\ cm^{-2}$ at the cathodes and $j = 7.58\ mA\ cm^{-2}$ at the anode. *Modified from W.-P. Ting, M.-C. Lu, Y.-H. Huang, The reactor design and comparison of Fenton, electro-Fenton and photoelectro-Fenton processes for mineralization of benzene sulfonic acid (BSA), J. Hazard. Mater. 156 (2008) 421–427. Copyright 2007. Elsevier.*

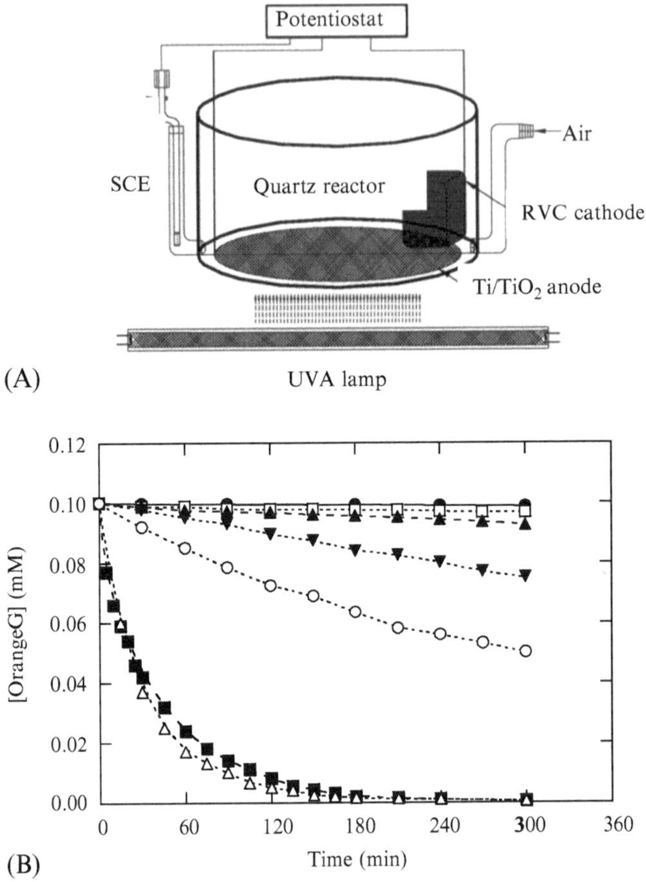

FIG. 7 (A) Experimental setup of the aerated three-electrode undivided quartz cell with 5 cm^2 electrodes used for the photoassisted electrochemical oxidation of 30 mL of a 0.1 mM Orange G solution with an 8 W UVA lamp. In photoelectrocatalysis (PEC) the bias potential applied to the Ti/TiO_2 anode was 0.71 V/SCE. A potential of −0.54 or −0.71 V/SCE was applied to the reticulated vitreous carbon (RVC) for H_2O_2 electrogeneration. (B) Dye concentration decay by: (●) direct photolysis, (□) EO with TiO_2/Pt cell, (▲) TiO_2 photocatalysis, (▼) PEC with TiO_2/Pt cell, (○) PEC with a TiO_2/RVC cell, (▪) PEF with 17.6 mg L^{-1} Fe^{2+} at pH 3.0 with a Fe/RVC cell, and (△) PEC coupled to EF with 17.2 mg L^{-1} Fe^{2+} at pH 3.0 using a TiO_2/RVC cell. *Based on Y.B. Xie, X.Z. Li, Interactive oxidation of photoelectrocatalysis and electro-Fenton for azo dye degradation using TiO2–Ti mesh and reticulated vitreous carbon electrodes, Mater. Chem. Phys. 95 (2006) 39–50. Copyright 2006. Elsevier.*

$$TiO_2 + h\nu \rightarrow h^+ + e^-_{CB} \tag{22}$$

Fig. 7B shows that Orange G was not directly photolyzed under UVA irradiation, being only slightly destroyed (about 3%) by EO in a TiO_2/Pt cell at $E_{anod} = 0.71$ V/SCE. TiO_2 photocatalysis behaved slightly better

(8%), as expected if small amounts of oxidizing species ($^•OH$ and/or holes) are formed at the TiO_2 surface upon illumination. PEC with a TiO_2/Pt cell enhanced the dye removal to 25% by the more efficient separation of charge carriers due to connection of the photoanode with a Pt cathode, whereas up to 50% dye removal was found using an RVC cathode (instead of Pt) at $E_{cat} = -0.54\,V/SCE$ because it was also oxidized with H_2O_2. In contrast, Orange G disappeared completely by PEF with a Fe/RVC cell using $17\,mg\,L^{-1}$ Fe^{2+} at pH 3.0 and $E_{cat} = -0.71\,V/SCE$, and its decolorization rate slightly increased when EF was coupled to PEC with the TiO_2/RVC cell, owing to the additional formation of $^•OH$ from Fenton's reaction (8) and/or photo-Fenton reaction (10). Similarly, Peralta-Hernández et al. [50] also described a strong improvement of the decolorization efficiency and TOC removal for 250 mL of O_2-saturated solutions with $2\,mg\,L^{-1}$ Direct Yellow 52 azo dye in 0.05 M Na_2SO_4 and 0.50 mM Fe^{2+} at pH 2.0 upon treatment by PEC + EF compared to single EF (see Table 3).

Finally, it is worth mentioning the recent work by Scialdone et al. [72], who dealt with the simultaneous generation of electricity and the treatment of wastewater contaminated with Acid Orange 7 by means of reverse electrodialysis (RED) with salinity gradients coupled to EF. Fig. 8A illustrates a scheme of the RED-EF cell utilized and the main processes involved for an effluent with 0.085 M KCl. Fig. 8B highlights the total decolorization obtained in 60 min and 56% COD reduction in 180 min for the treatment of $150\,mg\,L^{-1}$ of dye effluent using a stack of 40 cells with a Ti/RuO_2-IrO_2 anode and a carbon-felt cathode, which additionally supplied a power density of $48\,W\,m^{-2}$.

TWO-STEP PROCESSES

A large variety of sequential processes involving EC and/or EAOPs have been devised to enhance the remediation of synthetic and real wastewater containing organic contaminants. This subsection describes their characteristics and applicability.

Sequential Treatments Involving Electrocoagulation

Several kinds of two-step processes have been proposed using EC as a pretreatment followed by (i) degradation of the supernatant via chemical oxidation or EAOPs after sludge removal, or (ii) further separation by membrane filtration or adsorption. Table 4 summarizes relevant results obtained under these conditions.

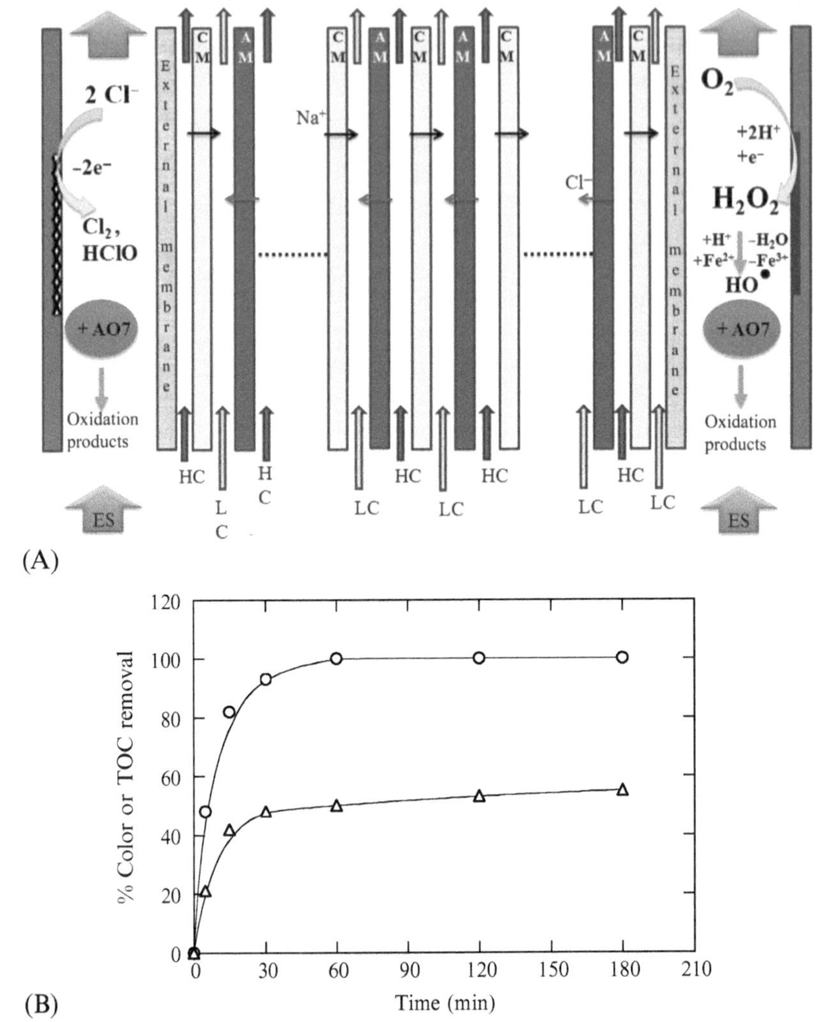

FIG. 8 (A) Scheme of an RED-EF stack showing the main reactions in the electrode compartments along with concentrated (HC), diluted (LC) and electrodic solutions (ES) flowing through the stack and the ionic flux generated by the salt gradients. Anion-exchange membranes (AMs) and cation-exchange membranes (CMs) selectively drive the flow of positive ions toward the cathode and negative ions toward the anode. (B) Percentage of (○) color and (△) TOC removals vs. time achieved in a stack equipped with 40 cell pairs with a $100\,cm^2$ Ti/RuO_2-IrO_2 anode and a $100\,cm^2$ carbon felt cathode. The system was fed with 0.5 M NaCl as HC and 0.01 M NaCl as LC at $190\,mL\,min^{-1}$. The ES was $150\,mg\,L^{-1}$ of Acid Orange 7 in 0.085 M KCl with 0.5 mM Fe^{2+} at pH 2 at $75\,mL\,min^{-1}$. *Modified from O. Scialdone, A. D'Angelo, A. Galia, Energy generation and abatement of Acid Orange 7 in reverse electrodialysis cells using salinity gradients, J. Electroanal. Chem. 738 (2015) 61–68. Copyright 2015. Elsevier.*

TABLE 4 Selected Two-Step Processes Based on EC for the Treatment of Several Polluted Effluents

Effluent	Experimental conditions	Removal	Ref.
US + EC/CHEMICAL OXIDANTS			
Winery wastewater	8 L of wastewater (10,240 mg L^{-1} COD, pH = 3.7), addition of 2 g L^{-1} NaCl		[18]
	EC (tank reactor, 10 SS electrodes of 187 cm^2 area, I = 65 A) + US (20 W, 42 kHz), 310 min, followed by:	55% COD	
	O_3 (2.5 g h^{-1}) + 9 W UVC + H_2O_2, 240 min	77% COD	
EC/MEMBRANE FILTRATION			
Pharmaceutical wastewater	EC: 1.7 L of wastewater (6535 mg L^{-1} COD, pH = 7.0), with 5 mg L^{-1} NaCl, stirred Al/Al cell, electrodes of 22 cm^2 area, j = 11.47 mA cm^{-2}, 20 min, followed by:	91% COD	[73]
	Nanofiltration (Nanomax 50 membranes)	98% COD	
Oil-in-water emulsion	EC: 1.8 L of solution (3200 mg L^{-1} COD, pH = 6.5), stirred Al/Al cell (2 anode-cathode pairs of 135 cm^2 area, j = 1.85 mA cm^{-2}, 6 min, followed by:	96% color, 92% COD	[74]
	Reverse osmosis (flow rate = 20 L h^{-1} m^{-2})	100% color and COD	
EC/EAOPs			
Carwash wastewater	EC: 300 mL of wastewater (3200 mg L^{-1} COD, pH = 6.4), Fe/SS tank reactor (27.5 cm^2 area), j = 2 mA cm^{-2}, 6 min	75% COD	[75]
	EO: 300 mL of treated wastewater, BDD/SS cell (50 cm^2 area), liquid flow rate = 300 L h^{-1}, j = 10 mA cm^{-2}, 100 min	98% COD	
Industrial wastewater	EC: 50 mL of wastewater (890 mg L^{-1} COD, pH = 8.0), Fe/Fe cell (25 cm^2 area), j = 80 mA cm^{-2}, 30 min	52% COD	[76]
	EO: 50 mL of treated wastewater, BDD/Fe tank reactor (0.50 cm^2 area), j = 80 mA cm^{-2}, 90 min	100% COD	
Tannery wastewater	EC: 250 mL of wastewater (1800 mg L^{-1} TOC, pH = 4.0), Fe/Fe cell (7.6 cm^2 area), j = 111 mA cm^{-2}, 180 min	65% TOC	[77]
	PEF: 250 mL of treated wastewater with 3.0 mM Fe^{2+}, pH = 3.0, BDD/BDD tank reactor (7.6 cm^2 area),6 W UVA, j = 111 mA cm^{-2}, 180 min	90% TOC	

Continued

TABLE 4 Selected Two-Step Processes Based on EC for the Treatment of Several Polluted Effluents—cont'd

Effluent	Experimental conditions	Removal	Ref.
Tartrazine dye	130 mL of 278 mg L^{-1} Tartrazine treated as shown in Fig. 9A, $I = 200$ mA for each method		[78]
	EC (15 min), followed by:	57% TOC	
	EO with BDD (300 min)	91% TOC	
	EF with Pt (300 min)	86% TOC	
	EF with BDD (300 min)	94% TOC	
	PEF with Pt (300 min)	88% TOC	
	PEF with BDD (300 min)	97% TOC	

Orescanin et al. [18] proposed a complex sequential treatment for winery wastewater with 10,240 mg L^{-1} COD upon addition of 2 g L^{-1} NaCl (see Table 4). The effluent was first degraded by EC with SS electrodes combined with US for 310 min, achieving 55% COD reduction. Subsequent treatment was performed with a mixture of chemical oxidants such as O_3, H_2O_2, and UVC irradiation for 240 min to obtain 73% COD abatement. The final effluent did not fulfill the legal requirements for discharge, indicating that more powerful post-treatments should be applied. Hence, better performance was found using membrane filtration [73,74,79], adsorption [80], and photo-Fenton [81,82] as post-treatments. For example, Table 4 shows that the sequential EC/membrane filtration process is able to yield total color removal and $\geq$98% COD decay for pharmaceutical wastewater [73] and oil-in-water emulsions [74].

Other papers reported the use of EC followed by EO [75,76,83] and also by EF and PEF [77,78,84] as effective two-step processes. For carwash wastewater with 3200 mg L^{-1} COD, Panizza and Cerisola [75] found 75% COD decay by EC with a Fe/SS tank reaction, which increased up to 98% when EO was subsequently applied with a BDD/SS cell (see Table 4). Similar excellent results were obtained by Linares-Hernández et al. [76] for the degradation of industrial wastewater with 890 mg L^{-1} COD by an analogous sequential EC/EO process. In this case, only 52% COD was removed by EC whereas the remaining pollutants were completely mineralized by EO with a BDD anode (see Table 4). On the other hand, Isarain-Chávez et al. [77] obtained 90% TOC abatement using a PEF post-treatment with a BDD/BDD cell after EC of tannery wastewater of 1800 mg L^{-1} TOC (see Table 4). The comparative post-treatment by

EO, EF, and PEF has been recently clarified in the works of Thiam et al. [78,84]. Fig. 9A highlights the change of TOC with electrolysis time for 278 mg L^{-1} Tartrazine solutions with 0.05 M NaCl treated by EC with Fe/SS cell and different consecutive EAOPs at $I = 200$ mA [78]. As can be seen, EC was limited to 15 min because a maximum TOC abatement was achieved. Increasing TOC removal was further reached in the order:

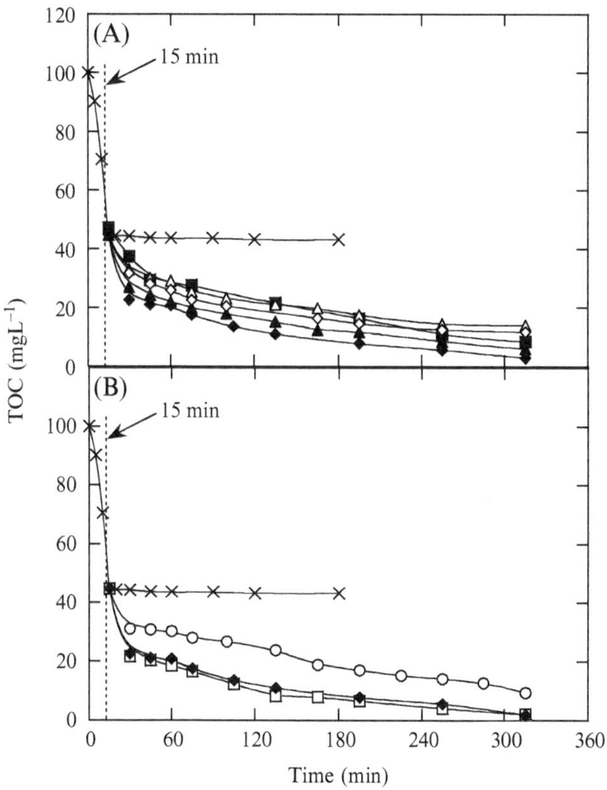

FIG. 9 TOC removal with electrolysis time for the treatment of 130 mL of 278 mg L^{-1} Tartrazine solutions treated by various two-step EC/EAOPs processes. (×) EC was performed with a stirred Fe/SS cell by adding 0.05 M NaCl at natural pH 6.3, applying $I = 200$ mA for 15 min. (A) Each consecutive EAOP was made at pH 3.0 and $I = 200$ mA: (■) EO with a stirred BDD/SS cell, (△) EF with Pt, (▲) EF with BDD, (◇) PEF with Pt, and (◆) PEF with BDD. An air-diffusion cathode was used in the four latter processes. (B) EC/PEF treatments under the same conditions, but with I of: (○) 100 mA, (◆) 200 mA, and (□) 300 mA in PEF step. Curve (×) at times >15 min shows the TOC removal by EC alone. *Based on A. Thiam, M. Zhou, E. Brillas, I. Sirés, Two-step mineralization of Tartrazine solutions: study of parameters and by-products during the coupling of electrocoagulation with electrochemical advanced oxidation processes, Appl. Catal. B: Environ. 150–151 (2014) 116–125. Copyright 2014. Elsevier.*

EF with Pt < PEF with BDD < EO with BDD < EF with BDD < PEF with BDD (see Table 4), informing about the prevalence of BDD over Pt anode and the superiority of PEF. Fig. 9B reveals that the best process (i.e., PEF with BDD) attained its best performance (97% TOC decay) at $I = 200\,mA$ without reaching any notable improvement at higher applied current.

Sequential Treatments Involving Electrochemical Advanced Oxidation Processes

EO has been successfully employed for the regeneration of adsorbents previously contaminated with organic pollutants [85–90]. As an example, Bouaziz et al. [90] studied the adsorption of phenol onto two kinds of adsorbents, softwood sawdust with a BET surface of $400\,m^2\,g^{-1}$ and commercial activated carbon with a BET surface of $980\,m^2\,g^{-1}$. These adsorbents were saturated with the target molecule attaining contents of 0.4 and $100\,mg\,g^{-1}$, respectively. Then, 2g of each loaded adsorbent were introduced into 200mL of 0.1M Na_2SO_4 contained in a stirred tank reactor with a $7\,cm^2$ BDD anode and a $67.5\,cm^2$ Pt cathode and electrolyzed at $j_{anod} = 21.5\,mA\,cm^{-2}$. While overall regeneration of softwood sawdust was easily reached after 60min of electrolysis due to the fast desorption and efficient EO of phenol and its breakdown products (i.e., catechol, hydroquinone, and *p*-benzoquinone), only 60% of the activated carbon was regenerated, indicating that a large proportion of phenol and its products remained strongly adsorbed onto its surface. This means that the nature of the adsorbent has to be well characterized in order to concentrate organics onto its surface for subsequent full regeneration by EO.

Recent research has shown good performance of sequential biological treatment and EAOPs (Bio/EAOPs) for the enhanced remediation of real wastewater [91]. EO [92–96] and SPEF [97,98] have been the preferred EAOPs to improve pollutant degradation. Table 5 summarizes the removal of COD or TOC achieved for selected trials of the above two-step processes. As can be seen, different anaerobic and aerobic biological treatments were performed, usually in continuous flow plants to attain an acceptable removal of organics contained in wastewater. The subsequent treatment by either EO or SPEF strongly increased the mineralization of the remaining biorecalcitrant pollutants and final COD or TOC abatement ≥99% was obtained, for example, for cheese whey [95] and winery [98] wastewater, thus demonstrating the effectiveness of Bio/EAOPs.

Other researchers examined the EO or EF degradation of wastewater previously treated by nanofiltration [99–101] or reverse osmosis [101–110]. Salazar et al. [101] pretreated acid ($499\,mg\,L^{-1}$ TOC, pH2.5) and alkaline

TABLE 5 Degradation of Selected Polluted Effluents by Bio/EAOPs

Effluent	Experimental conditions	Removal	Ref.
Textile wastewater	Biological treatment of wastewater (643 mg L^{-1} COD, pH = 9) with *Bacillus cereus*, 3 day, followed by:	47% COD	[92]
	EO of 300 mL of resulting wastewater, stirred quartz tank reactor with Ti/TiO_2-RuO_2-IrO_2 mesh anode (52 cm^2) upon UV/Vis irradiation, j = 20 mA cm^{-2}, 15 min	68% COD	
Carbamazepine	6 L of synthetic wastewater (393 mg L^{-1} COD, pH = 7.0) in mineral medium degraded by membrane bioreactor system with activated sludge in continuous, followed by:	11% drug 97% COD	[93]
	EO of 1 L of pretreated wastewater, flow PbO_2/Ti cell (12 cm diameter), liquid flow rate = 14 L h^{-1}, I = 1.37 A, 101 min	100% drug 98% COD	
Pulp and paper wastewater	6 L of wastewater (3300 mg L^{-1} COD, pH = 6.9) in thermophilic submerged aerobic membrane bioreactor (air bubbling = 3.2 L min^{-1}), hydraulic retention time = 1.1 day, continuous mode, followed by:	89% COD	[94]
	EO of 100 mL of pretreated wastewater, stirred tank reactor with Ti/SnO_2-Sb_2O_5-IrO_2 anode (5 cm^2 area), j = 18 mA cm^{-2}, 360 min	96% COD	
Cheese whey wastewater	550 L of wastewater (6 g L^{-1} COD, pH = 6.8–8.2) degraded in pilot-plant up-flow anaerobic sludge blanket (UASB) reactor, hydraulic retention time = 3 day, continuous mode, followed by:	94% COD	[95]
	EO of 10 L of pretreated wastewater, BDD/BDD Diacell (70 cm^2 area), with 3.3 g L^{-1} NaCl, I = 18 A, 180 min	100% COD	
Landfill leachate concentrate	Effluent pretreated by nitrification/denitrification process, ultrafiltration (UF) and DTRO system. Then, 200 mL of the resulting wastewater (3778 mg L^{-1} COD, pH = 5.2) was degraded by EO in BDD/SS cell (5 anodes, 24 cm^2 area each), liquid flow rate 6 L h^{-1}, j = 50 mA cm^{-2}, 360 min	88% COD	[96]
Winery wastewater	Aerobic biological oxidation of 45 L of wastewater (4300 mg L^{-1} TOC, pH = 3.5) in a batch immobilized biological reactor (IBR), air flow rate = 20 L min^{-1}, flow rate = 6.6 L min^{-1}, 10 day, followed by:	97% TOC	[97]

Continued

TABLE 5 Degradation of Selected Polluted Effluents by Bio/EAOPs—cont'd

Effluent	Experimental conditions	Removal	Ref.
	SPEF of 2.2 L of above wastewater (35 mg L^{-1} Fe^{2+}, pH = 2.8), solar flow plant with a BDD/air-diffusion MicroFlowCell (10 cm^2 area) and CPC photoreactor, j = 100 mA cm^{-2}, 240 min	99% TOC	
Sanitary landfill leachate	12 L of wastewater (1400 mg L^{-1} TOC, pH = 8.2–9.0) degraded biologically, by coagulation ($FeCl_3$) and aeration (flow rate 4000 L h^{-1}), followed by:	70% TOC	[98]
	SPEF of 2.2 L of above wastewater (60 mg L^{-1} Fe^{2+}, pH = 2.8), solar flow plant with a BDD/air-diffusion MicroFlowCell (10 cm^2 area) and CPC photoreactor, j = 200 mA cm^{-2}, 300 min, with clarification and sedimentation	89% TOC	

(594 mg L^{-1} TOC, pH 10.5) cellulose bleaching effluents by ultrafiltration, nanofiltration, and reverse osmosis. The latter method led to the greatest loss of organic load, attaining 113 and 52 mg L^{-1} TOC with pH values of 2.6 and 6.9 for acid and alkaline concentrates, respectively. However, subsequent degradation by EO-H_2O_2 of 250 mL of both permeates in a stirred BDD/air-diffusion cell of 3 cm^2 area at E_{cell} = 12 V for 9 h only allowed about 35%–37% TOC reduction, which was ascribed to the formation of highly recalcitrant chlorinated products from active chlorine generated from reactions (15)–(17). The production of chlorinated organic compounds has been confirmed by Bagastyo et al. [108], who degraded 5 L of a reverse osmosis concentrate (39 mg L^{-1} DOC, pH 7.72, 1391 mg L^{-1} Cl^-) from a municipal wastewater treatment plant by EO using undivided and divided flow BDD/BDD cells of 40.8 cm^2 area at constant I = 510 mA and liquid flow rate of 7.2 L h^{-1}. These authors found a poor DOC abatement of 72% and 69% after 120 h for the above arrangements, respectively, corroborating the inhibitory effect of Cl^- on the EO process. On the other hand, Zhou et al. [110] applied the EF process to a reverse osmosis concentrate from a steel plant with 150 mg L^{-1} COD, pH 7.4, very high conductivity of 22,300 μS cm^{-1} due to the presence of Cl^-, Ca^{2+}, and Mg^{2+} ions and biorefractory organics such as linear short-chain alkanes (eicosane, docosane, and tetracosane). Trials were performed with a stirred Pt/carbon-felt tank reactor containing 200 mL of an O_2-saturated effluent adjusted at optimum pH 3.0 with 0.2 mM Fe^{2+} and applying an E_{cat} = − 0.72 V/SCE. Under these conditions, 62% COD removal with an energy consumption of about 60 kWh (kg COD)$^{-1}$ was obtained after 180 min of electrolysis, which met the local wastewater discharge requirement regarding the organic load.

CONCLUSIONS

Conventional EC is an old, quite robust technology particularly well suited for the removal of large amounts of organic matter. Hybrid systems have not achieved a significant enhancement in terms of pollutant decay or solution mineralization. In contrast, interesting novel one-step treatments involving EO combined with membrane filters employed as anodes have demonstrated great potential for the treatment of certain wastewater effluents. The use of US or UV irradiation, however, does not seem to provide positive results so far. Regarding the evolution of EF process in recent years, SPEF is the most promising alternative for the scale up of technology for treating acidic effluents whereas PC is available for alkaline solutions. Lately, RED stacks have been devised for the cogeneration of energy coupled to self-sustained EF treatment of organic pollutants. Among two-step treatments, EC/PEF is especially interesting because EC pretreatment allows the coagulation of large percentages of organic matter along with the generation of iron ions whereas subsequent PEF serves as a tertiary method to completely degrade the refractory organic molecules. In the case of biodegradable effluents and aiming at implementing electrochemical technology on existing facilities, Bio/EAOPs have shown their ability to completely degrade organic pollutants. Finally, highly saline retentate fractions generated in membrane-based separation units can be effectively treated by EO.

From our point of view, the next steps should mainly address the critical factors that exert a larger influence on the performance of the electrochemical technologies, because less systematic studies have been reported compared to their chemical counterparts. On one hand, more research is needed on the development of advanced electrocoagulation reactors and their application to industrial effluents as well as on the comparison with conventional chemical coagulation in terms of capital and operating expenses. On the other hand, the main interests in EAOPs are related to new electrode materials with multiple abilities for enhancing pollutant removal as well as to the water-energy nexus with a focus on new processes that allow water decontamination with concomitant energy cogeneration.

References

[1] M. Panizza, G. Cerisola, Direct and mediated anodic oxidation of organic pollutants, Chem. Rev. 109 (2009) 6541–6569.

[2] E. Brillas, I. Sirés, M.A. Oturan, Electro-Fenton process and related electrochemical technologies based on Fenton's reaction chemistry, Chem. Rev. 109 (2009) 6570–6631.

[3] I. Sirés, E. Brillas, Remediation of water pollution caused by pharmaceutical residues based on electrochemical separation and degradation technologies: a review, Environ. Int. 40 (2012) 212–229.

[4] I. Sirés, E. Brillas, M.A. Oturan, M.A. Rodrigo, M. Panizza, Electrochemical advanced oxidation processes: today and tomorrow. A review, Environ. Sci. Pollut. Res. 21 (2014) 8336–8367.
[5] E. Brillas, C.A. Martínez-Huitle, Decontamination of wastewaters containing synthetic organic dyes by electrochemical methods. An updated review, Appl. Catal. B: Environ. 166-167 (2015) 603–643.
[6] C.A. Martínez-Huitle, M.A. Rodrigo, I. Sirés, O. Scialdone, Single and coupled electrochemical processes and reactors for the abatement of organic water pollutants: a critical review, Chem. Rev. 115 (2015) 13362–13407.
[7] G. Pliego, J.A. Zazo, P. Garcia-Muñoz, M. Munoz, J.A. Casas, J.J. Rodriguez, Trends in the intensification of the Fenton process for wastewater treatment: an overview, Crit. Rev. Environ. Sci. Technol. 45 (2015) 2611–2692.
[8] S.O. Ganiyu, E.D. van Hullebusch, M. Cretin, G. Esposito, M.A. Oturan, Coupling of membrane filtration and advanced oxidation processes for removal of pharmaceutical residues: a critical review, Sep. Purif. Technol. 156 ((2015) 891–914.
[9] C.E. Barrera-Díaz, G. Roa-Morales, P. Balderas Hernández, C.M. Fernandez-Marchante, M.A. Rodrigo, Enhanced electrocoagulation: new approaches to improve the electrochemical process, J. Electrochem Sci. Eng. 4 (2014) 285–296.
[10] M.J.M. Vianney, K. Muthukumar, Studies on dye decolorization by ultrasound assisted electrocoagulation, Clean 44 (2016) 231–238.
[11] E. Bocos, E. Brillas, M.A. Sanromán, I. Sirés, Electrocoagulation: simply a phase separation technology? The case of bronopol compared to its treatment by EAOPs, Environ. Sci. Technol. 50 (2016) 7679–7686.
[12] A.R.A. Aziz, P. Asaithambi, W.M.A.B.W. Daud, Combination of electrocoagulation with advanced oxidation processes for the treatment of distillery industrial effluent, Process. Saf. Environ. Prot. 99 (2016) 227–235.
[13] C. Gong, Z. Zhang, J. Zhang, S. Li, The addition of hydrogen peroxide in the electrocoagulation treatment for improving toxic organic matters removal: a comparative study, Sep. Sci. Technol. 52 (2017) 1404–1411.
[14] H.P. de Carvalho, J. Huang, M. Zhao, G. Liu, X. Yang, L. Dong, X. Liu, Parametric and energy consumption optimization of basic red 2 removal by electrocoagulation/egg shell adsorption coupling using response surface methodology in a batch system, Water Sci. Technol. 73 (2016) 2572–2582.
[15] B.G. Padilla-Robles, A. Alonso, S.A. Martínez-Delgadillo, M. González-Brambila, U.J. Jaúregui-Haza, J. Ramírez-Muñoz, Electrochemical degradation of amoxicillin in aqueous media, Chem. Eng. Process. 94 (2015) 93–98.
[16] A. Raschitor, C.M. Fernandez, I. Cretescu, M.A. Rodrigo, P. Cañizares, Sono-electrocoagulation of wastewater polluted with Rhodamine 6G, Sep. Purif. Technol. 135 (2014) 110–116.
[17] S. Cotillas, J. Llanos, O.G. Miranda, G.C. Díaz-Trujillo, P. Cañizares, M.A. Rodrigo, Coupling UV irradiation and electrocoagulation for reclamation of urban wastewater, Electrochim. Acta 140 (2014) 396–403.
[18] V. Orescanin, R. Kollar, K. Nad, I.L. Mikelic, S.F. Gustek, Treatment of winery wastewater by electrochemical methods and advanced oxidation processes, J. Environ. Sci. Health, Part A 48 (2013) 1543–1547.
[19] M.A. García-Morales, G. Roa-Morales, C. Barrera-Díaz, B. Bilyeu, M.A. Rodrigo, Synergy of electrochemical oxidation using boron-doped diamond (BDD) electrodes and ozone (O_3) in industrial wastewater treatment, Electrochem. Commun. 27 (2013) 34–37.
[20] G. Ramírez, F.J. Recio, P. Herrasti, C. Ponce-de-León, I. Sirés, Effect of RVC porosity on the performance of PbO_2 composite coatings with titanate nanotubes for the electrochemical oxidation of azo dyes, Electrochim. Acta 204 (2016) 9–17.

[21] F.J. Recio, P. Herrasti, I. Sirés, A.N. Kulak, D.V. Bavykin, C. Ponce-de-León, F.C. Walsh, The preparation of PbO_2 coatings on reticulated vitreous carbon for the electro-oxidation of organic pollutants, Electrochim. Acta 56 (2011) 5158–5165.
[22] M.C. Santos, Y.A. Elabd, Y. Jing, B.P. Chaplin, L. Fang, Highly porous Ti_4O_7 reactive electrochemical water filtration membranes fabricated via electrospinning/ electrospraying, AICHE J. 62 (2016) 508–524.
[23] C.D. Vecitis, G. Gao, H. Liu, Electrochemical carbon nanotube filter for adsorption, desorption, and oxidation of aqueous dyes and anions, J. Phys. Chem. C 115 (2011) 3621–3629.
[24] M.J. Martín de Vidales, C. Sáez, P. Cañizares, M.A. Rodrigo, Removal of triclosan by conductive-diamond electrolysis and sonoelectrolysis, J. Chem. Technol. Biotechnol. 88 (2013) 823–828.
[25] F.L. Souza, C. Sáez, M.R.V. Lanza, P. Cañizares, M.A. Rodrigo, Removal of herbicide 2,4-D using conductive-diamond sonoelectrochemical oxidation, Sep. Purif. Technol. 149 (2015) 24–30.
[26] H. Rubí-Juárez, S. Cotillas, C. Sáez, P. Cañizares, C. Barrera-Díaz, M.A. Rodrigo, Use of conductive diamond photo-electrochemical oxidation for the removal of pesticide glyphosate, Sep. Purif. Technol. 167 (2016) 127–135.
[27] M.J. Martín de Vidales, M. Millán, C. Sáez, P. Cañizares, M.A. Rodrigo, Irradiated-assisted electrochemical processes for the removal of persistent pollutants from real wastewater, Sep. Purif. Technol. 175 (2017) 428–434.
[28] F. Duan, Y. Li, H. Cao, Y. Wang, J.C. Crittenden, Y. Zhang, Activated carbon electrodes: electrochemical oxidation coupled with desalination for wastewater treatment, Chemosphere 125 (2015) 205–211.
[29] A. Xue, Z.-W. Yuan, Y. Sun, A.-Y. Cao, H.-Z. Zhao, Electro-oxidation of perfluorooctanoic acid by carbon nanotube sponge anode and the mechanism, Chemosphere 141 (2015) 120–126.
[30] W. Wu, Z.-H. Huang, Z.-T. Hu, C. He, T.-T. Lim, High performance duplex-structured SnO_2-Sb-CNT composite anode for bisphenol a removal, Sep. Purif. Technol. 179 (2017) 25–35.
[31] A.M. Zaky, B.P. Chaplin, Porous substoichiometric TiO_2 anodes as reactive electrochemical membranes for water treatment, Environ. Sci. Technol. 47 (2013) 6554–6563.
[32] Y. Liu, J. Xie, C.N. Ong, C.D. Vecitis, Z. Zhou, Electrochemical wastewater treatment with carbon nanotube filters coupled with in situ generated H_2O_2, Environ. Sci. Water Res. Technol. 1 (2015) 769–778.
[33] P. He, L. Wang, J. Xue, Z. Cao, Electrolytic treatment of methyl orange in aqueous solution using three-dimensional electrode reactor coupling ultrasonics, Environ. Technol. 31 (2010) 417–421.
[34] H. Zhao, J. Gao, G. Zhao, J. Fan, Y. Wang, Y. Wang, Fabrication of novel SnO_2-Sb/carbon aerogel electrode for ultrasonic electrochemical oxidation of perfluorooctanoate with high catalytic efficiency, Appl. Catal. B: Environ. 136-137 (2013) 278–286.
[35] F.L. Souza, C. Sáez, P. Cañizares, A.J. Motheo, M.A. Rodrigo, Sonoelectrolysis of wastewaters polluted with dimethyl phthalate, Ind. Eng. Chem. Res. 52 (2013) 9674–9682.
[36] M. Shestakova, M. Vinatoru, T.J. Mason, M. Sillanpää, Sonoelectrocatalytic decomposition of methylene blue using Ti/Ta_2O_5–SnO_2 electrodes, Ultrason. Sonochem. 23 (2015) 135–141.
[37] V. López-Grimau, M.C. Gutiérrez, Decolourisation of simulated reactive dyebath effluents by electrochemical oxidation assisted by UV light, Chemosphere 62 (2006) 106–112.
[38] M.J. Martín de Vidales, S. Cotillas, J.F. Perez-Serrano, J. Llanos, C. Sáez, P. Cañizares, M. A. Rodrigo, Scale-up of electrolytic and photoelectrolytic processes for water reclaiming: a preliminary study, Environ. Sci. Pollut. Res. 23 (2016) 19713–19722.

[39] E.V. dos Santos, C. Sáez, P. Cañizares, C.A. Martínez-Huitle, M.A. Rodrigo, UV assisted electrochemical technologies for the removal of oxyfluorfen from soil washing wastes, Chem. Eng. J. 318 (2017) 2–9.
[40] J.M. Aquino, D.W. Miwa, M.A. Rodrigo, A.J. Motheo, Treatment of actual effluents produced in the manufacturing of atrazine by a photo-electrolytic process, Chemosphere 172 (2017) 185–192.
[41] M.J. Martín de Vidales, S. Barba, C. Sáez, P. Cañizares, M.A. Rodrigo, Coupling ultraviolet light and ultrasound irradiation with conductive-diamond electrochemical oxidation for the removal of progesterone, Electrochim. Acta 140 (2014) 20–26.
[42] F.L. Souza, C. Sáez, P. Cañizares, A.J. Motheo, M.A. Rodrigo, Coupling photo and sono technologies to improve efficiencies in conductive diamond electrochemical oxidation, Appl. Catal. B: Environ. 144 (2014) 121–128.
[43] C. Lizama-Bahena, A. Álvarez-Gallegos, J.A. Hernandez, S. Silva-Martinez, Elimination of bio-refractory chlorinated herbicides like atrazine, alachlor, and chlorbromuron from aqueous effluents by Fenton, electro-Fenton, and peroxi-coagulation methods, Desalination Water Treat. 55 (2015) 3683–3693.
[44] F. Ghanbari, M. Moradi, A comparative study of electrocoagulation, electrochemical Fenton, electro-Fenton and peroxi-coagulation for decolorization of real textile wastewater: electrical energy consumption and biodegradability improvement, J. Environ. Chem. Eng. 3 (2015) 499–506.
[45] G. Coria, I. Sirés, E. Brillas, J.L. Nava, Influence of the anode material on the degradation of naproxen by Fenton-based electrochemical processes, Chem. Eng. J. 304 (2016) 817–825.
[46] J.R. Steter, E. Brillas, I. Sirés, On the selection of the anode material for the electrochemical removal of methylparaben from different aqueous media, Electrochim. Acta 222 (2016) 1464–1474.
[47] A. Thiam, I. Sirés, E. Brillas, Treatment of a mixture of food color additives (E122, E124 and E129) in different water matrices by UVA and solar photoelectro-Fenton, Water Res. 81 (2015) 178–187.
[48] T. Pérez, I. Sirés, E. Brillas, J.L. Nava, Solar photoelectro-Fenton flow plant modeling for the degradation of the antibiotic erythromycin in sulfate medium, Electrochim. Acta 228 (2017) 45–56.
[49] Y.B. Xie, X.Z. Li, Interactive oxidation of photoelectrocatalysis and electro-Fenton for azo dye degradation using TiO_2–Ti mesh and reticulated vitreous carbon electrodes, Mater. Chem. Phys. 95 (2006) 39–50.
[50] J.M. Peralta-Hernández, Y. Meas-Vong, F.J. Rodríguez, T.W. Chapman, M.I. Maldonado, L.A. Godínez, In situ electrochemical and photo-electrochemical generation of the Fenton reagent: a potentially important new water treatment technology, Water Res. 40 (2006) 1754–1762.
[51] M. Zarei, A. Niaei, D. Salari, A.R. Khataee, Removal of four dyes from aqueous medium by the peroxi-coagulation method using carbon nanotube–PTFE cathode and neural network modeling, J. Electroanal. Chem. 639 (2010) 167–174.
[52] Y. Yavuz, E. Öcal, A. Savas Koparal, Ü.B. Ögütveren, Treatment of dairy industry wastewater by EC and EF processes using hybrid Fe-Al plate electrodes, J. Chem. Technol. Biotechnol. 86 (2011) 964–969.
[53] X. Zhao, B. Zhang, H. Liu, F. Chen, A. Li, J. Qu, Transformation characteristics of refractory pollutants in plugboard wastewater by an optimal electrocoagulation and electro-Fenton process, Chemosphere 87 (2012) 631–636.
[54] R. Daghrir, P. Drogui, Coupled electrocoagulation–electro-Fenton for efficient domestic wastewater treatment, Environ. Chem. Lett. 11 (2013) 151–156.
[55] E. Brillas, A review on the degradation of organic pollutants in waters by UV photoelectro-Fenton and solar photoelectro-Fenton, J. Braz. Chem. Soc. 25 (2014) 393–417.

[56] I. Sirés, J.A. Garrido, R.M. Rodríguez, P.L. Cabot, F. Centellas, C. Arias, E. Brillas, Electrochemical degradation of paracetamol from water by catalytic action of Fe^{2+}, Cu^{2+}, and UVA light on electrogenerated hydrogen peroxide, J. Electrochem. Soc. 153 (2006) D1–D9.
[57] F. Kaplan, A. Hesenov, B. Gözmen, O. Erbatur, Degradations of model compounds representing some phenolics in olive mill wastewater via electro-Fenton and photoelectro-Fenton treatments, Environ. Technol. 32 (2011) 685–692.
[58] A. Thiam, E. Brillas, F. Centellas, P.L. Cabot, I. Sirés, Electrochemical reactivity of Ponceau 4R (food additive E124) in different electrolytes and batch cells, Electrochim. Acta 173 (2015) 523–533.
[59] O. García-Rodríguez, J.A. Bañuelos, A. El-Ghenymy, L.A. Godínez, E. Brillas, F.J. Rodríguez-Valadez, Use of a carbon felt-iron oxide air-diffusion cathode for the mineralization of malachite green dye by heterogeneous electro-Fenton and UVA photoelectro-Fenton processes, J. Electroanal. Chem. 767 (2016) 40–48.
[60] N. Flores, P.L. Cabot, F. Centellas, J.A. Garrido, R.M. Rodríguez, E. Brillas, I. Sirés, 4-Hydroxyphenylacetic acid oxidation in sulfate and real olive oil mill wastewater by electrochemical advanced processes with a boron-doped diamond anode, J. Hazard. Mater. 321 (2017) 566–575.
[61] C. Flox, J.A. Garrido, R.M. Rodríguez, P.L. Cabot, F. Centellas, C. Arias, E. Brillas, Mineralization of herbicide mecoprop by photoelectro-Fenton with UVA and solar light, Catal. Today 129 (2007) 29–36.
[62] V.S. Antonin, S. Garcia-Segura, M.C. Santos, E. Brillas, Degradation of Evans blue diazo dye by electrochemical processes based on Fenton's reaction chemistry, J. Electroanal. Chem. 747 (2015) 1–11.
[63] B. Garza-Campos, E. Brillas, A. Hernández-Ramírez, A. El-Ghenymy, J.L. Guzmán-Mar, E.J. Ruiz-Ruiz, Salicylic acid degradation by advanced oxidation processes. Coupling of solar photoelectro-Fenton and solar heterogeneous photocatalysis, J. Hazard. Mater. 319 (2016) 34–42.
[64] B.A. Roe, A.T. Lemley, Treatment of two insecticides in an electrochemical Fenton system, J. Environ. Sci. Health B 32 (1997) 261–281.
[65] K. Pratap, A.T. Lemley, Fenton electrochemical treatment of aqueous atrazine and metolachlor, J. Agric. Food Chem. 46 (1998) 3285–3291.
[66] A. Altin, An alternative type of photoelectro-Fenton process for the treatment of landfill leachate, Sep. Purif. Technol. 61 (2008) 391–397.
[67] W.-P. Ting, M.-C. Lu, Y.-H. Huang, The reactor design and comparison of Fenton, electro-Fenton and photoelectro-Fenton processes for mineralization of benzene sulfonic acid (BSA), J. Hazard. Mater. 156 (2008) 421–427.
[68] W.-P. Ting, M.-C. Lu, Y.-H. Huang, Kinetics of 2,6-dimethylaniline degradation by electro-Fenton process, J. Hazard. Mater. 161 (2009) 1484–1490.
[69] W.H. Leng, W.C. Zhu, J. Ni, Z. Zhang, J.Q. Zhang, C.N. Cao, Photoelectrocatalytic destruction of organics using TiO_2 as photoanode with simultaneous production of H_2O_2 at the cathode, Appl. Catal. A Gen. 300 (2006) 24–35.
[70] B.X. Zhao, X.Z. Li, P. Wang, 2,4-Dichlorophenol degradation by an integrated process: photoelectrocatalytic oxidation and E-Fenton oxidation, Photochem. Photobiol. 83 (2007) 642–646.
[71] X.Z. Li, B.X. Zhao, P. Wang, Degradation of 2,4-dichlorophenol in aqueous solution by a hybrid oxidation process, J. Hazard. Mater. 147 (2007) 281–287.
[72] O. Scialdone, A. D'Angelo, A. Galia, Energy generation and abatement of acid Orange 7 in reverse electrodialysis cells using salinity gradients, J. Electroanal. Chem. 738 (2015) 61–68.
[73] T. Chaabane, S. Zaidi, A. Darchen, R. Maachi, Treatment of pharmaceutical effluent by electrocoagulation coupled to nanofiltration, Desalination Water Treat. 51 (2013) 4987–4997.

[74] J. Rodrigues, P. da Silva, F. Merçon, L.F. da Silva, A.A. Cerqueira, P. Braz Ximango, M.R. da Costa Marques, Evaluation of electrocoagulation as pre-treatment of oil emulsions followed by reverse osmosis, J. Water Process. Eng. 8 (2015) 126–135.
[75] M. Panizza, G. Cerisola, Applicability of electrochemical methods to carwash wastewaters for reuse. Part 2: electrocoagulation and anodic oxidation integrated process, J. Electroanal. Chem. 638 (2010) 236–240.
[76] I. Linares-Hernández, C. Barrera-Díaz, B. Bilyeu, P. Juárez-García Rojas, E. Campos-Medina, A combined electrocoagulation–electrooxidation treatment for industrial wastewater, J. Hazard. Mater. 175 (2010) 688–694.
[77] E. Isarain-Chávez, C. de la Rosa, L.A. Godínez, E. Brillas, J.M. Peralta-Hernández, Comparative study of electrochemical water treatment processes for a tannery wastewater effluent, J. Electroanal. Chem. 713 (2014) 62–69.
[78] A. Thiam, M. Zhou, E. Brillas, I. Sirés, Two-step mineralization of Tartrazine solutions: study of parameters and by-products during the coupling of electrocoagulation with electrochemical advanced oxidation processes, Appl. Catal. B: Environ. 150–151 (2014) 116–125.
[79] P.S.A. de Souza, A.A. Cerqueira, M.M. Rigo, J.L. de Paiva, R.S.P. Couto, F. Merçon, D.V. Perez, M.R.C. Marques, Oilfield water treatment by electrocoagulation–reverse osmosis for agricultural use: effects on germination and early growth characteristics of sunflower, Environ. Technol. 38 (2017) 1151–1159.
[80] E. Bazrafshan, M.R. Alipour, A.M. Mahvi, Textile wastewater treatment by application of combined chemical coagulation, electrocoagulation, and adsorption processes, Desalination Water Treat. 57 (2016) 9203–9215.
[81] A.N. Módenes, F.R. Espinoza-Quiñones, F.H. Borba, D.R. Manenti, Performance evaluation of an integrated photo-Fenton—electrocoagulation process applied to pollutant removal from tannery effluent in batch system, Chem. Eng. J. 197 (2012) 1–9.
[82] P. Krystynik, P. Kluson, D.N. Tito, Water treatment process intensification by combination of electrochemical and photochemical methods, Chem. Eng. Process. 94 (2015) 85–92.
[83] G. Bahkar Raju, M. Thalamadai Karuppiah, S.S. Latha, S. Parvathy, S. Prabhakar, Treatment of wastewater from synthetic textile industry by electrocoagulation–electrooxidation, Chem. Eng. J. 144 (2008) 51–58.
[84] A. Thiam, M. Zhou, E. Brillas, I. Sirés, A first pre-pilot system for the combined treatment of dye pollutants by electrocoagulation/EAOPs, J. Chem. Technol. Biotechnol. 89 (2014) 1136–1144.
[85] S.N. Hussain, E.P.L. Roberts, H.M.A. Asghar, A.K. Campen, N.W. Brown, Oxidation of phenol and the adsorption of breakdown products using a graphite adsorbent with electrochemical regeneration, Electrochim. Acta 92 (2013) 20–30.
[86] S.N. Hussain, H.M.A. Asghar, H. Sattar, N.W. Brown, E.P.L. Roberts, Free chlorine formation during electrochemical regeneration of a graphite intercalation compound adsorbent used for wastewater treatment, J. Appl. Electrochem. 45 (2015) 611–621.
[87] S.N. Hussain, H.M.A. Asghar, H. Sattar, N.W. Brown, E.P.L. Roberts, Chlorinated breakdown products formed during oxidation of adsorbed phenol by electrochemical regeneration of a graphite intercalation compound, J. Ind. Eng. Chem. 30 (2015) 212–219.
[88] D. Liu, E.P.L. Roberts, A.D. Martin, S.M. Holmes, N.W. Brown, A.K. Campen, N. de las Heras, Electrochemical regeneration of a graphite adsorbent loaded with acid violet 17 in a spouted bed reactor, Chem. Eng. J. 304 (2016) 1–9.
[89] H. Sun, T. Chen, L. Kong, Q. Cai, Y. Xiong, S. Tian, Potential of sludge carbon as new granular electrodes for degradation of acid Orange 7, Ind. Eng. Chem. Res. 54 (2015) 5468–5474.
[90] I. Bouaziz, M. Hamza, A. Sellami, R. Abdelhedi, A. Savall, K. Groenen-Serrano, New hybrid process combining adsorption on sawdust and electroxidation using a BDD anode for the treatment of dilute wastewater, Sep. Purif. Technol. 175 (2017) 1–8.

[91] O. Ganzenko, D. Huguenot, E.D. van Hullebusch, G. Esposito, M.A. Oturan, Electrochemical advanced oxidation and biological processes for wastewater treatment: a review of the combined approaches, Environ. Sci. Pollut. Res. 21 (2014) 8493–8524.
[92] P. Aravind, V. Subramanyan, S. Ferro, R. Gopalakrishnan, Eco-friendly and facile integrated biological-cum-photo assisted electrooxidation process for degradation of textile wastewater, Water Res. 93 (2016) 230–241.
[93] C. García-Gómez, P. Drogui, B. Seyhi, P. Gortáres-Moroyoqui, G. Buelna, M.I. Estrada-Alvgarado, L.H. Álvarez, Combined membrane bioreactor and electrochemical oxidation using Ti/PbO_2 anode for the removal of carbamazepine, J. Taiwan Inst. Chem. Eng. 64 (2016) 211–219.
[94] A. Katsoni, D. Mantzavinos, E. Diamadopoulos, Coupling digestion in a pilot-scale UASB reactor and electrochemical oxidation over BDD anode to treat diluted cheese whey, Environ. Sci. Pollut. Res. 21 (2014) 12170–12181.
[95] X. Qu, W.J. Gao, M.N. Han, A. Chen, B.Q. Liao, Integrated thermophilic submerged aerobic membrane bioreactor and electrochemical oxidation for pulp and paper effluent treatment-towards system closure, Bioresour. Technol. 116 (2012) 1–8.
[96] F.C. Moreira, R.A.R. Boaventura, E. Brillas, V.J.P. Vilar, Remediation of a winery wastewater combining aerobic biological oxidation and electrochemical advanced oxidation processes, Water Res. 75 (2015) 95–108.
[97] F.C. Moreira, J. Soler, A. Fonseca, I. Saraiva, R.A.R. Boaventura, E. Brillas, V.J.P. Vilar, Incorporation of electrochemical advanced oxidation processes in a multistage treatment system for sanitary landfill leachate, Water Res. 81 (2015) 375–387.
[98] B. Zhou, Z. Yu, Q. Wei, H. Long, Y. Xie, Y. Wang, Electrochemical oxidation of biological pretreated and membrane separated landfill leachate concentrates on boron doped diamond anode, Appl. Surf. Sci. 377 (2016) 406–415.
[99] V. Buscio, M. García-Jiménez, M. Vilaseca, V. López-Grimau, M. Crespi, C. Gutiérrez-Bouzán, Reuse of textile dyeing effluents treated with coupled nanofiltration and electrochemical processes, Materials 9 (2016) 490–501.
[100] Y. Lan, C. Coetsier, C. Causserand, K. Groenen-Serrano, Feasibility of micropollutants treatment by coupling nanofiltration and electrochemical oxidation: case of hospital wastewater, Int. J. Chem. React. Eng. 13 (2015) 153–159.
[101] C. Salazar, I. Sirés, R. Salazar, H.D. Mansilla, C.A. Zaror, Treatment of cellulose bleaching effluents and their filtration permeates by anodic oxidation with H_2O_2 production, J. Chem. Technol. Biotechnol. 90 (2015) 2017–2026.
[102] A. Pérez-González, A.M. Urtiaga, R. Ibáñez, I. Ortiz, State of the art and review on the treatment technologies of water reverse osmosis concentrates, Water Res. 46 (2012) 267–283.
[103] K. Van Hege, M. Verhaege, W. Verstraete, Indirect electrochemical oxidation of reverse osmosis membrane concentrates at boron-doped diamond electrodes, Electrochem. Commun. 4 (2002) 296–300.
[104] E. Dialynas, D. Mantzavinos, E. Diamadopoulos, Advanced treatment of the reverse osmosis concentrate produced during reclamation of municipal wastewater, Water Res. 42 (2008) 4603–4608.
[105] G. Pérez, A.R. Fernández-Alba, A.M. Urtiaga, I. Ortiz, Electro-oxidation of reverse osmosis concentrates generated in tertiary water treatment, Water Res. 44 (2010) 2763–2772.
[106] A.Y. Bagastyo, J. Radjenovic, Y. Mu, R.A. Rozendal, D.J. Batstone, K. Rabaey, Electrochemical oxidation of reverse osmosis concentrate on mixed metal oxide (MMO) titanium coated electrodes, Water Res. 45 (2011) 4951–4959.
[107] A.Y. Bagastyo, D.J. Batstone, I. Kristiana, W. Gernjak, C. Joll, J. Radjenovic, Electrochemical oxidation of reverse osmosis concentrate on boron-doped diamond anodes at circumneutral and acidic pH, Water Res. 46 (2012) 6104–6112.

[108] A.Y. Bagastyo, D.J. Batstone, I. Kristiana, B.I. Escher, C. Joll, J. Radjenovic, Electrochemical treatment of reverse osmosis concentrate on boron-doped electrodes in undivided and divided cell configurations, J. Hazard. Mater. 279 (2014) 111–116.
[109] A.M. Urtiaga, G. Pérez, R. Ibáñez, I. Ortiz, Removal of pharmaceuticals from a WWTP secondary effluent by ultrafiltration/reverse osmosis followed by electrochemical oxidation of the RO concentrate, Desalination 331 (2013) 26–34.
[110] M. Zhou, Q. Tan, Q. Wang, Y. Jiao, N. Oturan, M.A. Oturan, Degradation of organics in reverse osmosis concentrate by electro-Fenton process, J. Hazard. Mater. 215-216 (2012) 287–293.

Further Reading

[1] M. Zarei, D. Salari, A. Niaei, A. Khataee, Peroxi-coagulation degradation of C.I. Basic yellow 2 based on carbon-PTFE and carbon nanotube-PTFE electrodes as cathode, Electrochim. Acta 54 (2009) 6651–6660.

CHAPTER

12

Microbial Fuel Cells and Wastewater Treatment

Sidney Aquino Neto, Valéria Reginatto, Adalgisa R. De Andrade

Chemistry Department, Faculty of Philosophy, Sciences and Letters at Ribeirão Preto, Universidade de São Paulo, Ribeirão Preto, SP, Brazil

OUTLINE

INTRODUCTION

In this chapter, we describe how anaerobic biodegradation of organic waste compounds is being revisited so that it will become a competitive strategy to treat wastewater and to produce an added-value product

https://doi.org/10.1016/B978-0-12-813160-2.00012-2

(H_2) simultaneously. The practical application of electrochemical hydrogen production (EHP) is being delayed due to the unfavorable energy balance of the whole process: the oxygen evolution reaction ($E > 1.4$ V vs. RHE) has large overpotential, so the amount of energy that is necessary to cleave water is almost the same as or even lower than the amount of energy that can be recovered from hydrogen in fuel cell devices [1]. An intelligent way to reduce the costs involved in the production of electricity and thus make the whole system sustainable is to explore the cogeneration of energy and hydrogen at microbial electrolysis cells (MECs), which could become an alternative approach for decentralized hydrogen production coupled to a cheap and efficient process for wastewater treatment. Here, we will present the bases for wastewater treatment by microbial fuel cells (MFCs) and the application of the resulting energy in hydrogen production. The main advantage of this system is that it can self-feed with the energy necessary to produce an added-value product, H_2, which boosts the whole system. A critical evaluation of the scientific production in this area can be conducted by analyzing the publications retrieved from the Web of Science database from 2000 until today. Fig. 1 shows a compilation of the search for microbial fuel cell (MFC), microbial electrolysis cell (MEC), and their respective subdivisions such as wastewater treatment and hydrogen production, respectively. The interest of the scientific

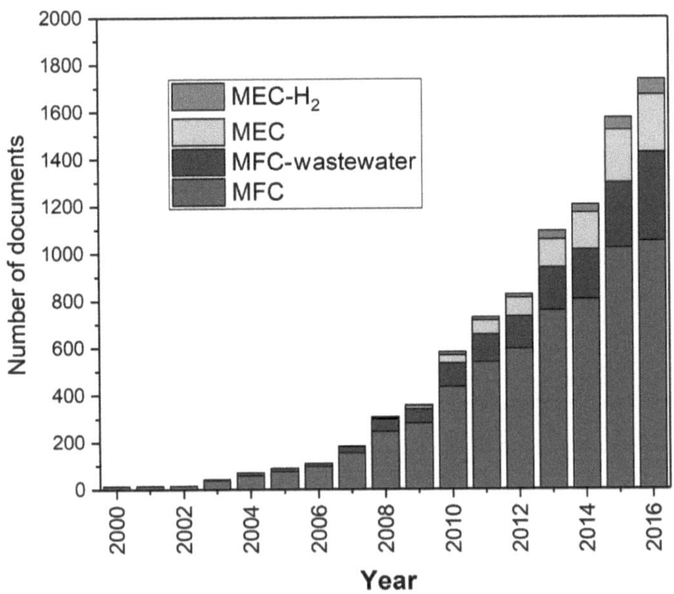

FIG. 1 Literature data search at the Web of Science database for microbial fuel cell (MFC) and wastewater treatment as well as microbial electrolysis cell (MEC) and hydrogen production.

community in these subjects increased along the analyzed period. Interest in MFC and MEC, which emerged as a relatively new subject after 2006, grew exponentially.

Fig. 2 shows the analysis of publications dealing with biohydrogen production in microbial electrolysis cells (MEC) and with wastewater treatment in microbial fuel cells (MFCs) over the past 20 years, retrieved from the Web of Science database. It can be observed that >50% are held in China and the United States.

The use of biocatalysts for either energy production (MFC) or electrolytic generation of valuable products by microorganisms (MEC) has gained increasing attention. Microorganisms have the ability to

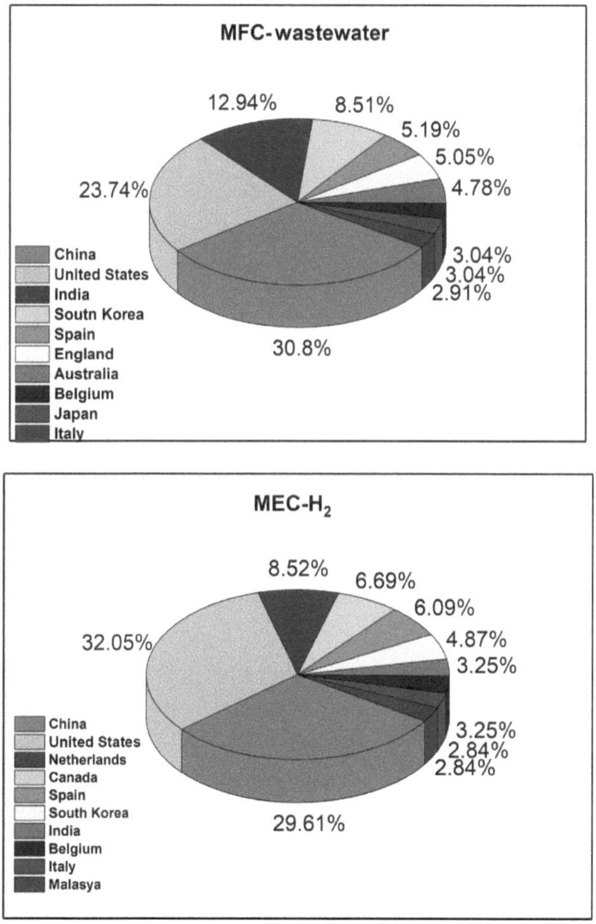

FIG. 2 Analysis of publications dealing with wastewater treatment in microbial fuel cells (MFCs) and biohydrogen production in microbial electrolysis cells (MEC) with over the past 20 years retrieved from the Web of Science database.

self-generate, to adapt to different environments easily, and to be flexible in terms of substrate as compared to other biosystems such as organelles and enzymes. Nevertheless, part of the substrate is used to keep the system alive and to maintain microbial growth. Control of this tenuous line calls for an interdisciplinary view and demands knowledge of biochemistry, microbiology, electrochemistry, engineering, and materials science. There are some good literature reviews discussing the main challenges and advantages of MEC and MFC [2–8] systems with respect to electron transfer and cell arrangement. Rabaey and Rozendal [6] have come up with the concept of biorefinery depicted in Fig. 3. In this concept, organic matter present in the wastewater can be used to produce energy in a

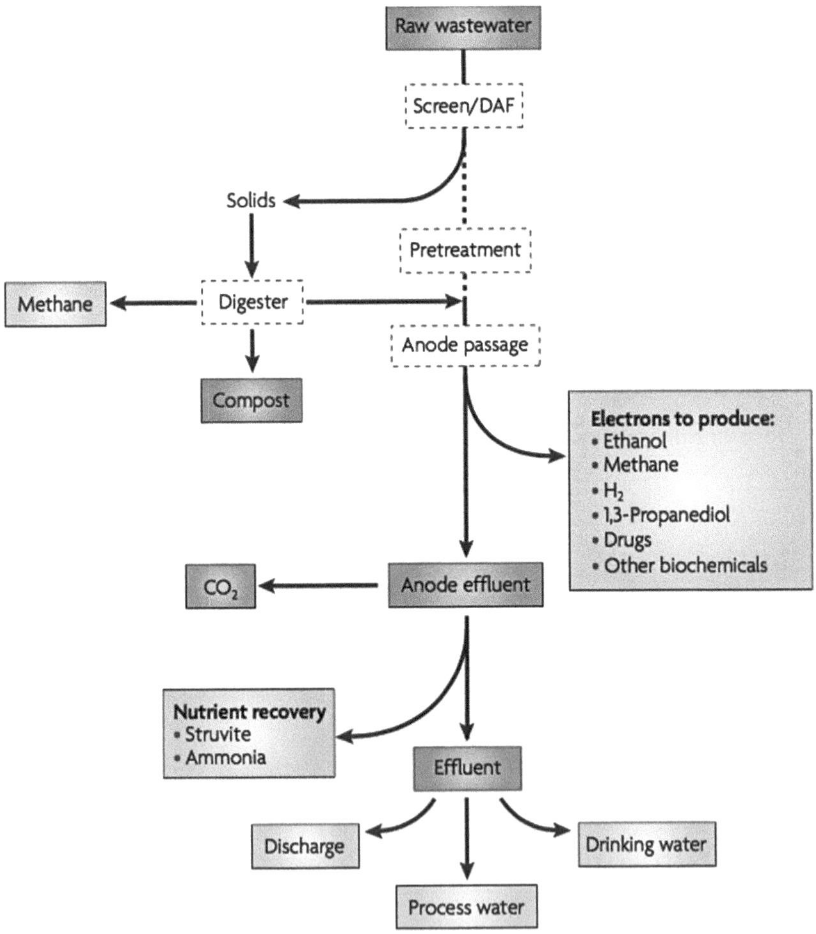

FIG. 3 Biorefinery concept proposed by Rabaey and Rozendal [6]. *DAF*, dissolved air flotation.

biodigestor. The biodigestor can be coupled to a bioelectrolytic system to generate several added-value products such as methane, ethanol, hydrogen, and 1,3-propanediol, among others. Control of the microbial groups is essential to ensure generation of the desired products and to prevent undesirable side reactions that will produce methane via the methanogenic route or lead to energy loss [9]. In 2007, Logan's group published another breakthrough paper in biohydrogen production in which they reported hydrogen production from different substrates [10].

Here we will deal with the concepts of organic matter degradation mediated by different microorganisms and review the current knowledge of MFCs and MECs.

BIOLOGICAL PROCESSES: ANAEROBIC DIGESTION

The anaerobic biodegradation of organic matter, or anaerobic digestion (AD), is probably one of the oldest biological processes to be described. In 1776, Alessandro Volta concluded that the amount of decaying organic matter directly correlated with the amount of produced flammable gas. Only in 1808 methane was detected among the gases produced during AD of cattle manure [11]. The first AD plant was built in India in 1859. Nowadays, AD is a commercial technology that is applied to convert municipal and industrial organic wastewater streams into renewable energy in the form of methane-rich biogas. Although AD recovers energy as methane, it offers advantages such as highly efficient organic matter removal, production of little sludge in excess, and need for small spaces [11].

Biochemically, AD is a process through which organic matter undergoes a series of sequential oxidations mediated by distinct groups of microorganisms to culminate in the formation of methane and carbon dioxide. This process occurs in the absence of inorganic acceptors of electrons, such as O_2, NO_3^-, NO_2^-, Fe^{+3}, Mn^{+4}, SO_4^{-2}, $S_2O_3^{-2}$, SO, and SO_3^{-2} [12].

Fig. 4 illustrates the three basic steps in the sequence of AD carried out by a wide range of microorganisms: (1) hydrolysis and fermentation; (2) acetogenesis; and (3) methanogenesis. This successive chain of biochemical reactions must occur in an orderly manner because the products of one reaction are the substrates of the next step. Finally, the electrons are transferred to CO_2, the final electron acceptor in this trophic chain of microorganisms, to yield methane [13].

These steps occur in the same anaerobic reactors used for wastewater treatment or in natural methanogenic ecosystems. In some cases, the first stages of biodigestion can be separated to recover H_2, as will be described later.

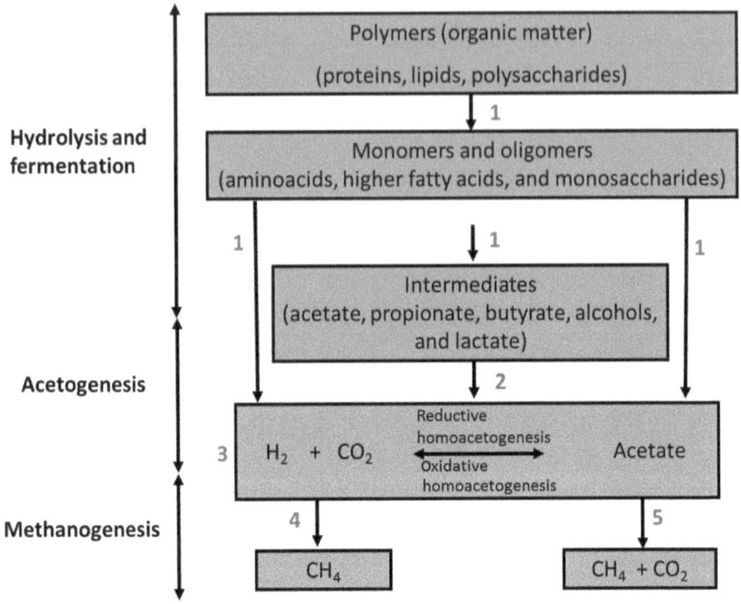

FIG. 4 Steps of the anaerobic degradation of organic matter mediated by different microorganisms. (1) Hydrolytic-fermentative bacteria; (2) hydrogen-producing acetogenic bacteria; (3) homoacetogenic bacteria; (4) hydrogen-consuming methanogenic archaea (hydrogenotrophic); and (5) acetotrophic methanogenic archaea. *Modified from B. Demirel, P. Scherer, The roles of acetotrophic and hydrogenotrophic methanogens during anaerobic conversion of biomass to methane: a review, Ver. Env. Sci. Biotechnol. 7 (2008) 173–190.*

Anaerobic Digestion Steps: Hydrolysis and Fermentation

AD begins with an enzyme-mediated transformation of insoluble organic material and compounds with higher molecular mass present in wastewater and waste materials to simpler compounds, followed by their assimilation by microorganisms. In this phase, complex organic materials, such as polysaccharides, fats, and proteins, are transformed into soluble organic materials, such as monosaccharides, organic acids, and amino acids, which are suitable energy and carbon sources for the cells. The hydrolytic-fermentative bacteria employ the soluble organic compounds to generate short-chain organic acids, alcohols, lactic acid, CO_2, H_2, NH_3, and H_2S, depending on the composition of the wastewater (Fig. 1). Some hydrolytic-fermentative microorganisms can directly ferment monomers to acetate (acetogenesis), as shown in Fig. 4.

The fermentation step is carried out by a diverse group of bacteria, the majority of which is obligatorily anaerobic. However, some species are facultative and can metabolize the organic material by the oxidative route; that is, they use oxygen as the final electron acceptor. These bacteria are important because they remove the dissolved oxygen that is possibly present in the reaction medium, ensuring the condition of anaerobiosis that is essential for the other anaerobic treatment steps to occur [14].

After hydrolysis of the complex materials, fermentative microorganisms are the first to act in the sequence of AD reactions. The most common fermentation products arising from organic material, represented by glucose, are acetic, propionic, and butyric acids, according to Eqs. (1)–(3) in Table 1, respectively. Oxidation of organic matter by fermentative bacteria is the most energetically favorable AD reaction (more negative ΔG), which translates into higher microorganism growth rate as compared to the other groups of microorganisms involved in AD (Table 1). In this way, the fermentation step will only be limiting if the material to be degraded is not easily hydrolyzed [14].

Fermentation of 1 mol of glucose produces 2 mol of pyruvic acid. Because pyruvic acid is more oxidized than glucose, this process also generates 2 mol of NADH, which is the reduced form of the most common electron carrier in the oxidation of carbohydrates (NAD^+). In the absence of oxygen, fermentative bacteria reoxidize NADH by transferring its electrons to the pyruvate molecule, to afford different reduced compounds such as propionate, butyrate, lactate, hydrogen, ethanol, propanol, and formate, depending on the microorganism and on the conditions of the medium [14]. Fig. 5 displays the main fermentation products and the

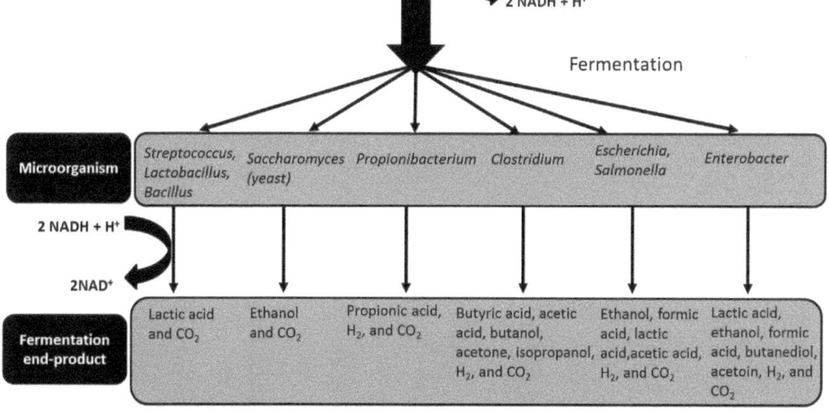

FIG. 5 Main fermentation products and the microorganisms involved in the process.

TABLE 1 Main Reactions of the Different Stages of Anaerobic Digestion and Free Standard Energy of the Reactions in the Biodigestors

Step	Eq.	Reaction	$\Delta G'^{\circ}$	ΔG°
Fermentation	(1)	Glucose → acetate $C_6H_{12}O_6 + 2H_2O \leftrightarrow 2CH_3COO^- + 2CO_2 + 2H^+ + 4H_2$	−206	−319
	(2)	Glucose → propionate $C_6H_{12}O_6 + 2H_2 \leftrightarrow 2CH_3CH_2COO^- + 2H_2O + 2H^+$	−358	
	(3)	Glucose → butyrate $C_6H_{12}O_6 \leftrightarrow CH_3CH_2CH_2COO^- + 2CO_2 + H^+ + 2H_2$	−255	−284
Acetogenesis	(4)	Propionate → acetate $CH_3CH_2COO^- + 3H_2O \leftrightarrow CH_3COO^- + H^+ + HCO_3^- + 3H_2$	+76.2	−5.5
	(5)	Butyrate → acetate $CH_3CH_2CH_2COO^- + 2H_2O \leftrightarrow 2CH_3COO^- + H^+ + 2H_2$	+48.1	−17.6
	(6)	Ethanol → acetate $CH_3CH_2OH + H_2O \leftrightarrow CH_3COO^- + H^+ + 2H_2$	+9.6	−37
	(7)	Bicarbonate and hydrogen → acetate $2HCO_3^- + 4H_2 \leftrightarrow CH_3COO^- + 4H_2O$	−104.6	−7.1
Hydrogenotrophic methanogenesis	(8)	Bicarbonate and hydrogen → methane $HCO_3^- + 4H_2 + H^+ \leftrightarrow CH_4 + 3H_2O$	−135.6	−3.2
Acetoclastic methanogenesis	(9)	Acetate → methane $CH_3COO^- + H_2O \leftrightarrow HCO_3^- + CH_4$	−31.0	−24.7
Global reaction	(10)	Glucose → methane $C_6H_{12}O_6 + 3H_2O \leftrightarrow 3CH_4 + 3HCO_3^- + 3H^+$	−403.4	−398.3

Standard conditions ($\Delta G'^{\circ}$, kJ/reaction): solutes = 1 mol/L, gases = 1 atm, and 25°C. Ambiental conditions (ΔG°, kJ/reaction), concentrations of the reactants in typical conditions of an anaerobic digestor: organic acids = 1 mmol/L; HCO_3^- = 20 mmol/L; CH_4 = 0.6 atm; $H_2 = 10^{-4}$ atm, at 37°C and pH 7.0.

Modified from S.F. Aquino, C.A.L. Chernicharo, Acúmulo de ácidos graxos voláteis (AGV) em reatores anaeróbios sob estresse: causas e estratégias de controle, Eng. Sanit. Ambient. 10 (2005) 152–161.

microorganisms involved in the process. The organic acids originating from fermentation are usually referred to as acetate, butyrate, and propionate because at the pH of AD, which is almost neutral, the acids occur mostly in their nonprotonated form. In the same way, CO_2 is represented as HCO_3^- in the AD reaction (Table 1)—CO_2 dissociates in aqueous medium, to form the bicarbonate ion.

Anaerobic Digestion Steps: Acetogenesis

Because methanogenic microorganisms can only assimilate acetate, CO_2, and H_2 among all the products formed during fermentation, acetogenesis is a preparation phase for methanogenesis. During acetogenesis, organic acids with more than two carbon atoms and the alcohols, fermentation products, are converted to acetate, CO_2, and H_2 by a group of specific fermenting bacteria, called acetogenic bacteria (Eqs. 4–7, Table 1) [13].

Based on metabolism, two groups of acetogenic bacteria can be distinguished. The first group, known as proton-reducing acetogenic bacteria or H_2-forming bacteria, can generate acetate, CO_2, and H_2 from various substrates, such as intermediate organic acids (propionate and butyrate), alcohols, or even longer-chain organic acids (valerate, isovalerate, and palmitate). The other group, known as acetate-oxidizing bacteria or homoacetogens, catalyzes acetate conversion to CO_2 and H_2. They also catalyze the reverse reaction to generate acetate from CO_2 and H_2 via the so-called Wood-Ljungdahl pathway, a CO_2-fixing process [15].

The proton-reducing acetogenic bacteria can transfer reduction equivalents (electrons) only to H^+, to give H_2. These reactions are not energetically favorable, as judged from the positive $\Delta G'^\circ$ value under standard conditions (1 mol/L of reagents, 1 atm for gases, 298 K) (Eqs. 4–6, Table 1). When the standard free energy is positive, metabolic processes are only thermodynamically favorable under environmental conditions if the reaction products are kept at low concentrations ($\Delta G^\circ < 0$).

Accordingly, given that $\Delta G'^\circ > 0$, the acetogenic H^+-reducing reactions are not thermodynamically favorable under standard conditions. However, these reactions occur naturally in anaerobic reactors or in anaerobic ecosystems ($\Delta G^\circ < 0$), as a result of the action of H_2-consuming microorganisms, mainly hydrogenotrophic methanogenic microorganisms (Eq. 8, Table 1).

Most methanogenic environments maintain an H_2 concentration below 10^{-4} atm, which is sufficient to stimulate the activity of proton-reducing acetogenic bacteria [14]. The association between proton-reducing acetogenic bacteria and H_2-consuming methanogenic microorganisms is called syntrophic association, and the transfer of reduction equivalents through H_2 is designated H_2 transfer between species [12].

The second group of acetogens is called homoacetogens to differentiate them from the proton-reducing acetogenic bacteria. Homoacetogens can convert acetate to H_2 and CO_2 (oxidative homoacetogenesis); they can also use the reverse reaction to produce acetate from H_2+CO_2 (reductive homoacetogenesis) (Eq. 7, Table 1). Indeed, Zinder suggested the term "reversible acetogenesis" in 1994. High H_2 concentrations ($\geq$500 Pa) favor reductive homoacetogenesis or methanogenesis from H_2+CO_2, whereas acetate is oxidized to CO_2 and H_2 at low H_2 concentrations ($\leq$40 Pa) [15]. The metabolic activity of homoacetogens does not depend on the syntrophic relationships mentioned above. Nevertheless, like hydrogenotrophic methanogens, homoacetogens can maintain low H_2 pressure in the medium [12].

Anaerobic Digestion Steps: Methanogenesis

Organic matter degradation finishes during methanogenesis. Among all the stages of AD, methanogenesis is the only one that microorganisms from the domain Archaea can perform. In contrast to the eubacteria responsible for the previous steps, Archaea are the most ancient microorganisms from an evolutionary viewpoint. Methane is formed by two distinct mechanisms. The first consists of methane production from acetate, known as acetoclastic methanogenesis (Eq. 9, Table 1) [15]. This reaction, mediated by the methanogenic acetoclastic microorganisms of the microbial consortium, is important because effective organic matter removal from the liquid phase depends on acetate conversion, during which about 70% of the original electrons from the substrate are channeled into methane [16].

The other methane-forming pathway is CO_2 reduction by H_2, known as hydrogenotrophic methanogenesis (Eq. 8, Table 1). Hydrogenotrophic methanogens use hydrogen as an electrons acceptor to form methane. Many methanogens can also use formate as an electron donor to reduce CO_2 to CH_4 [15].

In summary, considering that glucose is the organic matter to be degraded, it is possible to generate 3 mol of methane from 1 mol of glucose according to Eq. (10) (Table 1). In other words, the consumption of each gram of glucose produces 0.27 g of methane, which corresponds to 0.4 L of methane as calculated by the equation of ideal gases under standard conditions.

In general, microorganisms grow more slowly in the anaerobic environment where AD occurs as compared to aerobic bacteria present in activated sludge systems for wastewater treatment (2–10 d^{-1}). In particular, methanogenic microorganisms grow at a very low rate, at 0.08–0.7 d^{-1} [17]. In practice, during the treatment of effluents, AD produces a small amount of sludge. This contrasts with the large amount of residual sludge generated during aerobic processes of wastewater treatment. Therefore, besides producing methane that can be used as fuel, biodigestion of anaerobic wastewater oxidizes organic matter in effluents while generating a smaller amount of residual sludge as compared to aerobic processes.

Fermentative H_2 Production: A New Approach to Anaerobic Digestion

Besides methane, a traditional anaerobic digestion product, H_2, has raised the interest of a large number of researchers in recent years [18–20]. H_2 fuel appears as an attractive alternative to methane (50 kJ/g); its direct combustion produces a significant amount of energy (120 kJ/g) and releases only water as a by-product rather than CO_2 [21].

Contrary to the current processes used to generate H_2, which all employ fossil raw material with large energy consumption, fermentative H_2 production uses renewable materials, such as carbohydrate-rich wastes and wastewater, at low temperature and pressure, which makes the process even more advantageous [18].

Fermentative H_2 production by mixed culture (sludge) shares several points with the AD of organic matter; that is, the first steps, hydrolysis, fermentation, and acetogenesis, are common to H_2 and CH_4 production (Eq. 1–6, Table 1).

However, successful fermentative H_2 production by an anaerobic sludge poses some challenges. A typical anaerobic sludge cannot produce high H_2 concentrations because this gas will be readily consumed to produce methane [20]. Successful H_2 production by a mixed bacterial culture, such as anaerobic sludge, requires supression of the bacteria involved in the latest step of anaerobic digestion; that is, methanogenesis [20,22]. In addition, H_2 pressures must be low in the reactors so that $\Delta G°$ within the biodigestor remains negative and allows H_2-producing reactions to occur.

Suppression of the H_2-consuming steps during DA, which are mediated mainly by the hydrogenotrophic methanogens that produce methane from CO_2 and H_2 (Eq. 8, Table 1) and to a lesser extent by the homoacetogens that convert CO_2 and H_2 to acetate (Eq. 7, Table 1), have been successfully performed by applying pretreatments to the mixed cultures (sludge) used as inoculum during H_2 production [19]. The efficiency of the pretreatment of the inoculum for fermentative H_2 production is mostly based on H_2-producing bacteria, which are spore-forming bacteria (spores are structures that resist extreme environmental conditions) of the genus

Clostridium. Therefore, pretreatments create adverse conditions to eliminate the most sensitive bacteria and to favor the development of sporulating bacteria that produce H_2 [19,20].

Among the techniques available to pretreat inoculum, chemical treatments are worth highlighting, mainly the use of 2-bromoetanosulfonato-BESA (specific inhibitor of methanogenic archaea), acid treatment ($pH < 3$), and alkaline ($pH > 10$) treatment. Heat treatment (at 80°C or 100°C for a few minutes) is also worthy of note [19,22].

Other approaches have been attempted to ensure low partial H_2 pressure in the reactors and hence negative $\Delta G°$ for acetogenesis under environmental conditions. These procedures include continuous sparkling of inert gas into biodigesters and membrane-adsorption technologies to remove H_2 from a mixture of gases [20].

However, these practices are laborious and do not seem necessary because appropriate reactor configuration combined with low hydraulic retention times (HRT) to eliminate the methanogenic and thermophilic conditions (above 45°C) suffice to decrease H_2 solubility in the liquid [20,23].

Unfortunately, low yield still limits the use of fermentative H_2 production, which gives a maximum of 4 mol of H_2 per mole of glucose when acetate is the by-product, and 2 mol of H_2 when butyrate arises (Eqs. 1, 3, Table 1, respectively). Nonetheless, reactions in the presence of a mixed culture and real wastewater (brew, milk, and paper industrial wastewaters) rarely exceed 2 mol of H_2 per mole of glucose [20]. In addition, fermentative H_2 production by interrupted AD degrades only part of the organic matter (about 33%–50%); acetate and butyrate are the by-products.

Therefore, it does not seem likely that fermentative H_2 production will replace AD as the main technology to convert organic-rich waste and wastewater streams into energy. Nevertheless, the ability of fermentative H_2 production to generate an effluent rich in organic acids makes their combination with bioelectrochemical processes attractive [24,25], as represented in Fig. 6.

Bioelectrochemical systems can also be used to produce energy directly from the organic materials present in residues/effluents, to recover the energy contained in the by-products generated (organic acids) during fermentation [24,26,27]. MBCs are the most often used bioelectrochemical systems to generate energy. In turn, electricity can be applied to MECs to convert organic acids, typical side products of H_2 fermentation, to more H_2 (Fig. 3). Both these technologies will be further discussed in the next section of this chapter.

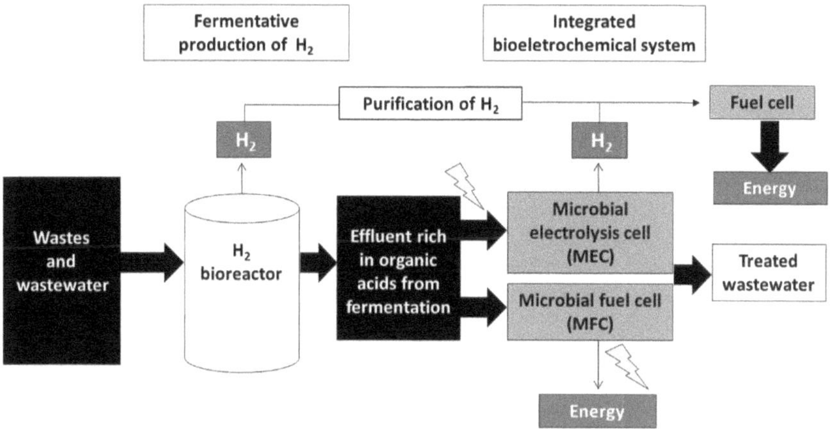

FIG. 6 Fermentative-bioelectrochemical system to generate energy and treat wastewater.

MICROBIAL FUEL CELLS

Fundamentals

The growing demand for sustainable energy associated with different types of environmental damage, especially contamination of water sources, has led to the development of several efficient and eco-friendly energy conversion systems. These alternative technologies seek to associate wastewater treatment with electric power generation and production of value-added compounds [28,29]. Among the several alternative energy conversion processes that have been presented in the specialized literature in recent decades, much attention has been given to the development of fuel cells, which afford higher conversion efficiency and power density as compared to other processes [30]. This device can be viewed as an electrochemical system that can directly convert intrinsic chemical energy in fuels into electrical energy [31]. A particular case of traditional fuel cells is the biological fuel cell or biofuel cell, which works almost in the same way as the traditional chemical fuel cell but uses biological entities as catalysts rather than noble metal-based catalysts [32]. When enzymes are employed to achieve electrode activity, we have the so-called enzymatic biofuel cell [33]; when microorganisms are responsible for the bioelectrocatalysis, we have the microbial fuel cell (MFC) [34]. Through the use of clean and renewable catalysts, MFCs provide a means to obtain renewable and sustainable energy and to treat wastewater, which is generally employed as the carbon source for the electrochemical system.

The connection between biology and electricity dates back to 1912, when M.C. Potter at the University of Durham was the first to demonstrate that microorganisms could generate electrical energy [35]. However, only after a certain period of time (in the early 1990s) this technology received more attention and the number of studies involving this technology increased. MFC is a device that directly converts chemical energy contained in organic compounds, such as sugars and alcohols, into electrical energy [34,36,37]. Such device is advantageous over conventional fuel cell technology: it operates under mild conditions (operation at ambient temperature and in physiological pH) and uses inexpensive materials.

Below is an example of MFC in which glucose is oxidized at the anodic side (assuming the electrolyte is acid) according to the reaction:

$$C_6H_{12}O_6 + 6H_2O \rightarrow 6CO_2 + 24H^+ + 24e^- \quad E^0 = 0.014\,V \qquad (11)$$

On the cathodic side, an oxidant (usually oxygen) is reduced according to the reaction:

$$6O_2 + 24H^+ + 24e^- \rightarrow 12H_2O \quad E^0 = 1.23\,V \qquad (12)$$

The resulting electrochemical reaction creates an electron flow, while protons originate from fuel oxidation [38]. Fig. 7 shows a schematic representation of how MFCs operate.

Considering the anodic side of MFCs, the bacteria used as catalysts can transfer electrons from the catalytic process directly (through specific bacteria that possess clusters in the outer membrane and are thus capable of transferring electrons to the electrode surface) or through a mediator molecule (a redox compound that can transfer electrons from the microorganism transport chain to the electrode surface is added to the catalyst surface) [39]. The electroactive groups, that is, the electron carriers present

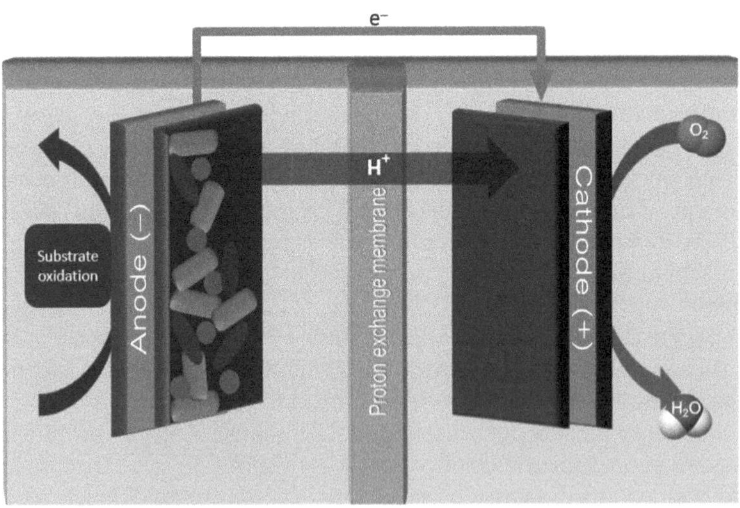

FIG. 7 Schematic representation of how a microbial fuel cell operates, based on oxygen reduction at the cathodic side.

in the microorganisms, are usually located within the prosthetic groups, which hinders the direct electron transfer process [38]. Therefore, the redox mediator species are important to improve the overall bioelectrode efficiency and must have the following characteristics: (i) its oxidized form must easily penetrate through the bacterial membrane; (ii) the redox potential of the mediator must be close to the redox potential of the metabolite; (iii) the oxidation state of the mediator should not interfere with metabolic processes; (iv) the oxidized form of the mediator should be promptly released from the bacterial membrane; (v) the redox mediator must have a high turnover number; and (vi) both redox forms of the mediator must be stable [40,41]. It is worth highlighting that the use of mediators in MFCs favors the electron transfer process and allows the microorganisms contained in the bulk solution to participate in the catalytic process. Various organic dyes, thionines, phenazine derivatives, and ferrocene derivatives exhibit these desired characteristics and provide efficient electron transfer rates [36]. In this context, mixed bacterial cultures (such as sewage sludge) generally afford greater electron transfer rates as compared to individual cultures because they can produce soluble mediator species (e.g., pyocyanin) [42].

Regarding the cathodic side of the cell, air oxygen is the most obvious choice of oxidant, although other compounds such as hydrogen peroxide, hexacyanoferrate $[Fe(CN)_6]^{-3}$, and permanganate, among others, are efficient electron acceptors. Nonetheless, such compounds need continuous replacement. The more reliable oxygen reduction reaction is the most used in microbiological cells [8]. Platinum-based materials are the most commonly employed electrocatalysts for this purpose, although other compounds have also been successfully used, such as porphyrins, phthalocyanines, many carbon materials, and nonprecious metallic species, among others [43,44]. An aqueous solution containing dissolved oxygen can also be applied in microbial fuel cells, but such a system requires the continuous addition of gas, and therefore gas diffusion electrodes are generally preferred [45].

Materials and Performances

The electrode support employed to construct fuel cells is crucial to good performance. In the case of MFCs, their composition, morphology, and surface properties will directly affect biofilm formation, electron transfer, and substrate oxidation. In this context, another key point in the development of an economically viable device is cost. Considering the overall fuel cell device, the electrode support and membranes account for most of the costs involved in electrode construction. One of the main advantages of MFCs is that they require relatively low-cost materials at both the anodic and cathodic sides. On the anodic side, carbon-based materials such as paper, cloth, and more recently carbon fiber brushes are mainly used as electrode support for biofilm formation [46]. Also, many malleable and large-scale applicable materials have been efficiently employed in this field, including stainless steel and titanium [47]. Bearing the practical features of materials used in MFCs in mind, characteristics such as high electrical conductivity and surface area, biocompatibility, and chemical and physical stability are mandatory for their application in large-scale systems. To achieve more efficient supports, researchers have targeted different possibilities such as nanostructured materials to obtain 3D biofilms with higher amount of biocatalyst and hence higher catalytic efficiency with lower overpotentials along the system [48–50]. Materials such as carbon nanotubes, graphene, and nickel foam display excellent electronic properties, such as high electron transfer rate and large active surface area, which make them quite a fit for MFCs. Along with their excellent biocompatibility, an increasing number of nanostructured materials are being developed for MFCs. Table 2 summarizes some of the electrode materials that are the most often employed in MFC technology.

Another crucial component of MFC technology is the membrane or separator, which separates the cathodic and anodic compartments physically. This material must allow efficient transport of protons (generated during substrate oxidation) from the anode to the cathode. Perfluorosulfonic acid-based polymers are most commonly applied as a fuel cell proton exchange membrane [69] because they give rise to a membrane with highly hydrophilic sulfonate groups that attract water molecules. When hydrated, these sulfonate groups form channels through which cations (e.g., protons) can move rapidly. Nafion membrane (Dupont), Flemion (Asahi Glass), and Aquivion (Solvay) are among the main commercially available membranes, and they vary essentially in terms of chain size and thickness.

The performance of bioelectrochemical systems based on microorganisms depends on several factors, including substrate, temperature, microorganism species or community, electrolyte composition, anodic and cathodic materials, and separators. In addition, parameters such as cell geometry and electrode size and arrangement influence the overall performance of the system [29]. In open circuit conditions, when no current flows, the typical open circuit potential values obtained in MFCs are in the range of 0.8 V. When electric current flows, this value decreases to around 0.6 V due to losses related mainly to activation and polarization. Increasing the electrode surface area, using more efficient biocatalysts, and increasing the temperature reduce these losses. To obtain higher voltages, the anode potential must be as negative as possible, but it cannot be so low as to inhibit the electron transport that results in fermentation [38]. The overall performance of these systems can be described in several ways and is represented in terms of coulombic efficiency, power density, and

TABLE 2 Some Examples of the Electrode Material Supports That Are the Most Often Employed in Microbial Fuel Cell Technology

Substrate	Electrode support material	Reference
Acetate	Graphite fiber brush	[51]
	Recycled tire crumb rubber coated with graphite paint	[52]
	Multiwalled carbon nanotubes	[50]
	Carbon brush	[46]
	Nonwaterproof carbon cloth	[53]
	Air-breathing gas diffusion activated carbon	[54]
	Polyaniline electrochemically synthesized on graphite felt	[55]
	Graphene	[56]
	TiO_2 nanoparticles decorating carbon nanotubes	[57]
Glucose	Electrochemically oxidized carbon mesh	[58]
	TiO_2(rutile)-C(semigraphitic)/C (semigraphitic) dual nanofiber	[59]
	Polyaniline-modified glassy carbon	[60]
Domestic wastewater	Graphite cylinder	[61]
	Nanostructured polyurethane/graphite/polypyrrole	[62]
Glucose and glutamate from artificial wastewater	Graphite felt	[63]
Protein-rich wastewater	Graphite rods	[64]
Sodium fumarate	Stainless steel	[65]
Sludge	Graphite electrodes covered with Co-nanostructured carbon	[49]
Sucrose	Nanostructure carbon based on 3D macroporous carbon with mesoporous carbon	[66]
1,2-Dichloroethane	Graphite plate anode and graphite granules cathode	[67]
Brewery wastewater	Carbon cloth	[37]
Ferricyanide	Flexible 3D graphene-nickel foam	[48]
Lactate	Graphite anodes decorated with Au and Pd nanoparticles	[68]

amount of hydrogen that is formed in the case of the electrolysis cell. The coulombic efficiency of the process will be calculated from the energy produced by the cell in relation to the theoretical charge from fuel oxidation, usually by employing the chemical oxygen demand parameter:

$$\mathrm{CE} = \frac{M \int_0^t I\,dt}{F n V \Delta \mathrm{DQO}}$$

where M is the oxygen molar mass, I is the current, F is Faraday's constant, $n=4$ is the number of electrons transferred per mole of oxygen, V is the volume of the liquid in the anodic side, and DQO is the chemical oxygen demand parameter.

If the system is employed to treat wastewater, parameters such as total organic carbon and chemical and biochemical oxygen demand are also used to control water quality [69].

Wastewater Treatment

The undue disposal of wastewater generated by various activities, such as domestic, agricultural, and industrial activities, causes much environmental damage. Such damage includes hypertrophication, hypoxia, and algae proliferation, which all deplete drinking water sources [70]. The currently available wastewater treatment processes require high initial investments and demand a huge amount of electrical energy. Moreover, the employed systems do not generate any revenue or commercially valuable compounds. Wastewater treatment requires about 0.5–2 kWh/m^3 depending on the process and on the quality of the treated water. Interestingly, these wastes contain about 3–10 times the energy required for their treatment [71]. In this context, the use of MFCs and MECs to perform the dual task of degrading an effluent and generating electrical energy as well as value-added products presents many advantages from the energetic, environmental, economic, and operational points of view, including:

(i) Direct generation of electrical energy, elimination of the aeration process, and centralized and decentralized applications;
(ii) High amount of water recovery, low carbon emission, and generation of a lower volume of sludge;
(iii) Profit through generation of energy and value-added products, low operational costs, and self-generation of microorganisms.

Bioelectrochemical systems based on microorganisms have been successfully used to treat organic waste ranging from pure compounds to very complex wastewater [47]. In this context, various substrates have been tested, and the efficient removal of total organic carbon, such as glucose, acetate, butyrate, cysteine, proteins, and lignocellulose, among others, has been achieved (Table 2). In the specialized literature there are works that seek to generate electrical energy and to produce hydrogen directly from complex organic residues, such as residues of food production, residues from breweries, domestic waste, chemical industry waste, animal waste, dye industry waste, pulp industry waste, and waste from sanitary landfills, for instance [72,73].

MICROBIAL ELECTROLYSIS CELLS

Fundamentals

The electrolyzer system represents an opposite situation to the fuel cell: the former system affords a product of interest through application of an electric potential. In the current scenario of sustainable technologies, electrolytic processes can be an attractive alternative for wastewater decontamination and generation of value-added products, such as hydrogen [38]. The production of hydrogen via an electrolytic process based on a microbial electrolysis cell (MEC) was first accomplished in 2005 [74]. In this system, biohydrogen can be directly produced at the cathodic side by reducing protons (generated on the anodic side through oxidation of organic compounds by microorganisms) in the presence of either inorganic catalysts or electroactive bacteria and low voltages [75].

In MEC, the hydrogen evolution reaction occurs when protons or water molecules are reduced on the electrode surface. As a result of its environmentally friendly behavior, this methodology has gained increasing attention. MECs produce hydrogen in a clean manner and are an alternative to the current methods that generate hydrogen from fossil fuels and consequently produce large amounts of carbon dioxide and consume high amounts of energy [5]. The microbial electrolytic process, that is, the conversion of electrical energy into chemical energy and the simultaneous production of hydrogen, can be called electrohydrogenesis or microbial electrolysis [10]. In other words, MEC is a reactor that produces hydrogen whereas MFC is a system that generates electricity, and both MEC and MFC consume organic matter. The cathodic reduction reaction is fueled by the protons and electrons generated in the anodic bioelectrocatalysis and by some extra electrical energy required for the occurrence of the cathodic reaction (Fig. 8) [76].

In the case of MFCs, the electrons flow from the anodic to the cathodic side, where they are consumed in the oxygen reduction reaction with the production of electrical current. In the case of MECs, which operate in the absence of oxygen, the electron flow is not spontaneous; however, it can be forced by application of an external potential between the anode and the

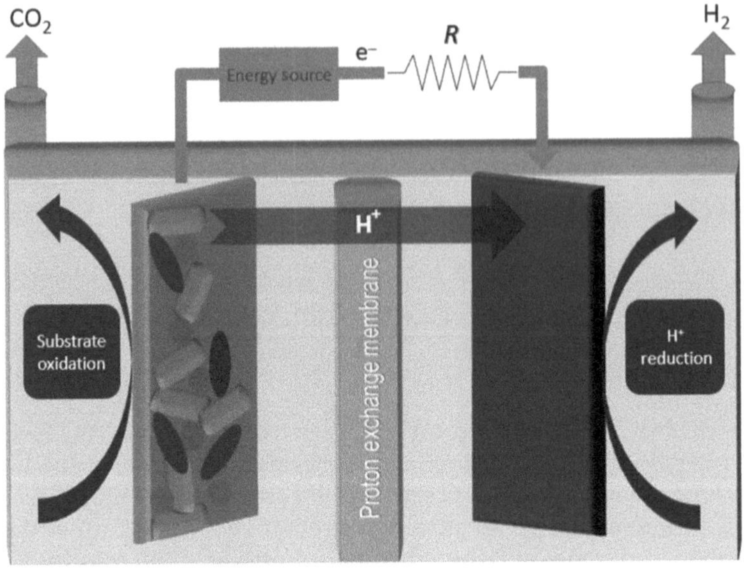

FIG. 8 Schematic representation of the operation of a microbial electrolysis cell based on the reduction of protons to hydrogen production.

cathode, and hydrogen gas is produced from the reduction of protons. To overcome the thermodynamic barrier limiting the hydrogen evolution reaction, the applied voltage must be greater than $\Delta G°/nF$ of each target reaction. Taking acetate as an example of substrate, the standard value would be −0.14 V. Nevertheless, the various losses that occur in an electrolysis system lead to a value that is typically above 0.3 V. Such value is quite low when compared to the potential required to produce hydrogen via electrolysis of water in a neutral medium (which can reach up to 1800 mV considering the overpotentials involved in the reaction) [74]:

Anode:

$$C_2H_4O_2 + 2H_2O \rightarrow 2CO_2 + 8e^- + 8H^+ \quad (3)$$

Cathode:

$$8H^+ + 8e^- \rightarrow 4H_2 \quad (4)$$

The possibility of obtaining pure hydrogen at low overpotentials without employing high-cost metal electrocatalysts, such as the many platinum-based materials, has dramatically raised interest in the bioelectrocatalytic production of hydrogen in recent years [5,72,77,78]. In terms of biological catalysts, wastewater or sludge from treatment plants displays a high concentration of exergonic bacteria and has generally been used as an inoculum in several works [79].

Association of hydrogen production technology by MEC with the traditional fermentative hydrogen production technology can boost the overall process efficiency [80]. To date, no microorganism capable of exceeding the Thauer limit (4 mol of H_2 mol^{-1} glucose) has been documented, and this limit is well below the theoretical limit of the overall reaction, which is 12 mol of H_2 mol^{-1} glucose [81]. Nonetheless, studies have shown that the association of these methodologies offers an interesting solution to convert waste generated during traditional fermentative biohydrogen production and consequently increase the overall amount of produced biofuel [82]. Fig. 9 shows a representative scheme in which organic acids that originated as byproducts of fermentative biohydrogen production (mostly acetic and butyric acids) can be later oxidized in MEC to give electrons and protons that will serve the electrolytic microbial cell during electrocatalytic hydrogen production.

The configuration of MEC resembles the configuration of MFC: both produce electricity only. However, there are some differences between MEC and MFC, especially in terms of achieving completely anaerobic conditions and using appropriate gas collectors. In terms of biological catalysts, wastewater sludge is the obvious source of inoculum because it contains high concentrations of exoelectrogenic bacteria. To obtain the biofilm, it is necessary to operate the system in MFC configuration until the biofilm emerges; after that, the bioanode is transferred to a MEC configuration. This methodology guarantees biofilm formation and preselects a community of exoelectrogenic bacteria for operation of the hydrogen-producing reactor.

To overcome the thermodynamic barrier limiting the hydrogen evolution reaction, a voltage greater than $\Delta G°/nF$ of each reaction must be

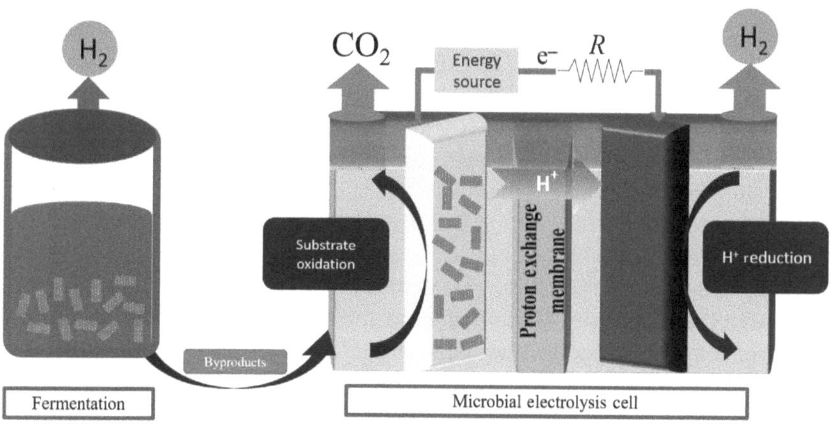

FIG. 9 Schematic representation of the association of traditional fermentative hydrogen production with a microbial electrolysis cell for hydrogen production.

applied. Different anodic and cathodic materials are being developed to reduce energy losses and the voltage applied to the system and to increase the efficiency of hydrogen production. All the environmental friendly materials described in the previous section are being targeted as electrode supports for the hydrogen producing side (cathode) microbial electrolysis systems.

Carbon-based supports such as papers, clothes, and brushes as well as composites obtained from the modification of these materials with conductive polymers such as polypyrrole and polyaniline are one of the most often-employed electrode materials. Considering the cathodic counterpart, where the hydrogen evolution reaction takes place, the importance of the electrode material goes further than stability and cost. For this specific purpose, researchers have focused on different catalytic materials that can lower the overpotential related with the desired reaction. Platinum-based electrodes are the standard material for hydrogen evolution with negligible overpotentials and excellent reaction kinetics. However, the high cost and lower abundance of platinum as compared to many earth-abundant metallic elements motivate the development of efficient and at the same time environmentally friendly materials. Nonnoble metal sulfides, phosphides, selenides, nitrides, Ni, Fe, Mo, and Co, among other earth-abundant particles and alloys, are in constant development [83]. Considering the performance parameters of the electrically driven hydrogen evolution process, coulombic efficiency is the most important factor as in the case of MECs. Nonetheless, the hydrogen yield is also crucial to overall efficiency.

PERSPECTIVES

The use of microorganisms as electrode catalysts has increased exponentially in the last decade. MEC is a relatively new strategy to produce biohydrogen. To make this technology commercially available for wastewater treatment, it is necessary to understand and control the microorganism community, which entails understanding how methanogenis (generation of methane) can be inhibited and how exoelectrogen microorganism (generation of energy at the MFC or generation of H_2 at the MEC) can be increased.

On the basis of the most recent studies on MFCs and MECs, the development of these techniques for practical application requires better understanding of the interactions between the microorganisms and the electrode surfaces. This should maximize biocatalyst loading and lead to significant improvement in electronic transfer processes. Therefore, the development of nanostructured materials and methodologies that boost the electron transfer efficiency in the case of both direct and mediated electron transfer processes will pave the way for future breakthroughs.

Improvement of this technology still requires that the following issues be tackled:

1. Efficient analysis of the microorganism communities.
2. Understanding the mechanism of bacterial electron transfer (direct electron transfer, mediated electron transfer, or wire communications).
3. Knowledge of the system architecture to diminish the internal system resistance.
4. Further studies must be performed to understand the anode and cathode material efficiency.
5. System scaling up to make the technology viable for wastewater treatment and energy generation.

References

[1] M. Carmo, D.L. Fritz, J. Mergel, D. Stolten, A comprehensive review on PEM water electrolysis, Int. J. Hydrogen Energy 38 (2013) 4901–4934.

[2] P. Clauwaert, P. Aelterman, T.H. Pham, L. De Schamphelaire, M. Carballa, K. Rabaey, W. Verstraete, Minimizing losses in bio-electrochemical systems: The road to applications, Appl. Microbiol. Biotechnol. 79 (2008) 901–913.

[3] H.S. Lee, W.F.J. Vermaas, B.E. Rittmann, Biological hydrogen production: Prospects and challenges, Trends Biotechnol. 28 (2010) 262–271.

[4] B.E. Logan, Scaling up microbial fuel cells and other bioelectrochemical systems, Appl. Microbiol. Biotechnol. 85 (2010) 1665–1671.

[5] B.E. Logan, D. Call, S. Cheng, H.V.M. Hamelers, T. Sleutels, A.W. Jeremiasse, R. A. Rozendal, Microbial electrolysis cells for high yield hydrogen gas production from organic matter, Environ. Sci. Technol. 42 (2008) 8630–8640.

[6] K. Rabaey, R.A. Rozendal, Microbial electrosynthesis—revisiting the electrical route for microbial production, Nat. Rev. Microbiol. 8 (2010) 706–716.

[7] R.A. Rozendal, H.V.M. Hamelers, K. Rabaey, J. Keller, C.J.N. Buisman, Towards practical implementation of bioelectrochemical wastewater treatment, Trends Biotechnol. 26 (2008) 450–459.

[8] H.M. Wang, Z.Y.J. Ren, A comprehensive review of microbial electrochemical systems as a platform technology, Biotechnol. Adv. 31 (2013) 1796–1807.

[9] M. Hasany, M.M. Mardanpour, S. Yaghmaei, Biocatalysts in microbial electrolysis cells: a review, Int. J. Hydrogen Energy 41 (2016) 1477–1493.

[10] S. Cheng, B.E. Logan, Sustainable and efficient biohydrogen production via electrohydrogenesis, Proc. Natl. Acad. Sci. 104 (2007) 18871–18873.

[11] T. Abbasi, S.M. Tauseef, S.A. Abbasi, A brief history of anaerobic digestion and Abbasi (2011). A brief history of anaerobic digestion and "biogas" in: Biogas Energy, SpringerBriefs in Environmental Science, vol. 2, 2011, p. 11.

[12] D.B. Archer, B.H. Kirsop, The microbiology and control of anaerobic digestion, in: A. Whetley (Ed.), Anaerobic Digestion: A Waste Treatment Technology, 1990, pp. 43–91.

[13] A.C. Van Haandel, G. Lettinga, Anaerobic Sewage Treatment. A Practical Guide for Regions With a Hot Climate, John Willey & Sons, New York, 1994.

[14] S.F. Aquino, C.A.L. Chernicharo, Acúmulo de ácidos graxos voláteis (AGV) em reatores anaeróbios sob estresse: causas e estratégias de controle, Eng. Sanit. Ambient. 10 (2005) 152–161.

[15] B. Demirel, P. Scherer, The roles of acetotrophic and hydrogenotrophic methanogens during anaerobic conversion of biomass to methane: a review, Rev. Environ. Sci. Biotechnol. 7 (2008) 173–190.
[16] P.L. McCarty, J. Bae, J. Kim, Domestic wastewater treatment as a net energy producer—can this be achieved? Environ. Sci. Technol. 45 (2011) 7100–7106.
[17] S.G. Pavlostathis, E. Giraldo-Gomez, Kinectics of anaerobic treatment: a critical review, Crit. Rev. Environ. Control. 13 (1991) 23–33.
[18] I.K. Kapdan, S. Alparslan, Application of anaerobic-aerobic sequential treatment system to real textile wastewater for color and COD removal, Enzym. Microb. Technol. 36 (2005) 273–279.
[19] S.V. Mohan, G. Velvizhi, J.A. Modestra, S. Srikanth, Microbial fuel cell: critical factors regulating biocatalyzed electrochemical process and recent advancements, Renew. Sustain. Energy Rev. 40 (2014) 779–797.
[20] I. Valdez-Vazquez, H.M. Poggi-Varaldo, Hydrogen production by fermentative consortia, Renew. Sustain. Energy Rev. 13 (2009) 1000–1013.
[21] F.M. Santos, F.A. Santos, O combustível "hidrogênio" Educação Ciência e Tecnologia (2007) 252–270.
[22] F.d.C. Lamaison, P.A.M. de Andrade, A.D. Bigaton, F.D. Andreote, R.V. Antônio, V. Reginatto, Long-term effect of acid and heat pretreatment of sludge from a sugarcane vinasse treatment plant on the microbial community and on thermophilic biohydrogen production, Int. J. Hydrog. Energy 40 (2015) 14124–14133.
[23] J.-S. Chang, K.-S. Lee, P.-J. Lin, Biohydrogen production with fixed-bed bioreactors, Int. J. Hydrog. Energy 27 (2002) 1167–1174.
[24] K.Y. Show, D.J. Lee, J.H. Tay, C.Y. Lin, J.S. Chang, Biohydrogen production: current perspectives and the way forward, Int. J. Hydrog. Energy 37 (2012) 15616–15631.
[25] D. Puyol, D.J. Batstone, T. Hülsen, S. Astals, M. Peces, J.O. Krömer, Resource recovery from wastewater by biological technologies: opportunities, challenges, and prospects, Front. Microbiol. 7 (2017) 2106.
[26] L.T. Angenent, K. Karim, M.H. Al-Dahhan, B.A. Wrenn, R. Domíguez-Espinosa, Production of bioenergy and biochemicals from industrial and agricultural wastewater, Trends Biotechnol. 22 (2004) 477–485.
[27] A. Schievano, T.P. Sciarria, K. Vanbroekhoven, H. De Wever, S. Puig, S.J. Andersen, K. Rabaey, D. Pant, Electro-fermentation merging electrochemistry with fermentation in industrial applications, Trends Biotechnol. 34 (2016) 866–878.
[28] O. Ellabban, H. Abu-Rub, F. Blaabjerg, Renewable energy resources: Current status, future prospects and their enabling technology, Renew. Sustain. Energy Rev. 39 (2014) 748–764.
[29] F.J. Hernandez-Fernandez, A.P. de Los Rios, M.J. Salar-Garcia, V.M. Ortiz-Martinez, L. J. Lozano-Blanco, C. Godinez, F. Tomas-Alonso, J. Quesada-Medina, Recent progress and perspectives in microbial fuel cells for bioenergy generation and wastewater treatment, Fuel Process. Technol. 138 (2015) 284–297.
[30] F. Barbir, Introduction, in PEM Fuel Cells, Academic Press, Burlington, 2005, pp. 1–16. (Chapter 1).
[31] B.C.H. Steele, A. Heinzel, Materials for fuel-cell technologies, Nature 414 (2001) 345–352.
[32] R.A. Bullen, T.C. Arnot, J.B. Lakeman, F.C. Walsh, Biofuel cells and their development, Biosens. Bioelectron. 21 (2006) 2015–2045.
[33] M. Rasmussen, S. Abdellaoui, S.D. Minteer, Enzymatic biofuel cells: 30 years of critical advancements, Biosens. Bioelectron. 76 (2016) 91–102.
[34] B.E. Logan, J.M. Regan, Microbial fuel cells—challenges and applications, Environ. Sci. Technol. 40 (2006) 5172–5180.
[35] M.C. Potter, Electrical effects accompanying the decomposition of organic compounds, Proc. R. Soc. London, Ser. B, Contain. Pap. Biol. Charact. 84 (1911) 260–276.

[36] A.S. Mathuriya, J.V. Yakhmi, Microbial fuel cells—applications for generation of electrical power and beyond, Crit. Rev. Microbiol. 42 (2016) 127–143.
[37] Y. Feng, X. Wang, B.E. Logan, H. Lee, Brewery wastewater treatment using air-cathode microbial fuel cells, Appl. Microbiol. Biotechnol. 78 (2008) 873–880.
[38] B.E. Logan, B. Hamelers, R. Rozendal, U. Schröder, J. Keller, S. Freguia, P. Aelterman, W. Verstraete, K. Rabaey, Microbial fuel cells: methodology and technology, Environ. Sci. Technol. 40 (2006) 5181–5192.
[39] Y. Qiao, S.J. Bao, C.M. Li, Electrocatalysis in microbial fuel cells-from electrode material to direct electrochemistry, Energy Environ. Sci. 3 (2010) 544–553.
[40] K. Rabaey, W. Verstraete, Microbial fuel cells: novel biotechnology for energy generation, Trends Biotechnol. 23 (2005) 291–298.
[41] Z. Du, H. Li, T. Gu, A state of the art review on microbial fuel cells: a promising technology for wastewater treatment and bioenergy, Biotechnol. Adv. 25 (2007) 464–482.
[42] K. Scott, An introduction to microbial fuel cells, in: Microbial Electrochemical and Fuel Cells, Woodhead Publishing, Boston, 2016, pp. 3–27.
[43] E.H. Yu, S. Cheng, B.E. Logan, K. Scott, Electrochemical reduction of oxygen with iron phthalocyanine in neutral media, J. Appl. Electrochem. 39 (2009) 705–711.
[44] M.H. Shao, Q.W. Chang, J.P. Dodelet, R. Chenitz, Recent advances in electrocatalysts for oxygen reduction reaction, Chem. Rev. 116 (2016) 3594–3657.
[45] S. Mateo, M. Rodrigo, L.P. Fonseca, P. Cañizares, F.J. Fernandez-Morales, Oxygen availability effect on the performance of air-breathing cathode microbial fuel cell, Biotechnol. Prog. 31 (2015) 900–907.
[46] V. Lanas, Y. Ahn, B.E. Logan, Effects of carbon brush anode size and loading on microbial fuel cell performance in batch and continuous mode, J. Power Sources 247 (2014) 228–234.
[47] K. Scott, E.H. Yu, Microbial Electrochemical and Fuel Cells: Fundamentals and Applications (Woodhead Publishing Series in Energy), first ed., Elsevier Science, 2015.
[48] H. Wang, G. Wang, Y. Ling, F. Qian, Y. Song, X. Lu, S. Chen, Y. Tong, Y. Li, High power density microbial fuel cell with flexible 3D graphene-nickel foam as anode, Nanoscale 5 (2013) 10283–10290.
[49] G.A. Kovalenko, T.V. Chuenko, L.V. Perminova, N.A. Rudina, O.V. Sherstyuk, A. Y. Tyurin-Kuzmin, I.A. Smirnov, Synthesis of nanostructured carbon on graphite electrodes with a supported co catalyst for preparing anodes for microbial fuel cells, Kinet. Catal. 57 (2016) 104–112.
[50] J.E. Mink, M.M. Hussain, Sustainable design of high-performance microsized microbial fuel cell with carbon nanotube anode and air cathode, ACS Nano 7 (2013) 6921–6927.
[51] B. Logan, S. Cheng, V. Watson, G. Estadt, Graphite fiber brush anodes for increased power production in air-cathode microbial fuel cells, Environ. Sci. Technol. 41 (2007) 3341–3346.
[52] H. Wang, M. Davidson, Y. Zuo, Z. Ren, Recycled tire crumb rubber anodes for sustainable power production in microbial fuel cells, J. Power Sources 196 (2011) 5863–5866.
[53] L.J. Ren, X.Y. Zhang, W.H. He, B.E. Logan, High current densities enable exoelectrogens to outcompete aerobic heterotrophs for substrate, Biotechnol. Bioeng. 111 (2014) 2163–2169.
[54] C. Santoro, K. Artyushkova, S. Babanova, P. Atanassov, I. Ieropoulos, M. Grattieri, P. Cristiani, S. Trasatti, B. Li, A.J. Schuler, Parameters characterization and optimization of activated carbon (AC) cathodes for microbial fuel cell application, Bioresour. Technol. 163 (2014) 54–63.
[55] P. Wang, H.R. Li, Z.W. Du, Polyaniline synthesis by cyclic voltammetry for anodic modification in microbial fuel cells, Int. J. Electrochem. Sci. 9 (2014) 2038–2046.
[56] C.E. Zhao, W.J. Wang, D. Sun, X. Wang, J.R. Zhang, J.J. Zhu, Nanostructured graphene/ TiO_2 hybrids as high-performance anodes for microbial fuel cells, Chem. Eur. J. 20 (2014) 7091–7097.

[57] Z.H. Wen, S.Q. Ci, S. Mao, S.M. Cui, G.H. Lu, K.H. Yu, S.L. Luo, Z. He, J.H. Chen, TiO_2 nanoparticles-decorated carbon nanotubes for significantly improved bioelectricity generation in microbial fuel cells, J. Power Sources 234 (2013) 100–106.
[58] M. Zhou, M. Chi, H. Wang, T. Jin, Anode modification by electrochemical oxidation: a new practical method to improve the performance of microbial fuel cells, Biochem. Eng. J. 60 (2012) 151–155.
[59] N.A. Garcia-Gomez, I. Balderas-Renteria, D.I. Garcia-Gutierrez, H.A. Mosqueda, E. M. Sanchez, Development of mats composed by TiO_2 and carbon dual electrospun nanofibers: a possible anode material in microbial fuel cells, Mater. Sci. Eng. B Adv. Funct. Solid-State Mater. 193 (2015) 130–136.
[60] A. Mehdinia, M. Dejaloud, A. Jabbari, Nanostructured polyaniline-coated anode for improving microbial fuel cell power output, Chem. Pap. 67 (2013) 1096–1102.
[61] M.A. Rodrigo, P. Cañizares, J. Lobato, R. Paz, C. Sáez, J.J. Linares, Production of electricity from the treatment of urban waste water using a microbial fuel cell, J. Power Sources 169 (2007) 198–204.
[62] P. Perez-Rodriguez, V.M. Ovando-Medina, S.Y. Martinez-Amador, J.A. Rodriguez-de la Garza, Bioanode of polyurethane/graphite/polypyrrole composite in microbial fuel cells, Biotechnol. Bioprocess Eng. 21 (2016) 305–313.
[63] J.K. Jang, T.H. Pham, I.S. Chang, K.H. Kang, H. Moon, K.S. Cho, B.H. Kim, Construction and operation of a novel mediator- and membrane-less microbial fuel cell, Process Biochem. 39 (2004) 1007–1012.
[64] Z. Liu, J. Liu, S. Zhang, Z. Su, Study of operational performance and electrical response on mediator-less microbial fuel cells fed with carbon- and protein-rich substrates, Biochem. Eng. J. 45 (2009) 185–191.
[65] C. Dumas, R. Basseguy, A. Bergel, Microbial electrocatalysis with Geobacter sulfurreducens biofilm on stainless steel cathodes, Electrochim. Acta 53 (2008) 2494–2500.
[66] M. Liu, M. Zhou, L. Ma, H. Yang, Y. Zhao, Architectural design of hierarchically meso-macroporous carbon for microbial fuel cell anodes, RSC Adv. 6 (2016) 27993–27998.
[67] H. Pham, N. Boon, M. Marzorati, W. Verstraete, Enhanced removal of 1,2-dichloroethane by anodophilic microbial consortia, Water Res. 43 (2009) 2936–2946.
[68] Y. Fan, S. Xu, R. Schaller, J. Jiao, F. Chaplen, H. Liu, Nanoparticle decorated anodes for enhanced current generation in microbial electrochemical cells, Biosens. Bioelectron. 26 (2011) 1908–1912.
[69] S.E. Oh, B.E. Logan, Proton exchange membrane and electrode surface areas as factors that affect power generation in microbial fuel cells, Appl. Microbiol. Biotechnol. 70 (2006) 162–169.
[70] W.-W. Li, H.-Q. Yu, Z. He, Towards sustainable wastewater treatment by using microbial fuel cells-centered technologies, Energy Environ. Sci. 7 (2014) 911–924.
[71] V.G. Gude, Energy and water autarky of wastewater treatment and power generation systems, Renew. Sustain. Energy Rev. 45 (2015) 52–68.
[72] N. Montpart, L. Rago, J.A. Baeza, A. Guisasola, Hydrogen production in single chamber microbial electrolysis cells with different complex substrates, Water Res. 68 (2015) 601–615.
[73] H.M. Wang, J.D. Park, Z.J. Ren, Practical energy harvesting for microbial fuel cells: a review, Environ. Sci. Technol. 49 (2015) 3267–3277.
[74] H. Liu, S. Grot, B.E. Logan, Electrochemically assisted microbial production of hydrogen from acetate, Environ. Sci. Technol. 39 (2005) 4317–4320.
[75] W. Liu, S. Huang, A. Zhou, G. Zhou, N. Ren, A. Wang, G. Zhuang, Hydrogen generation in microbial electrolysis cell feeding with fermentation liquid of waste activated sludge, Int. J. Hydrog. Energy 37 (2012) 13859–13864.
[76] R.A. Rozendal, H.V.M. Hamelers, G.J.W. Euverink, S.J. Metz, C.J.N. Buisman, Principle and perspectives of hydrogen production through biocatalyzed electrolysis, Int. J. Hydrog. Energy 31 (2006) 1632–1640.

[77] W.W. Cai, W.Z. Liu, J.L. Han, A.J. Wang, Enhanced hydrogen production in microbial electrolysis cell with 3D self-assembly nickel foam-graphene cathode, Biosens. Bioelectron. 80 (2016) 118–122.
[78] V.J. Watson, M. Hatzell, B.E. Logan, Hydrogen production from continuous flow, microbial reverse-electrodialysis electrolysis cells treating fermentation wastewater, Bioresour. Technol. 195 (2015) 51–56.
[79] A. Escapa, R. Mateos, E.J. Martinez, J. Blanes, Microbial electrolysis cells: An emerging technology for wastewater treatment and energy recovery. From laboratory to pilot plant and beyond, Renew. Sustain. Energy Rev. 55 (2016) 942–956.
[80] G. Kumar, P. Bakonyi, T. Kobayashi, K.-Q. Xu, P. Sivagurunathan, S.-H. Kim, G. Buitrón, N. Nemestóthy, K. Bélafi-Bakó, Enhancement of biofuel production via microbial augmentation: the case of dark fermentative hydrogen, Renew. Sustain. Energy Rev. 57 (2016) 879–891.
[81] G. Kumar, P. Bakonyi, S. Periyasamy, S.H. Kim, N. Nemestóthy, K. Bélafi-Bakó, Lignocellulose biohydrogen: practical challenges and recent progress, Renew. Sustain. Energy Rev. 44 (2015) 728–737.
[82] E. Lalaurette, S. Thammannagowda, A. Mohagheghi, P.-C. Maness, B.E. Logan, Hydrogen production from cellulose in a two-stage process combining fermentation and electrohydrogenesis, Int. J. Hydrog. Energy 34 (2009) 6201–6210.
[83] A. Kundu, J.N. Sahu, G. Redzwan, M.A. Hashim, An overview of cathode material and catalysts suitable for generating hydrogen in microbial electrolysis cell, Int. J. Hydrog. Energy 38 (2013) 1745–1757.

PART II

APPLICATIONS OF ELECTROCHEMICAL TECHNOLOGIES FOR DECONTAMINATION AND DISINFECTION OF WATER

CHAPTER

13

Fine Chemical Industry, Pulp and Paper Industry, Petrochemical Industry and Pharmaceutical Industry

Marco Panizza
University of Genoa, Genoa, Italy

OUTLINE

Nomenclature

AO	anodic oxidation
AOP	advanced oxidation process
AOX	adsorbable organic xalogen compounds
BDD	boron-doped diamond
BIO	biological oxidation
BOD	biological oxygen demand
CF	carbon felt
COD	chemical oxygen demand

https://doi.org/10.1016/B978-0-12-813160-2.00013-4

DOC dissolved organic carbon
EAOP electrochemical advanced oxidation processes
EC electrocoagulation
ECF electrofloatation
EF electro-Fenton
E_{sp} energy consumption
GC glassy carbon
i applied current density
I applied current
PEF photo electro-Fenton
SE sonoelectrochemistry
SS stainless steel
TDS total dissolved solids
TOC total organic carbon
TS total solids
TSS total suspended solids
US ultrasound

INTRODUCTION

Many industrial processes generate effluents that contain different types of organic compounds that have to be removed prior to discharge or reuse of the waste flow. In this field, research efforts have been made in order to develop more effective technologies to totally remove organic pollutants from wastewaters, including physical methods, incineration, biological oxidation, chemical oxidation or advanced oxidation processes (AOPs). Electrochemical processes, thanks to their attractive characteristics such as versatility, energy efficiency, amenability of automation, and environmental compatibility, have recently gained increasing attention as a promising treatment of industrial effluents containing both organic and inorganic substances [1–6]. Key electrochemical processes include electrocoagulation (EC), electro-floatation (ECF), direct and indirect anodic oxidation (AO), electro-Fenton (EF), photoelectro-Fenton (PEF), and sonoelectrochemistry (SE). The fundamentals of electrochemical processes have been presented and discussed deeply in the previous chapters of this book, together with the application for the treatment of synthetic wastewaters. This chapter intends to provide exhaustive information on the remediation of diversified real wastewaters from fine chemical, pulp and paper, petrochemical, and pharmaceutical industries by electrochemical processes, alone or in combination with other technologies. The remediation of each type of wastewater is assessed separately, including many features such as wastewater composition, applied electrochemical process, configuration of an electrochemical reactor, anode and cathode characteristics, and operational parameters.

FINE CHEMICAL INDUSTRIES

Fine chemicals industries produce complex, single, pure chemical substances with a low volume (<1000 tons $year^{-1}$) and a high price (>10 \$ kg^{-1}). The fine chemicals segment covers a broad range of chemical products that serve pharmaceuticals, life sciences, agrochemicals, photography chemicals, and, to a growing extent, electronic chemicals. For this reason, the composition of the wastewaters of fine chemical industries is diversified and largely depends on the nature of the product and process synthesis; thus, there is not a universal treatment for all the effluents. Some examples of fine chemical wastewater treatment by electrochemical methods are reported in Table 1.

TABLE 1 Example of Electrochemical Treatment of Wastewaters From Fine Chemical Industries

Wastewater characteristics	Process	Electrodes	Comment	Ref.
COD 6000 mg dm^{-3} TOC 1600 mg dm^{-3} Aromatic and aliphatic compounds, ethanol, methanol	AO, Fenton, ozone	Anode: BDD Cathode: SS	Complete COD and TOC removal obtained only with BDD E_{sp}: 800 kWh m^{-3}	[7,8]
COD 13068 mg dm^{-3} Nitrogen 347 mgN dm^{-3} TSS 13574 mg dm^{-3}	AO, Fenton, ozone	Anode: BDD Cathode: SS	Recommended the use of BDD after coagulation and biological pretreatment	[9]
COD 1000 mg dm^{-3} Dyes (methylene blue and rhodamine B), solvents (monoethylene glycol, diethylene glycol) and surfactants	AO	Anode: BDD Cathode: SS	COD removal higher than those of TOC Chlorine media favors the oxidation process	[10]
Naphthalen sulphonic acid 285.3 mg dm^{-3} Anthraquinon sulphonic acid 172.8 mg dm^{-3} COD 1361 mg dm^{-3}	AO	Anode: Ti/Pt, $TiRuSnSbO_2$, and CF	Ti/Pt anodes had the highest electrocatalytic activity. Complete COD and color removal only in the presence of 5 g dm^{-3} NaCl	[11]
Naphthalen sulphonic acid 285.3 mg dm^{-3} Anthraquinon sulphonic acid 172.8 mg dm^{-3} COD 1361 mg dm^{-3}	EF	Anode: Pt Cathode: CF	COD removal and color fading of 87% and 89%	[12]

Continued

TABLE 1 Example of Electrochemical Treatment of Wastewaters From Fine Chemical Industries—cont'd

Wastewater characteristics	Process	Electrodes	Comment	Ref.
COD 1084 mg dm^{-3} TDS 2740 mg dm^{-3} Chloride 524 mg dm^{-3} Phenolics 374 mg dm^{-3}	AO	Ti/TiO_2-RuO_2-IrO_2 graphite	Final COD: 191 mg dm^{-3} Current efficiency: 12.5% E_{sp}: 102.1 kWh kg_{COD}^{-1} E_{sp}: 91.2 kWh m^{-3}	[13]
COD 12000 mg dm^{-3} Composition Acetone, *n*-butanol, ter-butanol, methylacrylate, other	AO	Anode: BDD Cathode: SS	Complete COD and TOC removal E_{sp}: 22 Wh kg_{COD}^{-1}	[14]
Cu 45–335 mg dm^{-3} Cr 44.5–193 mg dm^{-3} Ni 394–526 mg dm^{-3}	EC	Fe or Al	Fe-Fe and Fe-Al electrode combinations were more effective. Almost 100% removal of Cu, Cr, and Ni	[15]

Canizares et al. [7,8] compared the performance of anodic oxidation with BDD, ozonation, and Fenton oxidation for the treatment of actual industrial wastewater of a fine chemicals plant containing solvents (ketones and alcohols) with a high concentration of aromatic compounds coming from the raw materials, intermediates, and products of the different processes of the plant. They reported that BDD allowed the complete removal of COD and TOC with high current efficiencies. During anodic oxidation with BDD, the organic compounds were mainly oxidized by the reaction with $^{\bullet}OH$ radicals generated from water discharge:

$$H_2O \rightarrow {}^{\bullet}OH + H^+ + e^- \tag{1}$$

$$R + {}^{\bullet}OH \rightarrow CO_2 + H_2O \tag{2}$$

The global efficiency of the process depended strongly on the pH and temperature but not on the current density. The presence of large amounts of chlorides in the actual effluents favored the rate of the electrochemical processes and increased the mineralization efficiency due to the electrogeneration of active chloro species. Using Fenton oxidation, an important fraction of both refractory carbon and inorganic compounds remained at the end of the treatment (50% of the organic carbon). The ozonation process was able to treat the actual wastewaters with an efficiency similar to the electrochemical oxidation, although the energy consumption required by the electrochemical process to remove a fixed percentage of COD or TOC was significantly smaller than that of ozonation (Fig. 1).

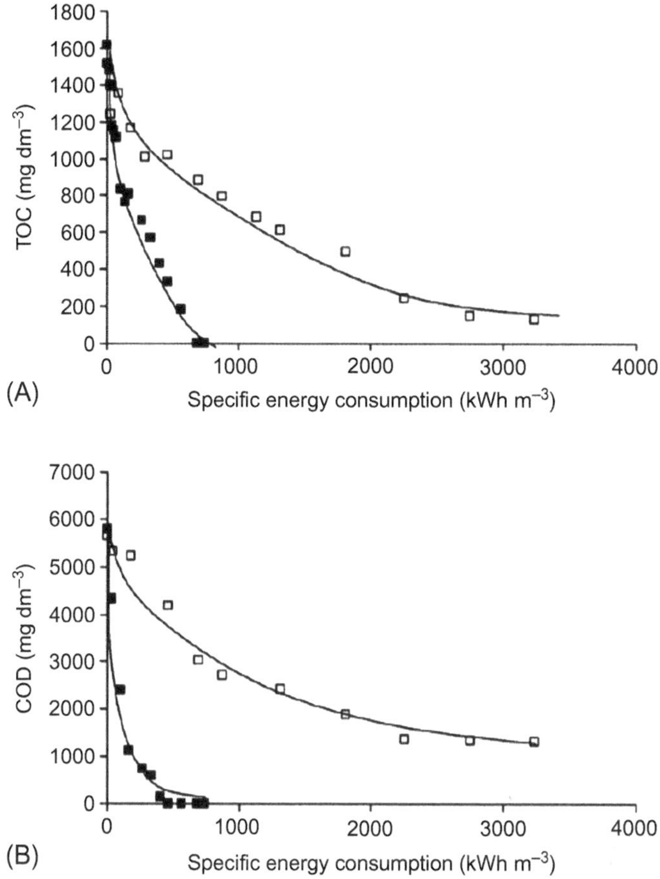

FIG. 1 Variation of the (A) TOC and (B) COD with specific energy consumption in the electrochemical oxidation with BDD (■) and ozonation (□) of fine chemical industry wastewater. Experimental conditions of BDD electrolysis: natural pH; T, 25°C; j, 300 A m^{-2}. Experimental conditions of ozonation: pH 12; T, 25°C; ozone production, 0.99 g h^{-1}. *Reproduced with permission from P. Canizares, R. Paz, J. Lobato, C. Saez, M.A. Rodrigo, Electrochemical treatment of the effluent of a fine chemical manufacturing plant. J. Hazard. Mater. 138 (2006) 173–181, Copyright 2006, Elsevier.*

Beteta et al. [9] investigated the improvement in conventional treatment consisting of coagulation-flocculation and biological oxidation of door manufacturing industry wastewater with three AOPs: anodic oxidation, ozonation, and the Fenton process. The wastewater produced in the gluing process was mainly composed of urea and formaldehyde with different additives. It presented a high concentration of COD (13,068 mg dm^{-3}), total nitrogen (347 mgN dm^{-3}), and suspended solids (13,574 mg dm^{-3}). The conventional treatment of coagulation-flocculation and biological oxidation reduced the COD up to 1500 mg dm^{-3}, which is over the discharge limits, and therefore a tertiary treatment was necessary to refine the quality of the effluent and reduce the retention time of the biological process. Thus, the effluent of the biological process was treated by anodic oxidation using a BDD anode, ozonation, and Fenton. All the processes obtained concentrations below the typical discharge limits; however, only the conductive-diamond electrochemical oxidation was able to mineralize the effluent while in the Fenton and ozonation processes there was a refractory COD. The estimated cost of the Fenton process and anodic oxidation to refine the quality of the biologically treated waste was under 1 € m^{-3}.

Disposal of industrial wastewater containing naphthalene- and anthraquinone-sulphonic acids by different electrochemical processes (i.e., anodic oxidation and electro-Fenton) was studied by Panizza et al. [11,12]. The wastewater derived from the regeneration of ion-exchange resin towers that constituted the final step in the treatment plant. These towers received the effluent from a biological treatment system and allowed the lowering of the COD value and color, as they hold the sulphonic compounds. During the regeneration, they released naphthalene- and anthraquinone-sulfonic acids, producing a dark-colored effluent with a COD of 1361 mg dm^{-3} that must be treated. The anodic oxidation of the effluent was performed using an undivided cell of 0.2 dm^{3} with three types of anodes (Ti/Pt, $TiRuSnSbO_2$, and carbon felt) [11]. The direct anodic oxidation did not give satisfactory results because only 40% of COD was removed and no relevant color fading was noticed. However, the addition of sodium chloride to the solution resulted in an increase in the organic oxidation rate. The pollutants were then destroyed in the bulk solution by oxidation reaction with hypochlorite/chlorine generated in situ from the oxidation of chloride ions:

$$2Cl^- \rightarrow Cl_2 + 2e^- \quad (3)$$

$$Cl_2 + H_2O \rightarrow HOCl + H^+ + Cl^- \quad (4)$$

$$HOCl \rightarrow H^+ + OCl^- \quad (5)$$

The best performance was obtained with a Ti/Pt anode that allowed complete COD removal and color disappearance (Fig. 2).

The same effluent was also treated by electro-Fenton methods using a graphite-felt cathode and a platinum anode in an undivided cell of 0.2 dm^{-3} [12]. In the EF process, the highly oxidizing $^{\bullet}OH$ was produced in the bulk of the polluted solution via Fenton's reaction (Eq. 6), which occurred by means of the electrogenerated Fenton's reagent:

$$Fe^{2+} + H_2O_2 \rightarrow Fe^{3+} + {}^{\bullet}OH + OH^- \quad (6)$$

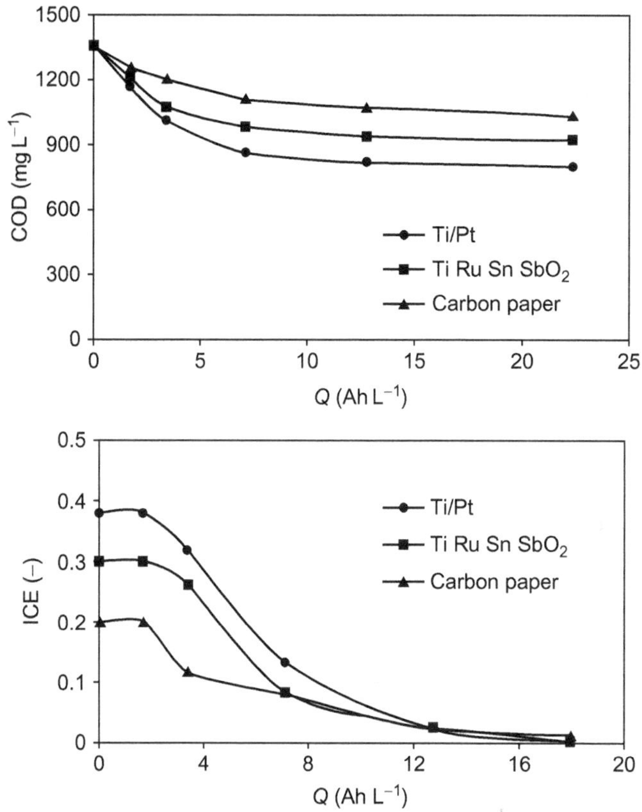

FIG. 2 Effect of anode materials on COD removal and ICE values for the indirect anodic oxidations of wastewater of a fine chemical industry. NaCl concentration = 5 g dm^{-3}, current density = 50 mA cm^{-2}. *Reproduced with permission from M. Panizza, C. Bocca, G. Cerisola, Electrochemical treatment of wastewater containing polyaromatic organic pollutants. Water Res. 34 (2000) 2601–2605, Copyright 2000, Elsevier.*

Thus, H_2O_2 was electrogenerated from the two-electron reduction of O_2 (Eq. 7) while Fe^{2+} was regenerated from Fe^{3+} reduction (Eq. 8), which prevented the Fe^{3+} accumulation in the medium and consequently minimized the production of iron sludge:

$$O_2 + 2H^+ + 2e^- \rightarrow H_2O_2 \quad (7)$$

$$Fe^{3+} + e^- \rightarrow Fe^{2+} \quad (8)$$

The authors reported that in the presence of 2 mM of iron, the maximum COD removal and color fading were 87% and 89%, respectively. The complete mineralization was not obtained because of the formation of intermediates that are more difficult to remove.

The anodic oxidation of industrial wastewater (BASF wastewater) containing acetone, *n*-butanol, ter-butanol, methylacrylate, and other compounds was studied by Ouattara et al. [14], using a BDD anode in an undivided flow-cell. The initial COD (12,000 mg dm^{-3}) and TOC (3000 mg dm^{-3}) were completely removed. They observed that the rate of COD elimination was higher than the theoretical value because monomeric materials present in the effluent polymerized and precipitated on the wall of the reactor. Operating at the limiting current throughout the treatment process, the specific energy consumption for the electrochemical treatment was strongly decreased to ~22 kWh kg_{COD}^{-1}.

PULP AND PAPER INDUSTRIES

The pulp and paper-making industry is a water-intensive industry, consuming >60 m^3 of freshwater per ton of paper produced. Furthermore, the effluents of these industries cause considerable damage to the receiving waters if discharged untreated because they have a high BOD, COD, blackish color, chlorinated compounds, suspended solids, fatty acids, tannins, resin acids, lignin and its derivatives, sulfur and sulfur compounds, etc. The pulp and paper wastewater is treated by a primary clarification, either sedimentation or flotation, followed by a secondary treatment, generally of a biological nature (e.g., aerobic or anaerobic digestion). In many cases, the effluent quality treated by conventional processes does not meet the regulation, making a tertiary process necessary to reduce parameters such as color and COD. The main tertiary treatments, which have been employed to date at a few sites in the industry, are membrane processes (ultrafiltration, nanofiltration, and reverse osmosis), advanced oxidation processes (ozonation, peroxidation, photocatalysis), adsorption, or coagulation. Over the past two decades, there has been growing interest in the use of electrochemical techniques such as electrocoagulation, electrofloatation, electrooxidation, etc., for the treatment of paper industrial effluent (Table 2). Among these methods, electrocoagulation emerges as one of the promising techniques for the disposal of this wastewater because it can be used either as a tertiary treatment to remove the residual COD and color or as a pretreatment to increase the biodegradability of the effluent.

For example, Mahesh et al. [16,17] studied the electrocoagulation of agri-based paper mill wastewater (black liquor) in a 2 dm^3 electrolytic batch reactor using Fe electrodes with multiple arrangements. The six-plate arrangement with current density of 55.56 A m^{-2} was found to be optimal, achieving a maximum COD and color removal of 80% and 90%, respectively. The addition of polyacrylamide (10 mg dm^{-3}) to the electrochemical reactor enhanced the COD removal rate with a very short treatment time

TABLE 2 Example of Electrochemical Treatment of Wastewaters From Pulp and Paper Industries

Wastewater characteristics	Process	Electrodes	Comment	Ref.
COD 2000–2100 $mg\,dm^{-3}$ BOD 615–670 $mg\,dm^{-3}$ TS 2100 $mg\,dm^{-3}$	EC	6 pair of Fe electrode	COD and color removal: 80% and 90% E_{sp}: from 6.64 to 5.73 $Wh\,kg_{COD}^{-1}$	[16,17]
COD 210–240 $mg\,dm^{-3}$ Conductivity 600–630 $mS\,cm^{-1}$ Color 1150–1350 U.C.	EC	Al or Fe anode	Al electrode: color and COD removal of 67% and 84% Fe electrode: color and COD removal 82% and 98%	[18]
COD 1620 $mg\,dm^{-3}$ BOD 450 $mg\,dm^{-3}$ TS 1336 $mg\,dm^{-3}$		Modified kaolin and graphite	96% COD removal in 40 min of at current density 30 $mA\,cm^{-2}$	[19,20]
COD 32000 $mg\,dm^{-3}$ BOD 8225 $mg\,dm^{-3}$ TS 8000 $mg\,dm^{-3}$	EC	Al or Fe anode	Al electrode: color and COD removal of 84% and 89% Fe electrode: color and COD removal 92% and 95%	[21]
TDS 1310–1560 $mg\,dm^{-3}$ COD 490–1114 $mg\,dm^{-3}$ BOD5 810–1095 $mg\,dm^{-3}$	EC	Fe anode	97% of color and 77% of COD removal operating cost of 0.29 $USD\,m^{-3}$	[22]
COD 925 $mg\,dm^{-3}$ BOD 102 $mg\,dm^{-3}$ TS 545 $mg\,dm^{-3}$	EC	Fe anode	COD and color removal: 55% and 87% E_{sp}: 1.19 $Wh\,kg_{COD}^{-1}$	[23]
COD 7960 $mg\,dm^{-3}$ Polyphenol 3220 $mg\,dm^{-3}$ Chloride 1207 $mg\,dm^{-3}$	EC	Al or Fe anode	COD removal 98%, polyphenols removal 92% and 99% of color removal	[24]
COD 620 $mg\,dm^{-3}$ BOD 210 $mg\,dm^{-3}$ TSS 1200 $mg\,dm^{-3}$	EC	Al anode	Color, COD, and BOD removal were 94%, 90%, and 87% respectively. E_{sp}: from 10.1 to 12.9 $kWh\,m^{-3}$	[25]
COD 1537 $mg\,dm^{-3}$ TOC 421 $mg\,dm^{-3}$ BOD 410 $mg\,dm^{-3}$	EC +UV	Fe anode	EC, EC+UV/ persulfate and EC+UV/ peroxymonosulfate removed 61%, 76.6%, and 80.5% of COD	[26]

Continued

TABLE 2 Example of Electrochemical Treatment of Wastewaters From Pulp and Paper Industries—cont'd

Wastewater characteristics	Process	Electrodes	Comment	Ref.
TOC 1275 mg dm^{-3} Total phenols 37 mg dm^{-3} AOX 60 mg dm^{-3}	Photo EC	$Ti/TiO_2/RuO_2$	Removal of color AOX, COD, and BOD was 90%, 25%, 30%, and 35%	[27]
COD 1410 mg dm^{-3} TSS 85 mg dm^{-3} TDS 2380 mg dm^{-3}	EC + AO	Fe anode and graphite cathode	Removal of PCP was very rapid (~100% removal in 10 min)	[28]
COD 1357 mg dm^{-3}	AO	$Ti/Co/SnO_2$-Sb_2O_5 + GC	COD removal of 90% and color removal of 75%	[29]
BOD 300–400 mg dm^{-3} COD 5000–6000 mg dm^{-3} TDS 4000–4500 mg dm^{-3}	AO	Pb sheet	COD reduced from 5500 to 160 Color removal from 53% to 100% E_{sp}: 4–29 kWh m^{-3}	[30]
COD 1510 mg dm^{-3} TOC 423 mg dm^{-3} TSS 354 mg dm^{-3} Cl^- 923 mg dm^{-3}	AO, AOP	BDD anode	Precipitation CaO: 90% COD removal Fenton + US: 12% COD removal Photo-Fenton: 20% COD removal AO: COD removal 28%	[31]
COD 129600–137,600 mg dm^{-3} BOD 90000–112,000 mg dm^{-3} TDS 154833–178,540 mg dm^{-3}	AO	Ti/RuO_2 grids	Color, BOD, and COD levels were reduced by around 91%, 83%, and 86%	[32]
TSS 330–510 mg dm^{-3} COD 3343–4250 mg dm^{-3} Total nitrogen: 0.01–0.02 mg dm^{-3}	AO + Bio	Ti/SnO_2-Sb_2O_5-IrO_2	Integrated bioreactor and AO gave complete decolorization and COD removal of 96.2%–98.2%	[33]

with excellent sludge settleability. Specific energy consumption (E_{sp}) was reduced from 6.64 to 5.73 kWh kg$^{-1}_{COD\ removed}$ with the addition of NaCl (625 mg dm^{-3}). The posttreatment of electrochemically treated wastewater by chemical coagulation using aluminium (360 mg dm^{-3}) along with 20 mg dm^{-3} polyacrylamide further reduced COD values to <180 mg dm^{-3}

and a near 100% color removal. An overall COD removal of 91% and color removal of near 100% was achieved by electrochemical treatment followed by coagulation/flocculation by adding aluminium and polyacrylamide. The results strongly indicated that the electrochemical method can be adopted effectively for the treatment of black liquor from agri-based pulp and paper mills for complete decolorization and COD removal.

The influences of current density, pH, rotational speed, and NaCl concentration on electrocoagulation using aluminium electrodes was also examined in detail by Sridhar [25]. From the experimental results, $15\,mA\,cm^{-2}$ current density, pH of 7, $1\,g\,dm^{-3}$ NaCl, 100 rpm, 28°C temperature, and 3 cm electrode distance were found to be optimum for maximum reduction of color, COD, and BOD. The reduction of color, COD, and BOD under the optimum conditions was found to be 94%, 90%, and 87% respectively. The electrode energy consumption was calculated and found to vary from 10.1 to $12.9\,kWh\,m^{-3}$, depending on the operating conditions. Under optimal operating condition such as $15\,mA\,cm^{-2}$ current density, pH of 7, $1\,g\,dm^{-3}$ NaCl, 100 rpm, 28°C temperature, and 3 cm electrode distance, the operating cost was found to be $1.56\,US\,\$\,m^{-3}$, proving that electrocoagulation is a suitable method for treating bleaching plant effluents for reuse.

Kalyani et al. [21] studied an integrated process combining electrocoagulation followed by a sequential batch reactor for the treatment of an effluent collected from a nearby pulp and paper industry. The maximum COD and color removal obtained during EC were 95% and 92% for mild steel and 84% and 89% with an aluminium electrode, respectively. Moreover, the biodegradability index (BOD/COD) increased gradually during EC to >0.4 from the original value of 0.2 within 10 min of process time (Fig. 3). Because the BOD/COD value was increased to beyond 0.4, which is required for an effective treatment by a biochemical technique, they attempted to couple EC with a sequential batch reactor. The performance of the biological treatment of pulp and paper effluent was significantly increased when electrocoagulation is combined with the sequential batch reactor. In fact, the percentage of COD removal was 98% in the integrated process, compared to only 35% in a direct biochemical route without electrocoagulation pretreatment.

Electrocoagulation can be also combined with a UV/oxidant system to improve the removal of organic compounds from pulp and paper wastewater, as reported by Jaafarzadeh et al. [26]. The initial COD of $1537\,mg\,dm^{-3}$ was reduced by 61% from electrocoagulation alone in 33.7 min at natural pH and current density of $5.55\,mA\,cm^{-2}$. When the effluent after electrocoagulation was remediated by UV/persulfate and UV/peroxymonosulfate, the COD removal increased to 76.6% and 80.5%, respectively. Moreover, EC along with UV/oxidant significantly increased the biodegradability (BOD5/COD ratio) of the effluent.

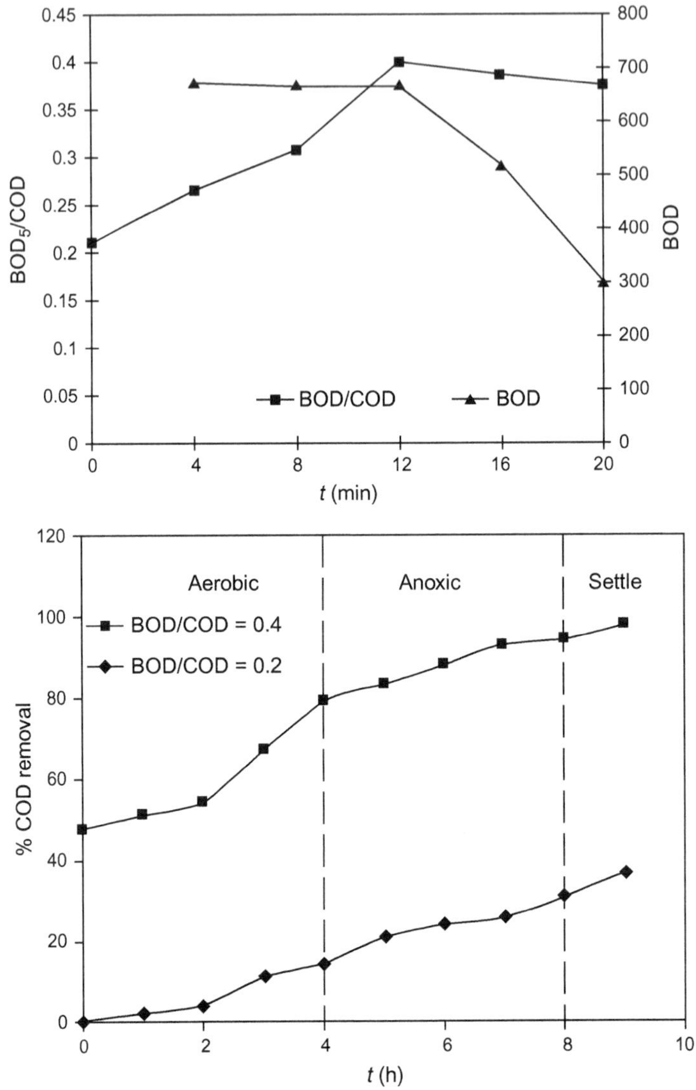

FIG. 3 Effect of electrocoagulation on biodegradability of pulp and paper effluent; anode: mild steel, pH 7; current density: $10\,mA\,cm^{-2}$; NaCl: $400\,mg\,dm^{-3}$. Evolution of COD removal during the treatment using a sequential batch reactor: effluent (▪) with EC pretreatment, (♦) without EC pretreatment. *Reproduced with permission from K.S.P. Kalyani, N. Balasubramanian, C. Srinivasakannan, Decolorization and COD reduction of paper industrial effluent using electro-coagulation. Chem. Eng. J. 151 (2009) 97–104, Copyright 2009, Elsevier.*

Ma et al. [19,20] evaluated a Fenton-like process involving electrocoagulation and electrochemical oxidation for the treatment of pulp and paper mill wastewater. The electrochemical degradation was performed in a batch cell with porous plate graphite electrodes with a molybdenum and phosphate (Mo-P) modified kaolin catalyst packed around the electrodes, forming a multiphase electrochemical oxidation packed bed. COD removal >96% was obtained when modified kaolin was loaded with Fe^{3+} while around 66% COD removal was obtained when the modified kaolin was loaded with Co^{2+} or Cu^{2+}. Electrochemical degradation with Mo-P modified kaolin loaded with Fe^{3+} took place via the mechanism of either electrocoagulation or electrochemical oxidation. The OH^- produced from a cathode reacted with iron ion and formed a number of iron species that favored the coagulation and precipitation of the pollutants. On the other hand, ferrous ions, reduced from ferric ions, initiated and catalyzed up the decomposition of H_2O_2 produced on the graphite electrodes, resulting in the generation of hydroxyl radicals according to Fenton's equation (Eq. 6).

A considerable increase in the COD removal rate was noted in the presence of chloride ions because the introduction of NaCl might have led to the formation of chlorinated species such as ClO^- or Cl_2 near the surface through the oxidation of chloride that participated in the removal of pollutants.

Currently, only a few studies have been performed on the electrochemical oxidation of pulp and paper effluents, probably because the high COD of the effluents resulted in high energy consumption. For example, El-Ashtoukhy et al. [30] treated paper mill effluent from Rakta's Pulp and Paper Company using a cylindrical vessel with lead sheet as the anode and stainless steel as the cathode. The results showed that anodic oxidation reduces the COD from a average value of 5500 to 160 and the percentage color removal ranged from 53% to 100% depending on the operating conditions. In particular, the addition of NaCl to the effluents favored the color removal thanks to the electrogeneration of hypochlorite. Energy consumption depended on the operating conditions and ranged from 4 to 29 $kWh\,m^{-3}$.

The performance of a photo-assisted electrolysis process for the degradation of organic pollutants in a bleach Kraft Mill Effluent was investigated by Pelegrini et al. [27]. Photocatalytic, electrochemical, and photoelectrochemical treatments were carried out in a single compartment, a glass cell with titanium coated with a TiO_2-RuO_2 anode using a 125 W mercury bulb lamp placed in the center of the cell. As shown in Fig. 4 there was a synergic effect of photo-electrochemical treatment. In fact, the electrochemical and photocatalytical processes, when used individually, reduced 50% and 20% of the solution color, respectively; however, 90% of discoloration was reached after 90 min of

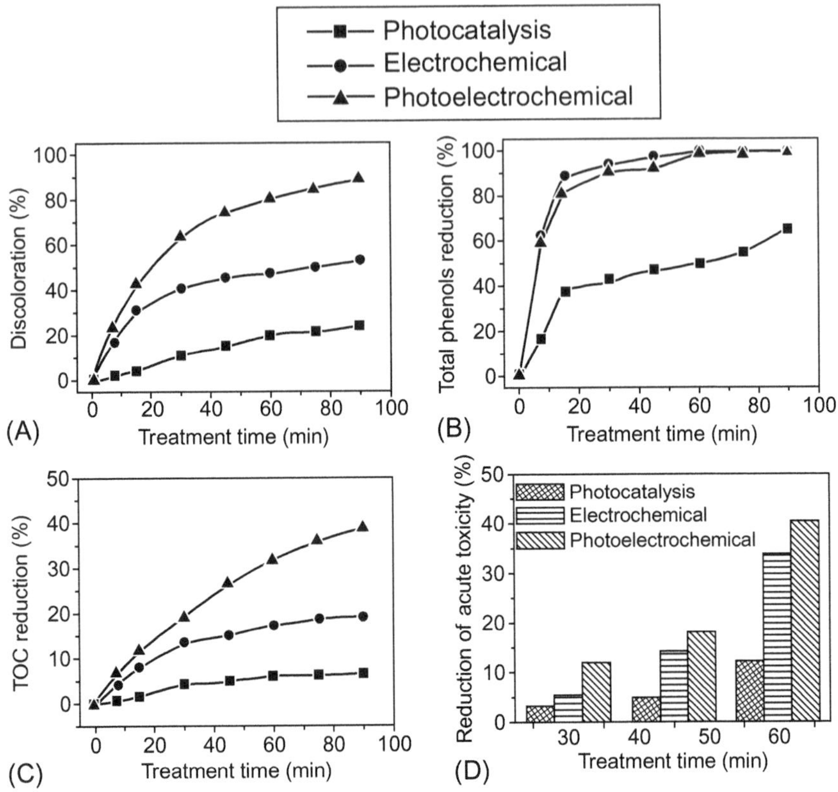

FIG. 4 Results for the E1 bleach Kraft mill effluent. Color (A), total phenols (B), TOC (C), and acute toxicity (D) reduction as a function of treatment time for the photocatalytic process (125 W mercury bulb), for the electrochemical process (20 mA cm^{-2}) and for the photoelectrochemical process (125 W, 20 mA cm^{-2}). *Reproduced with permission from R.T. Pelegrini, R.S. Freire, N. Duran, R. Bertazzoli, Photoassisted electrochemical degradation of organic pollutants on a DSA type oxide electrode: process test for a phenol synthetic solution and its application for the E1 bleach kraft mill effluent. Environ. Sci. Technol. 35 (2001) 2849–2853. Copyright (2001) American Chemical Society.*

photo-electrochemical treatment. Total phenols were practically eliminated in a short period of processing time, a reduction of AOX, COD, and BOD by 25%, 30%, and 35%, respectively, were also observed.

Eskelinen et al. [31] compared many processes for the treatment of pulp and paper mill effluents using chemical precipitation, electrochemical oxidation with BDD anode ultrasonic irradiation in combination with Fenton-like oxidation (Fe^{3+}/H_2O_2), or photo-Fenton degradation ($Fe^{2+}/H_2O_2/UV$). Among the treatments investigated, after 60 min, chemical precipitation using 5 g dm^{-3} of CaO at pH 12 achieved a higher COD removal (90%) than an integrated ultrasonic irradiation with Fenton-like oxidation (12%), photo-Fenton's oxidation (20%), or electrooxidation treatment (28%). As reflected by the slight decrease of COD concentrations after treatment, electrooxidative treatment did not effectively remove recalcitrant compounds from bleaching wastewater in a short time. However, this pretreatment facilitated an indirect removal of COD by improving the treatability of bleaching effluents for a subsequent biological process.

In this field, Qu et al. [33] studied a novel integrated thermophilic submerged aerobic membrane bioreactor coupled with anodic oxidation using a Ti/SnO_2-Sb_2O_5-IrO_2 for the treatment of thermomechanical pulping pressate effluent. The membrane bioreactor was able to achieve a COD removal efficiency of 88.6%–92.3% with an optimal hydraulic retention time of 1.1 days. The bioreactor permeate, however, contained color and residual COD and was thus further treated by EO. After 6 h of anodic oxidation, a complete decolorization was achieved and the COD removal efficiency increased to 96.2%–98.2%, producing a high-quality effluent that can be reused as process water for system closure in a pulp and paper mill.

PETROCHEMICAL INDUSTRIES

The wastewaters of petroleum refinery industries are very hazardous effluents because they have high COD (300–1000 mg dm^{-3}), phenols (20–200 mg dm^{-3}), benzene (1–100 mg dm^{-3}), heavy metals (0.1–100 mg dm^{-3}), suspended solids, surfactants, corrosion inhibitors, etc. Consequently, these effluents must be treated prior to their discharge or reuse. The traditional treatments are based on the mechanical and physicochemical methods such as separation and coagulation followed by biological treatment. However, these methods have shown various operational problems, such as generation of toxic gases, residual sludge production, demand for large territorial areas, and incomplete removal of refractory compounds. Many other innovative technologies for treating PW have been studied as an alternative to the physicochemical and bioremediation processes in order to comply with reuse and discharge limits.

Among them, different electrochemical technologies have been proposed as a promising alternative for the treatment of this kind of wastewater. They include electroflotation (ECF), electrocoagulation (EC), direct and indirect anodic oxidation (AO), and electro-Fenton (EF). Table 3 summarizes the most relevant results obtained for the degradation of petroleum hydrocarbons by electrochemical methods.

The efficiency of electroflotation and electrocoagulation applied to the purification of oil refinery wastewater was investigated by

TABLE 3 Example of Electrochemical Treatment of Wastewaters From Petroleum Refinery Industries

Wastewater characteristics	Process	Electrodes	Comment	Ref.
COD 257 mg dm^{-3} Phenol 3.17 mg dm^{-3} Hydrocarbons 11.72 mg dm^{-3} Grease 1.96 mg dm^{-3}	ECF EC	Al, Fe	COD and turbidity removal 72% and 83% for ECF, 80% and 88% for EC	[34]
Phenol content 13 mg dm^{-3} BOD 40.25 mg dm^{-3} COD 80–120 mg dm^{-3}	EC	Al	94.5% removal of phenol E_{sp}: 1.8 kWh g^{-1} phenol	[35]
TDS 34030 mg dm^{-3} COD 5800 mg dm^{-3} BOD 4590 mg dm^{-3}	EC +AO	Active metal-graphite-iron	Mixed EC and indirect AO COD and BOD reduced by over 90%, SS by 99%	[36]
TSS 0.03–0.04 g dm^{-3} COD 3970–4745 mg dm^{-3} Phenol 8–10 mg dm^{-3} o-Cresol 55–68 mg dm^{-3}	EC +BIO +Ads.	Al, Fe	96% COD removal after the adsorption column. Removal of phenol and cresols 100%	[37,38]
COD 19440 mg dm^{-3} BOD 298 mg dm^{-3} TSS 14900 mg dm^{-3}	EC +BIO	Al	Removal of 98% of the total petroleum hydrocarbon and 95% of COD	[39]
Phenol 192.9 mg dm^{-3} COD 590 mg dm^{-3}	EC	Fe	6.27% phenol and 2.26% COD removal	[40]
Petrobras (Brasil) COD 4980 mg dm^{-3}	AO	Ti/$Ru_{0.34}Ti_{0.66}O_2$	57% of COD removal	[41]
25–30 mg dm^{-3} of benzene, toluene, ethyl benzene, xylene 5 mg dm^{-3} of phenol	AO	RuO_2-TiO_2-SnO_2	Complete depollution in 0.5–2.5 h E_{sp}: 4.84–0.97 kWh m^{-3}	[42]
Chloride 51.4 g dm^{-3} COD 833.3 mg dm^{-3}	AO	Ti/RuO_2	COD removals 87% at 10 mA cm^{-2} Complete COD removal at 30 mA cm^{-2} Cost: US$ 38 kg$_{COD}^{-1}$	[43]

TABLE 3 Example of Electrochemical Treatment of Wastewaters From Petroleum Refinery Industries—cont'd

Wastewater characteristics	Process	Electrodes	Comment	Ref.
COD 901 mg dm^{-3} Benzene 311.2 μg dm^{-3} Toluene 1228.5 μg dm^{-3}	AO	Ti/Pt and Ti/IrO_2-Ta_2O_5 anodes	DOC and COD removal of 55.2% and 61.5% Cost: US$ 3.79 m^{-3}	[44]
COD 19842 mg dm^{-3} BOD5 475 mg dm^{-3} Phenol 59 mg dm^{-3} Oil and greases 56.7 g dm^{-3}	AO	PbO_2, BDD	COD removal of 85% and 96% with PbO_2 and BDD E_{sp} of 38 and 46 kWh m^{-3} with BDD and PbO_2	[45]
TDS 4.38 g dm^{-3} Phenol 2.7 mg dm^{-3} Oils and greases 15 mg dm^{-3} COD 1588 mg dm^{-3}	AO	Pt and BDD, tank cell	COD removal of 85% and 98% using Pt and BDD E_{sp}: 109 and 140 kWh dm^{-3} for BDD and Pt anodes	[46]
TDS 4.38 g dm^{-3} Oils and greases 15 mg dm^{-3} COD 2746 mg dm^{-3}	AO	Pt and BDD flow-cell	COD reduced to 200 mg dm^{-3} in 5 h E_{sp} of 56.2 kWh m^{-3} with BDD	[47]
COD 4753 mg dm^{-3} NH_3-N 3308.7 mg dm^{-3} Volatile phenol 146 mg dm^{-3}	AO +EF	Graphite + Fe particle	EC + Fe + Air + pH 3 COD value decreased to 479.4 mg dm^{-3} and NH_3-N decreased to 17.48 mg dm^{-3}	[48,49]
Phenol 192.9 mg dm^{-3} COD 590 mg dm^{-3} Conductivity 15.63 mS cm^{-1}	AO EC, EF	BDD, RuO_2 Fe electrode	Efficiency: EC < AO-RuO_2 < AO-BDD < EF With EF Phenol removal of 98.74% and COD removal of 75.71%	[40]

Dimoglo et al. [34] using laboratory-scale reactors, shown in Fig. 5. In the ECF unit, a graphite anode and a stainless steel mesh as cathode were used while in the EC unit, iron and aluminium were used simultaneously as materials for two blocks of alternating electrodes. The results have shown that EC, using Al anodes, removed the pollutants more effectively than ECF. COD and turbidity removal was estimated as 72% and 83% for ECF and 80% and 88% for EC, respectively.

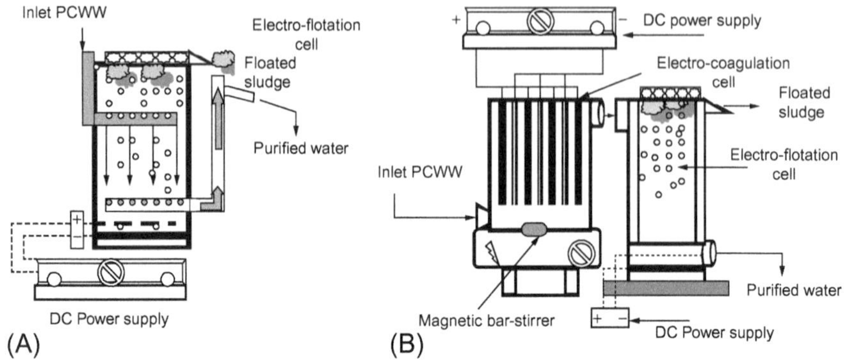

FIG. 5 The EF (A) and EC (B) reactors used in the laboratory experiments for EC and ECF of petroleum industry wastewater. *Reproduced with permission from A. Dimoglo, H.Y. Akbulut, F. Cihan, M. Karpuzcu, Petrochemical wastewater treatment by means of clean electrochemical technologies. Clean Technol. Environ. 6 (2004) 288–295, Copyright 2004, Springer.*

Abdelwahab et al. [35] explored the possibility of using electrocoagulation to remove phenol from oil refinery waste effluent using a cell with a horizontally oriented aluminium cathode and a horizontal aluminium screen anode. The removal of phenol was investigated in terms of various parameters, namely pH, operating time, current density, initial phenol concentration, and the addition of NaCl. Under optimal experimental conditions (current density of 23.6 mA cm^{-2} and pH 7) after 2 h of EC, 94.5% of the initial phenol concentration was removed from the petroleum refinery wastewater with energy consumption of 1.8 kWh g^{-1} phenol and electrode consumption of 0.091 g_{Al} per g of phenol.

Many studies indicated that EC alone is not able to achieve complete COD and phenol removal from petroleum refinery wastewaters, but this process can be efficiently used as pretreatment in a coupled plant with biological oxidation. For example, El-Naas et al. [37,38] proposed a three-step pilot plant for the treatment of highly contaminated petroleum refinery wastewater that consisted of an electrocoagulation unit, biological treatment in a spouted bed bioreactor using immobilized *Pseudomonas putida*, and an adsorption process using granular activated carbon in a packed column. Placing the EC unit as the pretreatment step resulted in the best performance because it reduced the COD and suspended solids and consequently enhanced the performance of the succeeding biodegradation and adsorption units. The COD removal was about 35% after EC and 55% after the biological unit; it further dropped by 96% after the adsorption column. The pilot plant was also able to reduce the concentrations of phenol and cresols by nearly 100%.

Anodic oxidation is the most popular electrochemical technology for removing organic pollutants from wastewaters. Consequently, many papers have been published concerning the application of AO for the treatment of petroleum industries (Table 3). In particular, due to the presence of high concentrations of NaCl in refinery wastewaters, free chlorine and/or chlorine-oxygen species can be electrogenerated on the anode surface. These chemicals can oxidize organic pollutants in the bulk until overall mineralization.

For example, Ramalho et al. [42] studied the anodic oxidation of produced water generated by petroleum exploration of the Petrobras plant in Brazil using an RuO_2-TiO_2-SnO_2 anode. Under galvanostatic conditions ($j = 89\,mA\,cm^{-2}$), it was observed that the performance of the electrode material was slightly influenced by flow rate. This fact suggests that the hydrocarbons are removed by direct oxidation on an anode surface and by indirect oxidation with other electrogenerated oxidants, such as active chlorine formed by NaCl oxidation. Almost complete elimination of pollutants was obtained after 2.5h of electrolysis with energy consumption values ranging from 4.84 to $0.97\,kWh\,m^{-3}$ and removal prices from US$ 0.14 to 0.61 (from 0.11 to 0.5 €).

The efficiency of the Ti/RuO_2 anode for the treatment of wastewaters from the Brazilian petroleum industry was investigated by Santos et al. [43] using an electrochemical flow reactor with a higher anode area. The COD removals after 120min of electrolysis, with a current density of $10\,mA\,cm^{-2}$, were above 96% for effluent after flotation with $712\,mg\,dm^{-3}$ COD and 87% for effluent before flotation with $833\,mg\,dm^{-3}$ COD. An almost complete COD removal from both effluents was achieved when current density was increased from 10 to $30\,mA\,cm^{-2}$. However, an increase of the current density also led to a higher specific energy consumption. At $30\,mA\,cm^{-2}$, the specific energy consumption for COD removal was around US$ $38\,kg_{COD}^{-1}$ and US$ $28\,kg_{COD}^{-1}$ for effluent before and after floatation, respectively.

Gargouri et al. [45] studied the feasibility of anodic oxidation for the removal of petroleum hydrocarbons from produced water generated by the petroleum exploration of the Petrobras plant in Tunisia, using PbO_2 and BDD electrodes. Compared with PbO_2, the BDD anode showed a better performance in removing petroleum hydrocarbon compounds from produced water. The COD removal was approximately 85% and 96% using PbO_2 and BDD reached after 11 and 7h, respectively. The average energy consumption values at the end of the electrochemical treatment were 38 and $46\,kWh\,m^{-3}$ by using BDD and PbO_2 electrodes, respectively. These results showed the high performance of BDD anodic oxidation in treating petrochemical wastewaters. Indeed, electrochemical oxidation using a BDD anode could be a feasible process for produced waters generated by the petrochemical industry in order to significantly reduce the cost and time of treatment.

Rocha et al. [46] investigated the anodic oxidation of produced water using platinum (Ti/Pt) and BDD anodes in an electrolytic batch cell. They also evaluated the influence of several operating parameters such as current, supporting electrolyte, agitation rate, and temperature on the COD removal and energy consumption. The COD removal during bulk electrolysis of wastewaters as received from the petrochemical industry at 25°C and 30 mA cm^{-2} was only 46.5% and 57.5% for Pt and BDD, respectively. However, the electrochemical treatment of the effluent in the presence of 5 g dm^{-3} of Na_2SO_4 at 60°C allowed an efficient degradation of these organic compounds due to the high amounts of effective hydroxyl radicals and peroxodisulfates generated from water oxidation. In these conditions, the BDD achieved higher COD removal and consumed less energy than the Pt anode. For example, 109 and 140 kWh dm^{-3} are consumed for BDD and Pt anodes, respectively, but 98% of COD removal was achieved at BDD while only 85% using the Pt electrode. These results pointed out the high performance of BDD anodic oxidation for treating petrochemical wastewaters. Although BDD oxidation is very effective, its energy consumption could be too high for practical application as the only treatment process, but it can be a feasible process for produced waters generated by the petrochemical industry as a pretreatment process.

On the basis of the promising results of Rocha et al. [46], dos Santos [47] investigated the scale up of an electrochemical flow system for treating petrochemical wastewater using Ti/Pt and BDD anodes. They demonstrated that flow cell was successfully scaled up because when it was compared with a batch mode by Rocha et al. [46], higher performances on organic matter removal were achieved. Even in the flow-cell, the BDD anode allows significantly faster COD removal than Pt, thanks to its greater oxidation ability with high reactivity of the strong oxidant species (hydroxyl radicals and peroxodisulfates) electrogenerated on this electrode (Fig. 6). Under suitable operating conditions, the COD of petrochemical wastewater was reduced from 2746 to 200 mg dm^{-3} within 5 h with an energy consumption of only 56.2 kWh m^{-3} in the scaled up BDD anode system. These results demonstrated that the anode flow system is very promising in practical biorefractory organic wastewater treatment.

Yan et al. [48,49] treated petroleum refinery wastewater by an electrochemical process with three-dimensional multiphase electrodes, which introduced the Fe particle and air into a traditional two-dimensional reactor. In the direct electrochemical process (EC), the COD value of the wastewater decreased from 4753 to 2040 mg dm^{-3} after 60 min treatment. When Fe particles were induced in the electrochemical process (EC + Fe), the COD value reduced to 1480 mg dm^{-3} and 68.86% of COD removal was achieved. This was attributed to the Fe particles that acted as charged microelectrodes in the presence of an electric field and then presented a high ratio of surface area to volume.

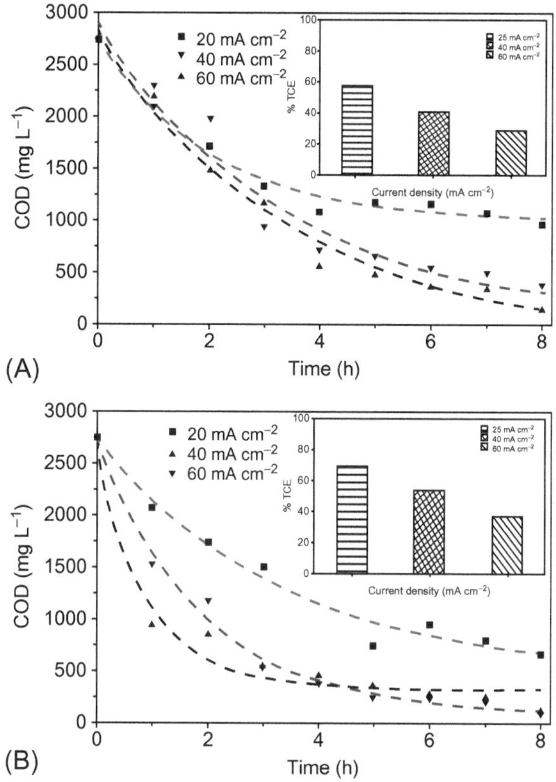

FIG. 6 Influence of applied current density on the COD removal as a function of time and total current efficiency (inset) during produced water anodic oxidation using (A) Ti/Pt and (B) BDD anodes. Operating conditions: applied current density = 20, 40, and 60 $mA\,cm^{-2}$, temperature = 25°C, flow rate, 151 $dm^3\,h^{-1}$. *Reproduced with permission from E.V. dos Santos, S.F.M. Sena, D.R. da Silva, S. Ferro, A. De Battisti, C.A. Martínez-Huitle, Scale-up of electrochemical oxidation system for treatment of produced water generated by Brazilian petrochemical industry. Environ. Sci. Pollut. Res. 21 (2014) 8466–8475, Copyright 2014, Springer.*

When air was pumped into the cell (EC+Air), some H_2O_2 was also electrogenerated from oxygen reduction. However, in this process the 77.91% of the COD removal was reached because H_2O_2 alone is a weak oxidant. In order to make Fe particles and air play a simultaneous role in removing pollutants in refinery wastewater, both Fe particles and air were introduced in the electrochemical process in the primary wastewater condition (EC+Fe+Air) and the pH value after adjusted to the value of 3 (EC+Fe+Air+pH 3). The COD removal was 83.65% and 89.91%, respectively. The largest COD removal of EC+Fe+Air+pH was attributed to the formation of $^{\bullet}OH$, a very powerful oxidizing agent from the reaction of H_2O_2 and Fe according to Fenton's reaction (Eq. 6).

The current efficiency in the EC + Fe + Air + pH 3 system was also much higher than that in other systems, indicating that it is a relatively low energy consumption and high efficiency system.

Yavuz et al. [40] compared different electrochemical processes for the treatment of petroleum refinery wastewater, including anodic oxidation by using BDD or ruthenium mixed metal oxide electrode and electro-Fenton and electrocoagulation by using an iron electrode (Fig. 7). Complete phenol and COD removal was achieved in almost all electrochemical methods except electrocoagulation, provided that the electrolysis time was

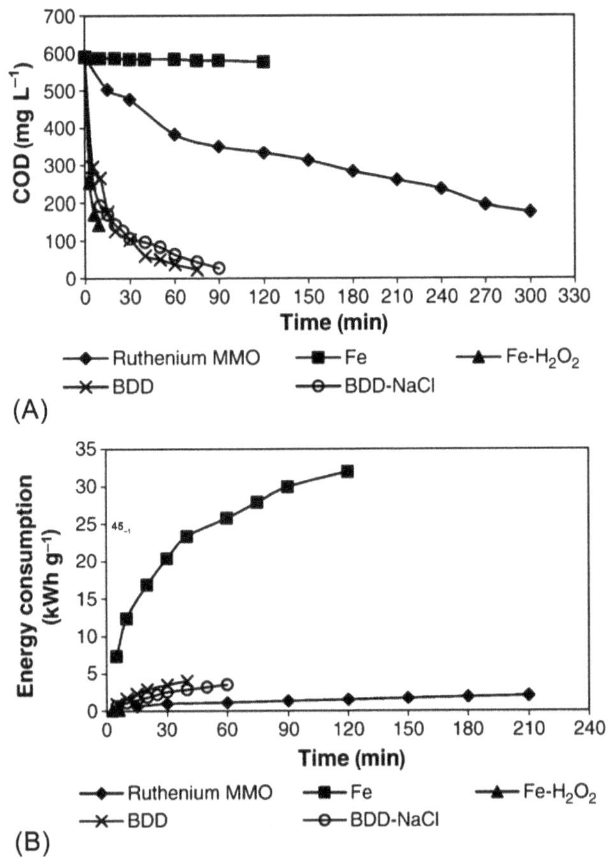

FIG. 7 Variation of (A) COD and (B) energy consumption during the treatment of petroleum refinery wastewater with different electrochemical processes. Experimental condition: AO with Ru-mixed metal oxides electrode (I = 7200 mA), EC with Fe (I = 176 mA), EF (Fe-H_2O_2) (1000 mg dm^{-3} H_2O_2, I = 176 mA), direct oxidation by using BDD (I = 882 mA), indirect oxidation by using BDD electrode (0.05 M NaCl, I = 530 mA). *Reproduced with permission from Y. Yavuz, A.S. Koparal, Ü.B. Öğütveren, Treatment of petroleum refinery wastewater by electrochemical methods. Desalination 258 (2010) 201–205, Copyright 2010, Elsevier.*

prolonged. The most efficient method was the electro-Fenton process followed by the electrochemical oxidation using a BDD anode. In EF, phenol removal of 98.74% was achieved in 6 min of electrolysis and COD removal of 75.71% was reached after 9 min of electrolysis. Average energy consumption values were 0.143 $kWh\,g^{-1}$ in the electro-Fenton method and 4.05 $kWh\,g^{-1}$ in direct electrochemical oxidation by using the BDD electrode. For other information on the utilization of electrochemical processes for the treatment of wastewater from refinery process units, please see also chapter 14.

PHARMACEUTICAL INDUSTRIES

The pharmaceutical industries are considered high-value, low-volume assets that produce a large variety of compounds. They use a wide range of chemicals as raw materials, including natural or synthetic organics, inorganics, solvents, catalysts, etc. This fact implies that wastewaters generated in pharmaceutical industries are generally highly toxic to biological life, contain a broad spectrum of substances, and are characterized by high BOD and COD values and a low BOD/COD ratio. Consequently, if they are not properly treated they cause the discharge to the environment of persistent organic pollutants that have negative effects on humans and other living organisms. AOPs are widely considered an efficient option to treat effluents containing toxic pollutants such as pharmaceutical industry wastewaters. In the AOPs, toxic and persistent pollutants are efficiently removed by the highly oxidizing hydroxyl radicals generated in situ. The most common AOPs proposed in the literature are H_2O_2 with UV radiation, ozone and ozone-based processes (O_3/UV, O_3/H_2O_2, and O_3/H_2O_2/UV), photocatalysis, Fenton, and photo-Fenton oxidation. In recent years, electrochemical advanced oxidation processes (EAOPs) have gained increasing attention as a promising class of AOPs because they provide several advantages for the prevention and remediation of pollution problems.

Up to now, many studies were performed on the application of EAOPs for the disposal of synthetic solutions containing pharmaceutical compounds (e.g., antibiotics or antimicrobial drugs, antiinflammatories, beta blockers, and endocrine-disrupting agents) [50], but only few studies dealt with the treatment of real effluents from the pharmaceutical industry (Table 4).

Domínguez et al. [51] investigated the application of an AO process to the treatment of wastewater from a pharmaceutical industry plant using a BDD anode and an SS cathode. The effluent exhibited a high organic content (DOC of 1600 $mg\,dm^{-3}$ and COD of 12,000 $mg\,dm^{-3}$) and contained high amounts of pharmaceuticals (aromatic and aliphatic compounds) as well as solvents such as methanol and ethanol. The design of experiments was used to optimize the process. The current density was found

TABLE 4 Example of Electrochemical Treatment of Wastewaters From Pharmaceutical Industries

Wastewater characteristics	Process	Electrodes	Comment	Ref.
COD 12000 mg dm^{-3} TOC 1600 mg dm^{-3} TS 5000 mg dm^{-3}	AO	BDD	100% TOC removal at 179.4 mA cm^{-2}	[51]
60 effluents: COD 100–500,000 mg dm^{-3} Conductivity 0.2–200 mS cm^{-1}	AO, Fenton	BDD	Complete mineralization with AO AO more efficient than Fenton in 80% of the effluents	[52]
TOC 110.1 mg dm^{-3} COD 326.9 mg dm^{-3} BOD5 13.3 mg dm^{-3}	SE	Nanocoated RuO_2-IrO_2/Ti	94% of COD removal at $i = 8$ mA cm^{-2} and an US frequency of 45 kHz	[53]
COD 4980 mg dm^{-3} TOC 1600 mg dm^{-3}	AO +UV +US	BDD	UV and US alone are not efficient Synergistic effect coupling AO +UV+US	[54]
COD 2518 mg dm^{-3} Phosphorous 27.65 mg dm^{-3} SS 170 mg dm^{-3} BOD 906 mg dm^{-3}	EC	Al	52% of COD removal 100% of total phosphorus removal	[55]
COD 4200 mg dm^{-3} TS 17.89 g dm^{-3} TDS 13.32 g dm^{-3} TSS 3.4 g dm^{-3}	EC	Al, SS	78.88% COD removal with SS 99.88% color removal with Al E_{sp}: 27.12 Wh kg_{COD}^{-1} with Al	[56]

to have a great positive influence on TOC removal while the flow was found to have a moderate negative influence; however, in the optimum conditions almost 100% of TOC content can be removed.

The efficiency in abatement of COD of 60 real effluents coming from different processes carried out in a pharmaceutical facility by Fenton oxidation and AO with a BDD anode was compared by Pérez et al. [52]. In 80% of the samples, AO was found to be more efficient than Fenton oxidation and in the remaining 20%, coagulation was found to exhibit a great significance in the COD abatement mechanism during Fenton, pointing out the effectiveness of the electrochemical technology. Results obtained in the separated tests were also confirmed by the treatment of the mixture of wastes collected in the homogenization tanks of the facility. For comparative purposes, COD removal was plotted against oxygen-equivalent chemical-oxidation capacity to quantify in arbitrary units the oxidants added to the waste [57]. As shown in Fig. 8, AO depleted completely the organic load while Fenton oxidation was only able to decrease COD down to 45% of the initial load.

The combination of AO with a BDD anode, photolysis, and sonolysis was evaluated by de Martín de Vidales et al. [54] for the treatment of an effluent coming from an organic-synthesis pharmaceutical industry. Results demonstrated that single irradiated technologies (photolysis and sonolysis) were inefficient in the oxidation but their combination with BDD improved significantly the efficiency of the process, showing clear synergistic effects. Complete mineralization of the organic compounds was attained with integrated process without the addition of any salt.

Yang et al. [53] developed a novel sonoelectrochemical catalytic oxidation-driven process using a nanocoated Ru-IrO_2 electrode for the disposal of biologically treated cephalosporin pharmaceutical wastewater. As shown in Fig. 9, sonolysis alone was ineffective for COD

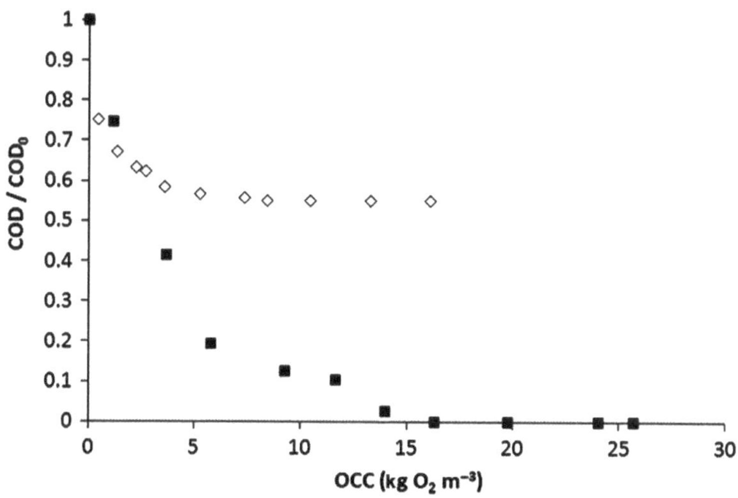

FIG. 8 Comparison of COD removal by AO with BDD (▪) and Fenton oxidation (◇) in a mixture of wastes collected in the homogenization tank. Wastewater (initial values): COD: 6000 $mg\,dm^{-3}$, TOC: 1650 $mg\,dm^{-3}$. Conditions: Fenton: H_2O_2/Fe 5:1 (mass units); AO: $i = 30\,mA\,cm^{-2}$. *Reproduced with permission from J.F. Pérez, J. Llanos, C. Sáez, C. López, P. Cañizares, M.A. Rodrigo, Treatment of real effluents from the pharmaceutical industry: a comparison between Fenton oxidation and conductive-diamond electro-oxidation. J. Environ. Manage. 195(Pt 2) (2017) 216–223, Copyright 2016, Elsevier.*

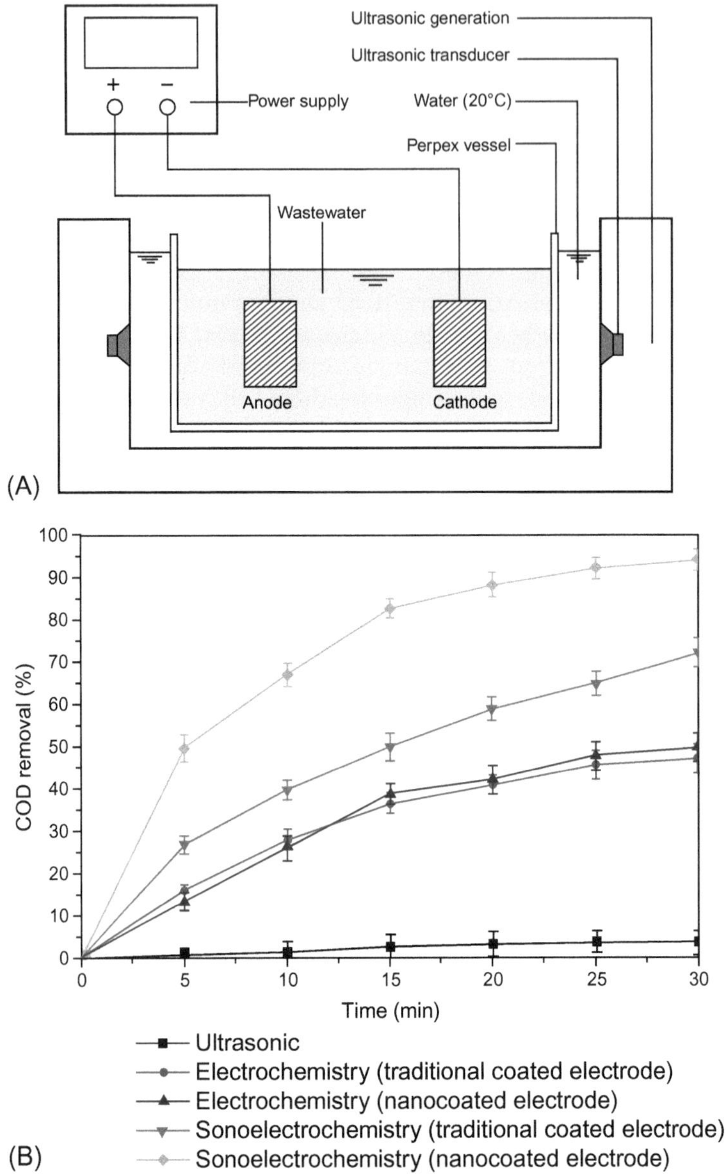

FIG. 9 (A) Experimental setup of ultrasound assisted electrochemical reactor. (B) COD removal via US, AO, and SE with traditional coated electrode and nanocoated electrode. Experimental conditions: current density, 8 mA cm^{-2}; US frequency, 45 kHz. *Reproduced with permission from B. Yang, J. Zuo, P. Li, K. Wang, X. Yu, M. Zhang, Effective ultrasound electrochemical degradation of biological toxicity and refractory cephalosporin pharmaceutical wastewater. Chem. Eng. J. 287 (2016) 30–37, Copyright 2016, Elsevier.*

removal but the use of US increased the nanocoated electrode activity to treat the wastewater. In the presence of ultrasonic waves, the mass transfer effects on the nanocoated electrode surface were enhanced, resulting in rapid diffusion of the generated hydroxyl radicals into the solution, and quickly reacted with organic pollutants. Compared with the traditional coated electrode, the effect of the nanocoated electrode used on the wastewater treatment process was more enhanced by ultrasound under the same conditions. The optimum operating conditions resulted in a 94% removal efficiency for COD and consisted of a current of density $8\,mA\,cm^{-2}$ and an ultrasound frequency of 45 kHz.

The treatment of Ayurveda pharmaceutical wastewater by the electrocoagulation method was recently studied by Singh et al. [56] with the aim of minimizing the energy consumption and maximizing the color and COD removals by using the stainless steel and aluminium electrodes. The maximum 78.88% COD removal was obtained with stainless steel electrodes whereas the maximum 99.88% color removal was obtained with Al electrodes. Moreover, the EC operating cost for both Al and SS electrodes was found to be around $6.64 and $4.93 (per kg of COD removed), respectively. Due to the presence of NaCl in the wastewaters, some active chlorine was produced during electrolysis and therefore the EC treatment mechanism was a combination of electrofloatation, electrooxidation, and electrocoagulation.

CONCLUSIONS

In recent decades, electrochemical technologies have reached a promising stage of development and can now be effectively used as an alternative to the traditional methods for wastewater remediation. The above provides an overview of the applications of electrochemical methods for the treatment of real industrial wastewaters belonging to the fine chemical, pulp and paper, petrochemical, and pharmaceutical industries.

Electrocoagulation and electroflotation are simple, effective, and economical processes for the treatment of pulp and paper or petrochemical effluents and in general, for the disposal of wastewaters containing suspended solids, organic, and inorganic pollutants. EC and ECF can be efficiently used as pretreatment to increase the efficiency of the biological oxidation.

The anodic oxidation and electro-Fenton processes proved to be a promising AOP for the removal of toxic organic pollutants commonly present in fine and pharmaceutical effluents. In order to reduce energy consumption, these processes can be used as tertiary treatment to refine the quality of the effluent and reduce the retention time of the biological process.

In conclusion, the literature results indicate that electrochemical methods could be applied as reliable alternative methods for the treatment of actual wastewaters.

References

[1] E. Brillas, I. Sirés, M.A. Oturan, Electro-Fenton process and related electrochemical technologies based on Fenton's reaction chemistry, Chem. Rev. 109 (2009) 6570–6631.
[2] G. Chen, Electrochemical technologies in wastewater treatment, Sep. Sci. Technol. 38 (2004) 11–41.
[3] C.A. Martínez-Huitle, E. Brillas, Decontamination of wastewaters containing synthetic organic dyes by electrochemical methods: a general review, Appl. Catal. B: Environ. 87 (2009) 105–145.
[4] M. Panizza, G. Cerisola, Application of diamond electrodes to electrochemical processes, Electrochim. Acta 51 (2005) 191–199.
[5] M. Panizza, G. Cerisola, Direct and mediated anodic oxidation of organic pollutants, Chem. Rev. 109 (2009) 6541–6569.
[6] I. Sires, E. Brillas, M.A. Oturan, M.A. Rodrigo, M. Panizza, Electrochemical advanced oxidation processes: today and tomorrow. A review, Environ. Sci. Pollut. Res. 21 (2014) 8336–8367.
[7] P. Canizares, R. Paz, J. Lobato, C. Saez, M.A. Rodrigo, Electrochemical treatment of the effluent of a fine chemical manufacturing plant, J. Hazard. Mater. 138 (2006) 173–181.
[8] P. Canizares, R. Paz, C. Saez, M.A. Rodrigo, Costs of the electrochemical oxidation of wastewaters: a comparison with ozonation and Fenton oxidation processes, J. Environ. Manag. 90 (2009) 410–420.
[9] A. Beteta, P. Canizares, M.A. Rodrigo, L. Rodriguez, C. Saez, Treatment of door-manufacturing factories wastewaters using CDEO and other AOPs: a comparison, J. Hazard. Mater. 168 (2009) 358–363.
[10] P. Canizares, B. Louhichi, A. Gadri, B. Nasr, R. Paz, M.A. Rodrigo, et al., Electrochemical treatment of the pollutants generated in an ink-manufacturing process, J. Hazard. Mater. 146 (2007) 552–557.
[11] M. Panizza, C. Bocca, G. Cerisola, Electrochemical treatment of wastewater containing polyaromatic organic pollutants, Water Res. 34 (2000) 2601–2605.
[12] M. Panizza, G. Cerisola, Removal of organic pollutants from industrial wastewater by electrogenerated Fenton's reagent, Water Res. 35 (2001) 3987–3992.
[13] D. Rajkumar, K. Palanivelu, Electrochemical treatment of industrial wastewater, J. Hazard. Mater. 113 (2004) 123–129.
[14] L. Ouattara, M.M. Chowdhry, C. Comninellis, Electrochemical treatment of industrial wastewater, New Diamond Front. Carbon Technol. (2004) 239–247.
[15] F. Akbal, S. Camcı, Treatment of metal plating wastewater by electrocoagulation, Environ. Prog. Sustain. Energy 31 (2012) 340–350.
[16] S. Mahesh, B. Prasad, I.D. Mall, I.M. Mishra, Electrochemical degradation of pulp and paper mill wastewater. Part 2. Characterization and analysis of sludge, Ind. Eng. Chem. Res. 45 (2006) 5766–5774.
[17] S. Mahesh, B. Prasad, I.D. Mall, I.M. Mishra, Electrochemical degradation of pulp and paper mill wastewater. Part 1. COD and color removal, Ind. Eng. Chem. Res. 45 (2006) 2830–2839.
[18] A.P. Buzzini, L.J. Patrizzi, A.J. Motheo, E.C. Pires, Preliminary evaluation of the electrochemical and chemical coagulation processes in the post-treatment of effluent from an upflow anaerobic sludge blanket (UASB) reactor, J. Environ. Manag. 85 (2007) 847–857.
[19] H. Ma, B. Wang, Y. Wang, Application of molybdenum and phosphate modified kaolin in electrochemical treatment of paper mill wastewater, J. Hazard. Mater. 145 (2007) 417–423.

[20] B. Wang, L. Gu, H. Ma, Electrochemical oxidation of pulp and paper making wastewater assisted by transition metal modified kaolin, J. Hazard. Mater. 143 (2007) 198–205.
[21] K.S.P. Kalyani, N. Balasubramanian, C. Srinivasakannan, Decolorization and COD reduction of paper industrial effluent using electro-coagulation, Chem. Eng. J. 151 (2009) 97–104.
[22] S. Khansorthong, M. Hunsom, Remediation of wastewater from pulp and paper mill industry by the electrochemical technique, Chem. Eng. J. 151 (2009) 228–234.
[23] P.A. Soloman, C. Ahmed Basha, M. Velan, N. Balasubramanian, P. Marimuthu, Augmentation of biodegradability of pulp and paper industry wastewater by electrochemical pre-treatment and optimization by RSM, Sep. Sci. Technol. 69 (2009) 109–117.
[24] M. Zaied, N. Bellakhal, Electrocoagulation treatment of black liquor from paper industry, J. Hazard. Mater. 163 (2009) 995–1000.
[25] R. Sridhar, V. Sivakumar, V. Prince Immanuel, J. Prakash Maran, Treatment of pulp and paper industry bleaching effluent by electrocoagulant process, J. Hazard. Mater. 186 (2011) 1495–1502.
[26] N. Jaafarzadeh, M. Omidinasab, F. Ghanbari, Combined electrocoagulation and UV-based sulfate radical oxidation processes for treatment of pulp and paper wastewater, Process Saf. Environ. 102 (2016) 462–472.
[27] R.T. Pelegrini, R.S. Freire, N. Duran, R. Bertazzoli, Photoassisted electrochemical degradation of organic pollutants on a DSA type oxide electrode: process test for a phenol synthetic solution and its application for the E1 bleach kraft mill effluent, Environ. Sci. Technol. 35 (2001) 2849–2853.
[28] U.D. Patel, S. Suresh, Electrochemical treatment of pentachlorophenol in water and pulp bleaching effluent, Sep. Sci. Technol. 61 (2008) 115–122.
[29] B. Wang, W. Kong, H. Ma, Electrochemical treatment of paper mill wastewater using three-dimensional electrodes with Ti/Co/SnO_2-Sb_2O_5 anode, J. Hazard. Mater. 146 (2007) 295–301.
[30] E.S.Z. El-Ashtoukhy, N.K. Amin, O. Abdelwahab, Treatment of paper mill effluents in a batch-stirred electrochemical tank reactor, Chem. Eng. J. 146 (2009) 205–210.
[31] K. Eskelinen, H. Särkkä, T.A. Kurniawan, M.E.T. Sillanpää, Removal of recalcitrant contaminants from bleaching effluents in pulp and paper mills using ultrasonic irradiation and Fenton-like oxidation, electrochemical treatment, and/or chemical precipitation: a comparative study, Desalination 255 (2010) 179–187.
[32] K. Chanworrawoot, M. Hunsom, Treatment of wastewater from pulp and paper mill industry by electrochemical methods in membrane reactor, J. Environ. Manag. 113 (2012) 399–406.
[33] X. Qu, W.J. Gao, M.N. Han, A. Chen, B.Q. Liao, Integrated thermophilic submerged aerobic membrane bioreactor and electrochemical oxidation for pulp and paper effluent treatment—towards system closure, Bioresour. Technol. 116 (2012) 1–8.
[34] A. Dimoglo, H.Y. Akbulut, F. Cihan, M. Karpuzcu, Petrochemical wastewater treatment by means of clean electrochemical technologies, Clean Technol. Environ. 6 (2004) 288–295.
[35] O. Abdelwahab, N.K. Amin, E.S.Z. El-Ashtoukhy, Electrochemical removal of phenol from oil refinery wastewater, J. Hazard. Mater. 163 (2009) 711–716.
[36] H. Ma, B. Wang, Electrochemical pilot-scale plant for oil field produced wastewater by M/C/Fe electrodes for injection, J. Hazard. Mater. 132 (2006) 237–243.
[37] M.H. El-Naas, M.A. Alhaija, S. Al-Zuhair, Evaluation of a three-step process for the treatment of petroleum refinery wastewater, J. Environ. Chem. Eng. 2 (2014) 56–62.
[38] M.H. El-Naas, R. Surkatti, S. Al-Zuhair, Petroleum refinery wastewater treatment: a pilot scale study, J. Water Process Eng. 14 (2016) 71–76.
[39] L.S. Pérez, O.M. Rodriguez, S. Reyna, J.L. Sánchez-Salas, J.D. Lozada, M.A. Quiroz, et al., Oil refinery wastewater treatment using coupled electrocoagulation and fixed film biological processes, Phys. Chem. Earth 91 (2016) 53–60.
[40] Y. Yavuz, A.S. Koparal, Ü.B. Öğütveren, Treatment of petroleum refinery wastewater by electrochemical methods, Desalination 258 (2010) 201–205.

[41] M.R.G. Santos, M.O.F. Goulart, J. Tonholo, C.L.P.S. Zanta, The application of electrochemical technology to the remediation of oily wastewater, Chemosphere 64 (2006) 393–399.
[42] A.M.Z. Ramalho, C.A. Martínez-Huitle, D.R.d. Silva, Application of electrochemical technology for removing petroleum hydrocarbons from produced water using a DSA-type anode at different flow rates, Fuel 89 (2010) 531–534.
[43] I.D. Santos, M. Dezotti, A.J.B. Dutra, Electrochemical treatment of effluents from petroleum industry using a Ti/RuO_2 anode, Chem. Eng. J. 226 (2013) 293–299.
[44] P.R.F. da Costa, D.R. da Silva, C.A. Martínez-Huitle, S. Garcia-Segura, Fuel station effluent treatment by electrochemical technology, J. Electroanal. Chem. 763 (2016) 97–103.
[45] B. Gargouri, O.D. Gargouri, B. Gargouri, S.K. Trabelsi, R. Abdelhedi, M. Bouaziz, Application of electrochemical technology for removing petroleum hydrocarbons from produced water using lead dioxide and boron-doped diamond electrodes, Chemosphere 117 (2014) 309–315.
[46] J.H.B. Rocha, M.M.S. Gomes, N.S. Fernandes, D.R. da Silva, C.A. Martínez-Huitle, Application of electrochemical oxidation as alternative treatment of produced water generated by Brazilian petrochemical industry, Fuel Process. Technol. 96 (2012) 80–87.
[47] E.V. dos Santos, S.F.M. Sena, D.R. da Silva, S. Ferro, A. De Battisti, C.A. Martínez-Huitle, Scale-up of electrochemical oxidation system for treatment of produced water generated by Brazilian petrochemical industry, Environ. Sci. Pollut. Res. 21 (2014) 8466–8475.
[48] L. Yan, H. Ma, B. Wang, Y. Wang, Y. Chen, Electrochemical treatment of petroleum refinery wastewater with three-dimensional multi-phase electrode, Desalination 276 (2011) 397–402.
[49] L. Yan, Y. Wang, J. Li, H. Ma, H. Liu, T. Li, et al., Comparative study of different electrochemical methods for petroleum refinery wastewater treatment, Desalination 341 (2014) 87–93.
[50] L. Feng, E.D. van Hullebusch, M.A. Rodrigo, G. Esposito, M.A. Oturan, Removal of residual anti-inflammatory and analgesic pharmaceuticals from aqueous systems by electrochemical advanced oxidation processes. A review, Chem. Eng. J. 228 (2013) 944–964.
[51] J.R. Domínguez, T. González, P. Palo, J. Sánchez-Martín, M.A. Rodrigo, C. Sáez, Electrochemical degradation of a real pharmaceutical effluent, Water Air Soil Pollut. 223 (2012) 2685–2694.
[52] J.F. Pérez, J. Llanos, C. Sáez, C. López, P. Cañizares, M.A. Rodrigo, Treatment of real effluents from the pharmaceutical industry: a comparison between Fenton oxidation and conductive-diamond electro-oxidation, J. Environ. Manag. 195 (Pt 2) (2017) 216–223.
[53] B. Yang, J. Zuo, P. Li, K. Wang, X. Yu, M. Zhang, Effective ultrasound electrochemical degradation of biological toxicity and refractory cephalosporin pharmaceutical wastewater, Chem. Eng. J. 287 (2016) 30–37.
[54] M.J. Martín de Vidales, M. Millán, C. Sáez, P. Cañizares, M.A. Rodrigo, Irradiated-assisted electrochemical processes for the removal of persistent pollutants from real wastewater, Sep. Sci. Technol. 175 (2017) 428–434.
[55] Ö. Gökkuş, Y.Ş. Yıldız, Application of electrocoagulation for treatment of medical waste sterilization plant wastewater and optimization of the experimental conditions, Clean Technol. Environ. 17 (2015) 1717–1725.
[56] S. Singh, S. Singh, S.L. Lo, N. Kumar, Electrochemical treatment of Ayurveda pharmaceuticals wastewater: optimization and characterization of sludge residue, J. Taiwan Inst. Chem. Eng. 67 (2016) 385–396.
[57] P. Cañizares, R. Paz, C. Sáez, M.A. Rodrigo, Costs of the electrochemical oxidation of wastewaters: a comparison with ozonation and Fenton oxidation processes, J. Environ. Manag. 90 (2009) 410–420.

CHAPTER

14

Application of Electrochemical Processes for Treating Effluents From Hydrocarbon Industries

*Alejandro Medel**, Florencio Lugo*[†], *Yunny Meas**

*Center for Research and Technological Development in Electrochemistry, Santiago de Querétaro, Mexico

[†]Autonomous National University of Mexico, Mexico City, Mexico

OUTLINE

Electrochemical Water and Wastewater Treatment
https://doi.org/10.1016/B978-0-12-813160-2.00014-6

INTRODUCTION

Through its different operations, the hydrocarbon industry (and particularly refineries), produces effluents containing highly toxic compounds (phenols and sulfurs, among others). Thus, in the transformation of crude oil, >2500 final products are produced such as liquefied petroleum gas, gasoline, gasoil, kerosene, jet fuel, diesel, lubricating oils, and petrochemical raw materials [1]. All of these are obtained through separation processes (initial desalination, atmospheric distillation, and vacuum distillation), conversion (catalytic and thermal cracking, catalytic and steam reforming, isomerization, alkylation, and lubricating oils production), and treatment (naphthas and diesel desulfurization, sour water, and catalyst regeneration) [2]. The production of fuel oil and gasoline, among others, leads to different configurations of plants and process units [3], which also allows for the classification of refineries as simple, complex, or very complex [4]. In this context, obtaining the different products leads to the production of effluents containing biorefractory compound that cannot be treated through biological methods. Because of this, the use of electrochemical processes has proven to be an excellent option. That is why, in this chapter, we will address the use of electrooxidation (EO) using metal oxide electrodes with Ti as a substrate (Ti/MO_x) as well as boron doped diamond (BDD) electrodes to eliminate phenol and phenolic compounds. Even though the different refinery process units (Fig. 1) can produce different types of effluents, in this chapter we will only discuss the main research reported in the literature as well as that obtained in our group regarding water produced, spent caustic, and desalter effluent.

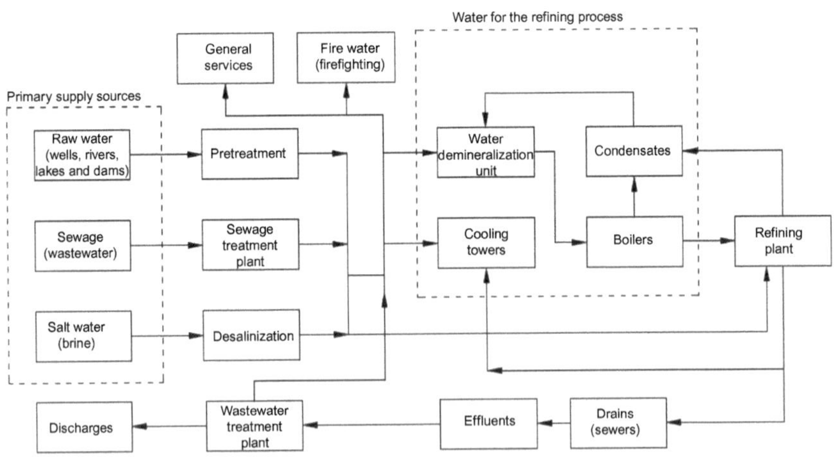

FIG. 1 Typical process flow diagram of crude oil refining.

WATER SYSTEM (AND EFFLUENTS) IN REFINERIES

The average water consumption in oil refining is 294 L per barrel of crude oil [5] while the volume of wastewater produced in these processes is approximately 0.4–1.6 times the volume of processed oil [6]. With the current processing of crude oil at 84 million barrels per day (MBD), it is estimated that 33.6 MBD of wastewater is produced globally [7]. Mexican Petroleums (Petróleos Mexicanos, PEMEX) reported that, in the National Refining System (Sistema Nacional de Refinación, SNR), an average of 1155 MBD were processed in 2014. This leads to the consumption of 340 million m^3 day^{-1} and the production of 462–1848 MBD of wastewater during the same time period [8]. Distillation, hydrotreatment, desalination, and cooling systems are among the oil refining processes that consume important amounts of water. Said waters contain fractions of aliphatic and aromatic hydrocarbons as well as halogenated organic substances [9]. The main sources of wastewater [5] with their respective pollutants and different process units are shown in Table 1.

In general, the wastewater produced in refineries can contain levels of biochemical oxygen demand (BOD_5) and chemical oxygen demand (COD) ranging from 150 to 350 and 300–800 mg L^{-1}, respectively; phenols at levels of 20–200 mg L^{-1}; greases and oils up to 3000 mg L^{-1}; suspended particles of >100 mg L^{-1}; benzene at levels of 1–100 mg L^{-1}; benzopyrene at levels of 1–100 mg L^{-1}; heavy metals at levels of 0.1–100 mg L^{-1} for chrome; and other organic and inorganic pollutants at levels of 0.2–10 mg L^{-1} [10]. Said waters are considered highly toxic due to the

TABLE 1 Main Oil Refining Processes [5]

Process unit	Pollutants produced
Crude desalting	Free oil, ammonia, sulfides, suspended solids
Crude oil distillation	Sulfides, ammonia, phenols, oil, chlorides, mercaptans
Thermal cracking	H_2S, ammonia, phenols
Catalytic cracking	Oil, sulfides, phenols, cyanide, ammonia
Hydrocracking	High in sulfides
Polymerization	Sulfides, mercaptans, ammonia
Alkylation	Spent caustic, oil, sulfides
Isomerization	Low level of phenols
Reforming	Sulfide
Hydrotreating	Ammonia, sulfides, phenol

presence of polycyclic aromatic compounds [4,11]. The amount of contaminated water produced by refineries as well as the elevated toxicity of the pollutants present justify the use of electrochemical processes for effective removal/oxidation of said pollutants.

Effluents Contaminated by Phenol and Phenolic Compounds

As previously stated, phenols and phenolic compounds are among the main pollutants produced by the hydrocarbon industry, especially from refineries. Phenols are aromatic compounds containing one or more hydroxyl groups attached to the ring. A phenol concentration around 5–25 $mg\,L^{-1}$ is as toxic for aquatic life as it is for humans [12,13]. The European Union (EU), in its Directive 80/778/EC, established a maximum allowable limit (MAL) of <0.0005 $mg\,L^{-1}$ for phenols in complete form in water for human consumption [12] due to the ability of said pollutants to cause cancer or to lead to death [14]. In turn, phenol and some phenolic compounds are considered as priority pollutants according to the Environmental Protection Agency (EPA) [12,13] due to their negative impact on aquatic life and human health. The MALs for discharge of waste effluents containing phenol range from 0.5 to 1.0 $mg\,L^{-1}$ [13]. Among the processes used in their treatment, the destructive [14–17] and nondestructive [12,13,18–21] methods stand out. In the latter category, physicochemical processes such as solvent extraction and adsorption with activated carbon are commonly used, with a disadvantage being the production of toxic residues. Within the destructive methods, biological oxidation is the most used process but it is sensitive to certain parameters such as pH, temperature, and the presence of other toxic compounds that can inhibit microbiological activity. Phenols act by precipitating the proteins from which its germicidal power and toxicity are deduced; for this reason, the effluents cannot be treated in conventional biological treatment plants. Phenols are not degraded to a concentration of >200 $mg\,L^{-1}$ [21], and the microorganisms are completely deactivated at a concentration of >3000 $mg\,L^{-1}$ [12]. In some processes, phenols can be combined with chlorine, allowing the formation of phenyl-polychlorinated compounds that are more toxic and more resistant to biodegradation than phenol itself. In turn, though the phenol molecule can be destroyed, the degradation of reaction byproducts is more difficult [16].

Given this situation, a very important alternative is the use of advanced oxidation processes (AOPs) [22], a category which includes all those processes that have the ability to produce the hydroxyl radical ($^{\bullet}OH$), which is a highly oxidizing species whose oxidation potential (2.8 V vs SHE) [23] allows the degradation of pollutants, transforming them into harmless species [24]. The AOPs include processes such as ozone (O_3) in an alkaline medium, hydrogen peroxide (H_2O_2) and its combination with ultraviolet light (UV), (UV/H_2O_2 and UV/O_3), the Fenton processes (with or without light), heterogeneous/homogeneous photocatalysis [25], and some electrochemical processes [26]. Considering the concentration of a pollutant, the combination of AOPs with biological processes has been proposed [27].

It has been demonstrated that the use of AOPs with O_3, H_2O_2, and UV light, followed by a biological process, represents an opportunity to increase the grade of biodegradability of highly toxic and biorefractory effluents [28]. Likewise, the combination of electrochemical and photochemical processes coupled with biological systems has also been proposed [29]. Even though the result of said couplings can be satisfactory, it is necessary to consider the type and concentration of the pollutants present. On the other hand, the electrochemical and electrochemical AOPs have been shown to be excellent options, depending on the material used as an electrode, that can directly (electrochemical oxidation (EO)/incineration) or indirectly (electro-Fenton, EF) promote a complete mineralization of pollutants up to CO_2 and water. Even though the electrochemically assisted Fenton processes are an excellent option, the formation of complex intermediates during the mineralization process (and the lack of a mechanically resistant cathode material) cause this application to have mineralization results at low levels [25], making it necessary to use UV-A light (photoelectro-Fenton process, PEF).

It is necessary to highlight that the efficiency of each process with the potential to be developed into an applied technology depends on the nature and contamination load to be treated. If the objective is total mineralization, an electrochemical AOP, such as EO (incineration), is the most favorable option. Special mention must be made to electrocoagulation (EC) as one of the electrochemical technologies with a high capacity to remove elevated organic charges, greases, oils, and suspended solids, among other pollutants. The nature of EC is nondestructive, meaning that it does not possess the ability to mineralize a pollutant in an efficient manner, at least not without the intervention of coupled chemical reactions. Among the main reaction mechanisms found in this technology, there are: (i) coagulation-flocculation-flotation, (ii) direct oxidation, (iii) mediated oxidation, and (iv) direct electronic transference (see the chapter "Electrocoagulation" in Part I, for additional information). Within these mechanisms, coagulation-flocculation-flotation is predominant. Depending on the material used as an anode, the initial removal process begins with the release of Al^{3+} or Fe^{2+} cations. The released ions possess the ability to remove any pollutant through chemical reactions, precipitation, or by adding materials that can precipitate or float. The colloidal particles, oils, or other pollutants move through the electrical field applied and can become ionized, undergoing electrolysis or hydrolysis reactions.

When using iron as the anode, the main reaction is oxidation of the Fe ion (Eq. 1), while the formation of H_2 occurs mainly in the cathode (Eq. 2). Depending on the medium's pH, hydroxide formation also takes place (Eq. 3).

$$Fe^0 \rightarrow Fe^{2+} + 2e^- \rightarrow Fe^{3+} + 1e^- \quad (1)$$

$$2H_2O + 2e^- \rightarrow H_{2(g)} + 2OH^- \quad (2)$$

$$Fe^{3+} + 3OH^- \rightarrow Fe(OH)_3 \quad (3)$$

In the case of an aluminum (Al) anode, the main reactions are represented by Eqs. (4)–(6).

$$Al \rightarrow Al^{3+} + 3e^- \quad (4)$$

$$2H_2O \rightarrow O_2 + 4H^+ + 4e^- \quad (5)$$

$$2H_2O + 2e^- \rightarrow H_{2(g)} + 2OH^- \quad (6)$$

Considering the elevated capacity of the EC process to remove organic pollutants, its use represents an excellent pretreatment for electrochemical AOPs, in this way achieving an improvement in current efficiency.

ELECTROCHEMICAL ADVANCED OXIDATION PROCESSES

Among the destructive processes, the use of electrochemistry has been shown to be an excellent option in the treatment of highly complex matrixes containing toxic organic compounds. Among the main advantages of electrochemical processes, there are: (a) versatility (redox processes in diluted or concentrated solutions), (b) automatization [easy handling of electrical variables (current and potential) and control for data acquisition], (c) environmental compatibility (use of the electron as a reagent), and (d) absence of muds. At present, the main application and with scaling potential is represented by EO, which is classified as electrochemical conversion and incineration [30] due to the ability of the materials used as anodes to partially (Eq. 7) or totally (Eq. 8) destroy the pollutants, respectively.

$$\text{Nonbiocompatibles} \rightarrow \text{Biocompatibles} + \text{Microorganisms} \rightarrow CO_2 + \text{Biomass} \quad (7)$$

$$R + M(^{\bullet}OH) \rightarrow M + mCO_2 + nH_2O + H^+ + e^- \quad (8)$$

Depending on the nature of the anode and its corresponding overpotential for the oxygen evolution reaction (η-O_2), there are electrodes with low or high $^{\bullet}OH$ production, according to its inner or outer sphere mechanism, respectively. If the η-O_2 is low, and the electrochemical performance of an anode is governed by an inner sphere mechanism, a low $^{\bullet}OH$ production is obtained. In contrast, if the η-O_2 is high and the reaction mechanism is in an outer sphere, a high production of $^{\bullet}OH$ is expected, leading to partial or total destruction, respectively. The commonly used anodes in EO are Pt, IrO_2, RuO_2, SnO_2, PbO_2, and BDD [31,32], with BDD the material with a greater η-O_2 as well as being more widely studied for electrochemical elimination of pollutants [33]. Fig. 2 shows the value of the η-O_2 for the evolution reaction of each of these materials.

The above graph was created based on the anodic overpotentials obtained in 0.5 M H_2SO_4 [34]. It is observed that the materials with a lower η-O_2 correspond to Pt, IrO_2, and RuO_2 with values of 0.37, 0.37, and 0.47 V vs SHE, respectively. In contrast, a greater η-O_2 is observed with electrodes such as PbO_2, SnO_2, and Ti/BDD with values of 0.67, 0.67, and 1.27 V vs SHE, respectively. Accordingly, materials such as Ti/SnO_2-Sb, Ti/PbO_2, and BDD represent the most suitable candidates in an electrochemical incineration process. However, the coating stability in the first two electrodes makes the BDD the number one candidate.

The BDD has won great acceptance due to the production of chemically active $^{\bullet}OH$ [35–37], which is responsible for the total destruction of pollutants (Eq. 8) [38]. Among the characteristics offered by the BDD are high chemical and electrochemical stability, high thermal conductivity, and extreme hardness as well as a wide interval of electrochemical potential

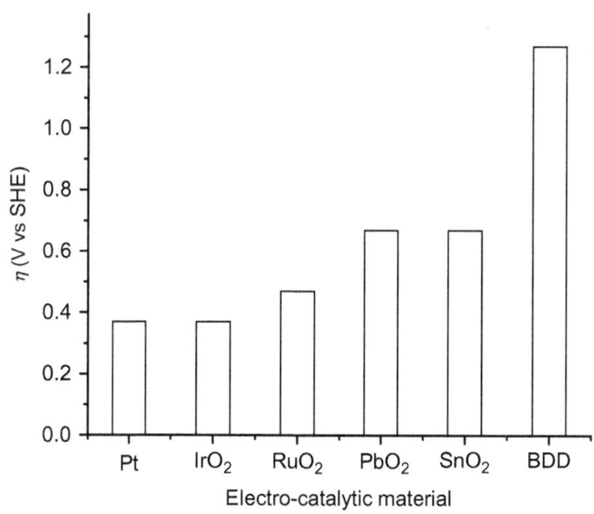

FIG. 2 Anodic overpotentials of anodes commonly used in EO. The values are expressed based on the thermodynamic potential 1.23 V vs SHE for Pt and using 0.5 M sulfuric acid (H_2SO_4) as a supporting electrolyte.

in both aqueous and nonaqueous mediums [32]. To improve the diamond's electrical properties for its use in electrochemistry, it must be previously doped with B, N, or P using Si, Nb, Ta, Mo, W, or Ti as substrates—these also being the most widely studied elements and commonly used substrates to obtain a diamond coating. Among said substrates, Ti represents the most ideal substrate for use in the field due to its mechanical resistance and low cost. However, said substrate has limitations in its use at a large scale due to adherence problems.

The reaction mechanisms in an EO process that lead to electrochemical conversion and incineration have been widely reported, considering the electrocatalytic performance of the material used as an anode. Thus, currently, there are two main classes of materials known: active [Pt, MOx (IrO_2 and RuO_2)] and nonactive [Ti/SnO_2-Sb, Ti/PbO_2, and X/BDD (X=Si, Nb, Ta, Mo, W, and Ti)] [30]. The "active" denomination is given when referring to the low chemical availability of the $^{\bullet}OH$ (*chemisorption*) and the even lower η-O_2, favoring the electrochemical conversion process. The "nonactive" denomination is given to electrodes with a high η-O_2 with *chemically available* $^{\bullet}OH$, as is the case with the BDD, leading to the electrochemical incineration process. A nonactive electrode does not participate in the anodic reaction and does not provide catalytic active sites for the adsorption of reagents and/or products of the aqueous medium. For this reason, the $^{\bullet}OH$ produced in the water oxidation process is subsequently implicated in the oxidation of organics, allowing their complete mineralization [38]. The electrochemical activity (related to η for oxygen evolution) and the chemical reactivity (oxidation rate of organics with electrogenerated $^{\bullet}OH$) of the adsorbed $^{\bullet}OH$ are strongly linked to the interaction strength of the M-$^{\bullet}OH$ complex. Generally, the weaker the interaction of the $^{\bullet}OH$ with the electrode surface, the higher the reactivity of the anode will be for oxidation of organics (fast chemical reaction). For correct electrochemical performance, in the case of the BDD, an accurate protocol of surface activation is required as well as an adequate selection of the material to be used as a cathode, which, when combined with the electrochemical parameters of interest [current density (j), supporting electrolyte, and pH], results in high $^{\bullet}OH$ production. Additional information on the process are given in Part I and III in the chapters "Electrochemical Oxidation" and "Advantages, Disadvantages, and Future Challenges of the Use of Electrochemical Technologies for Water and Wastewater Treatment."

Electrochemical Oxidation and Incineration With BDD Electrodes

The reaction mechanism that leads to the electrochemical incineration process with a BDD electrode is complex and entails the direct participation of the $^{\bullet}OH$ as the main oxidizing species (Eq. 9) as well as the secondary participation of H_2O_2 (Eq. 10) and O_3 (Eqs. 11 and 12), whose production is accentuated when mass transfer limitations exist. In the total absence of an organic substrate, corrosion of the BDD can also occur (Eq. 13).

$$H_2O \rightarrow {}^{\bullet}OH + H^+ + e^- \quad (9)$$

$$2{}^{\bullet}OH \rightarrow H_2O_2 \quad (10)$$

$${}^{\bullet}OH \rightarrow {}^{\bullet}O + H^+ + e^- \quad (11)$$

$${}^{\bullet}O + O_2 \rightarrow O_3 \quad (12)$$

$$C - sp^3 + 2{}^{\bullet}OH \rightarrow CO_2 + 2H^+ + 2e^- \quad (13)$$

In turn, the contribution of mediated oxidation from other oxidizing agents—such as peroxodisulfuric acid ($H_2S_2O_8$), peroxodiphosphoric acid ($P_2O_8{}^{-4}$), and percarbonate ($C_2O_6{}^{-2}$) [34,36] formed from the oxidation of sulfates, phosphates, and carbonates, respectively—to the overall oxidation process cannot be ruled out. Special mention must be made to the formation of active chlorine and oxychlorine radicals through chloride oxidation. It is important to always keep in mind that the ${}^{\bullet}OH$, due to a greater oxidation potential (2.8 V vs SHE), is the species responsible for the electrochemical incineration process. It is necessary to consider that a high production of the ${}^{\bullet}OH$ with BDD is strictly dependent on the elevated quality of the coating, doping level, reaction medium, and cathode material. Regarding the quality of the coating, it has been demonstrated that during the diamond synthesis (C-sp^3) process, graphite impurities (C-sp^2) can be produced [38], affecting the BDD's electrochemical performance.

In our work group, we have demonstrated that the elimination of C-sp^2 and the correct activation of the C-sp^3 surfaces can be achieved efficiently by applying current densities that favor the decomposition of the medium where the ${}^{\bullet}OH$ is electrogenerated, obtaining the complete mineralization of C-sp^2 with reproducible results and a constant production of the ${}^{\bullet}OH$ [39]. To demonstrate this, an ${}^{\bullet}OH$ analysis was performed using the spectroelectrochemical technique. Fig. 3 shows the experimental system used to detect the ${}^{\bullet}OH$ in situ. The analysis was done "qualitatively," using *N,N*-dimethyl-*p*-nitrosoaniline with a concentration of 2.5×10^{-5} M in 0.5 M H_2SO_4. In the analysis, different considerations and preliminary studies were carried out to obtain a reproducible method, also considering the reagent's disadvantages in determining the ${}^{\bullet}OH$.

Fig. 4A shows the analysis performed through Raman spectroscopy after carrying out the polarization of a Ti/BDD electrode contaminated with C-sp^2 while applying current densities of 0.45, 0.91, and 1.83 mA cm^{-2} in 0.5 M H_2SO_4. It is observed that at a reaction time (t_r) of 0 min, only the bands corresponding to C-sp^2 (1600 cm^{-1}) and C-sp^3

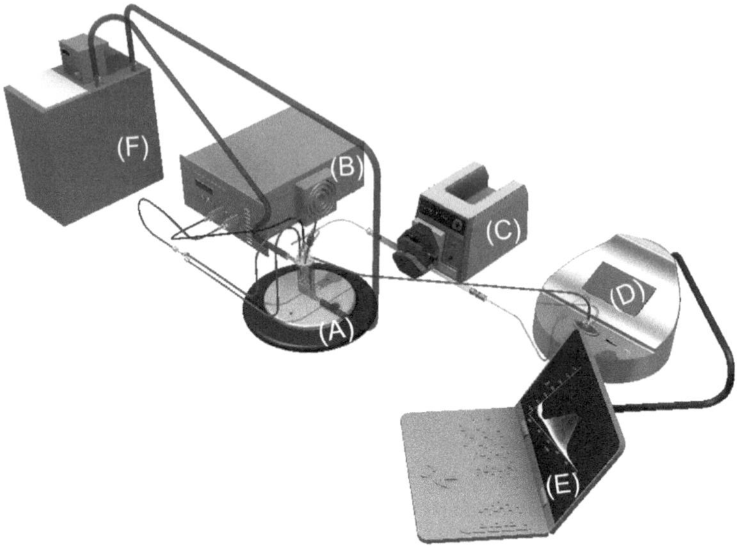

FIG. 3 Experimental system for in situ •OH analysis. (A) electrochemical cell, (B) rectifier, (C) peristaltic pump, (D) UV-Vis spectrophotometer, (E) data acquisition PC, and (F) heat exchanger. *Reproduced with permission from A. Medel, E. Bustos, L. Apátiga, Y. Meas, Surface activation of C-sp3 in boron-doped diamond electrode, Electrocatalysis 4 (2013) 189–195, with permission from Springer.*

(1313.24 cm^{-1}) were detected. When polarizing the electrode, a current density of 1.83 mA cm^{-2} was ideal to eliminate the C-sp^2 completely. In turn, one of the most notable changes was the activation of the Boron content (1200–1280 cm^{-1}), which was only detected after the elimination of the C-sp^2. The analysis of •OH production (Fig. 4B), performed through spectroelectrochemistry in situ, made it possible to link the participation of said species to the elimination of C-sp^2. The activation process developed in this study made it possible to obtain a constant production of the •OH as well as an excellent reproducibility of the analytical results.

Cathode Effect

Another interesting aspect of the use of electrochemical processes is the reaction that occurs in the cathode. In an EO process in an undivided cell, the reduction reactions of organic molecules compete with water reduction. Therefore, different products can be obtained during electrolysis due to the varying overpotential values in the hydrogen evolution reaction (η-H_2), depending on the cathode material used. In practice, a material with a low η-H_2 mainly produces hydrogen (H_2), and an electrode with a high η-H_2 is prone to reducing other reagents [40], which can affect the overall oxidation process. For example, during the phenol EO process,

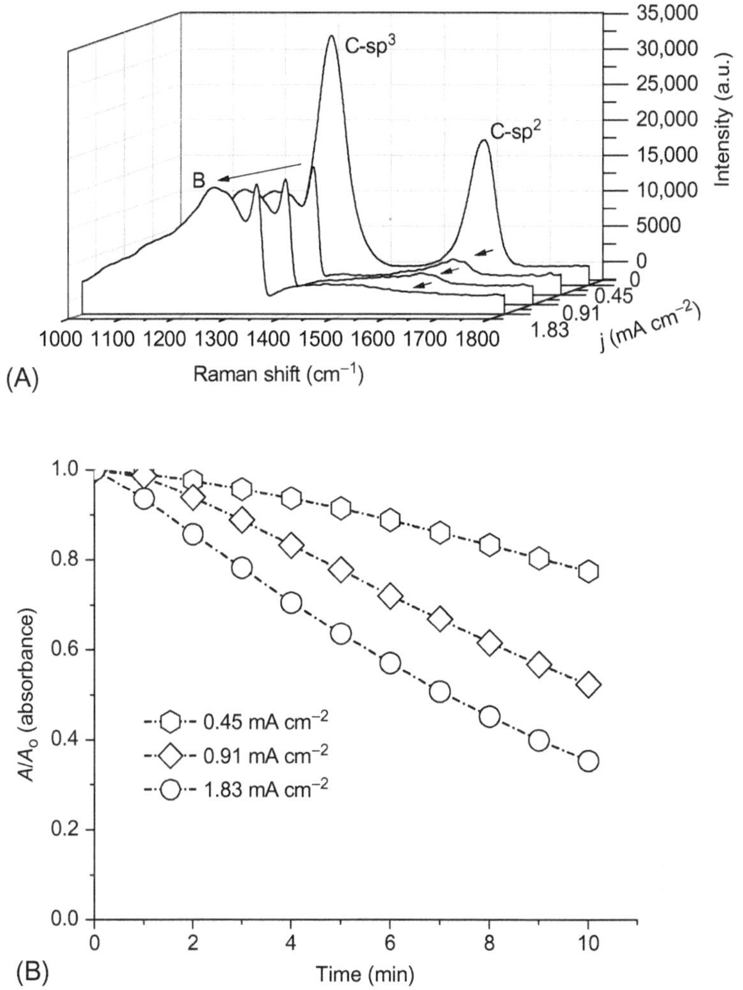

FIG. 4 (A) Raman spectrum of Ti/BDD polarized at current densities of 0.45, 0.91, and 1.83 $mA\,cm^{-2}$ for 10 min in 0.5 M H_2SO_4. (B). Absorbance normalized values of 2.5×10^{-5} M pNDA + 0.5 M H_2SO_4 solution, obtained at 1 min intervals during 10 min of galvanostatic electrolysis at 0.45, 0.91, and 1.83 $mA\,cm^{-2}$. *Reproduced with permission from A. Medel, E. Bustos, L. Apátiga, Y. Meas, Surface activation of C-sp3 in boron-doped diamond electrode, Electrocatalysis 4 (2013) 189–195, with permission from Springer.*

it has been suggested that compounds such as benzoquinone, produced as intermediates, can be reduced in the cathode to hydroquinone, in this way decreasing the process efficiency [41].

On the other hand, Azzam et al. [42] evaluated the EO of 4-clorophenol, according to the pH and the type of cathode material (Pb, Zn, Cu, and brass), using Ti/RuO_2 as the anode. The experimental system consisted of a cell with a working electrode (Ti/RuO_2) and two cathodes. The results obtained by the researchers showed that the nature of the cathode exerts a considerable influence on the destruction velocity of 4-clorophenol. Moreover, it was observed that the pH affects the destruction velocity versus the cathode material. At pH 12.6, the destruction velocity using Cu was higher than when using lead (Cu > Pb) while at pH 2.3 the destruction velocity was greater with Pb than with Cu (Pb > Cu). At pH 7, all the materials evaluated as cathodes showed the same electrocatalytic performance. At pH 12.6, the velocity was dependent on the cathode material in the following order: Cu > brass or Zn > Pb. Contrary to the behavior of the destruction velocity when Cu or brass was used, no difference was observed between Zn and brass, indicating that the presence of Cu in brass did not affect the behavior of Zn. The highest efficiency was obtained when Cu was used, with 37%, while brass, Zn, and Pb had efficiencies of 35%, 33%, and 32%, respectively. At pH 7, the concentration profiles were the same for the different cathodes. The change in pH from 12.6 to 7 caused a change in the destruction velocity. The decrease in pH decreased the degradation kinetics in the Cu, Zn, and brass cathodes while the destruction velocity increased for Pb. The decrease in the destruction velocity with the pH was attributed to the decrease of the η-H_2 and to the formation of other oxidizing species (active chlorine), in this way affecting the reaction mechanism. The current efficiencies with Pb, Zn, and brass were 41%, 21%, and 35%, respectively. At pH 2.3, Zn and brass showed different execution in comparison to pH 12.6 and 7, where the degradation profiles were similar. The destruction velocity using Pb was faster than Cu at an acidic pH, contrary to the behavior at pH 12.6 where Cu was faster than Pb.

Among other works, Borrás et al. [43] studied the EO of atrazine using BDD as the anode, and stainless steel and an electrode of gas diffusion layer [graphite impregnated with polytetrafluoroethylene (GDL)] as cathodes. The results showed that the mineralization process with BDD, in terms of total organic carbon (TOC), was improved when using stainless steel. The previous studies give an idea of the limited information that exists regarding the influence of the cathode in an EO process. It is important to highlight that, up to this point, reference has only been made to the electrochemical reduction of intermediates produced during the degradation process of a specific pollutant. However, Noël et al. [44] studied oxygen reduction on a Pt surface, considering possible $^{\bullet}OH$ production, but they did not carry out an analysis regarding the production of this species.

Other studies have also indicated that both O_2 [42,45] and H_2O_2 [46] can be reduced in the cathode, leading to $^{\bullet}OH$ production [41,45,46] without considering the disadvantages of the analytic method used. This subject has also been the object of different studies in our work group, and in collaboration with Dr. Ignasi Sirés of the University of Barcelona (UB), when evaluating the effect of the nature of the cathode material on $^{\bullet}OH$ production. Currently, the following are found among the main cathode materials used: GDL, graphite felt, titanium (Ti), Ti/Pt, Pt, 304 and 316 stainless steel, zirconium (Zr), graphite (C-sp^2), and diamond (C-sp^3), among others.

Conversion/Incineration of Phenolic Compounds

Electrolysis in synthetic samples. It has been previously stated that an EO process, depending on the type of anode used, can cause partial (*electrochemical conversion*) or total (*electrochemical incineration*) oxidation of pollutants. For this reason, in this section, both approaches have been considered and comparatively analyzed for the electrochemical destruction of phenol in synthetic samples. In comparison to Pt and Ti/IrO_2, it has been demonstrated that phenol can be oxidized up to CO_2 and water with high current efficiencies by using Ti/SnO_2 [30]. When doped with Sb, its performance improves, confirming its superiority with Pt and even with PbO_2 [47]. The mechanisms and the reaction efficiency have been dependent on the pH. Comninellis and Nerini [48] reported the EO of phenol with Pt, Ti/IrO_2, Ti/RuO_2, Ti/PbO_2, and Ti/SnO_2, indicating that the electrochemical performance of the material used as the electrode is influenced by the ionic composition of the solution and that in a medium containing chloride ions, phenol can be oxidized due to the production of hypochlorite ions (ClO^-). The degradation efficiencies obtained were relatively low with Pt, Ti/IrO_2, Ti/RuO_2, and Ti/PbO_2 electrodes whereas with Ti/SnO_2, the mineralization occurred in total form. With the latter, the reaction byproducts identified were mainly carboxylic acids (fumaric, maleic, and oxalic), which were subsequently oxidized to CO_2 [30]. The removal percentages in terms of TOC were of 90% and 38% for Ti/SnO_2 and Pt, respectively. In Ti/SnO_2, only traces of aromatic intermediates (hydroquinone, catechol, and benzoquinone) were detected whereas in Pt, great quantities were obtained. The aliphatic acids (fumaric, maleic, and oxalic) were quickly oxidized in Ti/SnO_2 and were electrochemically inactive in Pt.

It has been documented in other studies that the Ti/SnO_2 oxidizes a wide range of organic pollutants with an efficiency five times greater than that of Pt anodes [41]. The behavior of the SnO_2 is explained by a change in the chemical structure of the electrode's surface during the anodic polarization process. In other research by Feng and Li [49] on the electrocatalytic activity of Ti/Sb-Sn-RuO_2-Gd, Ti/Sb-Sn-RuO_2, Ti/RuO_2, Ti/β-PbO_2, and Pt on phenol destruction, they reported that the order of oxidation occurs as follows: Ti/β-PbO_2 > Ti/Sb-Sn-RuO_2-Gd > Ti/Sb-Sn-RuO_2 > Ti/RuO_2 > Pt. With Ti/β-PbO_2, the mineralization was carried out to CO_2. The intermediates detected, both aromatic and carboxylic acids, coincide with those reported in electrolysis with Ti/SnO_2 and PbO_2 [50].

In turn, Li et al. [51] evaluated the kinetics and paths of phenol degradation on Ti/SnO_2-Sb, Ti/RuO_2 and Pt. The results were presented in terms of TOC, phenol degradation percentage and quantification of the main reaction intermediates (Pt = hydroquinone, benzoquinone, and succinic, oxalic, and maleic acids; Ti/RuO_2 = hydroquinone, benzoquinone, and maleic, succinic, and oxalic acids; Ti/SnO_2-Sb = hydroquinone, benzoquinone, and oxalic, maleic, and succinic acids). The order of process effectiveness was as follows: Ti/SnO_2-Sb > Ti/RuO_2 > Pt. These intermediates were detected through high-performance liquid chromatography (HPLC) and gas chromatography-mass spectrometry (GC/MS), and it was observed that they were quickly oxidized in Ti/SnO_2-Sb but they accumulated in the cells of Ti/RuO_2 and Pt. It was also observed during electrolysis that dark colored polymeric compounds were formed and precipitated in the solutions obtained with Ti/RuO_2 and Pt, which did not occur with Ti/SnO_2-Sb. This reveals that the anodic property does not only affect the kinetic reaction but also alters the paths of phenol degradation in electrolysis. It was observed that surface treatment in Ti/SnO_2-Sb can exert an "apparent" catalytic function favoring the oxidation process. The best results were obtained with Ti/SnO_2-Sb, followed by Pt and Ti/RuO_2. It is worth mentioning that the complete removal of phenol was achieved in each one of the electrodes. However, some significant differences were found in the execution of each of these. For example, in Ti/SnO_2-Sb the complete phenol degradation was achieved in 5h (490 at 0 mg L^{-1}) and the degradation was slower in the other two anodes. The t_r for the complete removal of phenol was 18h for Pt and 36h for Ti/RuO_2. Even though the phenol degradation in the three electrodes occurred at 100%, the TOC removal percentage was in the following order: Ti/SnO_2-Sb, 100%, Pt, 40%, and RuO_2, 40%, with a t_r = 16, 48, and 48h, respectively. It is important to mention that much less energy was spent with the Ti/SnO_2-Sb than with the other materials for both phenol removal and TOC mineralization.

In this study, Na_2SO_4 was used as a supporting electrolyte and the pH changed during electrolysis. During electrolysis with Ti/SnO_2-Sb, a drop in the initial value was observed, followed by an increase during the first 16h. This can indicate that the species is completely degraded because a drop in the pH is caused by the production of organic acids during the electrolysis process. For the Pt and Ti/RuO_2 electrodes, the initial pH value of 5.3 fell to 4.0 and maintained a value around 3.5 until the end of electrolysis. Another important point is the change in color during electrolysis, turning yellow with Ti/SnO_2-Sb within the first 5h and disappearing when the concentration was almost zero (around 4h). This was explained due to the presence of benzoquinone, an aromatic compound formed during the initial phases of the oxidation process and responsible for the yellow color. For Pt and Ti/RuO_2, the color turned from dark yellow to brown when the phenol was totally removed and it did not change, even after 24h of electrolysis. The *hydroquinone* was also detected in high concentrations. The benzoquinone and the hydroquinone are known as an active redox couple in equilibrium in an aqueous solution. The hydroquinone was produced in greater concentration than the benzoquinone for Ti/SnO_2-Sb and the order of acid production during the first 5h of electrolysis was as follows: oxalic acid > maleic acid > succinic acid. In Pt, the hydroquinone concentration was much higher in comparison to that of benzoquinone and the order of carboxylic acid production was as follows: succinic acid > oxalic acid > maleic acid. In a similar manner, the benzoquinone and the hydroquinone were detected in Ti/RuO_2 and the residence time was much greater, approximately 8h for benzoquinone and 28h for hydroquinone, whose concentration was higher. The order of carboxylic acid production was as follows: maleic acid > oxalic acid > succinic acid. The authors concluded that the surface property of the anodes seems to affect the kinetics of several steps in the electrochemical reaction and that the anode could direct the path for oxidation of organic compounds. Furthermore, comparative studies with Si/BDD, Ti/SnO_2, Ta/PbO_2, and Pt found that at 20 $Ahdm^{-3}$, the TOC was reduced from an initial value of 1500–50 $mg L^{-1}$ in the Si/BDD electrode and around 300, 650, and 950 $mg L^{-1}$ in Ti/SnO_2, Ta/PbO_2, and Pt, respectively [52,53]. Some studies have also compared the behavior of the BDD with other electrodes such as SnO_2, PbO_2, and IrO_2 for the oxidation of organic pollutants, finding that the current efficiencies obtained with Ti/BDD in the oxidation of maleic acid, phenol, and colorants was of 1.6–4.3 times higher than that obtained with typical Ti/Sb_2O_5-SnO_2 electrodes [53–55].

In studies by Zhao et al. [55] regarding phenol EO with BDD and Pt in the presence of ultrasound (US), it was concluded that intermediates are formed in low proportions in BDD in comparison to Pt, not observing significant differences when coupled with US. The results obtained showed that the efficiency is greater in the US-BDD coupling in comparison to US-Pt. The degradation rates and the current efficiency in US-BDD increased to 301% and 100%, respectively, while for US-Pt only values of 51% and 49% were registered. This was attributed to mass transport limitations and the amount of adsorbed compounds. The kinetics research on the intermediates formed during the EO process with BDD showed that the intermediates were produced in lower quantity than with Pt. In the presence of US, there was no change in the amount of intermediates. The effect was more obvious for BDD than for Pt, indicating that the BDD is convenient for phenol degradation by EO assisted with US. The study focused fundamentally on the coupling with US, evaluating the effects of the material and the electrochemical character of the electrodes. Even though the majority of research reported in the literature has demonstrated the superiority of the BDD in comparison to conventional materials such as graphite, Pt, Au, stainless steel, Ni, Pd, and Ti/MOx (Ti/RuO_2, Ti/IrO_2, Ti/PbO_2, and Ti/SnO_2-Sb), these have been conducted on synthetic samples and very few studies have been reported on the use of the BDD on wastewater from refineries, meaning highly complex aqueous matrixes.

CONVERSION/INCINERATION IN REFINERY EFFLUENTS

The wastewater produced in the various process units of oil refining can have different names such as water produced, sour water, tank bottom draws, spent caustic, and desalter effluent. The composition and concentration of the pollutants vary widely; therefore the electrochemical parameters that control the EO process with BDD must be thoroughly studied, evaluating them for each type of effluent. This section will focus solely on the electrochemical treatment of water produced, spent caustic, and desalter effluent, presenting some of the main work reported in the literature as well as those conducted in our work group.

Water Produced

The effluents produced during the extraction of petroleum and gas are known as water produced. Among the most relevant researches regarding EO, there are those reported by Zanbotto Ramalho et al. [56], Yavuz et al. [10], and Bezerra Rocha et al. [57]. The work reported by Yavuz and Savaş Koparal [58] was carried out using a parallel plate reactor with Ti/RuO_2 as the anode. The main parameters evaluated were j, pH, temperature, concentration of the supporting electrolyte, flow rate, and initial phenol concentration. The study was performed on a synthetic sample using Na_2SO_4 as a supporting electrolyte. The best conditions were evaluated in a wastewater sample from a refinery without specifying the process unit. The results showed that the parameters of greater influence were the temperature and concentration of the supporting electrolyte in contrast with the pH and the flow rate, where no significant changes were observed. As conclusive results, removal efficiencies of 99.7% (phenol) and 85.4% (COD) and an energy consumption of $0.752\,kWh\,g^{-1}$ phenol were reported, considering the best conditions evaluated—synthetic phenol of $200\,mg\,L^{-1}$ equivalent to a concentration of $480\,mg\,L^{-1}$ of COD, $j = 15\,mA\,cm^{-2}$, $V = 500\,mL$, flow $= 24.83 \times 10^{-3}\,L\,min^{-1}$, room temperature ($\sim 20 \pm 5°C$, reaching values $\sim 50°C$), and $t_r = 50\,min$. Under these conditions, a current efficiency of 74.1% was obtained at a $t_r = 10$ min and 17.7% at a $t_r = 50$ min. In contrast to the refinery sample (phenol concentration of 192.9 mg L^{-1} and COD of 590 mg L^{-1}), the phenol removal percentages were 94.5% of phenol and 70.1% of COD ($j = 20$ mA cm^{-2}) at a $t_r = 3.5$ and 5 h, respectively. Here, the degradation time was much greater in comparison to the synthetic sample ($t_r = 50$ min). The differences observed are due to the extreme chemical composition of a real sample where both organic and inorganic compounds are found to be forming part of a complex matrix [17]. Here, reaction intermediates were not detected. In turn, Bezerra Rocha et al. [57] made a comparative analysis between BDD and Pt, treating samples produced from the exploration of petroleum (water produced) by the industry, *Petrobras en Brazil*.

The analysis focused on the influence of operation parameters, such as *j*, supporting electrolyte, stirring, and temperature, on the phenol removal efficiency in terms of COD. The results obtained showed that 98% of COD was removed with Si/BDD while with Pt the COD removal was 50% by applying 15–30 mA cm^{-2} and 80% at 60 mA cm^{-2}. These results were explained by the $^{\bullet}OH$ production and the formation of peroxodisulphates on the BDD as well as the active behavior of the Pt. In said study, when treating the sample without adding the supporting electrolyte, low COD removal efficiencies of 33.8% and 46.5% were achieved using Pt as electrode and applying $j = 15$ and 30 mA cm^{-2}, respectively, at 25°C; whereas when using Si/BDD under similar conditions, the removals were 50.3% and 57.5% at 10 h of treatment. This was attributed to the low electrical conductivity of the solution. It is necessary to mention that even though an improvement in the COD removal degree was obtained when increasing the values of *j*, a high consumption of charge was also achieved. Due to the existence of mass transfer limitations, the electrogenerated $^{\bullet}OH$ can participate in parallel reactions, reducing the degradation efficiency. This effect, added to the O_2 evolution, can result in elevated energy consumption. With Pt for values of $j = 15$ and 30 mA cm^{-2}, the total current efficiency (TCE) was 42% and 58%, respectively, while with BDD, efficiencies of 90% and 48%, respectively, were obtained. In the case of the BDD, the previous values confirm that a significant current is employed in the O_2 evolution after the first hours of reaction, decreasing the total efficiency of the electrochemical reaction.

In another study, the phenol degradation with Ti/TiO_2-RuO_2-IrO_2 from refinery samples showed only 74.75% and 48% removal of COD and TOC, respectively, although high quantities of chloride ions were employed [59]. On the other hand, the use of Ti/RuO_2-TiO_2-SnO_2 in the treatment of hydrocarbon industry discharge has recently been evaluated, but the removal efficiency under the evaluated conditions was very low, 20%–47% [60]. In turn, a comparative analysis was reported between EO with Ti/BDD and EF processes in actual refinery samples, indicating that the

EF process showed a better phenol degradation in comparison to BDD, although the better degradation of reaction intermediates (in terms of COD) occurred using BDD [10,17].

Spent Caustic

The aqueous wastes produced by the hydrocarbon industry, which involves the use of NaOH to carry out the elimination of sulfur compounds present in the different fractions of crude oil, are known as spent caustics and are classified as sulfidics (removal of H_2S and hydrocarbon mercaptans), naphthenics (removal of kerosene and diesel naphthenic acids), and cresylics (removal of organic, phenol, cresol, and xylenol acids). The typical composition of spent caustics includes a concentration of sulfide (S^{2-}, 1–4wt%), mercaptans (0.1–4wt%), phenols (0–2000 $mg\,L^{-1}$), TOC (6000–20,000 $mg\,L^{-1}$), COD (20,000–60,000 $mg\,L^{-1}$), BOD (5000–15,000 $mg\,L^{-1}$), and potentially toxic elements (PTEs) such as Cu, Ni, Cd, Pb, and Cr, among others [61]. Regarding the phenolic content, depending on the refining process, it can reach up to 30,600 $mg\,L^{-1}$ [62]. At present, electrochemical technologies for their treatment do not exist in the market. When these are confined and stored, they are mixed and the resulting composition is highly toxic.

Currently, spent caustics are treated through wet air oxidation (WAO), acid neutralization (AN), and biological oxidation (BO). Step 2 can be followed by steam stripping in which residual H_2S and mercaptans (which are highly perceivable, even in ppb) are removed. Even though the WAO process can treat concentrations of $<1000\,mg\,L^{-1}$ COD in 60min at 475K and 28bar, it is considered an expensive process [63]. A clear difference exists between water produced and spent caustic in what regards phenolic content. In spent caustics, the content of phenolic compounds can reach a value of 30,600 $mg\,L^{-1}$ [62] while in water produced or any other kind of wastewater this value can range between 20 and 200 $mg\,L^{-1}$ [10,18,59]. The EC has also been evaluated for treating spent caustics, proving to be effective in the removal of sulfides and organic material [61]. Considering that the nature of EC is nondestructive, this suggests its effectiveness as a pretreatment for a subsequent coupling process. This has been evaluated by our group during the treatment of spent caustics, finding a removal percentage of 8%–10% in terms of COD while working with Al anodes and different pH values (2, 5, and 7) and applying 1A. The evaluation of Ti/MO_x (IrO_2-RuO_2, IrO_2-Ta_2O_5) electrodes and graphite anodes has also been done in the EO of spent caustics, obtaining removal percentages of 90% and 60%–80%, respectively. These results are part of a subsequent publication in which studies will be presented regarding the evaluation of different cathode materials to improve the efficiency of the electrochemical processes evaluated. In this section, we will only focus on presenting the results corresponding to the EO of spent caustics using BDD.

In our group, we have evaluated the selective degradation of phenol and phenolic compounds found in a spent caustic diluted through an adequate chemical pretreatment, consisting of the optimization of the conversion process of phenols to phenolates using NaOH [17]. In this study, both the EO with Ti/BDD and PEF were comparatively evaluated. After the chemical pretreatment, two phases were obtained, one of high and the other of low density. An analysis performed through GC-MS showed that in the low density phase the main compounds were aliphatics, such as 4-methyldecane, undecane, nonane, tridecane, tetradecane, decane, dodecane, pentacosane, and tricosane. Here, the presence of phenolic compounds was not identified, showing the effectiveness of the chemical pretreatment. In contrast, the following phenolic compounds were identified in the high-density phase: 2,4-dimethyl-phenol, 2-methyl-phenol, *o*-cresol, 3-methyl-phenol, *m*-cresol, 1-methyl-phenol, *p*-cresol, 3,5-dimethyl-phenol, 3-ethyl-5-methyl-phenol, 4-benzyl-pyridacine, dibenzothiophene, 4-methyl-phenol, 2,4,6-trimethyl-phenol, 3-metoxy (methyltio)-phenol, and 1-methyl-3-phenoxy-benzene. The effective separation was explained because the chemical conversion of phenol to phenolates initiates at pH 10 and reaches a maximum efficiency at pH 12. When carrying out the electrolysis, and by using the best operating conditions (pH 1, 2 A, 298 K, $t_r = 2$ h) in an undivided cell and with a capacity of 100 mL, a removal efficiency of 99.5% of phenolic content and 97% of TOC was obtained. In contrast, the PEF process (undivided cell, 800 mL) allowed the removal of 99.9% of phenolic content and only a 25.69% removal of TOC.

In alkaline conditions, the results were not encouraging because the formation of CO_2 leads to the presence of carbonates and bicarbonates (Eqs. 14 and 15), which are known $^{\bullet}OH$ scavengers [64] as shown in Eqs. (16), (17) [65].

$$CO_2 + NaOH \rightarrow NaHCO_3 \tag{14}$$

$$NaHCO_3 + NaOH \rightarrow Na_2CO_3 + H_2O \tag{15}$$

$$^{\bullet}OH + CO_3^{2-} \rightarrow OH^- + CO_3^{\bullet -} \tag{16}$$

$$^{\bullet}OH + HCO_3^{-} \rightarrow H_2O + CO_3^{\bullet -} \tag{17}$$

The low degradation efficiency of phenol and phenolic compounds in an alkaline medium was also explained by the decrease of the η-O_2 [66]. The second study conducted by our work group consisted of the evaluation of the spent caustic mixtures [67], which are highly toxic, using Ti/BDD—polycrystalline boron [(B) = 1300 ppm] doped diamond film

(Ti/BDD) of 3 μm thickness—and activated in a special manner (elimination of C-sp^2). The study was done in an acidic medium (pH 1, H_2SO_4), observing an excellent electrochemical performance of the Ti/BDD. Before and after electrolysis, an intense physicochemical characterization, GC-MS, electrochemical tests, and analysis of $^{\bullet}OH$ production, were all performed to show the feasibility of using the EO process and its potential development into an applied technology for the destruction of spent caustics. As a chemical pretreatment, both HCl and H_2SO_4 were evaluated for (i) removing greases, oils, and other chemical compounds without altering the phenol concentration, and (ii) evaluating the $^{\bullet}OH$ production (through fluorescence spectroscopy) in a medium containing both $SO_4{}^{2-}$ and Cl^{-1} anions. This last analysis was performed through fluorescence spectroscopy using coumarin (3×10^{-5}M) as a probe compound to give way to the formation of the 7-hydroxycoumarin with a wavelength of maximum emission and excitation of 332 and 500 nm, respectively ($\lambda_{em} = 332$ nm, $\lambda_{ex} = 500$ nm). The equipment and the cell used to detect the 7-hydroxycoumarin are shown in Fig. 5. The electrochemical destruction process of the spent caustics mixture was performed in galvanostatic mode.

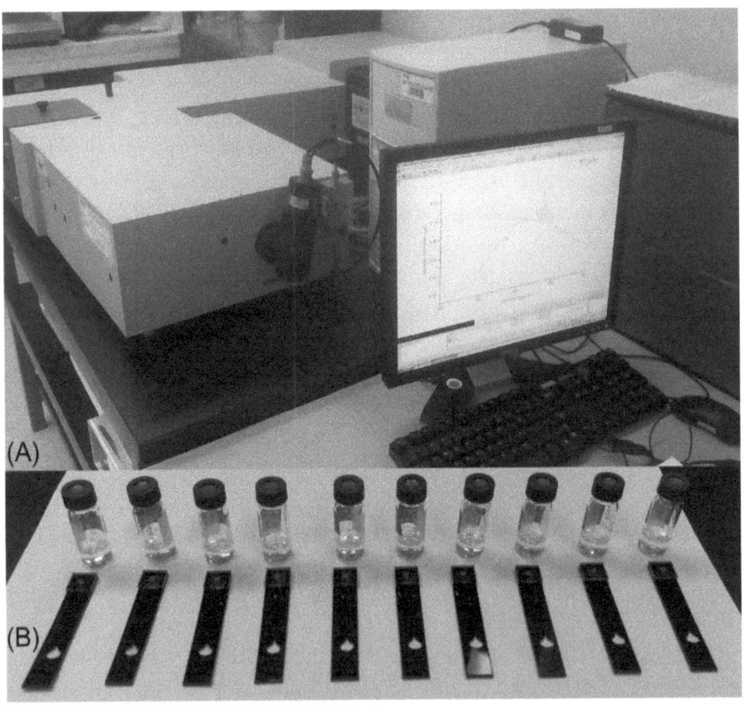

FIG. 5 Experimental system for: (A) hydroxyl radicals' analysis through fluorescence spectroscopy; and, (B) sample preparation using FluoroVettes.

In this process, a cell with a single compartment and a capacity of 120 mL was used [reaction volume of 100 mL, 8 rpm, 298 K, and Ti/Pt ($3\,cm^2$ as counter electrode)]. The area of the anode (prior selection) was $3\,cm^2$. Electrolysis with UV light ($\lambda = 254\,nm$) was performed under the same conditions indicated above. The composition of the spent caustic samples evaluated is shown in Table 2. The comparative table shows the content of a simple sample to demonstrate the complex composition that results from mixing different types of caustics.

It can be observed that the mixture includes a phenolic content of $11{,}041\,mg\,L^{-1}$ in comparison to $7270\,mg\,L^{-1}$ for the simple sample, meaning an elevated concentration, which implies that the mixture of spent caustics cannot be treated by biological processes, as is inferred from the value of BOD_5/COD with values of 0.09 and 0.07 for the simple sample and mixture, respectively. An elevated alkalinity of $52{,}266\,mg\,L^{-1}$ and values of pH 13–13.5 were also registered, which reflects the reductive nature of these residues and which can have a great impact on human health when selecting a specific treatment, meaning that some pollutants are fixed at an alkaline pH and liberated under oxidizing conditions. The mixture of caustics evaluated was classified as cresylic, also considering that sulfurs and cyanides were not detected. The analysis of anions through ion chromatography showed a content of chlorides and sulfates of 86 and $608\,mg\,L^{-1}$ for the simple sample while for the mixture it showed values of 54,900 and $1882\,mg\,L^{-1}$, respectively. In the analysis of volatile and semivolatile organic compounds performed through GC-MS for the mixture, 43 compounds for volatiles and 62 for semivolatiles were considered. The results showed the presence of 2,4-dimethylphenol, phenol, *m*-methylphenol, *p*-methylphenol, and *o*-methylphenol, all highly toxic compounds. Thus, this analysis also helped in identifying the type of spent caustics under evaluation. The main metallic and nonmetallic elements identified in the mixture of spent caustics were Na^+ and Fe with values of 19,148 and $1323\,mg\,L^{-1}$, respectively. Toxic elements such as Cu, Cr, Hg, Ni, and Pb exceeded the MAL established by Mexican regulations [68] with values of 6.89, 1.95, 0.07, 6.75, and $1.23\,mg\,L^{-1}$, respectively.

The main results obtained showed a greater efficiency in the chemical type of pretreatment process when using H_2SO_4 to carry out the process of acidification and removal of pollutants. This was explained by the effect of pH on the electrochemical response of the Ti/BDD. To demonstrate this, different voltammograms were obtained through the adjustment of 0.05 M NaOH with 0.5 M H_2SO_4 and HCl in a pH range of 12–1. In this analysis, the greater potential window was obtained using H_2SO_4, meaning that a greater η-O_2 was obtained in comparison to HCl. In NaOH, the potential window decreased considerably.

TABLE 2 Physicochemical Analysis of Spent Caustics [67]

Parameter	Value	
	Simple sample	**Mixture**
Total phenol ($mg\,L^{-1}$)	7270	11,041
Oil and grease ($mg\,L^{-1}$)	239	3399
BOD_5 ($mg\,L^{-1}$)	6930	7811
COD ($mg\,O_2\,L^{-1}$)	72,065	98,750–102,842
BOD_5/COD	0.09	0.07
Sulfides ($mg\,L^{-1}$)	<37.29	<37.29
Ammonia ($mg\,L^{-1}$)	–	–
Cyanides ($mg\,L^{-1}$)	<0.25	<0.25
pH	13.02	13.50
TOC ($mg\,L^{-1}$)	–	20,137
Hydrocarbons ($mg\,L^{-1}$)	–	–
Conductivity ($mS\,cm^{-1}$)	129.20	208.44
Alkalinity, $CaCO_3$ ($mg\,L^{-1}$)	52,266	–
Total dissolved solid ($mg\,L^{-1}$)	–	169,680
Total suspended solids ($mg\,L^{-1}$)	–	3940
Toxicity (%)	100	100
F^- ($mg\,L^{-1}$)	50	135
Cl^- ($mg\,L^{-1}$)	86	54,900
Br^- ($mg\,L^{-1}$)	12	<0.30
NO_3^- ($mg\,L^{-1}$)	11	<0.25
PO_4^{3-} ($mg\,L^{-1}$)	381	<0.30
SO_4^{2-} ($mg\,L^{-1}$)	608	1882
ORP (mV)	−334.20	–
Turbidity (NTU)	–	–

The voltammograms obtained for both HCl and H_2SO_4 (pH 1) showed an overpotential difference of 0.5 V vs $Hg|Hg_2SO_4$. This was congruent with the $^{\bullet}OH$ analysis, demonstrating a greater production in NaOH/H_2SO_4 (pH 1). Before conducting the degradation of the spent caustic mixture and to validate the correct performance of the BDD, a simple sample was electrolyzed. The electrolysis conditions were $j = 0.32\,A\,cm^{-2}$, pH 1 (adjustment with H_2SO_4), and $t_r = 12\,h$, finding that at pH 1 (H_2SO_4) the degradation occurred at 100% while for the mixture the destruction percentages of 90.40% and 85.21% were obtained for COD and TOC, respectively.

In our study, and considering the presence of chlorides, the use of UV light ($\lambda = 254\,nm$) was also evaluated to favor the high production of the $^{\bullet}OH$ (Eq. 18) through photochemical reactions. However, the results were not conclusive due to the complex composition of the spent caustic mixture.

$$HOCl + hv \rightarrow {}^{\bullet}OH + {}^{\bullet}Cl \quad (18)$$

Under the best operating conditions, a spent caustic mixture (100 mL) was treated at pH 1/H_2SO_4, $j = 0.32\,A\,cm^{-2}$, and $t_r = 16\,h$, obtaining similar results as those from the degradation of the simple sample. The TOC removal percentage reached was 88.46%. In this analysis, the removal was considered due to the chemical pretreatment (CAT) evaluated (Table 3).

Desalter Effluent

The effluents from the desalination process, known as desalter effluent, are characterized by having inorganic salts, sodium chlorides, magnesium, and calcium in their compositions [69]. All of these must be removed to avoid and prevent the plugging and fouling of process equipment by salt deposition, and to reduce corrosion caused by the formation of HCl from the chloride salts during crude processing. To carry out the desalination, chemical [69] and electrostatic [70] processes are used. However, the waters used can become contaminated by the different hydrocarbons contained in drilling muds. In the case of chemical treatment, the chemical surfactants (demulsifiers) can also contribute to the contamination. In both processes, other pollutants can be added, such as ammonia, caustic or acid for the pH adjustment. The cations concentrations in desalter effluents show values of 65,633, 8350, 1000, and 12.5 $mg\,L^{-1}$ for Na^+, Ca^{2+}, Mg^{2+},

TABLE 3 Electrochemical Incineration Efficiency of Spent Caustics Using Ti/BDD

Sample	TOC ($mg\,L^{-1}$)	TOC removal (%)	Total removal efficiency, IS-EI (%)
Initial sample (IS)	20,140	0	0
Chemical acidification treatment (CAT)	15,700	22	–
Electrochemical incineration (EI)	2320	85	88.5

and Fe^{2+}, respectively, while anions show values around 118,925, 153, and 216 mg L^{-1} for Cl^-, HCO_3^-, and SO_4^{2-}, respectively [69]. The values of COD range from 345 to 12,000 mg L^{-1}, TDS is 3475.94 mg L^{-1}, and total hardness ($CaCO_3$) of 510.33–9500 mg L^{-1} [71,72]. Among the methods used to treat waters from the desalination process of crude oil, there are API separators, hydrocyclones [73], membrane distillation [69], inverse ultrafiltration/osmosis [71], and nanofiltration [72].

It is important to emphasize that due to the quality of the crude processed and the nature of the desalination process, the physicochemical composition of the effluents completely differs from those identified in water produced and spent caustics. In our group, we conducted the EO of waters from the desalination process of crude oil, using Si/BDD as the anode and applying $j = 100$ mA cm^{-2} to different pH values (8.5, 7.0, and 3.0). Fig. 6 shows the different profiles of TOC in relation to the pH (normalized values, TOC_t/TOC_0).

It is clearly observed that there is no significant difference with regards to the pH in the long term ($t_r > 2$ h). Here, it was observed that the pH values changed drastically to acidic values during the first 2 min of electrolysis. Due to the high content of chlorides, a change in the reaction mechanism via indirect to mediated was observed. The efficiency obtained in the degradation process was related to the production of hypochlorous acid and oxychlorine radicals. It is important to mention that currently, we continue working on the electrochemical treatment of these types of effluents, applying EC as a pretreatment system and reaching removal percentages around 40% in terms of COD, using not only Fe

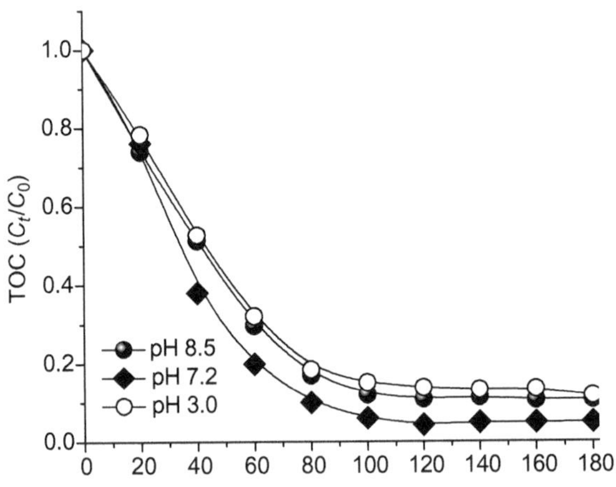

FIG. 6 Effect of pH and reaction time in electrolysis of waters from the desalination process of crude oil using Si/BDD at different pH values of 8.5, 7.2, and 3.0. Conditions: $j = 100$ mA cm^{-2}, 25°C, 6 rpm. The pH was adjusted with concentrated H_2SO_4, $t_r = 3$ h.

anodes but also Al anodes, both with different cathode materials. The results obtained in individual studies or coupled (EC-EO with BDD or Ti/MO_x) will be published at a later date [32].

CONCLUSIONS

In this chapter, results have been presented regarding the electrochemical treatment of spent caustics (diluted and concentrated) in simple samples and highly toxic mixtures as well as results of electrolysis in desalter effluent, which are two of the main effluents highly contaminated by hydrocarbons, phenolic compounds, and the presence of salts. It is important to mention that spent caustics possess an elevated hazard related to their physicochemical composition, which can exceed 11,000 $mg L^{-1}$ of phenolic content. The results presented prove the feasibility of the use of EO with BDD (incineration and mediated oxidation) to efficiently treat the effluents produced by the hydrocarbon industry.

References

[1] F. Benyahia, M. Abdulkarim, A. Embaby, M. Rao, (2006) The 7th Annual U.A.E. University Research Conference, Al Ain, UAE.
[2] P.L. Dold, Current practice for treatment of petroleum refinery wastewater and toxics removal, Water Poll. Res. J. Canada 24 (1989) 363–390.
[3] M. Al Zarooni, W. Elshorbagy, Characterization and assessment of Al Ruwais refinery wastewater, J. Hazard. Mater. A136 (2006) 398–405.
[4] H. Wake, Oil refineries: a review of their ecological impacts on the aquatic environment, Estuar. Coast. Shelf Sci. 62 (2005) 131–140.
[5] A. Alva-Argáez, A.C. Kokossis, R. Smith, The design of water-using systems in petroleum refining using a water-pinch decomposition, Chem. Eng. J. 128 (2007) 33–46.
[6] A. Coelho, A.V. Castro, M. Dezotti, G.L. Sant'Anna Jr., Treatment of petroleum refinery sourwater by advanced oxidation processes, J. Hazard. Mater. B137 (2006) 178–184.
[7] B.H. Diya'uddeen, W.M.A.W. Daud, A.R. Abdul Aziz, Treatment technologies for petroleum refinery effluents: a review, Process. Saf. Environ. Prot. 89 (2011) 95–105.
[8] Petróleos Mexicanos, PEMEX, Informe de Sustentabilidad, México, DF, (2014).
[9] P. Stepnowski, E.M. Siedlecka, P. Behrend, B. Jastorff, Enhanced photo-degradation of contaminants in petroleum refinery wastewater, Water Res. 36 (2002) 2167–2172.
[10] Y. Yavuz, A. Savaş Koparal, Ü. Bakır Öğütveren, Treatment of petroleum refinery wastewater by electrochemical methods, Desalination 258 (2010) 201–205.
[11] B. Mrayyan, M.N. Battikhi, Biodegradation of total organic carbons (TOC) in Jordanian petroleum sludge, J. Hazard. Mater. B120 (2005) 127–134.
[12] H. Jiang, Y. Fang, Y. Fu, Q.-X. Guo, Studies on the extraction of phenol in wastewater, J. Hazard. Mater. 101 (2003) 179–190.
[13] Y. Han, X. Quan, S. Chen, H. Zhao, C. Cui, Y. Zhao, Electrochemically enhanced adsorption of phenol on activated carbon fibers in basic aqueous solution, J. Colloid Interface Sci. 299 (2006) 766–771.

[14] A. Nuhoglu, B. Yalcin, Modelling of phenol removal in a batch reactor, Process Biochem. 40 (2005) 1233–1239.
[15] H. Ma, X. Zhang, Q. Ma, B. Wang, Electrochemical catalytic treatment of phenol wastewater, J. Hazard. Mater. 165 (2009) 475–480.
[16] Y. Liu, Simultaneous oxidation of phenol and reduction of Cr (VI) induced by contact glow discharge electrolysis, J. Hazard. Mater. 168 (2009) 992–996.
[17] A. Medel, E. Bustos, K. Esquivel, L.A. Godínez, Y. Meas, Electrochemical incineration of phenolic compounds from the hydrocarbon industry using boron-doped diamond electrodes, Int. J. Photoenergy 12 (2012) 1–6.
[18] H. Polat, M. Molva, M. Polat, Capacity and mechanism of phenol adsorption on lignite, Int. J. Miner. Process. 79 (2006) 264–273.
[19] A. Bódalo, J.L. Gómez, M. Gómez, G. León, A.M. Hidalgo, M.A. Ruíz, Phenol removal from water by hybrid processes: study of the membrane process step, Desalination 223 (2008) 323–329.
[20] A.V. Radushev, A.V. Plotnikov, V.N. Tyryshkina, Regeneration methods of decontamination of phenol-containing waste waters, Theor. Found. Chem. Eng. 42 (2008) 781–794.
[21] X. Hao, M. Pritzker, X. Feng, Use of pervaporation for the separation of phenol from dilute aqueous solutions, J. Membr. Sci. 335 (2009) 96–102.
[22] W.H. Glaze, J.W. Kang, D.H. Chapin, The chemistry of water treatment processes involving ozone, hydrogen peroxide and ultraviolet radiation, Ozone Sci. Eng. 9 (1987) 335–352.
[23] O. Legrini, E. Oliveros, A.M. Braun, Photochemical processes for water treatment, Chem. Rev. 93 (1993) 671–698.
[24] R.W. Matthews, Kinetics of photocatalytic oxidation of organic solutes over titanium dioxide, J. Catal. 111 (1988) 264–272.
[25] E. Brillas, I. Sirés, M.A. Oturan, Electro-Fenton process and related electrochemical technologies based on Fenton's reaction chemistry, Chem. Rev. 109 (2009) 6570–6631.
[26] I. Sirés, E. Brillas, M.A. Oturan, M.A. Rodrigo, M. Panizza, Electrochemical advanced oxidation processes: today and tomorrow: a review, Environ. Sci. Pollut. Res. 21 (2014) 8336–8367.
[27] P. Cañizares, A. Beteta, C. Sáez, L. Rodríguez, M.A. Rodrigo, Use of electrochemical technology to increase the quality of the effluents of bio-oxidation processes: a case studied, Chemosphere 72 (2008) 1080–1085.
[28] T. Rodríguez, D. Botelho, E. Cleto, Treatment of industrial effluents of recalcitrant nature using ozone, hydrogen peroxide and ultraviolet radiation, Rev. Fac. Ing. Univ. Antioquia 46 (2008) 24–38.
[29] M. Hassan, X. Wang, F. Wang, D. Wu, A. Hussain, B. Xie, Coupling ARB-based biological and photochemical (UV/TiO_2 and $UV/S_2O_8^{2-}$) techniques to deal with sanitary landfill leachate, Waste Manag. 63 (2016) 292–298.
[30] C. Comninellis, Electrocatalysis in the electrochemical conversion/combustion of organic pollutants for waste water treatment, Electrochim. Acta 39 (1994) 1857–1862.
[31] C.A. Martínez-Huitle, S. Ferro, Electrochemical oxidation of organic pollutants for the wastewater treatment: direct and indirect processes, Chem. Soc. Rev. 35 (2006) 1324–1340.
[32] A. Kraft, Doped diamond: a compact review on a new, versatile electrode material, Int. J. Electrochem. Sci. 2 (2007) 355–385.
[33] C.A. Martínez-Huitle, M.A. Quiroz Alfaro, Recent environmental applications of diamond electrode: critical review, J. Environ. Eng. Manag. 18 (2008) 155–172.
[34] G. Chen, Electrochemical technologies in wastewater treatment, Sep. Purif. Technol. 38 (2004) 11–41.
[35] D. Gandini, E. Mahé, P.-A. Michaud, W. Haenni, A. Perret, C. Comninellis, Oxidation of carboxylic acids at boron-doped diamond electrodes for wastewater treatment, J. Appl. Electrochem. 30 (2000) 1345–1350.

[36] P.-A. Michaud, M. Panizza, L. Ouattara, T. Diaco, G. Foti, C. Comninellis, Electrochemical oxidation of water on synthetic boron-doped diamond thin film anodes, J. Appl. Electrochem. 33 (2003) 151–154.
[37] J.-F. Zhi, H.-B. Wang, T. Nakashima, T.N. Rao, A. Fujishima, Electrochemical incineration of organic pollutants on boron-doped diamond electrode, evidence for direct electrochemical oxidation pathway, J. Phys. Chem. B 107 (2003) 13389–13395.
[38] B. Marselli, J. Garcia-Gomez, P.A. Michaud, M.A. Rodrigo, C. Comninellis, Electrogeneration of hydroxyl radicals on boron-doped diamond electrodes, J. Electrochem. Soc. 150 (2003) (2003) D79–D83.
[39] A. Medel, E. Bustos, L. Apátiga, Y. Meas, Surface activation of C–sp^3 in boron-doped diamond electrode, Electrocatalysis 4 (2013) 189–195.
[40] N. Sonoyama, T. Sakata, Electrochemical decomposition of CFC-12 using gas diffusion electrodes, Environ. Sci. Technol. 32 (1998) 375–378.
[41] S. Stucki, R. Kötz, B. Carcer, W. Suter, Electrochemical waste water treatment using high overvoltage anodes, Part II: anode performance and applications, J. Appl. Electrochem. 21 (1991) 99–104.
[42] M.O. Azzam, Y. Tahboub, M. Al-Tarazi, Effect of counter electrode material on the anodic destruction of 4-Cl phenol solution, IChemE 77 (1999) 219–226.
[43] N. Borrás, R. Oliver, C. Arias, E. Brillas, Degradation of atrazine by electrochemical advanced oxidation processes using a boron-doped diamond anode, J. Phys. Chem. A 114 (2010) 6613–6621.
[44] J.-M. Noël, A. Latus, C. Lagrost, E. Volanschi, P. Hapiot, Evidence for OH radical production during electrocatalysis of oxygen reduction on Pt surfaces: consequences and application, J. Am. Chem. Soc. 134 (2012) 2835–2841.
[45] B.-T. Zhang, L.-X. Zhao, J.-M. Lin, Study on superoxide and hydroxyl radicals generated in indirect electrochemical oxidation by chemiluminescence and UV-visible spectra, J. Environ. Sci. 20 (2008) 1006–1011.
[46] K. Imamura, Y. Tada, H. Tanaka, T. Sakiyama, K. Nakanishi, Removal of proteinaceous soils using hydroxyl radicals generated by the electrolysis of hydrogen peroxide, J. Colloid Interface Sci. 250 (2002) 409–414.
[47] R. Kotz, S. Stucki, B. Carcer, Electrochemical waste water treatment using high overvoltage anodes, Part I: physical and electrochemical properties of SnO_2 anodes, J. Appl. Electrochem. 21 (1991) 14–20.
[48] C. Comninellis, A. Nerini, Anodic oxidation of phenol in the presence of NaCl for wastewater treatment, J. Appl. Electrochem. 25 (1995) 23–28.
[49] Y.J. Feng, X.Y. Li, Electro-catalytic oxidation of phenol on several metal-oxide electrodes in aqueous solution, Water Res. 37 (2003) 2399–2407.
[50] E. Chatzisymeon, S. Fierro, I. Karafyllis, D. Mantzavinos, N. Kalogerakis, A. Katsaounis, Anodic oxidation of phenol on Ti/IrO_2 electrode: experimental studies, Catal. Today 151 (2010) 185–189.
[51] X.-Y. Li, Y.-H. Cui, Y.-J. Feng, Z.-M. Xie, J.-D. Gu, Reaction pathways and mechanisms of the electrochemical degradation of phenol on different electrodes, Water Res. 39 (2005) 1972–1981.
[52] F. Beck, W. Kaiser, H. Krohn, Boron doped diamond (BDD)-layers on titanium substrates as electrodes in applied electrochemistry, Electrochim. Acta 45 (2000) 4691–4695.
[53] M.A. Quiroz Alfaro, S. Ferro, C.A. Martínez-Huitle, Y. Meas Vong, Y. Boron, Doped diamond electrode for the wastewater treatment, J. Braz. Chem. Soc. 17 (2006) 227–236.
[54] X. Chen, G. Chen, F. Gao, P.-L. Yue, High-performance Ti/BDD electrodes for pollutant oxidation, Environ. Sci. Technol. 37 (2003) 5021–5026.
[55] G. Zhao, S. Shen, M. Li, M. Wu, T. Cao, D. Li, The mechanism and kinetics of ultrasound-enhanced electrochemical oxidation of phenol on boron-doped diamond and Pt electrodes, Chemosphere 73 (2008) 1407–1413.

[56] A.M. Zanbotto Ramalho, C.A. Martínez-Huitle, D. Ribeiro da Silva, Application of electrochemical technology for removing petroleum hydrocarbons from produced water using a DSA-type anode at different flow rates, Fuel 89 (2010) 531–534.
[57] J.H. Bezerra Rocha, M.M. Soares Gomes, N. Suely Fernandes, D. Ribeiro da Silva, C. A. Martínez-Huitle, Application of electrochemical oxidation as alternative treatment of produced water generated by Brazilian petrochemical industry, Fuel 96 (2012) 80–87.
[58] Y. Yavuz, A. Savaş Koparal, Electrochemical oxidation of phenol in a parallel plate reactor using ruthenium mixed metal oxide electrode, J. Hazard. Mater. 136 (2006) 296–302.
[59] D. Rajkumar, K. Palanivelu, Electrochemical treatment of industrial wastewater, J. Hazard. Mater. 113 (2004) 123–129.
[60] A. Dimoglo, H.Y. Akbulut, F. Cihan, M. Karpuzcu, Petrochemical wastewater treatment by means of clean electrochemical technologies, Clean Technol. Environ. 6 (2004) 288–295.
[61] I. Ben Hariz, A. Halleb, N. Adhoum, L. Monser, Treatment of petroleum refinery sulfidic spent caustic wastes by electrocoagulation, Sep. Purif. Technol. 107 (2013) 150–157.
[62] A. Olmos, P. Olguin, C. Fajardo, E. Razo, O. Monroy, Physicochemical characterization of spent caustic from the oximer process and sour waters from Mexican oil refineries, Energy Fuel 18 (2004) 302–304.
[63] S.-H. Sheu, H.-S. Weng, Treatment of olefin plant spent caustic by combination of neutralization and Fenton reaction, Water Res. 35 (2001) 2017–2021.
[64] A.L.N. Mota, L.F. Albuquerque, L.T.C. Beltrame, O. Chiavone-Filho, A. Machulek, C.A. O. Nascimento, Advanced oxidation processes and their application in the petroleum industry: a review, Braz. J. Petroleum Gas 2 (2008) 122–142.
[65] D. Wu, M. Liu, D. Dong, X. Zhou, Effects of some factors during electrochemical degradation of phenol by hydroxyl radicals, Microchem. J. 85 (2007) 250–256.
[66] D. Reyter, D. Bélanger, L. Roué, Nitrate removal by a paired electrolysis on copper and Ti/IrO_2 coupled electrodes—influence of the anode/cathode surface area ratio, Water Res. 44 (2010) 1918–1926.
[67] A. Medel, E. Méndez, J.L. Hernández-López, J.A. Ramírez, J. Cárdenas, R.F. Frausto, L. A. Godínez, E. Bustos, Y. Meas, Novel electrochemical treatment of spent caustic from the hydrocarbon industry using Ti/BDD, Int. J. Photoenergy 15 (2015) 1–18.
[68] Official Mexican Regulation, NOM-001-SEMARNAT-1996: Maximum Limits Allowed of Pollutants in Discharges of Wastewaters and National Resources, 1996.
[69] A. Pak, T. Mohammadi, Wastewater treatment of desalting units, Desalination 222 (2008) 249–254.
[70] E. Aryafard, M. Farsi, M.R. Rahimpour, S. Raeissi, Modeling electrostatic separation for dehydration and desalination of crude oil in an industrial two-stage desalting plant, J. Taiwan Inst. Chem. Eng. 58 (2016) 141–147.
[71] S. Norouzbahari, R. Roostaazad, M. Hesampour, Crude oil desalter effluent treatment by a hybrid UF/RO membrane separation process, Desalination 238 (2009) 174–182.
[72] S. Dadari, M. Rahimi, S. Zinadini, Crude oil desalter effluent treatment using high flux synthetic nanocomposite NF membrane-optimization by response surface methodology, Desalination 377 (2016) 34–46.
[73] Z.-S. Bai, H.-L. Wang, Crude oil desalting using hydrocyclones, Chem. Eng. Res. Des. 85 (2007) 1586–1590.

Further Reading

[1] P.K. Bachmann, D. Leers, H. Lydtin, Towards a general concept of diamond chemical vapour deposition, Diamond Relat. Mater. 1 (1991) 1–12.
[2] M. Panizza, G. Cerisola, Direct and mediated anodic oxidation of organic pollutants, Chem. Rev. 109 (2009) 6541–6569.

CHAPTER

15

Application of Electrochemical Processes for Treating Effluents From Landfill Leachate as Well as the Agro and Food Industries

Carolina Espinoza-Cisternas, Ricardo Salazar
University of Santiago of Chile, Santiago, Chile

OUTLINE

INTRODUCTION

On a worldwide basis, water at an industrial level is used as a raw material, coolant, solvent, transport agent, and/or energy transfer fluid. The large volumes of wastewater are always a critical problem in industrial manufacturing. Several wastewater treatment methods are available to ensure good quality effluent before disposal into municipal sewer systems or the environment. Wastewaters vary in terms of pollutant composition, depending on the origin of the industry. The composition of wastewater

Electrochemical Water and Wastewater Treatment
https://doi.org/10.1016/B978-0-12-813160-2.00015-8

originating from the agro and food industries as well as from landfill leachate varies widely in terms of the industrial processes (agro and foods) and the type of waste present in the leachate. It usually contains high concentrations of organic matter (BOD_5/COD ratio) due to the presence of proteins, sugars, oils, and greases, in addition to high pH, nonbiodegradable matter, toxic substances, detergents and soaps, sulfide, sodas, and heavy metals [1,2]. These kinds of wastewaters mostly have both dissolved and suspended material.

The disposal of such effluents in the environment will lead to surface and groundwater contamination, generating an increase in the chemical oxygen demand, eutrophication, ecosystem imbalance, and risks to human health [3,4]. For the elimination of organic wastes and content, biological treatment (aerobic digestion and anaerobic digestion) of industrial wastewaters is usually applied. In addition to these treatments, other conventional treatments are used at the industrial level for treating wastewaters such as adsorption onto inorganic and organic supports (usually with activated carbon), flocculation and sedimentation, flotation, membrane processes (micro-, ultra-, and nanofiltration), and coagulation with lime, aluminum, or iron salts [5]. Furthermore, chemical methods for water disinfection such as ozonation [6] and chlorination [7] are also used. The low efficiency of conventional water treatments for the degradation of many toxic and biorecalcitrant organic pollutants has promoted research on alternative methods. In this way, electrochemical technology has received great attention in the prevention of pollution problems, as previously reported in several books and reviews [8,9].

A potential way to obtain highly efficient decontamination of wastewater with organic pollutants consists of using electrochemical technologies. The main electrochemical procedures used for the remediation of organic pollutants in wastewaters are the so-called electrochemical advanced oxidation processes (EAOPs) [10], including anodic oxidation (AO) and electro-Fenton (EF) as well as photoassisted systems such as photoelectro-Fenton (PEF) and photoelectrocatalysis (PEC). On the other hand, the electrocoagulation (EC) process is not an oxidation method but an electrochemical process that allows higher removal of persistent organics from industrial wastewater. The chemical fundamentals of electrochemical technologies and their application in the decontamination of industrial wastewater from agro and food processes and from landfill leachate are presented below.

Anodic Oxidation (AO)

This methodology is called the "electrochemical combustion" method because organic pollutants are completely mineralized, that is, oxidized to CO_2, water, and inorganic salts, with physisorbed hydroxyl radical ($^{\bullet}OH$) electrogenerated from water discharge at the anode [11,12]:

$$M + H_2O \rightarrow M(^{\bullet}OH) + H^+ + e^- \quad (1)$$

This radical is the second strongest oxidant known after fluorine and has a high standard potential ($E^0 = 2.80$ V vs SHE) that ensures its fast reaction with most organics, producing dehydrogenated or hydroxylated derivatives up to conversion into CO_2.

Both the electrochemical and chemical reactivity of heterogeneous M($^{\bullet}$OH) are dependent on the nature of the electrode material [13]. Thus, a nonactive anode is required that interacts so weakly with $^{\bullet}$OH so as to allow the direct reaction of organics with M($^{\bullet}$OH) and yield fully oxidized reaction products such as CO_2:

$$aM(^{\bullet}OH) + R \rightarrow aM + mCO_2 + nH_2O + xH^+ + ye^- \quad (2)$$

where R is an organic compound with m carbon atoms and without any heteroatom, which needs $a = (2m + n)$ oxygen atoms to be totally mineralized to CO_2.

Among the anodes currently used in direct and indirect AO dimensionally, we can find stable anodes (DSA), titanium-supported platinum (Ti/Pt), PbO_2, and diamond electrodes [14,15]. Thin-film boron doped diamond (BDD) electrodes are preferred because the weak BDD-$^{\bullet}$OH interaction and higher O_2-overpotential leads to the generation of higher amounts of reactive physisorbed BDD($^{\bullet}$OH) radicals that mineralize aromatics and aliphatic contaminants more effectively than other anodes such as Pt and PbO_2 [16–18]. The BDD anode is the best nonactive electrode that exhibits this behavior, making it the preferable anode for treating organics by AO [19,20].

For an indirect AO, besides hydroxyl radical produced from water discharge, other stronger oxidants can be produced depending on the salts dissolved in the wastewater (solution). Peroxodisulphate is formed from the AO of bisulphate (or sulfate) by an indirect oxidation process where the $^{\bullet}$OH produced on the anode surface can react via homogeneous reaction with the dissolved sulfate salt [21], following mediated surface chemical steps that are responsible for the formation of persulfate ($S_2O_8^{2-}$):

$$HSO_4^- + {^{\bullet}OH} \rightarrow SO_4^{-\bullet} + H_2O \quad (3)$$

$$SO_4^{2-} + {^{\bullet}OH} \rightarrow SO_4^{-\bullet} + OH^- \quad (4)$$

$$SO_4^{-\bullet} + SO_4^{-\bullet} \rightarrow S_2O_8^{2-} \quad (5)$$

The contribution of the $^{\bullet}$OH present on the anode surface/reaction cage as well as the interaction of $^{\bullet}$OH with HSO_4^- and SO_4^{2-} anions favors the production of $S_2O_8^{2-}$ by the formation of $SO_4^{-\bullet}$. Both species act as strong oxidants for efficiently removing the organic matter in the solution by:

$$SO_4^{-\bullet} + \text{Organics} \rightarrow \text{Oxidation products} \rightarrow CO_2 + H_2O + SO_4^{2-} \quad (6)$$

$$S_2O_8^{2-} + \text{Organics} \rightarrow \text{Oxidation products} \rightarrow CO_2 + H_2O + 2SO_4^{2-} \quad (7)$$

In addition, active chlorine species can be generated from direct oxidation of chloride ion, promoting the subsequent organic oxidation steps. Chlorine and chlorine-oxygen species [22] have an oxidant power lower than that of $^{\bullet}OH$, but with an interesting activity. Between the main active chlorine species we can find chlorine and hypochlorite, produced by the following reactions [23]:

$$2Cl^- \rightarrow Cl_{2(aq)} + 2e^- \quad (8)$$

$$Cl_{2(aq)} + H_2O \leftrightarrows HClO_{(aq)} + H^+_{(aq)} + Cl^-_{(aq)} \quad (9)$$

$$HClO_{(aq)} \leftrightarrows ClO^-_{(aq)} + H^+_{(aq)} \quad (10)$$

Several types of wastewater from the agro and food industries and from landfill leachate have been treated by AO in recent years. The assays were performed at wastewater with different compositions, with different anodes, and in some cases without the addition of chemicals as supporting electrolytes or buffers (Table 1).

In all cases, COD or TOC decreased progressively with prolonged electrolysis time, and this decay was dependent on the nature of wastewater, applied current density, area of the electrode, and initial pH. For each type of wastewater, higher COD or TOC decay was achieved after 2 or 3 h of electrolysis. The presence of NaCl increased the efficiency of the removal of organics as in the case of Un et al. [32] and Cañizares et al. [33], achieving a complete mineralization of the organics from olive mill wastewater. Otherwise, it is possible to apply the anodic oxidation to industrial wastewater without the addition of supporting electrolytes as the industrial procedures included the presence of salts at different concentration scales. Anglada et al. [25] and Moraes et al. [26] reached >50% of TOC abatement during the AO of landfill leachate without the addition of any salts to the wastewater. The production of hydroxyl radicals and chlorine-active species reacts with the organic compounds contained in the wastewater, producing a decrease in COD or TOC and generating carbon dioxide and inorganic ions such as nitrate and ammonium.

Different electrochemical systems were used during the treatment of wastewater by anodic oxidation. For the treatment of landfill leachate, Moraes et al. [26] used a single compartment tubular flow reactor (Fig. 1A) formed by an anode of 1.0 m long titanium tube coated with an oxide layer composed of 7:3 TiO_2:RuO_2 (*w*/*w*). Concentric to the anode, a titanium tube cathode with a Ti screen was welded to increase the area and promote turbulence. The anode/cathode gap between electrodes was 3 mm. Panizza et al. [28] (Fig. 1B) performed the anodic oxidation of wastewater from the olive mill industry in a one-compartment flow cell with parallel plate electrodes. A mixed titanium and ruthenium oxide

TABLE 1 Percentage of COD or TOC Removal Together With Color or Turbidity Decay, Using Different Anode Surfaces for the Treatment of Real Industrial Wastewater

Wastewaters	Electrode	Water characteristics	Treatment time	Initial TOC or COD concentration (mg L^{-1})	% COD or TOC removal	% CR or TR	Ref
Distillery	Anode: $Ti/Ti_{0.7}$-$Ru_{0.05}$-$Ir_{0.25}O_2$, cathode: stainless steel	6 g L^{-1} NaCl, pH6.1	480 min	80,000–90,000	87	–	[24]
Landfill leachate	Anode and cathode: Si-BDD	No addition of supporting electrolyte, pH5	480 min	3358	51	84	[25]
Landfill leachate	Anode: Ti/Ti_{70}-$Ru_{30}O_2$, cathode: Ti	No addition of supporting electrolyte, natural pH (7–9.5)	180 min	1855	73	86	[26]
Landfill leachate	Anode: BDD, cathode: stainless steel	0.03 M Na_2SO_4, pH8.4	360 min	13,700	88	87	[27]
Landfill leachate	Anode: Si-BDD, cathode: stainless steel	No addition of supporting electrolyte, natural pH (8.2)	480 min	780	100	100	[28]
Olive mill	Anode: Ti/Ti-RuO_2, cathode: stainless steel	5 g L^{-1} NaCl, pH5	480 min	26,750	100	100	[29]
Olive mill	Anode and cathode: Ti	No addition of supporting electrolyte, natural pH (8.3)	600 min	65,000	87	55	[30]

Continued

TABLE 1 Percentage of COD or TOC Removal Together With Color or Turbidity Decay, Using Different Anode Surfaces for the Treatment of Real Industrial Wastewater—cont'd

Wastewaters	Electrode	Water characteristics	Treatment time	Initial TOC or COD concentration (mg L^{-1})	% COD or TOC removal	% CR or TR	Ref
Olive mill	Anode: Ti/IrO_2, cathode: Zr in a porous porcelain pot	0.3 g L^{-1} NaCl, natural pH (4.4)	120 min	1300	30	100	[31]
Olive mill	Anode and cathode: Ti/RuO_2	116 g L^{-1} NaCl, natural pH (4.6)	480 min	41,000	100	100	[32]
Olive mill	Anode: Si-BDD, cathode: stainless steel	No addition of supporting electrolyte, natural pH (6.0)	360 min	3000	100	–	[33]
Olive mill	Anode: Ti/RuO_2, cathode: Pt	1.2 g L^{-1} NaCl, no pH reported	24 h	40,000	54	100	[34]
Olive pomace	Anode and cathode: BDD	No addition of supporting electrolyte, natural pH7.6	360 min	3524	82	–	[35]
Almond	Anode: Ti/RuO_2, cathode: stainless steel	2 g L^{-1} NaCl, natural pH (9.0)	270 min	2000	70	–	[36]
Citric acid	Anode: Ti/RuO_2-IrO_2, cathode: stainless steel	3 g L^{-1} NaCl, natural pH (8.1)	180 min	349	85	100	[37]

COD, chemical oxygen demand; *CR*, color removal; *TOC*, total organic carbon; *TR*, turbidity removal.

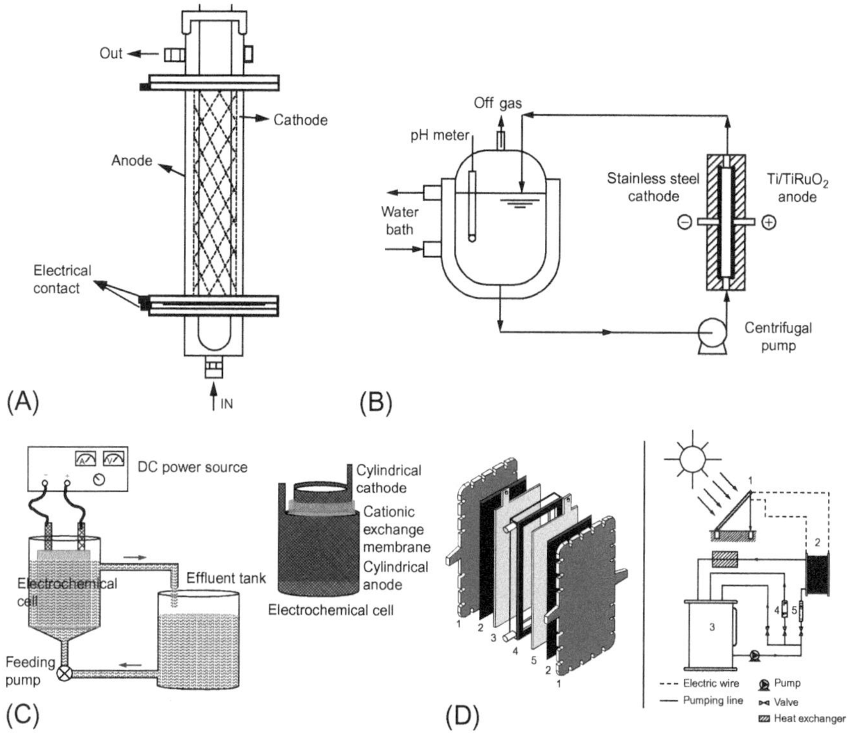

FIG. 1 Electrochemical reactors used during the treatments of wastewater by anodic oxidation. (A) Single compartment tubular flow reactor. (B) An one-compartment flow cell with parallel plate electrodes. (C) A cylindrical reactor operated in a flow mode. (D) Flow electrochemical reactor containing a nondivided electrochemical filter press reactor. *(A) Reproduced with permission from P. Moraes, R. Bertazzoli, Electrodegradation of landfill leachate in a flow electrochemical reactor, Chemosphere 58 (2005) 41–46 Copyright 2005. (B) from M. Panizza, C. Martínez-Huitle, Role of electrode materials for the anodic oxidation of a real landfill leachate—comparison between Ti–Ru–Sn ternary oxide, PbO_2 and boron-doped diamond anode, Chemosphere 90 (2013) 1455–1460 Copyright 2006. (C) from C. Belaid, M. Khadraoui, S. Mseddi, M. Kallel, B. Elleuch, J. Fauvarque, Electrochemical treatment of olive mill wastewater: treatment extent and effluent phenolic compounds monitoring using some uncommon analytical tools, J. Environ. Sci. 25 (2013) 220–230 Copyright 2013. (D) from D. Valero, V. García-García, E. Expósito, A. Aldaz, V. Montiel, Electrochemical treatment of wastewater from almond industry using DSA-type anodes: direct connection to a PV generator, Sep. Purif. Technol. 123 (2014) 15–22 Copyright 2014, with permission from Elsevier.*

($Ti/TiRuO_2$) was used as the anode and a stainless steel AISI 304 as the cathode. Both electrodes were square with a geometric area of 81 cm^2 each with an interelectrode gap of 0.5 cm. Wastewater totaling 400 mL was circulated through the electrolytic cell by a centrifugal pump with a flow rate of 180 $L\,h^{-1}$. Belaid et al. [30] treated 500 mL of olive mill wastewater by AO in a cylindrical reactor operated in a flow mode (Fig. 1C). The reactor cell consisted of two cylindrical platinized titanium N type electrodes separated by a cationic exchange membrane type IONAC MC 3470 (5 mm). For the treatment of almond industry wastewater, Valero et al. [36] employed a flow electrochemical reactor containing a nondivided electrochemical filter press reactor, as shown in Fig. 1B. The electrode geometric area was 63 cm^2.

Electro-Fenton (EF)

The electro-Fenton (EF) treatment of wastewaters involves the continuous generation of H_2O_2 from dissolved O_2 directly injected as pure gas or air [38], which is efficiently reduced at different cathodes (reticulated vitreous carbon, carbon felt, graphite-polytetrafluoroethylene (PTFE), carbon-PTFE O_2-diffusion, etc.). A small catalytic amount of Fe^{2+} is then added to the acidic contaminated solution to generate Fe^{3+} and $^{\bullet}OH$ according to the classical Fenton's reaction [39] (11) with a second order rate constant (k_2) of 63 $M^{-1}\ s^{-1}$:

$$Fe^{2+} + H_2O_2 \rightarrow Fe^{3+} + {}^{\bullet}OH + OH^- \qquad (11)$$

Reaction (11) can be propagated principally by the catalytic behavior of the Fe^{2+}/Fe^{3+} system. Thus, Fe^{2+} is continuously regenerated from the reduction of Fe^{3+} at the cathode ($E^0 = 0.77$ V vs SHE) from reaction (12).

$$Fe^{3+} + e^- \rightarrow Fe^{2+} \qquad (12)$$

EF technology utilizes the same electrochemical systems as the other electrooxidation methods, but differs in the additional use of O_2 gas or air which is bubbled through the solution or directly injected into the carbonaceous cathode for producing H_2O_2 from reaction (13) [40].

$$O_2(g) + 2H^+ + 2e^- \rightarrow H_2O_2 \qquad (13)$$

The initial pH of effluents treated by EF is always regulated to a value near 3.0, close to the optimum pH of 2.8 for Fenton's reaction (11), to ensure the fastest generation of homogeneous $^{\bullet}OH$.

In order to differentiate this process from the conventional Fenton process, which uses commercial H_2O_2, the term "electro-Fenton process (EF)" is applied; it is considered an EAOP. Also, the major advantages of the electro-Fenton process include [41]: (a) H_2O_2 can be continuously generated on-site whenever needed, which eliminates purchasing, shipment, and storage; (b) a dilute H_2O_2 solution enhances safety during material handling; (c) the production process can be carried out simply at ambient pressure and temperature; (d) Fe^{2+} can be electrochemically regenerated at the cathode, which minimizes the quantity of iron sludge; and (e) oxygen or air sparing enhances the mixing of the reaction solution. On the other hand, one disadvantage of this process is that H_2O_2 will accumulate at the cathode solution interface and may be partially decomposed. Additionally, protons at a high concentration may also compete for electrons, leading to hydrogen gas evolution. Both effects will reduce the current efficiency of H_2O_2 production. Therefore, in acidic solutions, cathodic potential and solution pH are two essential factors that control current efficiency. Finally, the $^{\bullet}OH$ formed in (3) reacts with the pollutants until complete mineralization to CO_2.

The simultaneous use of electrogenerated H_2O_2 in the presence of Fe^{2+} (EF conditions) and ultraviolet (UV) illumination of the solution to mineralize pollutants is known as the photoelectro-Fenton method (PEF) [42]. The action of this irradiation is complex and can be described by: (a) the production of a greater amount of hydroxyl radicals from photoreduction of $Fe(OH)^{2+}$ and (b) the predominant Fe^{3+} species in the acid medium by Eq. (14) and/or the photolysis of complexes of Fe(III) with generated carboxylic acids, as shown in Eq. (15)

$$Fe(OH)^{2+} + h\nu \rightarrow Fe^{2+} + {}^{\bullet}OH \tag{14}$$

$$2Fe(C_2O_4)_n(3-2n) + h\nu \rightarrow 2Fe^{2+} + (2n-1)C_2O_4{}^{2-} + CO_2 \tag{15}$$

The enhancement of the degradation rate of this process with respect to EF (when the solution is irradiated by UVA light) is well recognized [43].

In some cases, dyes may be degraded by the $^{\bullet}OH$ on the surface of the anode (AO) and in the electrolytic media generated by Fenton's reaction. A cost-effective alternative consists of the use of solar irradiation as a free and renewable UV/Vis source, in the so-called solar photoelectron-Fenton process (SPEF). Several works revealed the excellent results obtained for the removal of organic pollutants [44,45].

The electrolysis time of industrial wastewater by EF and PEF decreases with respect to the electrolysis time of the anodic oxidation process. With electrochemical Fenton processes, the organic pollutants present in agro or food wastewater and landfill leachate are destroyed by hydroxyl radicals generated in situ in the polluted water. In other words, a homogeneous and fast oxidation process occurs in the bulk, which reaches higher percentages of water decontamination, removing organic pollutants from different origins. The presence of iron ions plays an important role in EF processes as it is the catalyst for the formation of $^{\bullet}OH$ (Eq. 11). The concentration of Fe^{2+} is variable and depends on the initial characteristics of wastewater and the cathode nature. In Table 2, the range is between 0.07 and 38.0 mM. When solid cathodes are used in EF or PEF methods, it is very necessary that the solution contain a high concentration of dissolved oxygen or the air/oxygen be introduced near the cathode surface for efficient peroxide production. On the other hand, by using an air/oxygen diffusion cathode, the formation of peroxide occurs during the pass of the gas through the electrode when the gas comes in direct contact with the interface cathode/solution. TOC or COD abatement higher than 65% is reached by means of the EF process, with a complete decolorization

TABLE 2 Percentage of COD or TOC Removal, Together With Color or Turbidity Decay by EF and PEF Using Different Anode Surfaces for the Treatment of Real Industrial Wastewater

Wastewater	Process	Cathode electrode	Water characteristics	Treatment time	Initial TOC or COD concentration (mg L^{-1})	% COD or TOC removal	% CR or TR	Ref
Liquid organic fertilizer	EF	Fe	No addition of supporting electrolyte, 0.9 g L^{-1} of H_2O_2, pH3	45 min	435	91	99	[46]
Alcoholic	EF	Graphite electrodes coated by Fe_2O_3 nanoparticles	No addition of supporting electrolyte, 2.3 g L^{-1} H_2O_2, pH2.8	70 min	60,000–70,000	66	–	[47]
Distillery	EF	Graphite	No addition of supporting electrolyte, air introduced near of the cathode surface, 0.07 mM Fe^{2+}, pH3	240 min	98,400	–	44	[48]
Slaughterhouse	EF	Air diffusion	No addition of supporting electrolyte, 0.7 mM Fe^{2+}, pH3.5	60 min	405.6	84	–	[49]
Poultry	EF	Fe	No addition of supporting electrolyte, 29 g L^{-1}, H_2O_2 pH3	50 min	4650	97	93	[50]
Landfill	EF	Ti/RuO_2-IrO_2	No addition of supporting electrolyte, 38 mM Fe^{2+}, 11.6 g L^{-1} H_2O_2, pH3	100 min	5000	≈85	–	[51]

TABLE 2 Percentage of COD or TOC Removal, Together With Color or Turbidity Decay by EF and PEF Using Different Anode Surfaces for the Treatment of Real Industrial Wastewater—cont'd

Wastewater	Process	Cathode electrode	Water characteristics	Treatment time	Initial TOC or COD concentration (mg L^{-1})	% COD or TOC removal	% CR or TR	Ref
Landfill	EF	Fe	No addition of supporting electrolyte, 2 g L^{-1} H_2O_2 pH3	45 min	2350	72	90	[52]
Landfill	PEF	Fe	No addition of supporting electrolyte, 3 g L^{-1} H_2O_2, pH3	45 min	2350	94	97	[53]
Olive oil mill	PEF	Air diffusion	No addition of supporting electrolyte, 0.5 mM Fe^{2+}, pH3	540 min	598	80	–	[54]
Coffee	PEF	BDD	Air introduced near of the cathode surface, 7 g L^{-1} Na_2SO_4, 0.3 mM Fe^{2+}, pH3	240 min	10,821	80	90	[55]

COD, chemical oxygen demand; *CR*, color removal; *TOC*, total organic carbon; *TR*, turbidity removal.

and removal of the turbidity of the wastewater. If the electrolytic system is irradiated by UV lamps, the mineralization is faster than by EF without irradiation for large quantities of initial organic content.

Electrochemical systems with different configurations were employed for the treatment of industrial wastewater and landfill leachate by EF and PEF. As presented in Fig. 2A, Paramo Vargas et al. [49] used a batch reactor to treat 1.0 L of slaughterhouse wastewater by the EF process, with an agitation of 350 rpm at 23°C. The system included a carbon-felt cathode (6 cm × 10 cm × 1 cm) and three different anodes: DSA of Ti-RuO_2 DSA of Ti-$PdPtO_x$, and a BDD electrode with a contact area with the solution of 60 cm^2. The aeration of the system was carried out by using a fine bubbling from the bottom of the reactor and the inside of the carbon-felt cathode. Altin et al. [53] treated 1.0 L of landill leachate by the PEF process in a quartz glass reactor equipped with a magnetic stirrer (Fig. 2B). Two pairs of the cast iron anode and cathode electrodes (4.0 cm × 5.0 cm × 0.4 cm) were positioned ~1.0 cm separate from each other. UV radiation was provided by two low-pressure lamps of 4 W. The UV intensity of the lamps at 253.7 nm was 1.4 W m^{-2}.

Electrocoagulation (EC)

A traditional physicochemical treatment of phase separation for the decontamination of wastewater before discharge to the environment is chemical coagulation. Globally, the food and agro industries use coagulation as the first option (in many cases the only option) for the treatment of their residual wastewater [56].

Coagulation involves the addition of coagulating agents such as Fe^{3+} or Al^{3+} ions, usually in the form of chlorides, for the precipitation of dyes. The electrochemical technology can produce similar effects by means of the EC method [57,58]. This technique uses a current to dissolve Fe (or steel) or Al sacrificial anodes immersed in the polluted water [59], giving rise to the corresponding metal ions

$$Fe \rightarrow Fe^{2+} + 2e^- \ (\text{standard potential } E^0 = -0.44\,\text{V vs SHE}) \tag{16}$$

$$Al \rightarrow Al^{3+} + 3e^- \ (\text{standard potential } E^0 = 1.66\,\text{V vs SHE}) \tag{17}$$

that yield different Fe(II) (and/or Fe(III)) or Al(III) species with hydroxide ion depending on the medium pH [60]. Thus, Fe (or steel) or Al anodes produce insoluble $Fe(OH)_3$ and $Al(OH)_3$ flocs, removing dissolved dyes by surface complexation or electrostatic attraction [61]. The first mechanism considers that the organic pollutant can act as a ligand to bind a hydrous iron moiety of the floc, yielding a surface complex:

$$\text{Organic} - H + (HO)OFe_{(s)} \rightarrow \text{Organic} - OFe_{(s)} + H_2O \tag{18}$$

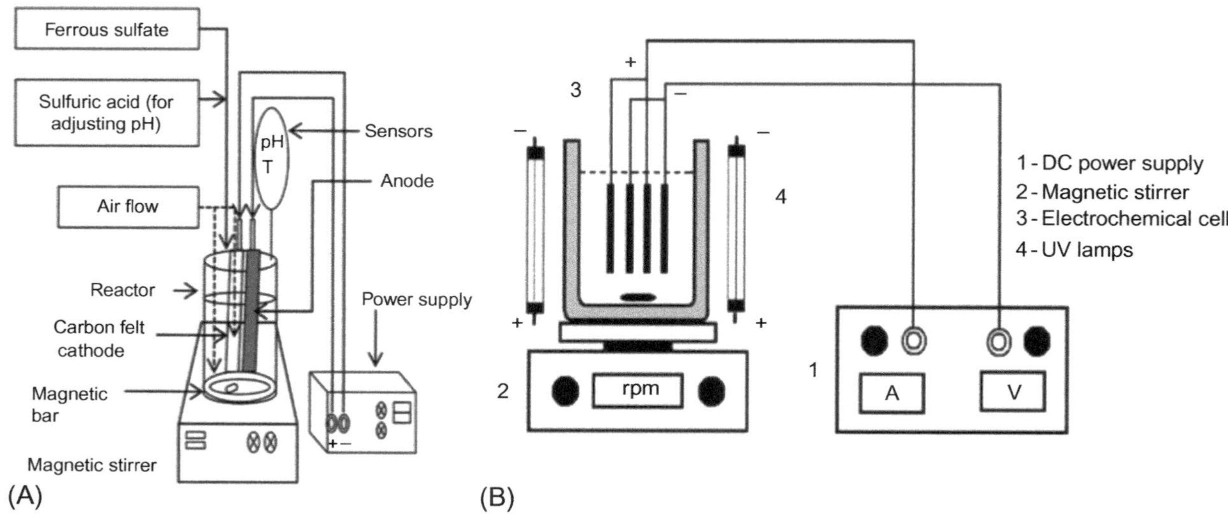

FIG. 2 Electrochemical reactors used during the treatments of industrial wastewater by the electro-Fenton process. *(A) Reproduced with permission from J. Paramo-Vargas, A. Camargo, S. Gutierrez-Granados, L. Godinez, J. Peralta-Hernandez, Applying electro-Fenton process as an alternative to a slaughterhouse effluent treatment, J. Electroanal. Chem. 754 (2015) 80–86 Copyright 2015. (B) from A. Altin, An alternative type of photoelectro-Fenton process for the treatment of landfill leachate, Sep. Purif. Technol. 61 (2008) 391–397 Copyright 2008, with permission from Elsevier.*

while the second mechanism assumes that $Fe(OH)_3$ flocs with surface complexes contain areas of apparent positive or negative charge that attract the opposite regions of the dyestuff. The coagulation of these flocs forms particles that are separated from the wastewater by sedimentation or electroflotation. Similarly to EC with Fe, the removal of pollutants from wastewaters using Al (or other metals such as Cu or alloys of Fe) can be explained by surface complexation and electrostatic attraction [62].

In summary, many advantages for EC have been reported [63,64]: (i) more effective and rapid organic matter separation than in coagulation; (ii) pH control is not necessary, except for extreme values; (iii) the amount of chemicals required is small; (iv) the amount of sludge produced is smaller when compared with coagulation. For example, the sludge formed in the EC method with Fe contains a higher content of dry and hydrophobic solids than that produced in coagulation by the action of $FeCl_3$ followed by the addition of NaOH or lime; and (v) the operating costs are much lower than in most conventional technologies.

It is not easy to establish a specific mechanism for the electrocoagulation process, even more so in the case of industrial wastewater. However, it is possible to describe the process steps in general for a given pollutant, as recently published by Zaleschi et al. for the electrocoagulation of an organic pollutant [79]. The common processes in EC are (i) the in situ formation of the coagulation agent, (ii) adsorption, (iii) complexation, (iv) flotation, and/or (v) precipitation. Finally, organic pollutants are fixed in the floc formed during electrolysis. The electrodes can be arranged in a monopolar or bipolar mode: (1) monopolar, with two electrodes in contact with the wastewater solution, cathode and anode connected to the power supply; or (2) bipolar, with three or more electrodes immersed in the solution, connected or not to the power supply.

In Table 3, the electrocoagulation of different industrial wastewater and landfill leachate is shown. Based on the results summarized there, almost a complete TOC or COD removal together with a complete decolorization and turbidity decay are achieved by means of the EC process after a short time of electrolysis. Both the presence and addition of sodium chloride (NaCl) to wastewater reduce the passivation of the working electrodes and increase the conductivity of the medium. In most works shown in Table 3, the TOC or COD decreased by >70% after 2 h of electrolysis at lower applied currents (0.5–10 mA cm^{-2}). With respect to the turbidity of the solution, during the first 20 min an increase in NTU is regularly found due to the formation of smaller flocs, as reported by Trompette et al. [64]. Then, turbidity decreases until it disappears almost entirely by the end of the electrolysis, which is in agreement with the results of decolorization and TOC and COD analysis. Finally, after the EC process, discolored water without organics (and organic matter) and with low COD and TOC loads is obtained.

TABLE 3 The Percentage of COD or TOC Removal, Together With Color or Turbidity Decay, Using Different Sacrificial Anode Surfaces for the Treatment of Real Industrial Wastewater by the EC Process

Wastewaters	Anode and cathode	Cell settings	Water characteristics	Treatment time	Initial TOC or COD concentration (mg L^{-1})	% COD or TOC removal	% CR or TR	Ref
Abattoir	Steel	Bipolar	No addition of supporting electrolyte, pH	60 min	1410	≈80	–	[65]
Dairy effluent	Al	Monopolar	4 g L^{-1} of commercial milk powder, 1.5 g L^{-1} NaCl, pH6.6	5 min	7560	80	96	[66]
Dairy effluent	Steel	Bipolar	0.3 g L^{-1} NaCl, pH7	20 min	25,300	98	–	[67]
Sugar	Copper plate/rod	Monopolar	7 g L^{-1} NaCl, pH6	120 min	3682	92	95	[68]
Sugar	Fe	Monopolar	No addition of supporting electrolyte, pH6	120 min	3682	84	86	[69]
Landfill leachate	Fe	Monopolar	2 g L^{-1} NaCl pH6.7	90 min	2566	≈50	–	[70]
Landfill leachate	Al	Monopolar	No addition of supporting electrolyte, pH8	210 min	260	56	80	[71]

Continued

TABLE 3 The Percentage of COD or TOC Removal, Together With Color or Turbidity Decay, Using Different Sacrificial Anode Surfaces for the Treatment of Real Industrial Wastewater by the EC Process—cont'd

Wastewaters	Anode and cathode	Cell settings	Water characteristics	Treatment time	Initial TOC or COD concentration (mg L^{-1})	% COD or TOC removal	% CR or TR	Ref
Rice grain-based distillery	Cu	Monopolar	No addition of supporting electrolyte, pH3.5	120 min	11,500	80	65	[72]
Rice grain-based distillery	Al	Monopolar	No addition of supporting electrolyte, pH8	120 min	13,800	93	87	[73]
Bio-digester effluent from distillery	Fe	Monopolar	No addition of supporting electrolyte, pH8	120 min	15,600	51	95	[74]
Cheese whey	Fe	Monopolar	No addition of supporting electrolyte, pH4.5	20 min	15,500	86	–	[75]
Slaughterhouse	Steel	Bipolar	No addition of supporting electrolyte 6.1–6.5	90 min	1290–1670	82	89	[76]
Poultry slaughterhouse	Al	Monopolar	No addition of supporting electrolyte, pH2	25 min	26,000–29,000	93	–	[77]
Yogurt	Fe	Monopolar	14 g L^{-1} Na_2SO_4, pH4.5	90 min	6500	≈80	–	[78]

COD, chemical oxygen demand; *CR*, color removal; *TOC*, total organic carbon; *TR*, turbidity removal.

Monopolar and bipolar configurations were employed to treat wastewater by the EC process. Sengil et al. [67] treated 0.65 L of dairy effluent in a bipolar electrochemical reactor with iron electrodes of 333 cm^2 (Fig. 3A). A stirrer was used in the electrochemical cell to maintain an unchanged composition and avoid the association of the flocs in the solution. Sahu et al. [68] treated sugar industry wastewater in an electrochemical open system in a monopolar configuration using aluminum anodes. The electrodes were spaced 20 mm for a good formation of flocs. Kobya et al. [77] decontaminated poultry slaughterhouse wastewater in a thermostated plexiglass electrocoagulator with four parallel monopolar electrodes: two anodes and two cathodes with dimensions of 46 mm × 55 mm × 3 mm, made of aluminum or iron plates. The total effective electrode area was 80 cm^2 and the spacing between electrodes was 11 mm.

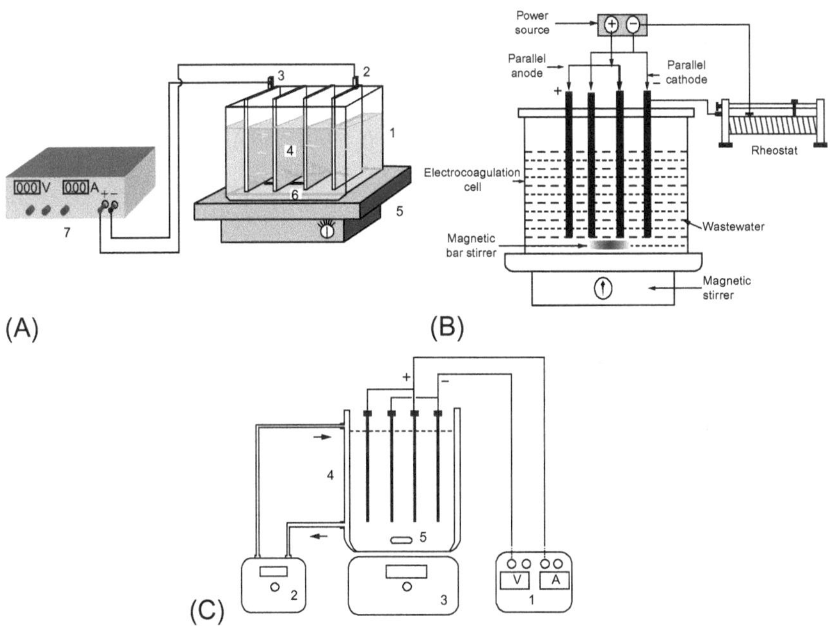

FIG. 3 Electrochemical systems used during the treatments of industrial wastewater by the electrocoagulation process. (A) Monopolar electrochemical reactor with iron electrodes. (B) Stirrer electrochemical open system in a monopolar configuration using aluminium anodes. (C) Thermostated plexiglass electrocoagulator with four parallel monopolar electrodes. *(A) Reproduced with permission from İ. Sengil, M. Özacar, Treatment of dairy wastewaters by electrocoagulation using mild steel electrodes, J. Hazard. Mater. 137 (2006) 1197–1205 Copyright 2009. (B) from O. Sahu, Performance of copper compounds in chemical and electro oxidation treatment of sugar industry waste water: batch reaction, J. Taiwan Inst. Chem. Eng. 65 (2016) 256–268 Copyright 2016. (C) from M. Kobya, E. Senturk, M. Bayramoglu, Treatment of poultry slaughterhouse wastewaters by electrocoagulation, J. Hazard. Mater. 133 (2006) 172–176 Copyright 2006, with permission from Elsevier.*

COUPLED PROCESSES

Worldwide, the vast majority of industrial water treatment plants use biological treatments to reduce the organic content in wastewater from agro or food manufacturing. For the process to be effective, microorganisms require oxygen to remove organic matter, transforming the initial organic compounds in carbon dioxide, methane, or organic compounds of low molecular weight. However, biological treatments are not capable of degrading residual bioresistant organic matter in wastewater. Furthermore, chemical processes such as oxidations or photolysis are used as tertiary treatments in the decontamination of landfill leachate or wastewater from the agro or food industries.

To reduce the times of traditional water processes or to increase the elimination of recalcitrant pollutants, in recent years the combination of traditional processes with electrochemical methods has become an interesting option for removing the organics in wastewater from the agro and food industries as well as landfill leachate.

As can be seen in Table 4, the application of electrochemical technologies before, after, or during an industrial wastewater treatment reduces the organic composition in high percentages (average > 70%). In the case of EAOPs such as AO, EF, or PEF, the mineralization of organic compounds occurs due to the generation of stronger oxidants such as hydroxyl radicals and chlorine-active species. These oxidants transform the pollutants into CO_2 and inorganic ions. When EC technology is used, the organic content is removed by the formation of flocs with a posterior separation step. In all cases, the combination between a traditional process and an electrochemical method increases water decontamination more than processes without electrochemical treatment. In the same way as electrochemical treatments previously described, the coupled processes could be effective without the addition of supporting electrolytes because the manufacturing or the previous water treatments contribute with ions such as sulfate or chloride.

Recently, Vidal et al. [80] studied the combination of an anaerobic digestion with SPEF process in the decontamination of wastewater from slaughterhouses. In this work, it was possible to remove >90% of COD through anaerobic treatment after 22 days of incubation with 90 mL of CH_4 production. After the anaerobic biological treatment, the water was treated by the SPEF process applying a current density of 30 mA cm^{-2}. The results of both processes are shown in Fig. 4. The colorless resulting water (postanaerobic digestion and SPEF) exhibited a 99% COD removal. Additionally, complete reduction of the turbidity of the solution and solids present was also achieved, showing that the combined treatment increased the COD removal. This high reduction in COD is explained

TABLE 4 Percentage of COD or TOC Removal, Together With Color or Turbidity Decay, Using Different Anode Surfaces for the Treatment of Real Industrial Wastewater by the Combination of Traditional Processes With Electrochemical Technologies

Processes	Wastewaters type	Electrode	Water characteristics	Time of electrochemical treatment	Initial TOC or COD concentration ($mg\ L^{-1}$)	% COD or TOC removal	% CR or TR	Ref.
Biological treatment-SPEF	Slaughterhouse	Anode: Si/BDD, cathode: carbon-PTFE air-diffusion	0.05 M Na_2SO_4, 1 mM Fe^{2+}, pH3	180 min	480	97	100	[80]
EC-AO	Landfill leachates	Anode and cathode: Fe (EC). Anode: Si-BDD, cathode: stainless Steel (AO)	No addition of supporting electrolyte, pH8.5	180 min for EC and then 300 for AO	21,700	95	–	[81]
Biological treatment-AO	Olive pomace leachate	Anode and cathode: Si-BDD	AO: 1.7 g L^{-1} NaCl, pH8	420 min	25,000–30,000	100	–	[82]
AO and UV irradiation	Landfill leachate	Anode: Ti/RuO_2 mesh, cathode: Ti mesh	No addition of supporting electrolyte, natural pH (7.5–8.0)	240 min	480–640	≈90	100	[83]

Continued

TABLE 4 Percentage of COD or TOC Removal, Together With Color or Turbidity Decay, Using Different Anode Surfaces for the Treatment of Real Industrial Wastewater by the Combination of Traditional Processes With Electrochemical Technologies—cont'd

Processes	Wastewaters type	Electrode	Water characteristics	Time of electrochemical treatment	Initial TOC or COD concentration (mg L^{-1})	% COD or TOC removal	% CR or TR	Ref.
AO and H_2O_2	Olive	Anode and cathode: BDD	No addition of supporting electrolyte, 0.5 g L^{-1} H_2O_2, pH3	840 min	5500	73	–	[84]
EC, EF and SBR	Landfill leachate	Anode and cathode: Fe	No addition of supporting electrolyte, 750 mg L^{-1} H_2O_2, pH4	30 min	2000	67	100	[85]
Biological treatment-SPEF	Winery	Anode: BDD, cathode: air diffusion	No addition of supporting electrolyte, 0.6 mM Fe^{2+}, pH2.8	240 min	12,000	86	–	[86]
Biological treatment-AO	Landfill leachate	Anode: Ti/Pt grid, cathode: cylindrical Ti/Pt mesh	No addition of supporting electrolyte, natural pH (7.0–7.8)	60 min	6500–8000	85	99	[87]

TABLE 4 Percentage of COD or TOC Removal, Together With Color or Turbidity Decay, Using Different Anode Surfaces for the Treatment of Real Industrial Wastewater by the Combination of Traditional Processes With Electrochemical Technologies—cont'd

Processes	Wastewaters type	Electrode	Water characteristics	Time of electrochemical treatment	Initial TOC or COD concentration (mg L^{-1})	% COD or TOC removal	% CR or TR	Ref.
AO-EF	Landfill leachate	Anode and cathode: BDD (AO), Anode: BDD and cathode: carbon felt (EF)	AO: No addition of supporting electrolyte, natural pH (8.8). EF: prebubbling with O_2 for 10 min, 0.02 g L^{-1} Fe^{2+} and pH3	24 h for AO and 8 h for EF	9900	60	–	[88]
Fenton oxidation-AO	Landfill leachate	Anode: Si/BDD, cathode: stainless steel	Fenton oxidation: pH3, 1 g L^{-1} Fe^{2+} and 5 g L^{-1} H_2O_2. AO: The same conditions before and without supporting electrolyte	400 min	4430	≈90	–	[89]

COD, chemical oxygen demand; *CR*, color removal; *TOC*, total organic carbon; *TR*, turbidity removal.

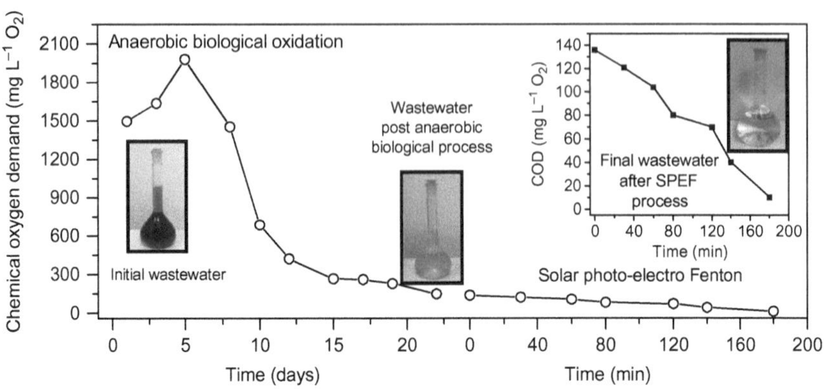

FIG. 4 COD decay under a combination of anaerobic digestion and solar photoelectron-Fenton treatments, with 30 days of anaerobic digestion followed by 180 min of treatment through the SPEF process. SPEF experimental conditions: 0.05 mol L^{-1} Na_2SO_4, pH=3.0, T=35°C, j=30 mA cm^{-2} [Fe^{2+}]=1.0 mM. The inset shows an enlargement of the SPEF treatment. *from J. Vidal, C. Huiliñir, R. Salazar, Removal of organic matter contained in slaughterhouse wastewater using a combination of anaerobic digestion and solar photoelectro-Fenton processes, Electrochim. Acta 210 (2016) 163–170 Copyright 2016, with permission from Elsevier.*

by the degradation of compounds of high molecular mass during biological processes, which transform them in lower compounds that are easier to oxidize by hydroxyl radicals during the SPEF process.

CONCLUSIONS

The versatility and efficiency of electrochemical technologies allow their application to treat wastewater with variable content and from different origins, as is the case of effluents from the agro and food industries as well as landfill leachate. The application of EAOPs such as anodic oxidation, electro-Fenton processes, and the electrocoagulation method has corroborated the viability of these technologies to remove many pollutants such as industrial chemicals. High mineralization and complete decolorization of turbidity decay were found with these environmentally friendly methods. It is possible to treat industrial wastewater from the agro and food industries as well as landfill leachate using these technologies without the addition of new chemical compounds to the wastewater and without adjusting pH. This makes the process simple, safe, and easy to scale up to industrial levels using recirculation flow plants or coupling to traditional processes.

The generation of hydroxyl radicals and other stronger oxidants by EAOPs, which react with the organics present in industrial wastewater, allows complete and effective mineralization to CO_2 and inorganic ions.

By electrocoagulation, an effective and fast removal of organics from industrial wastewater is achieved in a short time of electrolysis, thereby decreasing the cost of the process.

Acknowledgments

We are grateful to all authors included in this chapter as well as the support of FONDECYT Grant 1170352 and DICYT-USACh. PhD fellowship N° 21150338 awarded to L. Carolina Espinoza. Finally, we are grateful to "Proyectos Basales y Vicerrectoría de Investigación, Desarrollo e Innovación" 021742SG_PUBLIC Universidad de Santiago de Chile.

References

[1] C. Leea, M. Song, J. Ryu, C. Park, J. Choi, S. Lee, Application of carbon foam for heavy metal removal from industrial plating wastewater and toxicity evaluation of the adsorbent, Chemosphere 153 (2016) 1–9.
[2] K. Ma, Z. Qin, Z. Zhao, C. Zhao, S. Liang, Toxicity evaluation of wastewater collected at different treatment stages from a pharmaceutical industrial park wastewater treatment plant, Chemosphere 158 (2016) 163–170.
[3] C. Espinoza, N. Contreras, C. Berríos, R. Salazar, Degradation of a veterinary pharmaceutical product in water by electro-oxidation using a BDD anode, J. Chil. Chem. Soc. 59 (2014) 2507–2511.
[4] D. Han, M. Currell, Persistent organic pollutants in China's surface water, Sci. Total Environ. Syst. 580 (2017) 602–625.
[5] T. Robinson, G. McMullan, R. Marchant, P. Nigam, Remediation of dyes in textile effluent: a critical review on current treatment technologies with a proposed alternative, Bioresour. Technol. 77 (2001) 247–255.
[6] S. Pirgaliog, T.A. Ozbelge, Comparison of non-catalytic and catalytic ozonation processes of three different aqueous single dye solutions with respect to powder copper sulfide catalyst, Appl. Catal. A: Gen. 363 (2009) 157–163.
[7] G. Oliveira, E. Ferraz, F. Chequer, M. Grando, J. Angeli, M. Tsuboy, J. Marcarin, M. Mantovan, M. Osug, T. Lizier, M. Zanon, D. Oliviera, Chlorination treatment of aqueous samples reduces, but does not eliminate, the mutagenic effect of the azo dyes disperse red 1, disperse red 13 and disperse Orange 1, Mutat. Res. 703 (2010) 200–208.
[8] S. Raghu, C. Lee, S. Chellammal, S. Palanichamy, C. Basha, Evaluation of electrochemical oxidation techniques for degradation of dye effluents—a comparative approach, J. Hazard. Mater. 171 (2009) 748–754.
[9] C. Comninellis, G. Chen, Electrochemistry for the Environment, first ed., Springer, New York, 2009, pp. 1–23 (chapter 1).
[10] C.A. Martínez-Huitle, S. Ferro, Electrochemical oxidation of organic pollutants for the wastewater treatment: direct and indirect processes, Chem. Soc. Rev. 35 (2006) 1324–1340.
[11] O. Scialdone, Electrochemical oxidation of organic pollutants in water at metal oxide electrodes: a simple theoretical model including direct and indirect oxidation processes at the anodic surface, Electrochem. Acta 54 (2009) 6140–6147.
[12] I. Sirés, P.L. Cabot, F. Centellas, J.A. Garrido, R.M. Rodríguez, C. Arias, E. Brillas, Electrochemical degradation of clofibric acid in water by anodic oxidation: comparative study with platinum and boron-doped diamond electrodes, Electrochim. Acta 52 (2006) 75–85.
[13] M. Quiroz, J. Sánchez-Salas, S. Reyna, E. Bandala, J.M. Peralta-Hernández, C.A. Martínez-Huitle, Degradation of 1-hydroxy-2,4-dinitrobenzene from aqueous

solutions by electrochemical oxidation: role of anodic material, J. Hazard. Mater. 268 (2014) 6–13.

[14] J. Radjenovic, B. Escher, K. Rabaey, Electrochemical degradation of the β-blocker metoprolol by $Ti/Ru_{0.7}Ir_{0.3}O_2$ and Ti/SnO_2-Sb electrodes, Water Res. 45 (2011) 3205–3214.

[15] M. Quiroz, S. Reyna, C.A. Martínez-Huitle, S. Ferro, A. De Battisti, Electrocatalytic oxidation of p-nitrophenol from aqueous solutions at Pb/PbO_2 anodes, Appl. Catal. B Environ. 59 (2005) 259–266.

[16] A. Fabiánskaa, A. Bialk-Bielínskaa, P. Stepnowskia, S. Stoltea, E. Siedleckaa, Electrochemical degradation of sulfonamides at BDD electrode: kinetics, reaction pathway and eco-toxicity evaluation, J. Hazard. Mater. 280 (2014) 579–587.

[17] B. Marselli, J. García-Gómez, P.A. Michaud, M.A. Rodrigo, C. Comninellis, Electrogeneration of hydroxyl radicals on boron-doped diamond electrodes, J. Electrochem. Soc. 150 (2003) 79–83.

[18] G. Rodrigues de Oliveira, N. Suely Fernandesa, J. Vieira de Melo, D. Ribeiro da Silva, C. Urgeghe, C.A. Martinez-Huitle, Electrocatalytic properties of Ti-supported Pt for decolorizing and removing dye from synthetic textile wastewaters, Chem. Eng. J. 168 (2011) 208–214.

[19] M. Panizza, G. Cerisola, Application of diamond electrodes to electrochemical processes, Electrochim. Acta 51 (2005) 191–199.

[20] A. El-Ghenymy, F. Centellas, J.A. Garrido, R.M. Rodríguez, I. Sirés, P. Lluís Cabot, E. Brillas, Decolorization and mineralization of Orange G azo dye solutions by anodic oxidation with a boron-doped diamond anode in divided and undivided tank reactors, Electrochim. Acta 130 (2014) 568–576.

[21] A. Sánchez-Carretero, C. Sáez, P. Cañizares, M.A. Rodrigo, in: E. Brillas, C.A. Martínez-Huitle (Eds.), Synthetic Diamond Films: Preparation, Electrochemistry, Characterization and Applications, Wiley, New York, 2011 (Chapter 12).

[22] M. Wu, G. Zhao, M. Li, L. Liu, D. Li, Applicability of boron-doped diamond electrode to the degradation of chloride-mediated and chloride-free wastewaters, J. Hazard. Mater. 163 (2009) 26–31.

[23] L. Szpyrkowicz, C. Juzzolino, A comparative study on oxidation of disperse dyes by electrochemical process, ozone, hypochlorite and fenton reagente, Water Res. 35 (2001) 2129–2136.

[24] M. Susree, P. Asaithambi, R. Saravanathamizhan, M. Matheswaran, Studies on various mode of electrochemical reactor operation for the treatment of distillery effluent, J. Environ. Chem. Eng. 1 (2013) 552–558.

[25] A. Anglada, A. Urtiaga, I. Ortiz, D. Mantzavinos, E. Diamadopoulos, Boron-doped diamond anodic treatment of landfill leachate: evaluation of operating variables and formation of oxidation by-products, Water Res. 45 (2011) 828–838.

[26] P. Moraes, R. Bertazzoli, Electrodegradation of landfill leachate in a flow electrochemical reactor, Chemosphere 58 (2005) 41–46.

[27] A. Fernandes, M. Pacheco, L. Ciríaco, A. Lopes, Anodic oxidation of a biologically treated leachate on a boron-doped diamond anode, J. Hazard. Mater. 199–200 (2012) 82–87.

[28] M. Panizza, C.A. Martinez-Huitle, Role of electrode materials for the anodic oxidation of a real landfill leachate—comparison between Ti–Ru–Sn ternary oxide, PbO_2 and boron-doped diamond anode, Chemosphere 90 (2013) 1455–1460.

[29] M. Panizza, G. Cerisola, Olive mill wastewater treatment by anodic oxidation with parallel plate electrodes, Water Res. 40 (2006) 1179–1184.

[30] C. Belaid, M. Khadraoui, S. Mseddi, M. Kallel, B. Elleuch, J. Fauvarque, Electrochemical treatment of olive mill wastewater: treatment extent and effluent phenolic compounds monitoring using some uncommon analytical tolos, J. Environ. Sci. 25 (2013) 220–230.

[31] E. Chatzisymeon, A. Dimou, D. Mantzavinos, A. Katsaounis, Electrochemical oxidation of model compounds and olive mill wastewater over DSA electrodes: 1. The case of Ti/IrO_2 anode, J. Hazard. Mater. 167 (2009) 268–274.

[32] U. Un, U. Altay, A. Koparal, U. Ogutveren, Complete treatment of olive mill wastewaters by electrooxidation, Chem. Eng. J. 139 (2008) 445–452.
[33] P. Cañizares, J. Lobato, R. Paz, M.A. Rodrigo, C. Sáez, Advanced oxidation processes for the treatment of olive-oil mills wastewater, Chemosphere 67 (2007) 832–838.
[34] N. Papastefanakis, D. Mantzavinos, A. Katsaounis, DSA electrochemical treatment of olive mill wastewater on Ti/RuO_2 anode, J. Appl. Electrochem. 40 (2010) 729–737.
[35] M. Mavros, N.P. Xekoukoulotakis, D. Mantzavinos, E. Diamadopoulos, Complete treatment of olive pomace leachate by coagulation, activated-carbon adsorption and electrochemical oxidation, Water Res. 42 (2008) 2883–2888.
[36] D. Valero, V. García-García, E. Expósito, A. Aldaz, V. Montiel, Electrochemical treatment of wastewater from almond industry using DSA-type anodes: direct connection to a PV generator, Sep. Purif. Technol. 123 (2014) 15–22.
[37] X. Li, C. Wang, Y. Qian, Y. Wang, L. Zhang, Simultaneous removal of chemical oxygen demand, turbidity and hardness from biologically treated citric acid wastewater by electrochemical oxidation for reuse, Sep. Purif. Technol. 107 (2013) 281–288.
[38] M. Panizza, G. Cerisola, Electrochemical generation of H_2O_2 in low ionic strength media on gas diffusion cathode fed with air, Electrochim. Acta 54 (2008) 876–878.
[39] S. Sun, C.-J. Li, J.-H. Sun, S.-H. Shi, M.-H. Fan, Q. Zhou, Decolorization of an azo dye Orange G in aqueous solution by Fenton oxidation process: effect of system parameters and kinetic study, J. Hazard. Mater. 161 (2009) 1052–1057.
[40] E. Isarain-Chávez, C. Arias, P.L. Cabot, F. Centellas, R.M. Rodríguez, J.A. Garrido, E. Brillas, Mineralization of the drug beta-blocker atenolol by electro-Fenton and photoelectro-Fenton using an air-diffusion cathode for H_2O_2 electrogeneration combined with a carbon-felt cathode for Fe^{2+} regeneration, Appl. Catal. B: Environ. 96 (2010) 361–369.
[41] E. Brillas, I. Sirés, M. Oturan, Electro-Fenton process and related electrochemical technologies based on Fenton's reaction chemistry, Chem. Rev. 109 (2009) 6570–6631.
[42] R. Salazar, M.S. Ureta-Zañartu, Mineralization of triadimefon fungicide in water by electro-Fenton and photo electro-Fenton, Water Air Soil Pollut. 223 (2012) 4199–4207.
[43] A. Díez, O. Iglesias, E. Rosales, M. Sanromán, M. Pazos, Optimization of two-chamber photo electro Fenton reactor for the treatment of winery wastewater, Process Saf. Environ. Prot. 101 (2016) 72–79.
[44] S. Garcia-Segura, E. Brillas, L. Cornejo-Ponce, R. Salazar, Effect of the Fe^{3+}/Cu^{2+} ratio on the removal of the recalcitrant oxalic and oxamic acids by electro-Fenton and solar photoelectro-Fenton, Sol. Energy 124 (2016) 242–253.
[45] C. Espinoza, J. Romero, L. Villegas, L. Cornejo, R. Salazar, Mineralization of the textile dye acid yellow 42 by solar photoelectro-Fenton in a lab-pilot plant, J. Hazard. Mater. 319 (2016) 24–33.
[46] A. Akyol, O. Can, E. Demirbas, M. Kobya, A comparative study of electrocoagulation and electro-Fenton for treatment of wastewater from liquid organic fertilizer plant, Sep. Purif. Technol. 112 (2013) 11–19.
[47] R. Davarnejad, J. Azizi, Alcoholic wastewater treatment using electro-Fenton technique modified by Fe_2O_3 nanoparticles, J. Environ. Chem. Eng. 4 (2016) 2342–2349.
[48] C. David, M. Arivazhagan, F. Tuvakara, Decolorization of distillery spent wash effluent by electro oxidation (EC and EF) and Fenton processes: a comparative study, Ecotoxicol. Environ. Saf. 121 (2015) 142–148.
[49] J. Paramo-Vargas, A. Camargo, S. Gutierrez-Granados, L. Godinez, J. Peralta-Hernandez, Applying electro-Fenton process as an alternative to a slaughterhouse effluent treatment, J. Electroanal. Chem. 754 (2015) 80–86.
[50] K. Thirugnanasambandham, S. Kandasamy, V. Sivakumar, R. Kiran kumar, R. Mohanavelu, Modeling of by-product recovery and performance evaluation of electro-Fenton treatment technique to treat poultry wastewater, J. Taiwan Inst. Chem. Eng. 46 (2015) 89–97.

[51] H. Zhang, D. Zhang, J. Zhou, Removal of COD from landfill leachate by electro-Fenton method, J. Hazard. Mater. 135 (2006) 106–111.

[52] E. Atmaca, Treatment of landfill leachate by using electro-Fenton method, J. Hazard. Mater. 163 (2009) 109–114.

[53] A. Altin, An alternative type of photoelectro-Fenton process for the treatment of landfill leachate, Sep. Purif. Technol. 61 (2008) 391–397.

[54] N. Flores, P. Cabot, F. Centellas, J. Garrido, R. Rodríguez, E. Brillas, I. Sirés, 4-Hydroxyphenylacetic acid oxidation in sulfate and real olive oil mill wastewater by electrochemical advanced processes with a boron-doped diamond anode, J. Hazard. Mater. 321 (2017) 566–575.

[55] M. Villanueva-Rodríguez, R. Bello-Mendoza, D. Wareham, E. Ruiz-Ruiz, M. Maya-Treviño, Discoloration and organic matter removal from coffee wastewater by electrochemical advanced oxidation processes, Water Air Soil Pollut. 225 (2014) 2204–2215.

[56] C. Allegre, M. Maisseu, F. Charbit, P. Moulin, Coagulation–flocculation–decantation of dye house effluents: concentrated effluents, J. Hazard. Mater. 116 (2004) 57–64.

[57] M. Emamjomeh, M. Sivakumar, Review of pollutants removed by electrocoagulation and electrocoagulation/flotation processes, J. Environ. Manag. 90 (2009) 1663–1679.

[58] C. Yang, J. McGarrahan, Electrochemical coagulation for textile effluent decolorization, J. Hazard. Mater. 127 (2005) 40–47.

[59] I. Linares-Hernández, C. Barrera-Díaz, G. Roa-Morales, B. Bilyeu, F. Ureña-Núñez, Influence of the anodic material on electrocoagulation performance, Chem. Eng. J. 148 (2009) 97–105.

[60] J. Vidal, L. Villegas, J.M. Peralta-Hernández, R. Salazar, Removal of Acid Black 194 dye from water by electrocoagulation with aluminum anode, J. Environ. Sci. Health A Tox. Hazard. Subst. Environ. Eng. 51 (2016) 289–296.

[61] S. Tchamango, C. Nanseu-Njiki, E. Ngameni, D. Hadjiev, A. Darchen, Treatment of dairy effluents by electrocoagulation using aluminium electrodes, Sci. Total Environ. 408 (2010) 947–952.

[62] I. Sengil, M. Özcan, The decolorization of C.I. Reactive black 5 in aqueous solution by electrocoagulation using sacrificial iron electrodes, J. Hazard. Mater. 161 (2009) 1369–1376.

[63] P. Holt, G. Barton, C. Mitchell, The future for electrocoagulation as a localised water treatment technology, Chemosphere 59 (2005) 355–367.

[64] J.L. Trompette, H. Vergnes, On the crucial influence of some supporting electrolytes during electrocoagulation in the presence of aluminum electrodes, J. Hazard. Mater. 163 (2009) 1282–1288.

[65] P. Drogui, M. Asselin, S. Brar, H. Benmoussa, J. Blais, Electrochemical removal of pollutants from agro-industry wastewaters, Sep. Purif. Technol. 61 (2008) 301–310.

[66] K. Bensadok, N. El Hanafi, F. Lapicque, Electrochemical treatment of dairy effluent using combined Al and Ti/Pt electrodes system, Desalination 280 (2011) 244–251.

[67] İ. Sengil, M. Özacar, Treatment of dairy wastewaters by electrocoagulation using mild steel electrodes, J. Hazard. Mater. 137 (2006) 1197–1205.

[68] O. Sahu, Performance of copper compounds in chemical and electro oxidation treatment of sugar industry waste water: batch reaction, J. Taiwan Inst. Chem. Eng. 65 (2016) 256–268.

[69] O. Sahu, P. Chaudhari, Electrochemical treatment of sugar industry wastewater: COD and color removal, J. Electroanal. Chem. 739 (2015) 122–129.

[70] X. Li, J. Song, J. Guo, Z. Wang, Q. Feng, Landfill leachate treatment using electrocoagulation, Environ. Sci. 10 (2011) 1159–1164.

[71] C. Ricordel, H. Djelal, Treatment of landfill leachate with high proportion of refractory materials by electrocoagulation: system performances and sludge settling characteristics, J. Environ. Chem. Eng. 2 (2014) 1551–1557.

[72] A. Prajapati, P. Chaudhari, D. Pal, A. Chandrakar, R. Choudhary, Electrocoagulation treatment of rice grain based distillery effluent using copper electrode, J. Water Process Eng. 11 (2016) 1–7.
[73] A. Prajapati, P. Chaudhari, Electrochemical treatment of rice grain-based distillery effluent: chemical oxygen demand and colour removal, Environ. Technol. 35 (2013) 242–249.
[74] M. Kumar, F. Ponselvan, J. Malviya, V. Srivastava, I. Mall, Treatment of bio-digester effluent by electrocoagulation using iron electrodes, J. Hazard. Mater. 165 (2009) 345–352.
[75] U. Un, A. Kandemir, N. Erginel, S. Ocal, Continuous electrocoagulation of cheese whey wastewater: an application of response surface methodology, J. Environ. Manag. 146 (2014) 245–250.
[76] M. Asselin, P. Drogui, H. Benmoussa, J. Blais, Effectiveness of electrocoagulation process in removing organic compounds from slaughterhouse wastewater using monopolar and bipolar electrolytic cells, Chemosphere 72 (2008) 1727–1733.
[77] M. Kobya, E. Senturk, M. Bayramoglu, Treatment of poultry slaughterhouse wastewaters by electrocoagulation, J. Hazard. Mater. 133 (2006) 172–176.
[78] U. Un, E. Ozel, Electrocoagulation of yogurt industry wastewater and the production of ceramic pigments from the sludge, Sep. Purif. Technol. 120 (2013) 386–391.
[79] L. Zaleschi, C. Sáez, P. Cañizares, I. Cretescu, M.A. Rodrigo, Electrochemical coagulation of treated wastewaters for reuse, Desalin. Water Treat. 51 (2013) 3381–3388.
[80] J. Vidal, C. Huiliñir, R. Salazar, Removal of organic matter contained in slaughterhouse wastewater using a combination of anaerobic digestion and solar photoelectro-Fenton processes, Electrochim. Acta 210 (2016) 163–170.
[81] A. Fernandes, P. Spranger, A. Fonseca, M. Pacheco, L. Ciríaco, A. Lopes, Effect of electrochemical treatments on the biodegradability of sanitary landfill leachates, Appl. Catal. B Environ. 144 (2014) 514–520.
[82] A. Katsoni, D. Mantzavinos, E. Diamadopoulos, Sequential treatment of diluted olive pomace leachate by digestion in a pilot scale UASB reactor and BDD electrochemical oxidation, Water Res. 57 (2014) 76–86.
[83] S. Xiao, J. Peng, Y. Song, D. Zhang, R. Liu, P. Zeng, Degradation of biologically treated landfill leachate by using electrochemical process combined with UV irradiation, Sep. Purif. Technol. 117 (2013) 24–29.
[84] A. Deligiorgis, N. Xekoukoulotakis, E. Diamadopoulos, D. Mantzavinos, Electrochemical oxidation of table olive processing wastewater over boron-doped diamond electrodes: treatment optimization by factorial design, Water Res. 42 (2008) 1229–1237.
[85] S. Lin, C. Chang, Treatment of landfill leachate by combined electro-Fenton oxidation and sequencing batch reactor method, Water Res. 34 (2000) 4243–4249.
[86] F. Moreira, R. Boaventura, E. Brillas, V. Vilar, Remediation of a winery wastewater combining aerobic biological oxidation and electrochemical advanced oxidation processes, Water Res. 75 (2015) 95–108.
[87] F. Feki, F. Aloui, M. Feki, S. Sayadi, Electrochemical oxidation post-treatment of landfill leachates treated with membrane bioreactor, Chemosphere 75 (2009) 256–260.
[88] L. Labiadh, A. Fernandes, L. Ciríaco, M. Pacheco, A. Gadri, S. Ammar, A. Lopes, Electrochemical treatment of concentrate from reverse osmosis of sanitary landfill leachate, J. Environ. Manag. 181 (2016) 515–521.
[89] A. Urtiaga, A. Rueda, A. Anglada, I. Ortiz, Integrated treatment of landfill leachates including electrooxidation at pilot plant scale, J. Hazard. Mater. 166 (2009) 1530–1534.

CHAPTER

16

Practical Aspects on Electrochemical Disinfection of Urban and Domestic Wastewater

Simonetta Palmas, Michele Mascia*, Annalisa Vacca*, Laura Mais*, Simona Corgiolu*, Elisabetta Petrucci†*

*University of Cagliari, Cagliari, Italy
†University of Rome, La Sapienza, Roma, Italy

OUTLINE

Electrochemical Water and Wastewater Treatment
https://doi.org/10.1016/B978-0-12-813160-2.00016-X

INTRODUCTION

In view of sustainable water resources management, wastewater reclamation is a commonly acknowledged issue [1]. The origin of the wastewater (WW) and the kind and concentration of the different pollutants determine the combination of processes and technologies needed to reach the level of purification required by the final destination of the treated WW.

This chapter will be specifically devoted to the possible application of electrochemical techniques to the purification of civil and domestic WWs. Attention will be paid to tertiary treatments such as advanced oxidation processes (AOPs) and disinfection, which often represents the final treatment process to which the WW is submitted, either before its final discharge in a receptor basin or before its possible reuse.

Problems such as water scarcity and low water quality along with increasing water demand are becoming serious issues [2], especially in those regions where long periods of dryness are frequent or in remote lands where the infrastructures for water treatment and distribution are not sufficient or hardly available. As is well known, in many parts of the world it may not be possible to provide sewage facilities for all residents, now or in the future. In those cases, WW management is often addressed to decentralized WW treatment. Of note, it also has to be considered that most of the energy produced in those countries is used to guarantee drinkable water. Increasing costs have to be sustained depending on where the water is obtained: superficial water, ground water, WW, or seawater. If one considers that water itself is needed to produce energy (approximately 28–230 L/kWh) [3,4], it appears clear how important it is to manage the problem from both the economic and environmental point of view.

In this context, water recycling and reuse seem to be mandatory. Also, developing novel hybrid technologies able to simultaneously purify water and generate energy could be very helpful.

If recovery and reuse are considered, it is worth noting that in recent decades the development of new processes and the wide acceptance of membrane technologies in WW treatment have allowed recycled water applications to be broadened from nonpotable uses (e.g., irrigation, industry, environmental flow, residential use) to indirect and direct potable reuses [5,6].

However, if the possibility of producing recycled water of virtually any quality is already assessed from a theoretical point of view, many constraints still exist that limit practical application. Lack of infrastructure and funding, cost of transport, and the overall public objection still restrict the use of advanced treatment processes, a situation that can potentially cause health, environmental, and sanitation problems [7].

In this context, electrochemical technologies may be very useful, for example, in the disinfection process where a high concentration of oxidant can be generated in situ. The treatment only needs the application of an electric current, thus allowing easy process management and even remote control.

Classification of Wastewater on the Basis of Its Origin

Following a general classification, graywater refers to urban WW that includes domestic WW, that is, water from household sinks, dishwashers, showers, baths, hand basins, and laundry machines but not from toilets [8,9]. Rainwater is defined as rainfall precipitation collected from a roof catchment [10].

Municipal WWs were originally considered to be the collection of domestic and commercial sources.

A typical composition of municipal WW is reported in Table 1 [11].

Generally, graywater is less polluted and has a lower concentration of contaminating pathogens, nitrogen, suspended solids, and turbidity compared with municipal and industrial WWs. However, in countries where phosphorus-containing detergents are not banned, phosphorus concentrations in households can be as high as 45–280 mg/L. In some cases,

TABLE 1 Major Constituents of Typical Raw Municipal WW With a Minor Contribution of Industrial WW

	Concentration (mg/L)		
Constituent	**Strong**	**Medium**	**Weak**
COD total	1200	750	500
COD soluble	480	300	200
COD suspended	720	450	300
BOD	560	350	230
Volatile fatty acids	80	30	10
N total	100	60	30
Ammonia-N	75	45	20
Phosphorus (as P)	25	15	6
Total suspended solids	600	400	250
Volatile suspended solids	480	320	200

high biochemical oxygen demand (BOD) and chemical oxygen demand (COD) concentrations might also be observed. These are caused by chemical and pharmaceutical pollutants from soaps, detergents, and personal care products as well as food waste in kitchen sinks [12].

In this framework, monitoring the composition of the water to be treated becomes paramount either when the final aim is discharge into natural water streams or possible reuse. When possible, a diversification of water resources may be advisable. In fact, an optimization of the treatment process and of the related costs could be better obtained because high quality mains water is not required for nonpotable activities, which comprise more than 20% of a building's mains water demand.

When domestic and/or civil WWs are involved, the composition is obviously strongly dependent on the kind of urban area that generated the waste, the presence of possible production centers, and whether rainfall water is fed along with domestic WWs.

Due to the absence of separated sewage collection pipelines, sometimes separation of the water sources is not possible; in these cases, graywater, blackwater, industrial WW, and other waste streams from hospitals and commercial facilities are all discharged into municipal sewage systems. Hence, municipal WW often contains a broad spectrum of contaminants (e.g., organic matter, pathogens, and inorganic particles, along with a number of potentially toxic elements such as arsenic, cadmium, chromium, copper, lead, mercury, zinc, etc.), which can pose problems for both human health and the environment. For example, a strong residual salinity—determined by a high content of inorganic compounds as well as the presence of a high concentration of heavy metals—may constitute a problem when reused in irrigation.

On the other hand, when considering recycled water for potable reuse, even a very low concentration of organic pollutants such as active pharmaceutical compounds or other organic hydrophilic substances, which are scarcely adsorbed at the secondary treatment stage of activated sludge, may cause adverse biological effects on human health. Besides, from a microbiological aspect, the main pollution groups in municipal WW are excreted organisms and pathogens from human and animal origin where enteric viruses and protozoan pathogens are significantly more infectious than other bacterial pathogens.

Disinfection processes become, in this case, of paramount importance to prevent waterborne diseases. Such processes as chlorination, chloramination, chlorine dioxide, ozonation, ultraviolet (UV), and UV/H_2O_2 are mostly used to guarantee the absence of pathogens. However, while this may prevent the toxicity of pathogens, a second aspect has to be monitored carefully: the possible formation of harmful by-products, which can be even more toxic than the original compounds.

Attention must be paid, for example, to possible anthropogenic activities or such negative events as the infiltration of seawater into drinking water sources, which may be the origin of a high concentration of bromide and iodide. In these cases, bromide and iodide have to be removed before final disinfection. In fact, their presence can result in the formation of brominated and/or iodinated disinfection by-products (DBPs) upon exposure to natural organic matter (NOM) and disinfectant, which are often more toxic than their chlorinated analogues [13,14]. Actually, when disinfection is performed with chlorine [15] or chloramine [16], the formation of brominated/iodinated DBPs is expected while, when disinfection is performed by ozone [17,18] or with other advanced oxidation products (AOP), the production of bromate—a suspected carcinogen—is expected.

How Can Electrochemistry Be Useful for Disinfection?

Electrochemistry offers different approaches for water and WW disinfection based on physical, chemical, and physicochemical methods. However, the most commonly used method is the electrogeneration of biocides. The main advantage provided by electrochemical disinfection is the possibility of producing the species involved in situ, thus avoiding all the hazards correlated to the manipulation of highly concentrated oxidants, very often without a requirement of additional chemicals. In those cases, the treatment only needs the application of an electric current, thus allowing easy process management and even remote control.

Disinfectants action usually implies cell lysis, that is, the dissolution of the cell membrane of the target organism, thus causing a change in cell permeability and inhibition of enzyme activity. The added chemicals affect membrane functions—they change osmotic pressure, permeability, and the transport processes of molecules and ions through the membrane while also inhibiting metabolic processes, bio-oxidation, and cell division. An additional goal is represented by oxidation of the residual organic matter content in water, favoring the death of the surviving microorganisms due to a lack of nutrients. The action of in situ electrogenerated biocides is further enhanced by synergy with the electric field and in some cases with pH changes, thus resulting in fast and effective inactivation of several microorganisms.

In electrochemical disinfection processes, the production of oxidants (disinfectants) may occur directly by water discharge (i.e., hydroxyl radicals, ozone), dissolved species (i.e., active chlorine or hydrogen peroxide via oxygen reduction), or anode dissolution (i.e., ferrate) [18]. Types and concentrations of the formed oxidants strictly depend on the adopted operating conditions, mostly on the used electrode material.

ELECTRODE MATERIALS

Active and Nonactive Electrodes for the Anodic Production of Oxidants

According to a well-accepted hypothesis, the anodic materials can be classified as *active* or *nonactive*, depending on the mechanisms involved in the oxidation process.

The differences between the mechanisms involved using active or non-active electrodes can allow different choices depending on the composition of WW: the treatment of urban and municipal WW needs both the oxidation of pollutants and the removal of pathogens, which can be achieved by exploiting both direct and bulk oxidation.

Active electrodes consist of a metal support coated by a film of inorganic conductive oxide. Originally, the most used oxides in the preparation of these types of anodes were RuO_2 and IrO_2. However, the addition of such catalytically active components and oxides such as TiO_2, SnO_2, and Ta_2O_5 has also been studied to improve the stability and durability of these materials while maintaining their main catalytic characteristics [19].

With "active electrodes" such as Ti/RuO_2, Ti/Pt and Ti/IrO_2, which are commonly adopted in industrial electrochemical processes [20–23], the electrochemical oxidation is supposed to occur through the formation of higher metal oxides that can catalyze the transformation of organic substances to more oxidized compounds by several chemical and electrochemical steps [19]. However, in this case, low faradic efficiencies have generally been observed.

According to the following mechanism, the oxidation by active electrodes starts with the initial decomposition of water, producing hydroxyl radicals that, at these electrodes, are strictly linked to the active sites (M) on the electrode surface [23].

$$MO_x + H_2O \rightarrow MO_x(-OH) + H^+ + e^- \quad (1)$$

A possible transition of the oxygen from OH to the electrode surface forms a higher-state oxide, at a potential higher than that required for oxygen evolution (1.23 V vs RHE).

$$MO_x(^{\bullet}OH) \rightarrow MO_{x+1} + H^+ + e^- \quad (2)$$

The active species at the higher oxidation state at the surface electrode are consumed by two competing reactions (oxidation of aqueous compounds and oxygen evolution) that regenerate the electrode active site in a reversible mechanism (redox catalysis).

$$MO_{x+1} + R \rightarrow MO_x + RO \quad (3)$$

$$MO_{x+1} \rightarrow MO_x + \frac{1}{2}O_2 \quad (4)$$

The closer the reversible potential of the surface redox couple (MO_x/MO_x+1) to the potential of oxygen evolution, the higher the active character of the anode [24].

Several experimental results showed that if chloride ions are present in the solution, the faradic efficiencies may be strongly improved [25–28]. A synergetic effect involving surface and bulk reactions has been proposed in which the heterogeneous surface reaction is mediated by chlorohydroxyl radicals formed during the discharge of chloride and adsorbed at the electrode surface. The homogeneous bulk reaction occurs in the liquid phase, and involves electrogenerated $HClO/ClO^-$ [29].

The reaction is generally simplified in the following steps:

Chloride migration toward the anode, where it undertakes oxidation to gaseous chlorine

$$2Cl^-_{aq} \rightarrow Cl_{2(g)} + 2e^- \quad (5)$$

hydrolysis of chlorine with the formation of hypochlorous acid

$$Cl_2 + H_2O \rightarrow HClO + H^+ + Cl^- \quad (6)$$

pH-dependent equilibrium between hypochlorous and hypochlorite ion

$$HClO \rightarrow ClO^- + H^+ \quad (7)$$

The three involved species Cl_2 ($E_0 = +1.358$ V), HClO ($E_0 = +1.611$ V), and ClO^- ($E_0 = +0.81$ V), whose amount depends on pH and temperature conditions, form a powerful oxidizing mixture named *active* or *free chlorine*.

It is worth noting that not only is hypochlorous acid endowed with higher oxidizing power, but it also behaves as a stronger disinfectant than hypochlorite. In fact, with the cell wall of pathogenic microorganisms being naturally negatively charged, it can be entered more easily by the neutral hypochlorous acid than the negatively charged hypochlorite ion [30].

Due to its high reactivity and relatively low-cost production, active chlorine has been widely used as a biocide for the disinfection of potable, industrial, or agro-industrial water; surface sanitation in hospitals; and the sterilization of medical equipment.

Different concomitant effects, such as irreversible permeabilization of the cytoplasmic membrane caused by the electrical field, led to cell lysis by reaction of the biopolymers with electrogenerated oxidants, a release of highly reactive atomic oxygen [31] thus contributing to effectively inactivating a broad range of microorganisms [32,33].

$$HClO \rightarrow O + Cl^- + H^+ \quad (8)$$

$$ClO^- \rightarrow O + Cl^- \quad (9)$$

A paper also reports the removal of Staphylococcus aureus by active ruthenium-based electrodes in the absence of chloride [34].

Only a few species such as *Giardia* and *Cryptosporidium* are known to be chlorination-resistant [30].

The superior disinfecting power of electrochlorination with respect to chemical chlorination can be explained by considering that, during the electrolysis, a number of different oxidants (i.e., hydroxyl radical, hydrogen peroxide, ozone) can be easily electrogenerated by water discharge at the anode [35].

An issue of serious concern related to electrochlorination is the occurrence of side reactions leading to the formation of hazardous by-products. Although trihalomethanes (THMs) and haloacetic acids (HAAs) are considered as the main DBPs, hundreds of different chemicals have been identified in treated drinking water; for many of them, mutagenic or carcinogenic behavior has been verified [14].

Despite concerns raised by toxic by-product formation along with unpleasant odor and taste, active chlorine remains hardly replaceable due to its high reactivity and relatively low-cost production and mostly for its capacity to provide a long-term biocidal action.

With *nonactive electrodes,* typically fully oxidized metal oxides and boron doped diamond (BDD) anodes, the oxidation is mediated by physi-adsorbed $OH^{\bullet}$ radicals generated at the anode surface by water discharge [23].

In particular, diamond electrodes exhibit high overpotential for water oxidation, allowing the electrogeneration of hydroxyl radicals directly from an aqueous solution [36,37]. Because $OH^{\bullet}$ radicals are scarcely adsorbed at the BDD surface, they are quickly desorbed and may react either with oxidizable compounds or with each other to give oxygen. Moreover, other bulk oxidizing agents such as hydrogen peroxide and ozone can be produced. Thus, reactive oxygenated species (ROS) may be generated during water oxidation, which can be exploited for a chlorine-free disinfection process. The global process of OER at this electrode material can be sketched as [38]:

$$H_2O \rightarrow OH^{\bullet} + H^{+} + e^{-} \tag{10}$$

$$OH^{\bullet} \rightarrow O^{\bullet} + H^{+} + e^{-} \tag{11}$$

$$2O^{\bullet} \rightarrow O_2 \tag{12}$$

$$2OH^{\bullet} \rightarrow H_2O_2 \tag{13}$$

$$H_2O_2 \rightarrow O_2 + 2H^{+} + 2e^{-} \tag{14}$$

$$O_2 + O^{\bullet} \rightarrow O_3 \tag{15}$$

Due to their high reactivity, the radicals desorbed from the electrode surface remain confined near the anode in a thin layer of solution δ_r, (thickness of 10–40 nm) [39], where they rapidly give rise to oxygen (steps 12, 14) or other *reactive oxygen species* (ROS) such as ozone or hydrogen peroxide (steps 13, 15). The kinetics of the different series/parallel steps involved in the global process determine the dependence of the flux of ROS on the applied current.

It was also reported that, in the presence of sulfate, peroxidisulphate is produced [40]:

$$2SO_4{}^{2-} + 2OH^{\bullet} \rightarrow S_2O_8{}^{2-} + 2OH^- \quad (16)$$

In the presence of chloride and chlorine, hypochlorite/hypochlorous acid can be also produced. Moreover, the free chlorine produced can react with $OH^{\bullet}$ radicals to give chlorine dioxide and more oxidized chlorine species as chlorates. When the chemistry near the electrode is dominated by radical mechanisms characterized by rate constants in the range of 10^5–$10^{10}\ M^{-1}\ s^{-1}$, those mechanisms would replace the known reaction paths, based on the electrochemical oxidation of hypochlorite or chloride to chlorate [41,42].

Good oxidative performance is generally observed during disinfection and purification of water by electrolysis at BDD where the so-called mixed oxidant is formed, which shows a disinfecting ability higher than the addition of a single chemical [43,44]. High performance was achieved, thanks to the possible presence of short-life radical species [45]. Moreover, the high bactericidal effect of the electrochemical treatment was also attributed to the combined effect of mixed oxidants, the electric field, and pH gradients in the reactor [46–49].

Nevertheless, similarly to other treatments, the electrochemical disinfection may also lead to the formation of different by-products (DPBs) [50–52]. Among them, chlorates are of great concern, and the effect of these compounds on human health is still under study. They are suspected to have health risks so that the World Health Organization (WHO) recommends very low concentrations of $ClO_3{}^-$ and $ClO_4{}^-$ in drinking water [53]. A correct choice of anode material and operative conditions are key parameters to obtain a powerful mixture of disinfectant, limiting the production of DPBs.

Sacrificial Anodes

An emerging oxidant that is stirring great interest is Ferrate $[FeO_4]^{2-}$, an anion containing iron at its highest oxidation state (IV) whose use has long been limited by the difficulty of preparation [54].

Ferrate behaves as a powerfully strong oxidant at acidic pH ($E_0 = +2.20$ V) while it shows a milder oxidizing ability under alkaline conditions ($E_0 = +0.72$ V) [55].

Chemical synthesis implies the oxidation of ferric ions under a very alkaline solution with hypochlorite or ozone while electrochemical synthesis requires electrolysis with a sacrificial iron-based anode in a NaOH or KOH-rich electrolyte. The process performance mostly depends on anode material, applied current density, and electrolytes. The reactions involved are:

at the anode

$$Fe + 8OH^- \rightarrow FeO_4^{2-} + 4H_2O + 6e^- \quad (17)$$

and at the cathode

$$3H_2O + 6e^- \rightarrow 3H_2 + 6OH^- \quad (18)$$

Anode materials generally consist of iron, carbon steel, or cast iron. Porosity [56] and composition of electrodes greatly affect the ferrate electrogeneration [57]. For example, high and similar current yields were achieved by both Bouzek and Rousar [58] (65%) by using a cast iron anode containing 3.16% C as iron carbide, or by Denvir and Pletcher [59] (70%) by using a silver steel electrode with 0.9% carbon content. However, this last electrode's performance was drastically reduced to 12% when the carbon content was lowered to 0.08%.

Outstanding performance of ferrate (VI) relies on its multifunctional action: it can inactivate a variety of microorganisms [60] including some viruses while oxidizing a wide range of compounds (including emerging contaminants); moreover, due to the release of ferric ions, it also provides coagulant action [61].

For example, Jiang et al., indicate that 0.5 mg/L of ferrate (VI) is sufficient to oxidize many chemicals and to achieve complete disinfection, with the performance unaffected by the initial content of the microbial load [62,63].

Ferrate is generally considered a green oxidant whose decomposition only produces ferric ions and insoluble ferric hydroxides (III) [30]. In addition, it is unreactive toward bromide, thus avoiding the formation of bromate [64]. However, contradictory results have been reported about the production of possible mutagenic by-products by ferrate-involving reactions [65,66], so that the release of DBP of environmental and sanitary concern is still a pending issue. For this reason, the potential for ferrate(VI) to replace conventional disinfectants requires further study.

Carbon-Based Electrodes for Cathodic Production of Oxidants

The adoption of suitable electrode materials enables the generation of high oxidizing and biocidal species, also at the cathode compartment. This is the case of hydrogen peroxide, whose on-site production occurs via a reduction of the dissolved oxygen at carbon-based electrodes.

The electrochemical route involves several reactions with formation of different radical species such as superoxide radicals $O_2^{\bullet-}$ and hydroperoxide radicals $HOO^{\bullet}$ [67,68] with the following overall reaction:

$$O_2 + 2H_3O^+ + 2e^- \rightarrow H_2O_2 + 2H_2O \quad (19)$$

Major side reactions are [69]:

$$O_2 + 4H^+ + 4e^- \rightarrow 6H_2O \quad (20)$$

the decomposition of hydrogen peroxide

$$H_2O_2 + 2H^+ + 2e^- \rightarrow 2H_2O \quad (21)$$

and hydrogen gas evolution

$$2H_2O + 2e^- \rightarrow H_2 + 2OH^- \quad (22)$$

A broad range of materials has been successfully tested including graphite [68,70], and graphite felt [71], activated carbon fiber [72], carbon sponge [73], and modified electrodes [64,74].

The development of gas diffusion electrodes (GDEs) [75,76] have avoided the limitations related to oxygen solubility and transport from the bulk to the cathode surface, thus allowing for high production of hydrogen peroxide, also in airflow.

Current efficiencies close to theoretical values are generally observed, at least at the beginning of the electrolysis before the achievement of the steady state concentration.

An increased reactive area and porosity account for the excellent performance of materials such as carbon felt (CF) [77,78] and reticulated vitreous carbon (RVC) [67,79,80] that present as additional advantage of low cost and the possibility of being tailored in different shapes. In particular, the great electrocatalytic power has been explained by considering the high number of active sites that promote selectivity, reducing the occurrence of the side reactions leading to hydrogen evolution. It has been found that, due to higher reproducibility together with a decreased tendency to deactivation, RVC outperforms CF [81].

Unlike commercial H_2O_2 that is stabilized to prevent decomposition, electro-generated hydrogen peroxide is quite reactive and synergistically enhanced by all the different radical species that are simultaneously produced during oxygen reduction, thus resulting in improved oxidation and disinfection [82].

A relatively unexplored application of electrogenerated hydrogen peroxide is disinfection of ballast waters, which are responsible for introduction of exotic aquatic organisms through cargo transport, as well as of seawater for aquaculture and swimming pool.

In a divided reactor, *Escherichia coli* and the total bacterial load were effectively removed, even if the high salinity of the medium caused electrode fouling and increased cell resistance [83]. In a membrane-free reactor, the occurrence of an anodic oxidation and reaction with the species anodically generated (especially active chlorine), as well as reactions with species cathodically generated (O_2^-, $^{\bullet}OH$, $O_2H^{\bullet}$) and decomposition at the cathode, seriously affect the production yields, preventing hydrogen peroxide from accumulating [84].

A combined treatment for the disinfection of ballast water, involving a sequential anodic (on BDD) and cathodic (on GDE) process, was tested for the inactivation of two dinoflagellates responsible for algal blooms, *Alexandrium minutum* Halim and *Alexandrium taylori* Balech, and of the pathogenic marine bacterium *Pseudomonas aeruginosa* [85]. In that case, cathodically electrogenerated hydrogen peroxide contributed to the reduction of treatment time and the removal of residual species.

Of note, decomposition of hydrogen peroxide forms water and oxygen as unique compounds, thus producing one of the least environmental impacts among all the known oxidants [86]. However, the reactivity of hydrogen peroxide does not allow provision of the long-term effect required by current legislation.

ISSUES RELATED TO DIRECT ELECTROLYSIS

Direct electrolysis for purification and disinfection of water usually involves the application of an electrical field or, more frequently, direct current to solutions with low conductivity. This issue may lead to high energy consumption as well as to low effectiveness, due to the mass transfer limitations related to the low concentrations of such inorganic precursors as chlorides or sulphates.

As the operative current density is concerned, the values should be carefully selected, considering the electrode processes involved in the treatment. When water electrolysis is the main electrochemical reaction involved, a minimum value is required to reach the reaction potential.

However, the water to be treated usually contains chlorides, and relatively high values of current density may result in the formation of undesired byproducts such as perchlorates, chlorates, and chlorites [87–89].

It is worth observing that high chloride content (considering 250 ppm as maximum) limits the formation of perchlorate in the treated water, current density and residence time being the only parameters influencing the formation of perchlorates [43].

Moreover, high current densities may lead to a process unsustainable due to the energy cost. The optimum values of current density are very low if compared to the values usually found in industrial electrolytic processes: optimum values were found in the order of 20–200 A/m^2, which are values one or two orders of magnitude less than those of the chloro-alkali process [90,91].

Mass Transfer Limitation

Mass transfer study represents one of the most important points to be afforded in order to correctly design the electrochemical reactor, especially when dilute solutions have to be treated. This is a typical situation that occurs not only during disinfection but also, for example, in metal removal processes or in all those cases in which the electrochemical process is used as a final step in the whole treatment process. Low concentrations of organics and inorganic pollutants require the use of a high specific electrode area and high mass transfer rate to make the related recovery process sustainable.

Actually, three-dimensional electrodes have been extensively tested for removing metal ions from water [92] as well as to remove organic and inorganic pollutants [93,94].

Mass transfer at porous flow-through electrodes has been the subject of several studies [95,96], aimed at obtaining a high space-time yield of electrochemical reactors suitable for processing dilute solutions [97].

In the specific case of disinfection during electrolysis of low conductive water, only a fraction of the applied current is used for the generation of disinfectants, such as active chlorine or reactive oxygen species. Most of the current is used for oxygen evolution: the gas bubbles at the anode surface can reduce the electrochemically active area, decreasing the efficiency and increasing ohmic drops in the layer of electrolyte adjacent to the electrode surface [98,99]. High mass transfer rates, that is, high turbulence conditions, may overcome these drawbacks. Moreover, because most of the disinfectant is generated at the anode surface and diffuse toward the bulk of the solution, high mass transfer rates enhance the concentration of these species. However, in undivided cells this effect may be balanced by the cathodic reduction of oxidants, which is, in turn, a mass transfer controlled reaction. Moreover, high turbulence conditions may lead to high pressure drop.

As a very common practice in electrochemical reactor engineering, mass transfer study is generally performed through limiting current approach. Mass transfer coefficients are derived from the limiting current values obtained with the standard redox ferricyanide/ferrocyanide couple as redox probe [100–102]. The method is based on the reduction of ferricyanide to ferrocyanide ions, and it allows the (convective-diffusion controlled) limiting current density i_L to be measured over a range of flow rates of the electrolyte. The limiting current densities are measured for $K_3Fe(CN)_6$ reduction and in presence of an excess of $K_4Fe(CN)_6$ in order to ensure that the anodic reaction never became the limiting process. Different supporting electrolytes are used, although potassium carbonate provides much more stable limiting current versus time behavior [103].

The values of mass transfer coefficient (k_m) are derived from:

$$k_m = \frac{i_L}{zFC_b} \tag{23}$$

and used to obtain dimensionless Sherwood (Sh) versus Reynold (Re) correlations in the form:

$$\mathrm{Sh} = a\,\mathrm{Re}^b\mathrm{Sc}^{0.33} \tag{24}$$

where a and b are fitting parameters, and the 0.33 power of the Schmidt number (Sc) is assumed a priori.

The traditional configuration of electrochemical reactors is based on two parallel electrodes, with the electrolyte flowing between them. Because empty parallel plate cells usually present low mass transfer rates, enhancements are often requested, which can be obtained by introducing physical obstacles in the flow path [104]. New cell design and electrode geometries were proposed to enhance the mass transfer; most of the studies focused on the optimization of flow distribution or on the use of obstacles and turbulence promoters along the flow path [105–108]. Significant improvement can be obtained by introducing such turbulent promoters as net-type spacers and foams in the reaction region [109,110].

Although a parallel plate system does not present high pressure drops, the flow restriction imposed by the manifolds should be considered. The increase in the pressure drop across the internal channel is often negligible compared with the overall value [111].

Table 2 shows some examples of dimensionless correlations for parallel plate electrochemical reactors (as parameters of Eq. 24). Data in Table 2 refer to different lab scale cells proposed in the literature, commercial (A and B) or homemade (HM), realized for the particular investigation (C—F). The turbulence promoters (TP) are plastic grids with different

TABLE 2 Comparison Between Values of Parameters in Eq. 24 From the Literature

Cell		*a*	*b*	d_e (mm)	$L \times H$ (mm)	Ref.
A (empty)	FM01-LC	0.22	0.71	9.7	40 × 160	[102]
A (with TP)		0.74	0.62			
B (with TP)	Electro MP-Cell	0.49	0.51	19.2	120 × 180	[89]
C (empty)	Home made, University of Alicante (Spain)	1.08	0.61	21.8	40 × 40	[112]
C (with TP)		1.11	0.53			
C (with TP)		1.25	0.56			
D (empty)		0.84	0.63	24.7	70 × 90	
D (with TP)		1.29	0.44			
E (empty)		0.17	0.82	5.8	70 × 90	
E (with TP)		0.54	0.59			
E (with TP)		0.43	0.63			
F (empty)	Home made, University of Cagliari (Italy)	0.32	0.51	19.2	120 × 180	[113]
F (with TP)		0.61	0.54			
F (with TP)		0.63	0.48			
Fully developed laminar flow		0.023	0.4			
Fully developed turbulent flow		2.5	0.8			

meshes; details can be found in the relevant references. The equivalent diameter (d_e) for the parallel plate configurations compared is given by:

$$d_e = 2\frac{LS}{(L+S)} \tag{25}$$

where L is the horizontal length of the plate and S is the interelectrode gap.

The values are compared with the relevant ones of laminar and turbulent flow: the slopes are higher than the values for laminar flow and quite similar to the slope of turbulent flow, even if the operative range of Re is typical of a laminar regime. This behavior has been attributed to the effect of manifolds because, as a consequence of practical design, most of the filter-press cells do not have a calming zone [110,114].

Among other configurations, stirred tank cells and three-dimensional electrodes have been considered.

Three-dimensional electrodes have attracted much attention for electrochemical treatment of water and WW when high specific areas and high mass transfer rates are required [115]. Three-dimensional electrode configuration was adopted in traditional filter press cells in which the anode was constituted by titanium grids with different coatings [116,117] as well as in fixed-bed reactors [118].

Values of the parameters a and b in Eq. 24 of 1.62 and 0.4 have been obtained for fixed-bed reactors with two-dimensional meshed electrodes and an electric field parallel to the electrolyte flow, using the bed diameter as a characteristic length [119]. Values of $a = 0.076$ and $b = 0.25$ were obtained with parallel plate reactors with three-dimensional meshed electrodes; in this case, the characteristic length was the equivalent diameter of particles if three-dimensional meshes had a spherical geometry [115]. Because different Re definitions were used, the values are not directly comparable; however, both are in agreement with the values reported in the literature for laminar flow in packed-bed reactors.

Stirred tank reactors have also been proposed for disinfection of water with low conductivity [120]. A continuous stirred tank reactor (CSTR) is extensively used in electrochemical processes; it consists of a vessel equipped with an anode, cathode, and stirrer. The main advantage of CSTR is related to the high retention times. This may increase the effectiveness of the disinfection process but, in turn, it may originate undesired byproducts, mainly when the water contains chlorides, as mentioned for parallel plate reactors. Another drawback of CSTR is the real mixing achievable, which depends on the complexity of flow in most mixing applications [121].

Reactor Design

The design of reactors usually starts from a more or less close representation of the flow behavior. The simplest approach is based on the hypothesis of ideal flow conditions, with the two extreme behaviors of a completely stirred reactor (CSTR) and a perfect plug flow reactor (PFR).

The PFR model is based on the assumption of no longitudinal but complete radial mixing, resulting in an identical velocity profile and residence time distribution for all elements of the flow.

The CSTR model is characterized by identical composition of the fluid at the exit stream and within the reactor. However, the real flow pattern inside the treatment unit normally presents dispersion phenomena, dead or stagnant zones, bypass flows, exchanges between them, and recirculation zones that can affect the efficiency of the process [122]. To achieve the contact time suitable for disinfection, low volumetric flow rates may be required. Under these conditions, the effects of canalization and stagnant zones are important, and deviations from the ideal flow pattern should be considered [90].

In the electrochemical disinfection process, dead zones may result in long reaction times, favoring the bulk reactions to form undesired by-products. In electrolysis of water containing chlorides, a high residence time leads to the formation of chlorates by bulk reactions, such as reactions between two electrogenerated species, namely active chlorine and reactive oxygen species, or by disproportionation of chlorine dioxide [85].

The effects of bypass are contact times lower than those predicted by considering the hydraulic residence time. This leads to the overestimation of the disinfection efficiency, mainly when bulk disinfection is the main mechanism involved in the process.

To account for the nonideal flow behavior of electrochemical cells, two approaches were proposed: the first involves the use of traditional models, such as in series CSTR or dispersed plug-flow, and the second is based on computational fluid dynamics.

The hydrodynamic behavior of nonideal reactors can be interpreted using models consisting of a combination of plug-flow zones, ideally mixed zones, by-pass flow elements, and dead zones. However, an experimental characterization of the reactors is necessary in order to both obtain the parameters of the model and to validate the results of the model predictions. A characterization of the flow behavior of the cells can be obtained by the residence time distribution (RTD) curves [27,122].

A number of papers appeared in the literature on the performance of various reactors. RTDs in vessels were represented by a combination of different tanks or volumes [123], as two parallel flows dominated by a perfectly mixed volume corrupted by the presence of bypass and stagnant zones [124], by two CSTRs in parallel [121], or by a CSTR with dead zone [122].

An in series, a CSTR model was adopted for empty parallel plate cells [89] while a dispersed plug flow was more representative in the presence of turbulence promoters [125] and in a fixed bed with three-dimensional meshed electrodes [117].

A plug flow behavior with both axial and lateral dispersion phenomena was also observed for a filter press-type electrochemical reactor [126].

RTD curves are obtained by pulse-response tests. A sharp pulse of inert tracer is injected upstream of the reactor while the concentration of the tracer in the outlet stream is measured by a suitable detector. The response of the detector is the RTD.

The analysis of the experimental RTD may be done by evaluating the following parameters:

(1) Exit age distribution $E(t)$

$$E(t) = \frac{C(t)}{\int_0^\infty C(t)dt} \tag{26}$$

(2) Mean residence time

$$t_m = \frac{\int_0^\infty tC(t)dt}{\int_0^\infty C(t)dt} \tag{27}$$

(3) Variance

$$\sigma^2 = \frac{\int_0^\infty (t-\tau)^2 C(t)dt}{\int_0^\infty C(t)dt} \tag{28}$$

Depending on the system, different boundary conditions are needed to describe the inlet and outlet, which correspond to different relationship of the Peclet number (Pe $= \nu H/D$) as a function of mean residence time and variance. From Pe, such model parameters as the number of in-series CSTR or the axial dispersion coefficient can be calculated [126,127].

Table 3 shows some example of the results in terms of Pe as a function of the surface velocity for different cell configurations. A wide range of velocities has been used, and the corresponding Pe numbers are typical

TABLE 3 Pe as a Function for Different Cell Configurations

v (cm/s)	Pe	Cell	Ref.
0.06	12.7	Commercial EC Electro MP-Cell (with turbulence promoter)	[89]
0.1	14.1		
0.2	27.2		
0.02	5.5	Homemade fixed bed cell with three-dimensional meshed electrodes	[117]
0.08	5.9		
0.3	6.2		
10	5.4	Commercial ICI-INEOS FM01-LC (empty cell)	[126]
15	5.5		
20	6.1		
10	6.8	Commercial ICI-INEOS FM01-LC (with three-dimensional RVC electrode)	[126]
15	11.1		
20	12.4		

of moderate and highly dispersed systems, confirming the nonideal behavior of electrochemical reactors.

Computational fluid dynamics (CFD) is considered a feasible tool for reactor design purposes. CFD has been employed to model hydrodynamics of empty channels [127] as well as in electrochemical cells with different turbulence promoters. CFD offers the possibility for evaluating, under realistic conditions, the flow pattern and its effect within the reactor [128].

In modeling electrochemical disinfection processes, the numerical approach must consider the solution of transport governing equations of fluid dynamics, which must be combined with kinetic equations. The kinetic equations should represent both electrochemical processes at the electrodes and bulk reactions involving chemical species and/or microorganisms. This allows interpreting the effect of velocity distribution on the effectiveness of disinfection, obtaining the profiles of concentration of all the species involved in the process [129,130].

Implementation of the CFD model can be summarized in the following steps:

- Definition of the reactor geometry (shape, length, dimension, and geometry of inlet and outlet ports). This usually involves the drawing of the reactor in an appropriate graphic environment. The graphic environment is embedded in several commercial software programs of numerical simulation.
- Building of the different domains with the different flow conditions. In an electrochemical reactor, Navier Stokes equations for flow in porous media can be used, depending on the conditions.
- Definition of mesh pattern and mesh size.
- Definition of fluid properties and selection of boundary conditions at the inlet and outlet ports and at the walls.

Once the velocity and pressure profiles within the reactor are obtained, the fluid model should be combined with the kinetic model, which is constituted by the equation balance of each of the species involved with the corresponding kinetics [117]. The electrochemical reactions are usually expressed as boundary conditions at the electrode surface, as Faraday's law.

CONCLUSIONS

This chapter presented a general view of some aspects related to electrochemical WW treatment. Attention has been focused on the disinfection process, which represents the final treatment step of WW in the case of either reuse or discharge in the final receptor water sources.

Some aspects are mainly examined, which may be crucial in terms of costs of the process and/or pollutant removal. Of note, depending on the composition of the WW, problems of major concern are represented by: (i) the low conductivity of the treated water, which may lead to high current values needed for the electrochemical process, and (ii) the presence of chloride/bromide, which may lead to generation of by-products, which can be more dangerous than the original pollutants.

In the present work, first the possible electrode materials are considered for the production of active oxidizing agents via anodic oxidation, or via cathodic reduction. The behavior of active or nonactive oxides is considered, at which the electrochemical oxidation is supposed to occur through the formation of higher metal oxides or mediated by hydroxyl radicals generated at the anode surface by water discharge. In any case, the effect of the presence of chloride is investigated.

Sacrificial anodes are then taken into account as a possible way to obtain ferrate ions as a powerful oxidizing agent, which is able to inactivate a variety of microorganisms and some viruses and also provide a coagulant action.

Information is also given on the cathodic generation of oxidizing agent at carbon electrodes, which occurs via reduction of the dissolved oxygen leading to the formation of H_2O_2.

The reaction mechanism of pollutant demolition and inactivation of bacteria is considered.

The study is completed by the analysis and modeling of the fluid dynamics of the reactor, along with the study of the kinetics of the process, which are paramount to model the complex reactions that occur at the electrodes and in the bulk solution involving chemical species and/or microorganisms. This allows interpreting the effect of velocity distribution on the effectiveness of disinfection, obtaining the profiles of concentration of all the species involved in the process.

References

[1] T. Asano, A.D. Levine, Wastewater reclamation, recycling and reuse: past, present and future, Water Sci. Technol. 33 (10 – 11) (1996) 1–14.

[2] M. Falkenmark, A. Berntell, A. Jagerskog, J. Lundqvist, M. Matz, H. Tropp, On the Verge of a New Water Scarcity: A Call for Good Governance and Human Ingenuity, SIWI Policy Brief, SIWI, Sundbyberg, Sweden, 2007.

[3] M.E. Webber, Catch 22: water vs. energy, scientific American, Earth 3 (18) (2008) 34–41.

[4] V. Lazarova, K.-H. Choo, P. Cornel, Water–Energy Interactions in Water Reuse, first ed., IWA Publishing, London, 2012.

[5] G.K. Pearce, UF/MF pre-treatment to RO in seawater and wastewater reuse applications: a comparison of energy costs, Desalination 222 (2008) 66–73.

[6] C. Rodriguez, P.V. Buynder, R. Lugg, P. Blair, B. Devine, A. Cook, P. Weinstein, Indirect potable reuse: a sustainable water supply alternative, Int. J. Environ. Res. Public Health 6 (2009) 1174–1209.
[7] T. Asano, F.L. Burton, H.L. Leverenz, R. Tsuchihashi, G. Tchobanoglous, Water Reuse–Issues, Technologies and Applications, McGraw-Hill, New York, NY, 2007.
[8] E. Eriksson, K. Auffarth, M. Henze, A. Ledin, Characteristics of grey wastewater, Urban Water 4 (2002) 85e104.
[9] F.Y. Li, Review of the technological approaches for grey water treatment and reuses, Sci. Total Environ. 407 (2009) 3439–3449.
[10] P.R. Thomas, G.R. Greene, Rainwater quality from different roof catchments, Water Sci. Technol. 28 (1993) 291–299.
[11] M. Henze, Y. Comeau, in: M. Henze, M.C.M. van Loosdrecht, G.A. Ekama, D. Brdjanovic (Eds.), Wastewater Characterization in Biological Wastewater Treatment: Principles Modelling and Design, IWA Publishing, London, 2008. ISBN: 9781843391883.
[12] A. Morel, S. Diener, M. Alderlieste, A. Baumeyer, M.J. Bino, J. Burnat, S. Dallas, M. Hind, C. Martin, N. Priest, R.R. Shrestha, Greywater Management in Low and Middle Income Countries, Retrieved from, http://www.eawag.ch/forschung/sandec/publikationen/ewm/dl/MorelDiener Greywater 2006.pdf, 2006.
[13] R.S. Magazinovic, B.C. Nicholson, D.E. Mulcahy, D.E. Davey, Bromide levels in natural waters: its relationship to levels of both chloride and total dissolved solids and the implications for water treatment, Chemosphere 57 (2004) 329–335.
[14] S.D. Richardson, M.J. Plewa, E.D. Wagner, R. Schoeny, D.M. DeMarini, Occurrence, genotoxicity, and carcinogenicity of regulated and emerging disinfection by-products in drinking water: a review and roadmap for research, Mutat. Res. Rev. Mutat. Res. 636 (2007) 178–242.
[15] Y. Bichsel, U. von Gunten, Formation of iodo-trihalomethanes during disinfection and oxidation of iodide-containing waters, Environ. Sci. Technol. 34 (2000) 2784–2791.
[16] G. Hua, D.A. Reckhow, J. Kim, Effect of bromide and iodide ions on the formation and speciation of disinfection byproducts during chlorination, Environ. Sci. Technol. 40 (2006) 3050–3056.
[17] U. von Gunten, J. Hoigne, A. Bruchet, Bromate formation during ozonation of bromide-containing waters, Water Supply 13 (1995) 45–50.
[18] U. von Gunten, Ozonation of drinking water: part II. Disinfection and byproduct formation in presence of bromide, iodide or chlorine, Water Res. 37 (2003) 1469–1487.
[19] O. Simond, V. Schaller, C. Comninellis, Theoretical model for the anodic oxidation of organics on metal oxide electrodes, Electrochim. Acta 42 (13–14) (1997) 2009–2012.
[20] A.M. Polcaro, S. Palmas, Electrochemical oxidation of chlorophenols, Ind. Eng. Chem. Res. 36 (1997) 1791–1798.
[21] M. Gattrell, K. Kirk, Study of the oxidation of phenol at platinum and preoxidized platinum surfaces, J. Electrochem. Soc. 140 (6) (1993) 1534.
[22] Z.C. Wu, M. Zhou, Partial degradation of phenol by advanced electrochemical oxidation process, Environ. Sci. Technol. 35 (13) (2001) 2698.
[23] J.D. Rodgers, W. Jedral, N.J. Bunce, Electrochemical oxidation of chlorinated phenols, Environ. Sci. Technol. 33 (1999) 1453.
[24] G. Fóti, C. Comninellis, Ralph, E. White, B.E. Conway, C.G. Vayenas, M.E. Gamboa-Adelco (Eds.), Electrochemical oxidation of organics on iridium oxide and synthetic diamond based electrodes, in: Modern Aspects of Electrochemistry, vol. 37, Plenum, New York, NY, 2004, pp. 87–130
[25] F. Bonfatti, A. De Battisti, S. Ferro, G. Lodi, S. Osti, Anodic mineralization of organic substrates in chloride-containing aqueous media, Electrochim. Acta 46 (2000) 305–314.

[26] F. Bonfatti, S. Ferro, F. Lavezzo, M. Malacarne, G. Lodi, A. De Battisti, Electrochemical incineration of glucose as a model organic substrate. II. Role of active chlorine mediation, J. Electrochem. Soc. 147 (2000) 592–596.
[27] C. Comninellis, A. Nerini, Anodic oxidation of phenol in the presence of NaCl for wastewater treatment, J. Appl. Electrochem. 25 (1995) 23–28.
[28] L.C. Chiang, J.E. Chang, T.C. Wen, Indirect oxidation effect in electrochemical oxidation treatment of landfill leachate, Water Res. 29 (1995) 671–678.
[29] A.M. Polcaro, M. Mascia, S. Palmas, A. Vacca, Kinetic study on the removal of organic pollutants by electrochemical oxidation process, Ind. Eng. Chem. Res. 41 (12) (2002) 2874–2881.
[30] AWWA (Ed.), Water Chlorination/Chloramination Practices and Principles, second ed., AWWA Manual M20. AWWA catalog no. 30020. American Water Works Association, Denver, CO, 2011, p. 173.
[31] A. Kraft, Electrochemical water disinfection: a short review, Platin. Met. Rev. 52 (2008) 177–185.
[32] L.F. Trigueiro, L.M. Silva, L.A.B.D. Itto, T.M.B.F. Oliveira, A.J. Motheo, C.A. Martínez-Huitle, J.J.F. Alves, S.S.L. Castro, Inactivation, lysis and degradation by-products of Saccharomyces Cerevisiae by electrooxidation using DSA, Environ. Sci. Pollut. Res. 24 (2017) 6096–6105.
[33] I.C.P. Gusmão, P.B. Moraes, E.D. Bidoia, Studies on the electrochemical disinfection of water containing escherichia coli using a dimensionally stable anode, Braz. Arch. Biol. Technol. 53 (2010) 1235–1244.
[34] I.C.P. Gusmão, P.B. Moraes, E.D. Bidoia, A thin layer electrochemical cell for disinfection of water contaminated with Staphylococcus Aureus (Uma Célula eletroquímica de camada delgada para desinfecção de água contaminada com Staphylococcus Aureus), Braz. J. Microbiol. 40 (2009) 649–654.
[35] X. Huang, Y. Qu, C.A. Cid, C. Finke, M.R. Hoffmann, K. Lim, S.C. Jiang, Electrochemical disinfection of toilet wastewater using wastewater electrolysis cell, Water Res. 92 (2016) 164–172.
[36] B. Marselli, J. Garcia-Gomez, P.A. Michaud, M.A. Rodrigo, C. Comninellis, Electrogeneration of hydroxyl radicals on boron-doped diamond electrodes, J. Electrochem. Soc. 150 (2003) D79–D83.
[37] J. Iniesta, P.A. Michaud, M. Panizza, G. Cerisola, A. Aldaz, C. Comninellis, Electrochemical oxidation of phenol at boron-doped diamond electrode, Electrochim. Acta 46 (2001) 3573–3578.
[38] M.H.P. Santana, L.A.D. Faria, J.F.C. Boodts, Electrochemical characterisation and oxygen evolution at a heavily boron doped diamond electrode, Electrochim. Acta 50 (2005) 2017–2027.
[39] G. Patermarakis, E. Fountoukidis, Disinfection of water by electrochemical treatment, Water Res. 24 (1990) 1491–1496.
[40] K. Groenen-Serrano, P.A. Michaud, C. Comninellis, A. Savall, Electrochemical preparation of peroxodisulfuric acid using boron doped diamond thin film electrodes, Electrochim. Acta 48 (2002) 431–436.
[41] N. Ibl, D.J. Landolt, On the mechanism of anodic chlorate formation in dilute NaCl solutions, Electrochem Soc 115 (1968) 713–720.
[42] L.R. Czarnetzki, L.J.J. Janssen, Formation of hypochlorite, chlorate and oxygen during NaCl electrolysis from alkaline solutions at an RuO2/TiO2 anode, J. Appl. Electrochem. 22 (1992) 315–324.
[43] M.E.H. Bergmann, J. Rollin, T. Iourtchouk, The occurrence of perchlorate during drinking water electrolysis using BDD electrodes, Electrochim. Acta 54 (7) (2009) 2102–2107.
[44] M.E.H. Bergmann, A.S. Koparal, Studies on electrochemical disinfectants production using anodes containing RuO_2, J. Appl. Electrochem. 35 (2005) 1321–1329.

[45] M.E.H. Bergmann, A.T. Koparal, A.S. Koparal, F. Ehrig, The influence of products and by-products obtained by drinking water electrolysis on microorganisms, Microchem. J. 89 (2008) 98–107.

[46] G. Pérez, P. Gómez, R. Ibañez, I. Ortiz, A.M. Urtiaga, Electrochemical disinfection of secondary wastewater treatment plant (WWTP) effluent, Water Sci. Technol. 62 (2010) 892–897.

[47] D. Ghernaout, A. Badis, A. Kellil, B. Ghernaout, Application of electro coagulation in *Escherichia coli* culture and two surface waters, Desalination 219 (2008) 118–125.

[48] D. Gaskova, K. Sigler, B. Janderová, J. Plášek, Effect of high-voltage electric field pulses on yeast cell: factors influencing the killing efficiency, Bioelectrochem. Bioenerg. 39 (1996) 195–202.

[49] K.P. Drees, M. Abbaszadegan, R.M. Maier, Comparative electrochemical inactivation of bacteria and bacteriophage, Water Res. 37 (2003) 2291–2300.

[50] B. Hamm, Disinfection by-product reduction using on-site generated mixed oxidants in groundwater treatment, Wat Cond Purif 44 (2002) 24–27.

[51] H. Son, M. Cho, J. Kim, B. Oh, H. Chung, J. Yoon, Enhanced disinfection efficiency of mechanically mixed oxidants with free chlorine, Water Res. 39 (2005) 721–727.

[52] A.M. Polcaro, A. Vacca, M. Mascia, S. Palmas, J. Rodriguez Ruiz, Electrochemical treatment of waters with BDD anodes: kinetics of the reactions involving chlorides, J. Appl. Electrochem. 39 (2009) 2083–2092.

[53] World Health Organization (WHO), Guidelines for Drinking Water Quality, third ed., WHO Press, Geneva, 2008.

[54] J.-Q. Jiang, B. Lloyd, Progress in the development and use of ferrate(VI) salt as an oxidant and coagulant for water and wastewater treatment, Water Res. 36 (2002) 1397–1408.

[55] R.H. Wood, The heat, free energy and entropy of the ferrate (VI) ion, J. Am. Chem. Soc. 80 (1958) 2038–2041.

[56] M. De Koninck, D. Belanger, The electrochemical generation of ferrate at pressed iron powder electrodes: comparison with foil electrode, Electrochem. Acta 48 (2003) 1435–1442.

[57] M. Alsheyab, J.-Q. Jiang, C. Stanford, On-line production of ferrate with an electrochemical method and its potential application for wastewater treatment—a review, J. Environ. Manage. 90 (3) (2009) 1350–1356.

[58] K. Bouzek, I. Rousar, Influence of anode material on current yields during ferrate (VI) production by anodic iron dissolution. Part 1: Current efficiency during anodic dissolution of grey cast iron to ferrate (VI) in concentrated alkali hydroxide solutions, J. Appl. Electrochem. 26 (9) (1996) 919–924.

[59] A. Denver, D. Pletcher, Electrochemical generation of ferrate. 2. Influence of anode composition, J. Appl. Electrochem. 26 (1996) 823–827.

[60] V. Sharma, F. Kazama, H. Jiangyong, A. Ray, Ferrates (iron (VI) and iron (V)): environmentally friendly oxidants and disinfectants, J. Water Health 3 (2005) 45–58.

[61] B.J. Yates, R. Zboril, V.K. Sharma, Engineering aspects of ferrate in water and wastewater treatment-a review, journal of environmental science and health—part a toxic/hazardous substances and, Environ. Eng. 49 (14) (2014) 1603–1614.

[62] J.-Q. Jiang, H.B.P. Durai, R. Winzenbacher, M. Petri, W. Seitz, Drinking water treatment by in situ generated ferrate(VI), Desalin. Water Treat. 55 (3) (2015) 731–739.

[63] E. Gombos, T. Felfoldi, K. Barkacs, C. Vertes, B. Vajna, G. Zaray, Ferrate treatment for inactivation of bacterial community in municipal secondary effluent, Bioresour. Technol. 107 (2012) 116–121.

[64] N.A. Kruusenberg, K. Tammeveski, The pH-dependence of oxygen reduction on multiwalled carbon nanotube modified glassy carbon electrodes, Carbon 47 (2009) 651–658.

[65] S.J. DeLuca, A.C. Chao, C. Smallwood, Removal of organic priority pollutants by oxidation-coagulation, J. Environ. Eng. 109 (1983) 36–46.

[66] J.-Q. Jiang, Z. Zhou, S. Patibandla, X. Shu, Pharmaceutical removal from wastewater by ferrate(VI) and preliminary effluent toxicity assessments by the zebrafish embryo model, Microchem. J. 110 (2013) 239–245.

[67] G.A. Alvarez, D. Pletcher, The removal of low level organics via hydrogen peroxide formed in a reticulated vitreous carbon cathode cell. Part 1. The electrosynthesis of hydrogen peroxide in aqueous acidic solutions, Electrochim. Acta 44 (1998) 853–861.

[68] Z. Qiang, C.P. Huang, Electrochemical generation of hydrogen peroxide from dissolved oxygen in acidic solutions, Water Res. 36 (2002) 85–94.

[69] A. Da Pozzo, L. Di Palma, C. Merli, E. Petrucci, An experimental comparison of a graphite electrode and a gas diffusion electrode for the cathodic production of hydrogen peroxide, J. Appl. Electrochem. 35 (2005) 413–419.

[70] E. Peralta, R. Natividad, G. Roa, R. Marin, R. Romero, T. Pavon, A comparative study on the electrochemical production of H2O2 between BDD and graphite cathodes, sustain, Environ. Res. 23 (2013) 259–266.

[71] J. Miao, H. Zhu, Y. Tang, Y. Chen, P. Wan, Graphite felt electrochemically modified in H_2SO_4 solution used as a cathode to produce H_2O_2 for pre-oxidation of drinking water, Chem. Eng. J. 250 (2014) 312–318.

[72] A. Wang, J. Qu, H. Liu, J. Ru, Mineralization of an azo dye acid red 14 by photoelectro-Fenton process using an activated carbon fiber cathode, Appl. Catal. Environ. 84 (2008) 393–399.

[73] A. Özcan, Y. Şahin, A.S. Koparal, M.A. Oturan, Carbon sponge as a new cathode material for the electro-Fenton process: comparison with carbon felt cathode and application to degradation of synthetic dye basic blue 3 in aqueous medium, J. Electroanal. Chem. 616 (2008) 71–78.

[74] E.P. Randviir, C.E. Banks, The oxygen reduction reaction at graphene modified electrodes, Electroanalysis 26 (2014) 76–83.

[75] E. Brillas, R.M. Bastida, E. Llosa, J. Casado, Electrochemical destruction of aniline and 4-chloroaniline for wastewater treatment using a carbon-PTFE O_2-fed cathode, J. Electrochem. Soc. 142 (1995) 1733–1741.

[76] T. Harrington, D. Pletcher, The removal of low level of organics from aqueous solutions using Fe(II) and hydrogen peroxide formed in situ at gas diffusion electrodes, J. Electrochem. Soc. 146 (1999) 2983–2989.

[77] I. Sirés, E. Guivarch, N. Oturan, M.A. Oturan, Efficient removal of triphenylmethane dyes from aqueous medium by in situ electrogenerated Fenton's reagent at carbon-felt cathode, Chemosphere 72 (2008) 592–600.

[78] H. Lin, H. Zhang, X. Wang, L. Wang, J. Wu, Electro-Fenton removal of Orange II in a divided cell: reaction mechanism, degradation pathway and toxicity evolution, Sep. Purif. Technol. 122 (2014) 533–540.

[79] E.L. Gyenge, C.W. Oloman, Electrosynthesis of hydrogen peroxide in acidic solutions by mediated oxygen reduction in a three-phase (aqueous/organic/gaseous) system part I: emulsion structure, electrode kinetics and batch electrolysis, J. Appl. Electrochem. 33 (2003) 655–663.

[80] Q. Li, C. Batchelor-McAuley, N.S. Lawrence, R.S. Hartshorne, C.J.V. Jones, R. G. Compton, A flow system for hydrogen peroxide production at reticulated vitreous carbon via electroreduction of oxygen, J. Solid State Electrochem. 18 (2014) 1215–1221.

[81] E. Petrucci, A. Da Pozzo, L. Di Palma, On the ability to electrogenerate hydrogen peroxide and to regenerate ferrous ions of three selected carbon-based cathodes for electro-Fenton processes, Chem. Eng. J. 283 (2016) 750–758.

[82] A. Da Pozzo, E. Petrucci, C. Merli, Electrogeneration of hydrogen peroxide in seawater and application to disinfection, J. Appl. Electrochem. 38 (2008) 997–1003.

[83] G.R. Agladze, G.S. Tsurtsumia, B.–.I. Jung, J.S. Kim, G. Gorelishvili, Comparative study of hydrogen peroxide electro-generation on gas-diffusion electrodes in undivided and membrane cells, J. Appl. Electrochem. 37 (2007) 375–383.

[84] E. Petrucci, D. Montanaro, L. Di Palma, A feasibility study of hydrogen peroxide lectrogeneration in seawater for environmental remediation, Chem. Eng. Trans. 28 (2012) 91–96.
[85] E. Petrucci, L. Di Palma, E. De Luca, G. Massini, Biocides electrogeneration for a zero-reagent on board disinfection of ballast water, J. Appl. Electrochem. 43 (2) (2013) 237–244.
[86] P. Drogui, S. Elmaleh, M. Rumeau, C. Bernard, A. Rambaud, Oxidising and disinfecting by hydrogen peroxide produced in a two-electrode cell, Water Res. 35 (2001) 3235.
[87] M.E.H. Bergmann, J. Rollin, Product and by-product formation in laboratory studies on disinfection electrolysis of water using boron-doped diamond anodes, Catal. Today 124 (2007) 198–203.
[88] M. Mascia, A. Vacca, S. Palmas, A.M. Polcaro, Kinetics of the electrochemical oxidation of organic compounds at BDD anodes: modeling of surface reactions, J. Appl. Electrochem. 37 (2007) 71–76.
[89] A.M. Polcaro, A. Vacca, M. Mascia, F. Ferrara, Product and by-product formation in electrolysis of dilute chloride solutions, J. Appl. Electrochem. 38 (2008) 979–984.
[90] A. Vacca, M. Mascia, S. Palmas, A. Da Pozzo, Electrochemical treatment of water containing chlorides under non-ideal flow conditions with BDD anodes, J. Appl. Electrochem. 41 (2011) 1087–1097.
[91] M. Mascia, A. Vacca, A.M. Polcaro, S. Palmas, J. Rodriguez Ruiz, A. Da Pozzo, Electrochemical treatment of phenolic waters in presence of chloride with boron doped diamond (BDD) anodes: Experimental study and mathematical model, J. Hazard. Mater. 174 (2010) 314–322.
[92] B.K. Ferreira, Three-dimensional electrodes for the removal of metals from dilute solutions: a review, Miner. Process. Extract. Metall. Rev. Int. J. 29 (2008) 330–371.
[93] A.M. Polcaro, S. Palmas, F. Renoldi, M. Mascia, Three-dimensional electrodes for the electrochemical combustion of organic pollutants, Electrochim. Acta 46 (2000) 389–394.
[94] H. Cui, Y. Qian, H. An, C. Sun, J. Zhai, Q. Li, Electrochemical removal of fluoride from water by PAOA modified carbon felt electrodes in a continuous flow reactor, Water Res. 46 (2012) 3943–3950.
[95] G.H. Sedahmed, Y.A. El-Taweel, A.H. Konsowa, M.H. Abdel-Aziz, Mass transfer intensification in an annular electrochemical reactor by an inert fixed bed under various hydrodynamic conditions, Chem. Eng. Process. 50 (2011) 1122–1127.
[96] E.A. Soltan, S.A. Nosier, A.Y. Salem, I.A.S. Mansour, G.H. Sedahmed, Mass transfer behaviour of a flow-by fixed bed electrochemical reactor under different hydrodynamic conditions, Chem. Eng. J. 91 (2003) 33–44.
[97] A.A. Mobarak, M.S.E. Abdo, M.S.M. Hassan, G.H. Sedahmed, Mass transfer behaviour of a flow-by fixed bed electrochemical reactor composed of a vertical stack of screens under single and upward two phase flow, J. Appl. Electrochem. 30 (2000) 1269–1276.
[98] H. Vogt, The rate of gas evolution at electrodes I. An estimate of efficiency of gas evolution from the supersaturation of electrolyte adjacent to a gas-evolving electrode, Electrochim. Acta 19 (1984) 167–173.
[99] I.L.T. Albuquerque, E.B. Cavalcanti, E.O. Vilar, Mass transfer study of electrochemical processes with gas production, Chem. Eng. Process. 48 (2009) 1432–1436.
[100] W.M. Taama, R.E. Plimey, K. Scott, Mass transfer rates in a DEM electrochemical cell, Electrochim. Acta 41 (1996) 543–551.
[101] T.R. Ralph, M.L. Hitchman, J.P. Millington, F.C. Walsh, Mass transport in an electrochemical laboratory filter-press reactor and its enhancement by turbulence promoters, Electrochim. Acta 41 (1996) 591–599.
[102] M. Griffiths, C. Ponce de Leon, F.C. Walsh, Mass transport in the rectangular channel of a filter-press electrolyzer (the FM01-LC reactor), AICHE J. 51 (2005) 681–687.

[103] W.M. Taama, E.R. Plimey, K. Scott, Influence of supporting electrolyte on ferricyanide reduction at a rotating disc electrode, Electrochim. Acta 41 (1996) 549–551.
[104] J.L.C. Santos, V. Geraldes, S. Velizarov, J.G. Crespo, Characterization of fluid dynamics and mass-transfer in an electrochemical oxidation cell by experimental and CFD studies, Chem. Eng. J. 157 (2010) 379–392.
[105] C.F. Oduoza, A.A. Wragg, Local mass-transfer distribution in the channels of a serpentine flow baffled parallel plate cell, Chem. Eng. J. 85 (2002) 119–126.
[106] A. Djati, M. Brahimi, J. Legrand, B. Saidani, Entrance effect on mass transfer in a parallel plate electrochemical reactor, J. Appl. Electrochem. 31 (2001) 833–837.
[107] C.F. Oduoza, A.A. Wragg, M. Patrick, The effects of a variety of wall obstructions on local mass transfer in a parallel plate electrochemical flow cell, Chem. Eng. J. 68 (1997) 145–155.
[108] C. Bengoa, A. Montillet, P. Legentilhomme, J. Legrand, Characterization and modeling of the hydrodynamic behavior in the filter-press-type FM01-LC electrochemical cell by direct flow visualization and residence time distribution, Ind. Eng. Chem. Res. 39 (2000) 2199–2206.
[109] A. Wragg, A. Leontaritis, Local mass transfer and current distribution in baffled and unbaffled parallel plate electrochemical reactors, Chem. Eng. J. 66 (1997) 1–10.
[110] T.R. Ralph, M.L. Hitchman, J.P. Millington, F.C. Walsh, Mass transport in an electrochemical laboratory filter-press reactor and its enhancement by turbulence promoters, Electrochim. Acta 41 (1996) 591–603.
[111] J. Gonzalez-Garcia, A. Frias, E. Exposito, V. Montiel, A. Aldaz, J.A. Conesa, Characterization of an electrochemical pilot-plant filter-press reactor by hydrodynamic and mass transport studies, Ind. Eng. Chem. Res. 39 (2000) 1132–1142.
[112] A. Frías-Ferrer, J. González-García, V. Sáez, C. De Ponce León, F.C. Walsh, The effects of manifold flow on mass transport in electrochemical filter-press reactors, AICHE J. 54 (3) (2008) 811–823.
[113] M. Mascia, A. Vacca, A.M. Polcaro, S. Palmas, A. Da Pozzo, Electrochemical treatment of simulated ground water containing MTBE and BTEX with BDD anodes, J. Chem. Technol. Biotechnol. 86 (2011) 128–136.
[114] P. Trinidad, F.C. Walsh, Hydrodynamic behaviour of the FM01-LC reactor, Electrochim. Acta 41 (1996) 493–502.
[115] L. Wei, S. Guo, G. Yan, C. Chen, X. Jiang, Electrochemical pretreatment of heavy oil refinery wastewater using a three-dimensional electrode reactor, Electrochim. Acta 55 (2010) 8615–8620.
[116] F.F. Rivera, M.R. Cruz-Díaz, E.P. Rivero, I. González, Analysis and interpretation of residence time distribution experimental curves in FM01-LC reactor using axial dispersion and plug dispersion exchange models with closed–closed boundary conditions, Electrochim. Acta 56 (2010) 361–371.
[117] M. Cruz-Díaz, F.F. Rivera, E.P. Rivero, I. González, The FM01-LC reactor modeling using axial dispersion model with a reaction term coupled with a continuous stirred tank (CST), Electrochim. Acta 63 (2012) 47–54.
[118] N.R. Neti, R. Misra, Efficient degradation of reactive blue 4 in carbon bed electrochemical reactor, Chem. Eng. J. 184 (2012) 23–32.
[119] M. Mascia, A. Vacca, S. Palmas, Fixed bed reactors with three dimensional electrodes for electrochemical treatment of waters for disinfection, Chem. Eng. J. 211–212 (2012) 479–487.
[120] A.M. Polcaro, A. Vacca, M. Mascia, S. Palmas, R. Pompei, S. Laconi, Characterization of a stirred tank electrochemical cell for water disinfection processes, Electrochim. Acta 52 (2007) 2595–2602.

[121] R. Saravanathamizhan, R. Paranthaman, N. Balasubramanian, C. Ahmed Basha, Residence time distribution in continuous stirred tank electrochemical reactor, Chem. Eng. J. 142 (2) (2008) 209–216.

[122] V.G. Tzatchkov, S.G. Buchberger, A. Martin-Dominguez, Water treatment units residence-time distributions as probabilistic functions, J. Environ. Eng. 135 (2009) 944–952.

[123] S.P. Ferro, R.J. Principe, M.B. Goldschmit, A new approach to the analysis of vessel residence time distribution curves, Metall. Mater. Trans. B Process Metall. Mater. Process. Sci. 32B (2001) 1185–1193.

[124] L.R.P. de Andrade Lima, D. Hodouin, Residence time distribution of an industrial mechanically agitated cyanidation tank, Miner. Eng. 18 (2005) 613–621.

[125] P. Trinidad, C. Ponce de León, F.C. Walsh, The application of flow dispersion models to the FM01-LC laboratory filter-press reactor, Electrochim. Acta 52 (2006) 604–613.

[126] C. Bengoa, A. Montillet, A. Legentilhomme, J. Legrand, Flow visualization and modelling of a filter-press type electrochemical reactor, J. Appl. Electrochem. 27 (1997) 1313–1322.

[127] E. Pellerin, E. Michelistsch, K. Darcovich, S. Lin, C.M. Tam, Turbulent transport in membrane modules by CFD simulation in two dimensions, J. Membr. Sci. 100 (1995) 139–153.

[128] L. Vázquez, A. Alvarez-Gallegos, F.Z. Sierra, C. Ponce de León, F.C. Walsh, CFD evaluation of internal manifold effects on mass transport distribution in a laboratory filter-press flow cell, J. Appl. Electrochem. 43 (4) (2013) 453–465.

[129] M. Mascia, A. Vacca, S. Palmas, Electrochemical treatment as a pre-oxidative step for algae removal using Chlorella Vulgaris as a model organism and BDD anodes, Chem. Eng. J. 219 (2013) 512–519.

[130] M. Mascia, S. Monasterio, A. Vacca, S. Palmas, Electrochemical treatment of water containing Microcystis Aeruginosa in a fixed bed reactor with three-dimensional conductive diamond anodes, J. Hazard. Mater. 319 (2016) 111–120.

PART III

CHEMICAL AND TECHNOLOGICAL ADVANTAGES AND DISADVANTAGES OF ELECTROCHEMICAL APPROACHES

CHAPTER

17

Advantages, Disadvantages, and Future Challenges of the Use of Electrochemical Technologies for Water and Wastewater Treatment

Brian P. Chaplin

University of Illinois at Chicago, Chicago, IL, United States

OUTLINE

Electrochemical Water and Wastewater Treatment
https://doi.org/10.1016/B978-0-12-813160-2.00017-1

INTRODUCTION

This chapter will provide an overview of the advantages, disadvantages, and emerging challenges associated with the use of electrochemical technologies for water treatment. A specific focus will be given to electrochemical advanced oxidation processes (EAOPs). These are electrochemical methods that produce hydroxyl radicals ($OH^{\bullet}$) from the oxidation of water. In order to provide a focused review on the topic, only these methods will be discussed; hybrid electrochemical methods, such as electro-Fenton, photo-electrochemical oxidation, and sono-electrochemical oxidation, will not be reviewed. Numerous review articles on these topics can be found in the literature [1–11]. There are also a number of excellent review articles that focus on electrochemical oxidation for water treatment [12–22]. This chapter draws from these reviews to concisely summarize the knowledge in the field and identify advantages and disadvantages of EAOPs in order to define key challenges that should be addressed that will facilitate the use of EAOP technologies in water treatment.

The chapter is organized in the following manner. First, an overview of EAOPs is given, which is followed by a review of common electrode types. Next, the advantages and key innovations in the area of EAOPs are discussed, with a specific focus on the evolution from two-dimensional to three-dimensional porous electrodes. This discussion is followed by a careful review of the key challenges associated with EAOP implementation, and finally recommendations for future research and development to overcome the challenges are provided.

ELECTROCHEMICAL ADVANCED OXIDATION PROCESSES

Overview

EAOPs have been extensively studied for their ability to destroy a wide range of synthetic and real waste streams [20]. The major appeal of this technology is that electrochemical cells do not need the addition of chemical reagents and instead produce oxidizing species by the reactions that occur at the anode surface. These reactions mainly involve the oxidation of water at an appropriate anode to produce $OH^{\bullet}$, as shown in Eq. (1).

$$H_2O \rightarrow OH^{\bullet} + H^+ + e^- \quad (1)$$

The $OH^{\bullet}$ species is very reactive with many organic and inorganic contaminants and often reacts at diffusion-limited rates [23]. The production of $OH^{\bullet}$ is a defining requirement of EAOPs. Although $OH^{\bullet}$ is considered the primary oxidant, the production of H_2O_2 (Eq. 2) and the reaction of inorganic anions at the anode can also produce a number of other oxidants that have been shown to increase the oxidation rates of contaminants [24–28]. The most reactive oxidants include Cl_2 and $S_2O_8^{2-}$, which can be formed by direct electron transfer reactions as shown in Eqs. (3)–(6).

$$2OH^{\bullet} \rightarrow H_2O_2 \quad (2)$$

$$Cl^- \rightarrow Cl^{\bullet} + e^- \quad (3)$$

$$2Cl^{\bullet} \rightarrow Cl_2 \quad (4)$$

$$SO_4^{2-} \rightarrow SO_4^{\bullet -} + e^- \quad (5)$$

$$2SO_4^{\bullet -} \rightarrow S_2O_8^{2-} \quad (6)$$

The Cl^- and SO_4^{2-} anions can also react with $OH^{\bullet}$, as shown in Eqs. (7), (8).

$$SO_4^{2-} + OH^{\bullet} \rightarrow SO_4^{\bullet -} + OH^- \quad (7)$$

$$Cl^- + OH^{\bullet} \rightarrow ClOH^{\bullet -} \quad (8)$$

While H_2O_2 (hydrogen peroxide) and Cl_2 (chlorine) are commonly used disinfectants in water treatment, $S_2O_8^{2-}$ is a chemical oxidant that has been primarily used for aquifer remediation [29]. The standard reduction potential for $S_2O_8^{2-}$ is $E^o = 2.01\,V$. This E^o value is greater than that for H_2O_2 ($E^o = 1.8\,V$) and Cl_2 ($E^o = 1.48\,V$). A recent study has also shown that $S_2O_8^{2-}$ is an efficient disinfectant against the bacterial species *Escherichia coli* and *Enterococcus faecalis* [30]. The $Cl^{\bullet}$, $ClOH^{\bullet -}$, and $SO_4^{\bullet -}$ species have also been shown to be active oxidants for contaminant destruction [26,27].

In addition to this mix of oxidants that is present for contaminant oxidation, contaminants may also react on the anode surface via a direct electron transfer reaction, as shown in Eq. (9).

$$R \rightarrow R^{\bullet +} + e^- \quad (9)$$

This direct oxidation pathway is an important rate-determining step for the oxidation of contaminants that are unreactive with $OH^{\bullet}$ (e.g., fluorinated organics) [31–33]. The participation of all these oxidants during contaminant transformations makes EAOPs an exciting yet complicated system to understand, and as will be discussed, only a few electrode materials are effective as EAOP anodes.

Electrode Types

Electrode Classification

Electrodes for water treatment are either classified as inactive or active electrodes, based on the stability of the electrode surface atoms to redox cycling. Active electrodes (e.g., Pt, IrO_2) continually cycle oxidation states during electrochemical oxidation of substrates. The oxidation of H_2O is initiated by the following direct electron transfer reaction.

$$M^n[\,] + H_2O \rightarrow M^n[OH^\bullet] + H^+ + e^- \quad (10)$$

where $M^n[\,]$ is an electrode surface site with a given oxidation state n, and $M^n[OH^\bullet]$ represents the physically sorbed $OH^\bullet$ at this surface site [34]. The $M^n[OH^\bullet]$ sites on active electrode materials are further oxidized to a higher M^{n+2} oxide, as shown in Eq. (11).

$$M^n[OH^\bullet] \rightarrow M^{n+2}O + H^+ + e^- \quad (11)$$

Due to the strong interaction of $OH^\bullet$ with the electrode surface, the production of $OH^\bullet$ on active anodes is low, and the oxidation of substrates (R) primarily occurs via an oxygen transfer reaction, which restores the surface site to its original oxidation state (M^n), as shown in Eq. (12).

$$M^{n+2}O + R \rightarrow M^n[\,] + RO \quad (12)$$

Support for the above mechanism on inactive electrodes was obtained by the detection of ^{18}O on the surface of an IrO_2/Ti electrode from ^{18}O-enriched water [35]. The oxidation of formic acid during this study resulted in the transfer of ^{18}O from the electrode to the CO_2 oxidation product [35]. The evolution of O_2 (Eq. 13) is in competition with the oxygen transfer reaction (Eq. 12). The final step for O_2 evolution involves the associative desorption reaction of O atoms via surface diffusion.

$$2\left[M^{n+2}O\right] \rightarrow 2M^n[\,] + O_2 \quad (13)$$

Active electrodes typically are associated with high rates of O_2 evolution and thus active electrodes are characterized with a low potential for oxygen evolution (~1.5 V/SHE) relative to the standard redox potential of the H_2O/O_2 redox reaction ($E^o = 1.23$ V/SHE). The main disadvantage of inactive electrodes is that they do not generate a sufficient quantity of $OH^\bullet$ and result in only the partial oxidation of substrates [34,36–38].

By contrast, inactive electrode materials are those whose surface atoms do not cycle oxidation states during electrochemical reactions. Oxygen evolution on inactive electrodes is not completely understood, but is thought to involve the reaction between an additional H_2O molecule and the $M^n[OH^\bullet]$ site to form O_2 (Eq. 14) [34].

$$M^n[OH^\bullet] + H_2O \rightarrow M^n[\,] + O_2 + 3H^+ + 3e^- \quad (14)$$

It has been speculated that O_2 evolution forms H_2O_2 as an intermediate on some anodes [38,39]. The weak interaction of $OH^\bullet$ with inactive electrodes requires a high anodic potential for water oxidation (~2.0 V/SHE) [16,40], which require an electrode that is stable against anodic corrosion. The most effective inactive electrodes for EAOPs are doped-SnO_2, PbO_2, boron doped diamond (BDD), and substoichiometric and doped-TiO_2. The classification of these electrodes as inactive anodes is based primarily on their ability to form high yields of $OH^\bullet$. However, as will be shown in the discussion below, not all electrodes clearly divide into these two classifications, and some electrodes have characteristics of both anode types. A discussion of each common type of inactive electrodes follows, with careful consideration given to the advantages and disadvantages of each.

Doped-SnO_2 Electrodes

The conductivity of pure SnO_2 is low, and thus it must be doped to obtain sufficient conductivity to allow it to function as a water treatment electrode. The most common dopant for SnO_2 is Sb, which achieves high conductivity and a potential for O_2 evolution of ~1.9 V/SHE [16]. However, Sb is a toxic compound that is regulated in drinking water. For example, the Environmental Protection Agency's (EPA) drinking water limit for Sb is 6.0 μg L^{-1}. Because of concerns over toxic leaching of Sb, research has focused on using other dopants (e.g., Ar, B, Bi, F, Cl, P) [41–43]. Many studies have investigated the use of doped-SnO_2 electrodes in laboratory settings [33,44–54], but at the time of writing they were not commercially available, which is a direct result of their short service life [55,56]. Two mechanisms for deactivation of doped-SnO_2 electrodes have been proposed. Both mechanisms are associated with corrosion processes. The first mechanism is attributed to surface corrosion, which results in the formation of a nonconductive Sn hydroxide layer on the anode surface [42,57,58], The second corrosion mechanism is attributed to corrosion of the underlying Ti substrate that results in doped-SnO_2 film delamination [42,58]. The Sn hydroxide surface layer was proposed to form due to hydration of the SnO_2 surface [42,59], and can be largely mitigated by doping with Pt [58,59]. Passivation of the Ti support can be minimized by depositing an anodically stable interlayer between the Ti substrate and doped-SnO_2 film. Such an approach has been carried out using a thin IrO_2 interlayer, which resulted in a significant improvement in service life of the SnO_2-Sb_2O_5 coating [48,56]. The IrO_2 interlayer is conductive and stable under high anodic polarization while Ir atoms isomorphically substitute for both Ti and Sn atoms at the metal-metal interfaces [48,56].

Other coatings, for example F-doped SnO_2, have also resulted in improved electrode longevity [60]. In summary, it appears that doped-SnO_2 electrodes are promising from a general performance and cost standpoint but have the disadvantage of a short cycle life that is a current challenge being addressed in electrode fabrication research.

PbO_2 Electrodes

Undoubtedly the first EAOP electrodes were comprised of PbO_2 packed-bed reactors, which were prepared by oxidizing Pb pellets [61,62]. Later studies deposited PbO_2 and doped-PbO_2 films on to a substrate (e.g., Ta, Ti, stainless steel, Ebonex) [63–72]. In general, the service life of PbO_2 anodes is sufficient, but various dopants (e.g., Fe, Co, Bi, F) have been added to PbO_2 to increase the electroactivity. The dopant was primarily reported to affect electrochemical performance by reducing the crystal grain size and thus increasing the electroactive surface area [73]. Additionally, doping can result in formation of a more porous β-PbO_2, which has a higher specific surface area and thus higher oxidation rate then the less porous α-PbO_2 crystal structure [74].

PbO_2 anodes suffer from the distinct disadvantage that they have a very low tolerance for Pb ion leaching into solution when used for water treatment applications, which has limited the appeal of PbO_2 anodes. Although experiments have shown that PbO_2 electrodes are stable during anodic polarization and Pb leaching is minimal [75], the well known toxicity of Pb and the environmental regulations set forth by the EPA (15 $\mu g\,L^{-1}$ drinking water standard) and other regulatory agencies prevents the wide spread use of PbO_2 electrodes for water treatment applications.

Doped- and Substoichiometric TiO_2

Stoichiometric titanium dioxide (TiO_2) is a well-known broad band-gap semiconductor with an electrical conductivity of $\sim 10^{-9}\,S\,cm^{-1}$ [76]. Various methods of doping TiO_2 have been performed and result in a significant increase in conductivity and the generation of $OH^{\bullet}$. These methods include doping with group V elements [77,78], reduction to Magneli phase suboxides of TiO_2 (Ti_nO_{2n-1}, $4 \leq n \leq 10$) [76,79], and more recently the electrochemical reduction of TiO_2 nanostructured electrodes [80–82].

Doping of rutile phase TiO_2 with group V elements (e.g., V, Nb, or Ta) produces ceramic materials with high electrical conductivities. The most common dopant is Nb, and Nb-doped rutile (NDR) oxides with the general formula of $Ti_{1-x}Nb_xO_2$ ($0 \leq x \leq 0.8$) have been synthesized from sol gel methods and found to be highly conductive [77,78]. Doping involves direct substitution of Nb^{5+} for Ti^{4+}, which provides n-type doping behavior [83]. Doped oxides have the advantage of having negligible anion deficiencies [77], allowing them to be resistant to anodic passivation [78].

In fact, oxidative wear of NDR electrodes has been found to be reversible via polarization of the electrode as a cathode [78]. Due to their high anodic stability, NDR electrodes have been used in unitized regenerative fuel cells [78], fuel cell and battery supports [84,85], dye-sensitized solar cells [86–89], water splitting [90], and gas sensors [91,92]. Doping TiO_2 with other transition metals has also produced stable and conductive electrodes, which have been tested as O_2-evolving electrocatalysts [93]. Only one study could be found that used doped-TiO_2 electrodes for electrochemical oxidation [94]. Therefore, more studies are warranted to determine its effectiveness in water treatment applications, especially for use in EAOPs.

Creating oxygen deficiencies in the lattice structure of TiO_2 can also greatly enhance the electrical conductivity and electrochemical properties of the material. Creating oxygen vacancies can be accomplished at temperatures above 900°C and in the presence of a suitable electron donor. Conversion is typically performed in a H_2 atmosphere or in the presence of reduced metals (e.g., C, Ti) [76]. The process involves rearrangement to create a unique crystal structure [95], and results in suboxides of TiO_2 containing $Ti^{(III)}$ atoms, which provide n-type semiconductor behavior. These suboxides are known as Magneli phases (Ti_nO_{2n-1}, $4 \leq n \leq 10$), where the more reduced species are the most conductive compounds (e.g., Ti_4O_7 and Ti_5O_9), and thus are desired in high percentages after synthesis. For example, single-crystal Ti_4O_7 has an electrical conductivity of 166 S cm^{-1}, several orders of magnitude greater than TiO_2 [76]. The oxygen deficiencies in Magneli phases are due to edge sharing of TiO_6 octahedra in the crystallographic shear planes that produce a layered structure of rutile and corundum phases in Ti_4O_7 [95]. Over-reduction to Ti_3O_5 creates a significant oxygen-deficient structure that is brittle and has a lower conductivity than Ti_4O_7 [76]. Reduction methods have been tailored to synthesize materials consisting primarily of Ti_4O_7, as confirmed by XRD measurements [76,79,96]. Ceramic Magneli phase electrodes are commercially available, and are known by the trade name Ebonex [76]. However, Ebonex electrodes have been shown to vary significantly with respect to the Magneli phases that are present, and XRD measurements are necessary to confirm the exact composition [97]. Ebonex electrodes have been used for lab-scale water treatment applications [97–103], and recent work comparing Ebonex electrodes to BDD concluded that the $OH^{\bullet}$ produced on Ebonex were less abundant but more reactive than those formed on BDD [100]. The primary industrial application of Ebonex electrodes is for the cathodic protection of metal structures [104]. However, due to its corrosion resistance and electrochemical stability, there is growing interest in the synthesis and testing of high-purity Ti_4O_7 electrodes as supports for batteries, fuel cells, catalysts, and electrocatalysts [78,96,99,104–107].

The oxygen-deficient nature of substoichiometric TiO_2 makes it possible for the reincorporation of oxygen atoms into the crystal structure during anodic polarization, which may result in the formation of a passivating TiO_2 coating [108–110]. However, due to the crystal rearrangement associated with conversion of TiO_2 to Ti_4O_7, the passivation should not be feasible at room temperature [95]. Nevertheless, studies have shown through conductivity and electrochemical impedance spectroscopy (EIS) measurements that passivation can occur under mild electrochemical polarization (e.g., ~1.6 V/SHE) [108]. Some studies have reported that the passivation was reversible upon cathodic polarization [110] while other studies have found that the passivation was irreversible [108]. The discrepancies in reported results are not clear and may be due to the presence of different Magneli phase compositions in the as-prepared samples. Recent work with Ebonex electrodes showed that periodic polarity reversals were able to prevent electrode passivation [110]. These studies indicate that Magneli phase suboxides may show properties characteristic of both inactive and active anodes. The electrodes have been reported to form $OH^{\bullet}$ and therefore meet the requirement of an inactive anode, but may also be associated with a higher oxide formation similar to an active anode [110]. More detailed studies of the passivation and re-reduction of Magneli phase suboxides are needed in order to understand how to efficiently manage these anodes during water treatment applications.

Recent results indicate that TiO_2 nanostructures can be doped by cathodic reduction under mild conditions [80–82]. Most studies used TiO_2 nanotubes [111] as the precursor structure and reduced them with cathodic currents in the $mA\,cm^{-2}$ range for a few minutes [80–82]. The resulting electrode is bluish-black in color and is often referred to as "blue titania." These blue titania electrodes show high conductivity, suspected $OH^{\bullet}$ generation, and high oxidation rates of organic substrates [80–82]. However, they suffer from fairly rapid passivation during anodic polarization [82] and thus are not as stable as the Magneli phase TiO_2 suboxides previously discussed.

The exact mechanism responsible for the doping process to obtain blue titania is not fully understood. Blue titania electrodes have an almost identical XRD and XPS spectra to that of TiO_2 [80,81], indicating that substantial changes to the crystal structure do not occur. Doping by cathodically generated H_2 is one possible mechanism that has been proposed [112,113]. This mechanism requires cathodically generated H_2 to diffuse into the TiO_2 lattice and ionize at point defect sites. The electron delocalizes in the oxide conduction band, resulting in the formation of a highly conductive H-doped TiO_2 electrode [112,113]. Another possible mechanism is the intercalation of H^+ into the TiO_2 lattice [114]. This later mechanism may be responsible for the reactivation of passivated Magneli phase TiO_2

suboxides previously discussed [110]. The formation of blue titania is associated with a shift of the flat band potential to more negative potentials and a substantial increase in the donor density [82]. Upon anodic aging, the flat band potential increased and donor density decreased, confirming passivation [82]. It has been proposed that a periodic reverse polarization of the blue titania is necessary to maintain high conductivity [82], which can easily be incorporated into a water-treatment strategy. However, a bigger disadvantage of this material is that it is extremely brittle and does not offer any distinct advantage over Magneli phase anodes.

The water treatment potential of doped and substoichiometric TiO_2 electrodes is very promising, and many recent studies related to water treatment have been performed [79,94,97,100,102,103,105,115–117]. Recent work has also taken advantage of the porous monolithic structure of Ti_4O_7 relectrodes to use them as porous flow-through electrodes and reactive electrochemical membranes [79,97,102,103,115–118]. Future research and development should focus on understanding the passivation/ reactivation mechanisms and incorporating strategies to implement these electrodes in water treatment.

Boron Doped Diamond Electrodes

The most widely studied electrode for EAOPs is BDD. BDD electrodes are produced by the chemical vapor deposition (CVD) method. The CVD method is relatively straightforward and has resulted in widespread interest in utilizing polycrystalline diamond films for industrial applications [15]. Boron is the most common dopant for diamond electrodes. Boron has a low charge carrier activation energy (0.37 eV), and thus boron atoms substitute for carbon atoms in the diamond film and behave as a p-type dopant by creating excess holes in the electrode [119,120]. At low doping levels ($\sim 10^{17}$ atoms cm^{-3}), the diamond has semiconductor properties and electrical conduction occurs via a hole hopping mechanism [15]. At high doping levels ($\sim 10^{20}$ atoms cm^{-3}), the diamond achieves semimetallic conductivity where electron conduction occurs through low energy impurity bands [121]. At high doping levels, metallic resistivities $<0.1\,\Omega$-cm are achieved [121]. BDD electrodes have been synthesized in different crystal sizes ranging from microcrystalline [121–123] to nanocrystalline [121–129]. The electrochemical and chemical properties have been linked to the crystal size and synthesis method [130]. The smaller crystal size is associated with a larger proportion of surface sites at the grain boundaries, and thus a higher sp^2 C content [126]. However, detailed studies regarding the effect of crystal size on oxidation rates of organic compounds are difficult due to continuous evolution of the surface during anodic conditions.

BDD electrodes have the distinct advantage over other electrodes that they are extremely stable under anodic polarization, due to the C atoms being in sp^3 hybridization. However, BDD film electrodes are still subject to failure, by either anodic wear at grain boundaries at high applied current densities (e.g., 1 A cm^{-2}) [129], or more commonly by film delamination from the substrate [128]. The higher proportion of sp^2 C at nanocrystalline BDD electrodes appears to enhance corrosion rates by the oxidation of the edge sites to CO_2 [129]. The most common substrate for BDD electrodes is p-silicon. This substrate is very fragile but highly conductive and film delamination is rare because Si has a relatively low electrochemical activity and forms a compact self-limiting oxide [128]. Industrial applications require a more robust substrate, which has led to studies with various different metal substrates for the BDD films, including Ta, Nb, W, Zr, C, Ti, and different interlayers on metal substrates [128,131–135]. Stability towards film delamination as a function of substrate type is difficult to assess between independent studies due to nonuniform synthesis conditions between studies. However, in a single study it was found that the stability of nanocrystalline BDD electrode films on various substrates followed the order of: $Ta > Si > Nb > W \gg Ti$ [128]. Film delamination could be prevented by matching the coefficient of thermal expansion (CTE) value for diamond ($CTE = 1.18 \times 10^{-6}\ K^{-1}$) [136] to that of the substrate. Large differences between the CTE value of diamond and the substrate cause the build-up of residual stress in the BDD film upon cooling to room temperature. This process leads to film defects that allow electrolyte permeation to the substrate and anodic corrosion of the substrate is possible. The most resistant substrates to BDD film delamination are those with CTE values that are lower for the metal oxide relative to the pure metal [128]. Therefore, the oxide that is formed by corrosion is more compact than the pure metal and film delamination is minimal (e.g., Si and Ta) [128]. An increase of the CTE value from the metal to the oxide causes a physical expansion of the oxide and BDD delamination often occurs (e.g., Nb, W, and Ti) [128]. Film adhesion can also be improved by roughening the substrate through bead blasting or other methods [128,137]. The deposition of interlayers between the BDD films and Ti substrates has been attempted to overcome the difference in CTE value. Studies have deposited interlayers consisting of Si [135], Ta [132], and nanocrystalline BDD [138]. Increasing the stability of BDD/Ti electrodes is an ongoing focus of research, with the ultimate goal of decreasing the cost and increasing the robustness of BDD electrodes.

Research has also focused on determining the location of active sites on BDD electrodes. The understanding of active site location evolved with increased accuracy of the characterization methods. Early work suggested that the active sites were at grain boundaries, crystal edges, and other defects, which was based on the spatial mapping of the boron dopant [139]. However, later it was shown through characterization of hydrogen-terminated BDD electrodes using conductive probe atomic force microscopy (CP-AFM) and scanning electrochemical microscopy (SECM) that a large fraction of the electrode was nonconductive and only discrete random areas showed high conductivity [140]. The electrochemical active surface area was strongly correlated with the boron doping level, but was not located at grain boundaries, crystal edges, or other defects [140]. More recently, the use of a higher sensitivity scanning electrochemical cell microscopy (SECCM) technique showed that the entire BDD surface was electroactive [141]. However, the heterogeneous rate constants measured for different redox couples varied on different crystal facets, and was also correlated to boron doping levels [141]. Evidence was not found for increased reactivity at grain boundaries [141], which did not support the initial hypothesis of active site location.

The surface functional groups on the BDD electrode surface have a significant effect on charge transfer at the electrode surface, and may be related to the findings of Patten et al. [141], discussed previously. The surface functional groups continuously evolve on the BDD surface. As shown in Fig. 1, freshly prepared electrodes are H-terminated (i.e., $=CH_2$, $\equiv C—H$), but anodic polarization (or other oxidative conditions) converts these H-terminated groups to various different oxygenated-functional groups (i.e., hydroxyl ($\equiv C—OH$), carbonyl ($=C=O$), carboxyl ($—COOH$)), all of which have been detected by X-ray photoelectron spectroscopy (XPS) [121,128,129,142–146]. Based on the coordination number of C atoms, the $\equiv C—OH$ groups form along the crystal planes and the $—C—OOH$ and $=C=O$ groups form at defect and edge sites. Therefore, the density of specific functional groups is a function of the crystal size, as supported by XPS measurements that showed a much higher density of $=C=O$ groups on nanocrystalline BDD electrodes [128,129], relative to microcrystalline BDD electrodes [147]. Density functional theory (DFT)

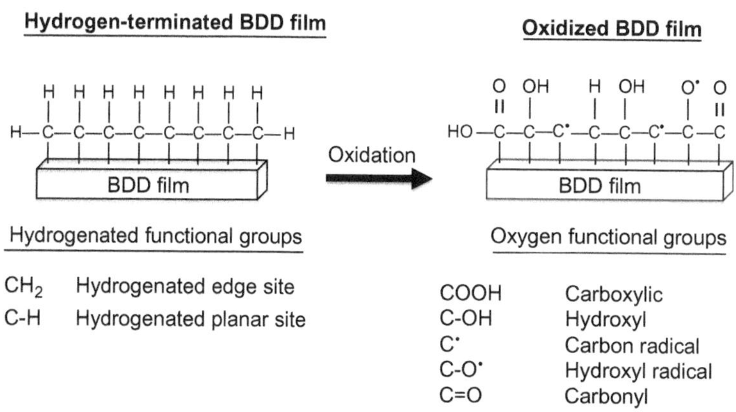

FIG. 1 Surface functional groups present on hydrogenated and oxidized BDD film electrodes.

calculations have proposed mechanisms for the formation of these various oxygenated functional groups, and ultimately to the corrosion of the electrode to CO_2 [129].

Intermediate to the corrosion process is the formation of carbon radical ($\equiv C^{\bullet}$) and deprotonated hydroxyl radical sites ($\equiv C—O^{\bullet}$), as shown in Eqs. (15), (16) [129,148].

$$\equiv C-H \rightarrow \equiv C^{\bullet} + H^{+} + e^{-} \tag{15}$$

$$\equiv C-OH \rightarrow \equiv C-O^{\bullet} + H^{+} + e^{-} \tag{16}$$

These radical sites were suggested to participate in compound oxidation, which was supported by DFT studies [129,148]. Experimental studies have also shown that BDD surface functional groups have a strong effect on charge transfer of both anodic and cathodic reactions [144,148–151]. Oxidized BDD electrodes inhibit some redox reactions while enhancing other types of redox reactions [148,150–152]. The surface functional group effect on direct electron transfer reaction rates was attributed to a combination of hydrophobic, dipole, and catalytic interactions [127,129,152]. However, it is difficult to determine the role of specific functional groups as the surface may change as a function of potential [19]. Therefore, it is important to characterize the effect of functional groups on target compound transformation at the appropriate electrode potentials of interest and to develop in situ characterization techniques to determine the functional groups present during compound transformation. Such experiments in combination with DFT calculations would greatly expand our understanding of oxidation mechanisms at BDD electrodes.

The mechanism responsible for $OH^{\bullet}$ production at BDD electrodes has also been linked to the surface functional groups present on the electrode [129]. Experimental results indicated that the overpotential for water oxidation decreased on an oxidized BDD surface relative to a freshly prepared H-terminated electrode [129]. DFT calculations supported these experimental findings, and the electrode potential necessary for the activationless water oxidation to $OH^{\bullet}$ decreased from 2.74 to 2.29 V/SHE when adding oxygenated functional groups ($=C=O$ and $\equiv C—O^{\bullet}$ sites) to DFT simulations [129]. These results illustrated a link between electrode structure and function. These findings are counter to the original belief that the BDD surface lacked adsorption sites, and therefore there may be opportunities for electrocatalysis at the BDD surface.

Characterization of Hydroxyl Radicals at Anodes

A typical benchmark of performance for EAOP anode materials is their relative production rate of $OH^{\bullet}$. The characterization of $OH^{\bullet}$ production is primarily accomplished by two methods, and both have distinct disadvantages that can lead to the misinterpretation of $OH^{\bullet}$ generation. The first method uses electron spin resonance (ESR) spectroscopy with a spin trap compound (R) that selectively reacts with $OH^{\bullet}$ to form an adduct species that is detectable by ESR. The second method uses probe molecules that selectively react with $OH^{\bullet}$ to either form a detectable adduct species or $OH^{\bullet}$ production is inferred by the disappearance of the probe compound.

It has been shown that both methods have shortcomings when used to characterize electrochemically produced $OH^{\bullet}$. Spin trap and probe molecules both can react on the electrode surface via direct electron transfer, followed by nucleophilic attack by water to form adducts that give a false positive detection of $OH^{\bullet}$ [153–159]. This method is known as inverted spin trapping, and is given by Eq. (17).

$$R \xrightarrow{-e^-} R^{\bullet +} \xrightarrow{+OH^-} R-OH \quad (17)$$

Yet another mechanism can result in false positive detection, which is known as the Forrester-Hepburn mechanism and is the reverse of inverted spin trapping. The Forrester-Hepburn mechanism involves nucleophilic attack of the spin trap by water followed by a direct electron transfer oxidation reaction (Eq. 18).

$$R \xrightarrow{+OH^-} R-OH^- \xrightarrow{-e^-} R-OH \quad (18)$$

Both these methods result in the production of a radical adduct (R—OH) that is identical to that formed by $OH^{\bullet}$ attack on the original spin trap.

A recent study by Jing and Chaplin [160] carefully evaluated the suitability of four commonly used $OH^{\bullet}$ probe compounds—coumarin (COU), *p*-chlorobenzoic acid (*p*-CBA), *p*-benzoquinone (*p*-BQ), and terephthalic acid (TA)—for the detection of $OH^{\bullet}$ produced during EAOPs with a BDD electrode. Experimental studies and DFT simulations were used to examine whether these four probe molecules were susceptible to the false detection pathways discussed above [160].

Results from linear sweep voltammetry experiments showed that both COU and *p*-CBA could react by direct electron transfer reactions. The numerical simulations of voltammetry results showed that the electrode potentials (E^o) for COU and *p*-CBA were 1.95 and 2.34 V/SHE, respectively, which were consistent with the values (2.04 and 2.48 V/SHE) determined by the DFT simulations. The activation energies of COU oxidation were calculated experimentally as a function of potential and experimental results compared well with those determined by DFT. These results indicated that both COU and *p*-CBA were not appropriate $OH^{\bullet}$ probe compounds, as they reacted by direct electron transfer reactions at potentials lower then that necessary to form $OH^{\bullet}$. Furthermore, a non-$OH^{\bullet}$ mediated and thermodynamically favorable pathway to form the COU adduct compound (e.g., R—OH) was determined by DFT simulations, which further complicates the use of this compound for $OH^{\bullet}$ detection. By contrast, both TA and *p*-BQ were found to be unreactive to direct electron transfer reactions, as determined by linear sweep voltammetry experiments, and DFT simulation yielded E^o values for TA and *p*-BQ of 2.76 and 3.21 V/SHE, respectively. Because these values are higher then the E^o value for $OH^{\bullet}$ ($E^o = 2.6$ V/SHE), both these compounds should be appropriate $OH^{\bullet}$ probes. However, *p*-BQ is only a useful $OH^{\bullet}$ probe in a divided cell reactor because it can be easily reduced to hydroquinone ($E^o = 0.26$ V vs SHE) [161]. DFT simulations were also used to investigate the possibility of these four probe molecules undergoing the Forrester-Hepburn mechanism. The results indicated that the nucleophilic addition/substitution of OH^- to COU, TA, and *p*-CBA were unlikely to be significant at room temperature. However, attention is needed when these probes are used in Fenton-type reactions because it has been reported that catalytic nucleophilic addition by metal ions in solution is possible (e.g., Fe, Cu, and Cr) [162,163]. It was also shown that *p*-BQ is likely to react via the Forrester-Hepburn mechanism, and thus the use of *p*-BQ in alkaline conditions should be used with caution.

The study by Jing and Chaplin [160] illustrates the importance of carefully choosing a probe compound for $OH^{\bullet}$ detection and also indicates that past work claiming $OH^{\bullet}$ formation should be viewed with scrutiny. Of the four major electrodes used for EAOPs (i.e., doped-SnO_2, PbO_2, substoichiometric TiO_2, and BDD) there are published studies that show $OH^{\bullet}$ formation using TA and/or *p*-BQ [68,79,101,164].

KEY ADVANTAGES AND INNOVATIONS OF EAOPs

Overview

The anodes discussed above have been used in various different electrochemical cells. An electrochemical cell consists of an anode and cathode, which may also contain a membrane separator (Fig. 2). Membrane separators allow for the control of the solution conditions in both the anode and cathode compartments, and may be desired when these solution conditions need to remain distinct. However, in order for electrical current to flow in the cell, the membrane separator must be permeable to ions. For this reason, ion exchange membranes are usually used as membrane separators in electrochemical cells. The most common membrane separator is Nafion, which is a fluoropolymer with high stability to acidic, alkali, and oxidizing conditions as well as a high selectivity to H^+. In most EAOP applications, a membrane separator is not used because it increases the cost of water treatment. Also, membranes can easily

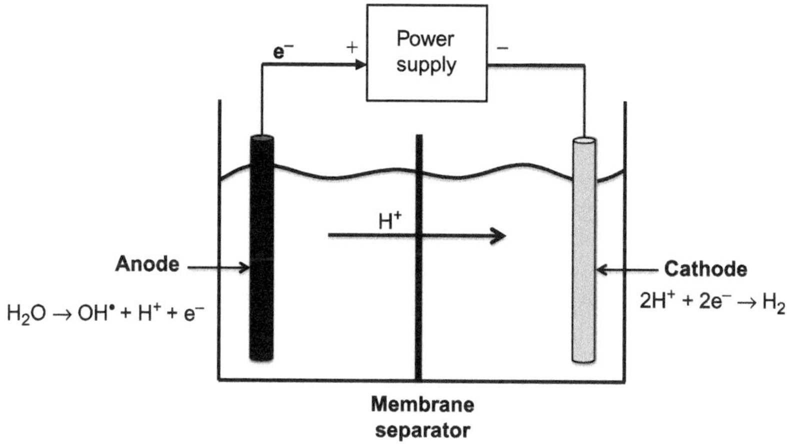

FIG. 2 Electrochemical cell with a membrane separator.

become fouled, which results in increased power consumption, increased maintenance, and diminished performance. Cathodic processes usually do not interfere with anodic processes because reactions between contaminants and electrochemically generated oxidants usually occur at or near the anode surface.

Advantages of using electrochemical technologies for water treatment, specifically EAOPs, are that they can operate at ambient temperatures and pressures; electrode materials have been developed that have a long operating life and are robust under acidic, alkali, and oxidizing solution conditions (e.g., BDD, substoichiometric TiO_2); and they are easily automated if coupled with the appropriate sensors which monitor system performance. Furthermore, electrochemical cells are modular in nature, much like a water treatment membrane, and treatment systems can be designed for either small or large flow volumes by configuring multiple electrochemical cells in series and/or parallel. The ability to fabricate compact electrochemical cells also makes EAOPs an attractive technology for point-of-use and point-of-entry water treatment, which thus far has received little attention [97,165]. The key advantages of EAOPs from an operational standpoint will be briefly discussed, followed by a more detailed discussion of how the evolution from two-dimensional (2D) to three-dimensional (3D) electrodes has the possibility of realizing the high reactivity of EAOP systems.

System Parameters

Electrochemical technologies offer a distinct advantage over competitive technologies in that they are effective at ambient temperatures and pressures. Although increasing the temperature increases the exchange current density (i_o) for electrode reactions, the reaction rate (or current) is exponentially linked to electrode potential, as shown by the Butler-Volmer kinetic expression (Eq. 19). Therefore overcoming the activation energy of a given reaction can conveniently be accomplished by adjusting the electrode potential [166].

$$i = i_o\left[\exp\left(\alpha\frac{nF}{RT}(E - E^o)\right) - \exp\left((1-\alpha)\frac{nF}{RT}(E - E^o)\right)\right] \tag{19}$$

where i is current, α is the electron transfer coefficient, E is the electrode potential, E^o is the standard electrode potential, n is the number of electrons transferred, F is Faraday's constant, and R and T have their usual meanings. Temperature also has a negligible effect on the temperature-dependent standard redox potential (E_T^o) of a given reaction, according to Eq. (20) [167].

$$E_T^o = E^o + (T - 298.15)\left(\frac{dE^o}{dT}\right) \tag{20}$$

The values for $\left(\frac{dE^o}{dT}\right)$, evaluated for approximately 1700 half reactions, are typically in the range of $\pm 0.0 - 2.0\,\text{mV}\,\text{K}^{-1}$, and thus temperature does not significantly affect E_T^o values. The temperature dependence of reaction rates reported in many electrochemical studies is small, which is primarily due to the fact that mass transport is limiting the reactions. Ambient pressure is also primarily used because most reactions involve condensed phases, and thus there is not an advantage to operating at elevated pressures when gaseous species are not involved.

The anodes used for EAOPs, which were discussed previously, are extremely robust under the full pH range and are very stable under corrosive conditions [12,76]. For this reason, EAOPs have been used to treat a wide range of waste streams, including but not limited to wastewaters from the pulp and paper industry, the olive oil industry, the pharmaceutical industry, the chemical industry, the dye industry, landfill leachate, perfluorinated organics, and water disinfection [12,31,33,168–177]. Additionally, when salt management is not an issue nontoxic salts (e.g., NaCl, Na_2SO_4, etc.) can be added to increase the conductivity of the waste stream and thus reduce the power consumption by reducing the ohmic losses due to solution resistance. Studies have also shown that the addition of NaCl or oxidation of saline wastewaters generates high concentrations of chlorine and other radicals that significantly reduce the time necessary for organic compound oxidation and contribute to ammonia removal [169,170,178]. However, as will be discussed later in this chapter, the formation of halogenated byproducts is of significant concern in some water treatment applications and should be monitored carefully.

3D Electrodes

Electrochemical reactions can either be limited by chemical kinetics or mass transport of contaminants to the electrode surface. Although reactions at the electrode surface, either through direct electron transfer or reactions with electrochemically produced oxidants, can vary by several orders of magnitude. Reaction time scales are generally much shorter then those for mass transport processes. This is confounded by the fact that many electrochemical reactors use flat 2D electrodes housed in parallel plate reactors. This type of configuration produces a relatively thick diffusion layer next to the electrode surface, where it is necessary for contaminants to diffuse through this layer before they can react at or near the electrode surface. The reaction of substrates with $OH^{\bullet}$ is often in the diffusion-limited range with second order rate constants commonly ranging from 10^9 to 10^{10} M^{-1} s^{-1} [23]. Experimental and modeling studies have also shown that, due to the high reactivity of $OH^{\bullet}$, it exists in only a narrow reaction zone adjacent to the electrode surface (<1.0 μm) [39,179,180]. As a result of these very fast kinetic processes, the use of 2D electrodes and small reaction zone volumes, the operation of electrochemical water treatment technologies become mass transport limited at relatively low applied current densities (e.g., <5 mA cm^{-2}) [180,181]. Most treatment systems are therefore operated under mass transport limited conditions where reaction rates are governed by the diffusion of contaminants to the electrode surface. Therefore, current research is focused on the development of high electroactive surface area electrodes that utilize strategies to maximize mass transport rates [79].

One strategy being explored to address mass transport limitations is the use of 3D electrodes operated in flow-through mode as opposed to the more commonly used 2D electrodes operated in flow-by mode. Fig. 3 shows a sketch of these two different operational flow modes. Traditional electrochemical reactors utilize 2D electrodes arranged in a parallel plate configuration and operated in flow-by mode (Fig. 3A). The electrodes are typically separated by narrow flow channels (mm to cm range), and the solution to be treated is pumped through these channels in either one-pass or recirculation mode. Reaction rate constants are governed by the diffusion of substrates through a thin stagnant diffusion boundary layer (δ ~100 μm) that forms at the electrode surface. The diffusion boundary layer thickness is a function of the cross-flow velocity and the channel spacing, which are both affected by the inclusion of flow distributors within the channels.

Mass transport processes in electrochemical reactors are usually analyzed by nondimensional groups, which include the Sherwood number (Sh), the Reynolds number (Re), and the Schmidt number (Sc).

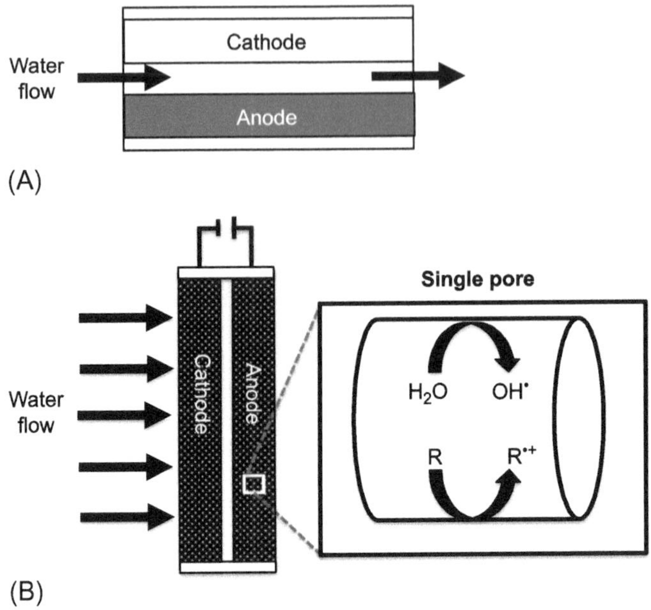

FIG. 3 Schematic of: (A) an undivided parallel plate electrochemical cell operated in flow-by mode; (B) porous electrodes operated in flow-through mode. Inset shows reactions on the pore wall within a single pore.

$$\mathrm{Sh} = \frac{k_m d}{D} \tag{21}$$

$$\mathrm{Re} = \frac{ud}{\upsilon} \tag{22}$$

$$\mathrm{Sc} = \frac{\upsilon}{D} \tag{23}$$

These nondimensional groups relate the mass transport rate constant (k_m) to the diffusivity (D), the equivalent diameter (d), the fluid velocity (u), and the kinematic viscosity of the fluid (ν). Additionally, a nondimensional shape factor (Le) is also included in order to account for the effects of the geometry of the cell on mass transport. The following relation can be used to calculate k_m as a function of the nondimensional groups.

$$\mathrm{Sh} = \frac{k_m d}{D} = a\,\mathrm{Re}^b \mathrm{Sc}^c \mathrm{Le}^d \tag{24}$$

where a, b, c, and d are typically determined empirically.

Various correlations have been developed in the form of Eq. (24). For flat-plate 2D electrodes in a parallel plate configuration, Eqs. (25), (26) have been proposed for laminar and turbulent conditions, respectively [181].

$$Sh = 2.54\ Re^{0.33} Sc^{0.33} Le^{0.33} \tag{25}$$

for $58 < Re < 2000$ and $2850 < Sc < 5140$

$$Sh = 0.023\ Re^{0.67} Sc^{0.33} Le^{0.25} \tag{26}$$

for $2000 < Re < 20000$ and $2850 < Sc < 5140$.

With the use of parallel plate electrodes, k_m values on the order of 10^{-6}–$10^{-5}\ m\,s^{-1}$ are attainable [181]. The use of high surface area 3D electrodes in parallel plate configurations only results in a small increase in k_m, as features of electrode roughness are smaller than the diffusion length and therefore become averaged into the diffusion field [166].

In order to overcome diffusional limitations in electrochemical reactors, recent research has focused on the use of 3D, monolithic, porous, flow-through electrodes, as shown in Fig. 3B. This reactor type utilizes a porous electrode in the form of a filter or membrane and water is convected through the pores that are on the order of 0.01–1.0 μm in diameter [79,102,182]. The advantage of this technique is that a high specific surface area of the electrode is active for electrochemical reactions [79,102], and the very small pore diameters prevent radial diffusion limitations [79,102,182].

Due to the recent introduction of 3D flow-through electrodes to the electrochemistry field, rigorous mass transport analysis and correlations such as those presented in Eqs. (25), (26) have not been determined for these systems. However, a simple mathematical analysis, assuming Hagen-Poiseuille flow in idealized straight pores, can provide insights into the effects of mass transport on measured reaction rates [183]. For example, evaluating the effect of the pressure drop (ΔP) across the 3D electrode on u allows for an estimation of the effective pore radius (r_p), according to Eq. (27) [79].

$$r_p = \left(8\eta u \frac{\Delta x}{\Delta P}\right)^{0.5} \tag{27}$$

where η is the dynamic fluid viscosity and Δx is the electrode thickness. It should be noted that the velocity u is the volume-averaged velocity in the pore. Obtaining a value for r_p by this method has been confirmed by Hg porosimetry analysis [79].

A rigorous analysis of reaction kinetics in 3D flow-through electrodes according to the classic Bultler-Volmer kinetics is complicated by the fact that the electrode potential is not constant but varies with electrode depth, due to the presence of solution and electrode resistances [184]. Rigorous mathematical analysis to account for this depth-dependent electrode potential has been presented by Newman and Thomas-Alyea [185], but these effects are difficult to validate experimentally. Alternatively, a simple semiempirical analysis is useful to determine if observed rate constants (k_{obs}) are limited by the kinetic rate constant (k_r), radial diffusive mass transport rate constant (k_d), or the convective mass transport rate constant (k_c) (Fig. 4). The relationship between k_{obs} and the other rate constants is given in Eq. (28).

$$k_{obs} = \left(\frac{C_{in} - C_{out}}{C_{in}}\right) k_c = \frac{k_c}{1 + \frac{k_c}{k_i}} \tag{28}$$

where k_i represents the limiting process (either k_d or k_r), and k_c is determined by Eq. (29).

$$k_c = \frac{Q}{A} \tag{29}$$

where Q is the volumetric flow rate and A is the projected surface area of the electrode. Deviation from the 1:1 relationship between k_{obs} and k_c is due to either kinetic limitations or radial diffusive mass transport limitations. Fig. 4A contains an example of the effect of k_c (dotted line, 1:1 with flux) on k_{obs} (solid line) where the asymptotic value for k_{obs} (k_i, shown as a dashed line) is a measurement of either k_r or k_d. These processes are depicted in a single pore in Fig. 4B. Estimates of k_d can be determined by Eq. (30) or using packed bed correlations [186].

$$k_d = \frac{D}{r_p} \tag{30}$$

The analysis discussed above has been applied to porous substoichiometric TiO_2 electrodes [79,86,102,103,115]. Limitations due to k_c [102,103], k_r [97], and k_d [115] have all been observed. The largest k_{obs} value measured in these porous electrodes was $1.4 \times 10^{-4}\ m\,s^{-1}$, which is the highest measured in a flow-through electrochemical reactor and was limited by k_r [79]. Current work in this area is focused on increasing the reactivity of the

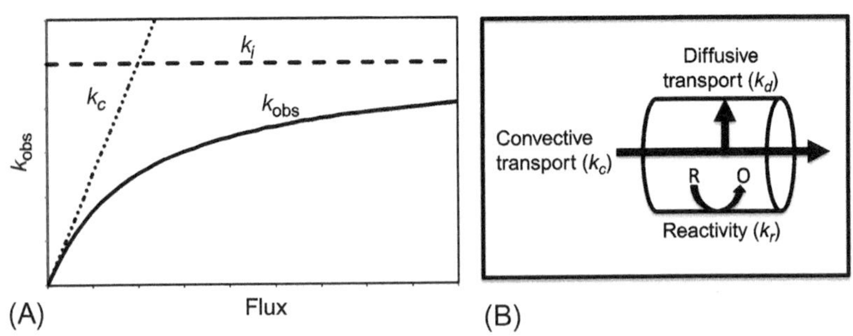

FIG. 4 (A) Relationship between rate constants as a function of fluid flux. (B) Schematic showing the different processes (kinetic and mass transport) in a single electrode pore.

porous electrodes to further take advantage of the enhanced mass transfer that is realized by flow-through operation.

CHALLENGES ASSOCIATED WITH EAOP IMPLEMENTATION

Overview

Before EAOPs can be considered a viable water treatment option, several challenges must be assessed. Important to this assessment is quantifying the performance of EAOPs in applied treatment settings and the negative environmental impacts that may arise from their use. To that end, an assessment of EAOPs to byproduct formation and matrix effects, and a comparison to traditional AOPs using performance and environmental measures is provided.

Halogenated Byproduct Formation

The formation of halogenated byproducts is a concern when using electrochemical oxidation for water and wastewater treatment. Byproducts include halogenated oxyanions (e.g., ClO_3^-, ClO_4^-, BrO_3^-) [148,187–193] and a wide variety of halogenated organic compounds [194–198]. Each byproduct type is discussed separately.

Inorganic Byproduct Formation

The formation of halogenated oxyanions, including ClO_3^-, ClO_4^-, and BrO_3^-, has been detected in numerous studies [148,187–193]. Of these oxyanions, ClO_4^- is especially problematic because it is a terminal oxidation product, its consumption has been linked to adverse health risks [199,200], and Cl^- is ubiquitous in natural and wastewater streams, making it especially challenging to control its formation. The EPA has issued a health advisory level for ClO_4^- of 15 $\mu g\,L^{-1}$ for drinking water sources [24], and Massachusetts and California have set even lower drinking water standards of 2 and 6 $\mu g\,L^{-1}$, respectively [201,202].

The formation of ClO_4^- is via a stepwise oxidation pathway starting from Cl^-, as shown in Eq. (31) [187].

$$Cl^- \rightarrow OCl^- \rightarrow ClO_2^- \rightarrow ClO_3^- \rightarrow ClO_4^- \tag{31}$$

The rate-determining step for the formation of ClO_4^- is the oxidation of ClO_3^- to ClO_4^- [188,203]. The conversion of ClO_3^- to ClO_4^- on BDD anodes has been shown to be a two-step process [148]. The first step is a direct electron transfer reaction of ClO_3^- at the electrode surface (Eq. 32) [148].

$$ClO_3^- \rightarrow ClO_3^\bullet + e^- \quad (32)$$

The second step is the homogeneous reaction between $ClO_3^\bullet$ and $OH^\bullet$ to form $HClO_4$ (Eq. 33) [148].

$$ClO_3^\bullet + OH^\bullet \rightarrow HClO_4 \quad (33)$$

Under conditions of kinetic control and the absence of other competing species (e.g., $OH^\bullet$ scavengers), it was found that the direct electron transfer reaction (Eq. (32)) was the rate-determining step for ClO_4^- formation on BDD anodes [148]. However, in the presence of organic compounds that can react with $OH^\bullet$ it was found that ClO_4^- formation was limited by the homogeneous reaction (Eq. (33)) [180], which will also be the case in applied treatment systems.

Studies focused on the formation of ClO_4^- on other EAOP electrodes under water treatment conditions could not be found. However, it was shown that ClO_4^- formation on BDD electrodes is approximately 50–100 times higher than on Pt and mixed-metal oxide electrodes [187]. These results are likely due to the fact that $OH^\bullet$ is not formed at high quantities on these active electrode materials. It is also possible that a different reaction mechanism is dominant on active electrodes. For example, the second reaction step may involve $ClO_3^\bullet$ reacting by a heterogeneous oxygen transfer reaction instead of a homogeneous reaction with $OH^\bullet$ (Eq. 33), but mechanistic studies focused on ClO_4^- formation on inactive electrodes could not be found in the literature.

Theoretical and experimental studies also suggested that the functional groups on the BDD electrode surface greatly influenced ClO_4^- formation [148,204,205]. DFT and experimental studies indicate that the surface radical sites can act as adsorption sites for $ClO_x^\bullet$ radicals [148,204], which can increase their lifetimes for further reaction with $OH^\bullet$ [204]. This mechanism may increase ClO_4^- formation. Studies have also focused on determining the effects of different oxygenated-functional groups on BDD-mediated ClO_4^- formation [205]. These studies showed that ClO_4^- formation varied greatly as a function of electrode aging, which was attributed to the continuous evolution of oxygenated-functional groups on the BDD surface [205]. Postreaction characterization of the BDD surface functional groups by XPS was not sufficient to clearly determine the role of surface groups on ClO_4^- formation rates [205]. It is likely that in situ studies are necessary to gain the necessary information to determine an appropriate link between surface functional groups and ClO_4^- formation.

It has been shown that the formation of ClO_4^- from ClO_3^- occurs significantly slower than the mass transfer rate, and therefore ClO_3^- concentrations at the electrode surface are near bulk solution concentrations [180]. Under these conditions, the mass transport rate of $OH^\bullet$ scavenger compounds (i.e., contaminants) has a substantial effect on lowering ClO_4^- formation rates. Results from reactive-transport modeling of the diffuse layer at the anode surface indicated that ClO_4^- formation was controlled by the competition for $OH^•$ between organic compounds and $ClO_3^•$ within a reaction zone (0.02–0.96 μm) adjacent to the anode surface [180]. Under kinetic-limited conditions, substrates reached the anode surface and those with higher $OH^•$ reaction rates inhibited ClO_4^- formation to a greater degree. When organic compound oxidation rates became mass transfer-limited, they were degraded by $OH^•$ before reaching the electrode surface, which prevent $OH^•$ scavenging at the anode surface. Such a process resulted in a minimal effect on ClO_4^- formation rates. These results indicated that controlling reactor operating conditions could potentially limit ClO_4^- formation during EAOP treatment of organic compounds.

Electrode modification is one strategy that has been explored to minimize ClO_4^- formation rates [205,206]. Modified electrodes were based on the concept of creating a thin blocking film on the BDD surface that would significantly slow the direct electron transfer reaction involving ClO_3^- (Eq. 32), but would allow for reaction of organic compounds with $OH^•$ diffusing through the film [205]. In order to meet the requirements of chemical stability, $—C_nF_{2n+1}$ chains were deposited on the BDD electrode surface by a variety of methods [205]. The most successful modification used the silanization method to covalently attach 1H, 1H, 2H, 2H perfluorodecyltrichlorosilane groups to the oxygen groups on the BDD surface [206]. Perchlorate formation was completely inhibited ($<1.0\ \mu g\,L^{-1}$ detection limit) while organic compound oxidation was moderately affected (decrease by 66%) [206]. Aging experiments determined that the silanized surface was stable over several experimental cycles before detectable levels of ClO_4^- began to form. These modified electrodes are promising but long-term stability of the coatings is still an issue.

The formation of BrO_3^- during electrochemical oxidation is also of concern, as the EPA has set a maximum contaminant level (MCL) of $0.01\ mg\,L^{-1}$ for drinking water. The Br^- content of natural waters is often at least an order of magnitude lower then Cl^-, but BrO_3^- formation has been shown in numerous electrochemical studies [189]. In some cases, BrO_3^- was further oxidized to BrO_4^- [191]. There have not been any mechanistic studies that focused on the electrochemical formation of BrO_3^- and BrO_4^-. However, initial studies have suggested that the formation of BrO_x^- species follows a stepwise pathway, which is similar to ClO_x^- formation [191].

$$Br^- \rightarrow OBr^- \rightarrow BrO_2^- \rightarrow BrO_3^- \rightarrow BrO_4^- \qquad (34)$$

Experimental results for this system appear to have similarities with ClO_4^- formation. That is, BrO_3^- accumulates in solution and slowly oxidizes to BrO_4^-, which suggests the rate-determining step for BrO_4^- formation is the oxidation of BrO_3^- [191]. However, further work is necessary to properly elucidate the mechanisms for BrO_x^- formation.

Halogenated Organic Compound Formation

The formation of halogenated-organic compounds (HOCs) has been detected during the application of EAOPs to a variety of solution conditions [194–198]. HOCs have been detected during the oxidation of organics in landfill leachate [194], reverse osmosis concentrates [197,198], and model aqueous systems [195,196,207]. HOC formation is attributed to addition and substitution reactions between organic compounds and in situ formed halogenated oxidants. These oxidants include active chlorine (e.g., Cl_2, OCl^-, $HOCl/OCl^-$) and bromine (e.g., Br_2, OBr^-, $HOBr/OBr^-$) species, chlorine radicals ($Cl^\bullet$, $Cl_2^{\bullet-}$), and bromine radicals ($Br^\bullet$, $Br_2^{\bullet-}$) [2]. These studies indicated that (1) HOCs were continuously formed during electrolysis while halogens and organics were both present, (2) HOCs were completely oxidized to inorganic end products (i.e., CO_2, ClO_3^-, ClO_4^-, BrO_3^-, BrO_4^-) after elimination of halogen ions or halogenated oxidants, and (3) the incorporation of halogens into organic compounds increased in the following order: $Cl < I < Br$.

The potential formation of individual HOCs is extremely diverse and therefore few studies have investigated the topic in detail. However, it appears that general classes of HOCs tend to form. The most comprehensive HOC characterization study was conducted by Anglada et al. [194]. The authors performed a detailed analysis of the HOCs formed during the treatment of landfill leachate using a BDD electrode flow-cell operated under mass transfer limited conditions in recycle mode. The landfill leachate had an initial COD concentration of $3385\,mg\,L^{-1}$, Cl^- concentration of $2587\,mg\,L^{-1}$, and Br^- concentrations of $9\,mg\,L^{-1}$ [194]. The formation of various HOCs was detected, including trihalomethanes (THMs), haloacetonitriles (HANs), haolketones (HKs), and 1,2-dichloroethane (DCA). Chloroform was the primary HOC formed, and represented on average 55% of the total HOC concentration. HANs (primarily dichloroacetonitrile and bromochloroacetonitrile) comprised 19%–25% of total HOCs. Some trends were observed for the formation of specific classes of HOCs. It was observed that the formation of HANs and HKs was enhanced at low pH. The formation of DCA increased when the pH, applied current density, and chloride concentration increased. The formation of THMs was relatively insensitive to changes to operational parameters. The concentration of HOCs, with the exception of DCA, increased with electrolysis time, and maximum concentrations of 1.9, 753, and $431\,mg\,L^{-1}$ were detected for THMs, HANs, and HKs, respectively, at an applied charge of $14.4\,Ah\,L^{-1}$. DCA was not detected at this applied charge, indicating that it was oxidized to other products. At applied charges $>14.4\,Ah\,L^{-1}$, HOC concentrations plateaued, which corresponded to ~50% COD removal. Unfortunately, trends of chloride and its inorganic oxidation products (Cl_2, OCl^-, ClO_3^-, ClO_4^-) were not reported by the authors, which would allow an assessment of the relationship between these species and HOC formation. HOCs are expected to eventually be mineralized, as prior studies have shown halogenated-organics are efficiently oxidized on EAOP electrodes [208,209], but these long residence times are likely not a cost-effective treatment strategy.

Studies have also characterized HOC formation during EAOP treatment of reverse osmosis concentrates [197,198]. Bagastyo et al. [197] investigated the formation of HOCs during the treatment of reverse osmosis concentrates using BDD anodes. Instead of characterization of individual HOCs, Bagastyo et al. [197] used the formation of adsorbable organic chlorine (AOCl), adsorbable organic bromine (AOBr), and adsorbable organic iodine (AOI) compounds as a measure of HOC formation. As expected, results were dependent on the initial halogen ion concentration. The relatively low concentration of Br^- (1.6 mg L^{-1}) and I^- (0.5 mg L^{-1}) in the RO concentrate resulted in low AOBr and AOI concentrations, which were degraded at higher applied charges. By contrast, the high chloride concentration (1386 mg L^{-1}) resulted in a continuous increase in AOCl concentration (~0.88 mM) until the end of the experiment (i.e., 10.9 Ah L^{-1}) [197]. The formation of total trihalomethanes (tTHMs) and total haloacetic acids (tHAAs) was also observed. Both these HOC groups were degraded with an increase in applied charge [197]. Final total concentrations between 1–4 (tTHMs) and 12–22 mM (tHAAs) were measured [197]. The detection of HOCs during the entirety of these experiments was attributed to the continual presence of residual free available chlorine (FAC) and DOC [197]. It appears that complete removal of FAC is necessary to completely eliminate chlorinated organic compound formation, which ultimately results in ClO_3^- and ClO_4^- formation.

Bagastyo et al. [197] also found that Br^- and I^- were converted to halogenated products at much greater proportions (~100% and 50% molar concentrations, respectively) compared to Cl^- (~3%). This result is likely related to the higher reactivity of $HOBr/OBr^-$ and HOI/OI^- species towards phenolic compounds [210], and the possibility of stripping of chlorine into the gas phase, which is not as significant for the less volatile bromine and iodine species.

The relative formation of THMs and HAAs as a function of electrode type (Pt-IrO_2/Ti, SnO_2-Sb/Ti, BDD) and electrolyte composition (NaCl, $NaNO_3$, and Na_2SO_4) was also investigated by Bagastyo et al. [198] Results showed that HAA and THM production was highest for the BDD electrode in the 0.05 M NaCl electrolyte, although Ti/Pt-IrO_2 was the most effective electrode for the production of free available chlorine [198]. These results may be attributed to the interaction of BDD surface sites with halogenated radical and organic species [26,27]. The oxygenated-functional groups on the BDD surface may stabilize these radicals, which would increase their relative lifetimes and increase their availability for reaction with organic compounds. The electrolyte composition played a significant role in HOC formation. It was found that the formation of THMs and HAAs on BDD electrodes was an order of magnitude higher in the Na_2SO_4 electrolyte compared to the $NaNO_3$ electrolyte ($Cl^- = 4$ mM for both) [198]. This result may be due to the production of $S_2O_8^{2-}$ or $SO_4^{-\bullet}$ at the BDD anode [211,212]. These in situ formed oxidants have a high electrode potential and can promote Cl^- oxidation and thus formation of HOCs. The BDD anode showed the lowest production of HAAs and THMs of the three electrodes in the $NaNO_3$ electrolyte [198]. The reason for this result was not discussed, but may be related to increased adsorption of NO_3^- at the BDD anode relative to the other electrodes, which has the possibility of blocking reaction sites for direct electron transfer reactions of Cl^- species.

Under more controlled laboratory conditions, the oxidation of different organic compounds (resorcinol, dimethyl phthalate, diethyl phthalate, bisphenol-A) at a concentration of 0.5 mM and in the presence of 10 mM Cl^- was used to evaluate the formation of chloroform under constant current (20 mA cm^{-2}) conditions [213]. Chloroform formation was approximately an order of magnitude higher in the presence of resorcinol compared to the other organic compounds [215]. The high production of HOCs during the electrolysis of resorcinol/Cl^- solutions is related to the high reactivity of the ortho-position carbon atom toward electrophilic addition of chlorine atoms [214]. These results are similar to those found for the chlorination of resorcinol containing water, where large quantities of monochloro-resorcinol and dichloro-resorcinol were initially formed followed by hydrolysis to form chloroform [215,216]. Due to the complexity of HOC formation in complex water sources, it should be evaluated on a case-by-case basis and appropriate indicator compounds should be identified that are representative of a larger set of HOCs.

Matrix Effects

When EAOPs are used in water treatment applications, substrate degradation rates are affected by solution conditions. The effects can be related to fouling at the electrode surface and scavenging of $OH^\bullet$ by natural organic matter (NOM) and other inorganic ions in solution. These factors can result in reduced reaction rates of the target substrate.

Fouling

The fouling of EAOP electrodes has been observed at low potentials, primarily due to the polymerization of phenolic compounds on the electrode surface [217,218]. However, it was found that increasing the anodic potential in the range of $OH^{\bullet}$ formation removed the polymers [218]. When porous flow-through electrodes are used, the anodic potential is not constant with depth into the porous electrode, which is primarily due to solution resistance. Therefore, there are areas within the electrode whose potentials are lower than that necessary to form $OH^{\bullet}$ and polymerization of phenolic compounds is favored. Such an effect was observed for a series of *p*-substituted phenolic compounds that were oxidized with a Ti_4O_7 porous electrode [103]. It was found that adsorption and pore blocking did occur from the polymer films that formed [103]. Fouling of the porous electrodes could be overcome by performing periodic backwash treatment while maintaining anodic potentials [116]. The foulants were flushed to areas of higher potential where the $OH^{\bullet}$ could react and remove the foulants [116].

An additional minor issue when treating hard waters is the scaling that forms on the cathode of the electrochemical cell. This scaling on the cathode can easily be dissolved by performing reverse polarity treatment. Therefore, it is useful to work with EAOP electrodes as both anode and cathode. A periodic polarity reversal can maintain both electrodes scale-free for the duration of treatment, and has the added advantage of rereducing any passivation that may have formed on the anode. Reverse polarity has been shown to reactivate BDD, substoichiometric TiO_2, and blue titania anodes [110], and may be necessary for maintaining blue titania electrode conductivity [82].

Scavenging of OH•

The reaction between $OH^{\bullet}$ and nontarget water constituents can have the effect of lowering target compound oxidation rates. A general expression for estimating the effect of these nontarget compounds in given by the following expression:

$$R_R = \frac{k_{R,OH^{\bullet}} C_R}{k_{R,OH^{\bullet}} C_R + \sum k_{i,OH^{\bullet}} C_i} \tag{35}$$

where R_R is the relative reaction rate with the target compound, $k_{R,\ OH^{\bullet}}$ is the second-order reaction rate constant of R with $OH^{\bullet}$, $k_{i,\ OH^{\bullet}}$ is the second-order reaction rate constant of the ith compound with $OH^{\bullet}$, and C_R and C_i are aqueous surface concentrations. Unless the surface concentrations are known, it is difficult to get an accurate result from Eq. (35). Additionally, the reaction of large NOM molecules can react at the electrode surface to form products that are also reactive with $OH^{\bullet}$. The reaction rate constants and surface concentrations of these species are generally unknown, which further complicates the process. A list of several common $OH^{\bullet}$ scavengers and their typical concentration ranges in natural waters and reported second-order rate constants with $OH^{\bullet}$

($k_{i,\ OH\bullet}$) are provided in Table 1A. In natural waters, carbonate species, chloride, and NOM have the largest effect on contaminant oxidation rates. Industrial waste streams can have substantially varying concentrations of organic compounds and salt concentrations. Typical ranges of Cl^- for different water and waste streams are also given in Table 1B. Eq. (35) should only be used as an order of magnitude estimate and detailed transport modeling is necessary to obtain accurate results. However, even seemingly simple systems are difficult to model properly due to the high reactivity of radical species that are formed at the anode. Several studies have focused on modeling such systems [39,180,219–222].

The fundamental difference between EAOPs and traditional AOPs (e.g., H_2O_2/O_3, H_2O_2/UV) is that EAOP reactions are happening near or at the anode surface and traditional AOPs involve homogeneous reactions. A disadvantage of EAOPS is that reaction rates are typically slower then AOPs because contaminants must diffuse to the surface in order to react.

TABLE 1 (A) Hydroxyl Radical Scavengers Found in Water and Waste Streams, (B) Representative Chloride Concentrations in Different Water and Waste Streams

Compound	Hydroxyl radical rate constant ($L\,mol^{-1}\,s^{-1}$)	Natural water range ($mg\,L^{-1}$)
(A)		
H_2CO_3	$<1.0\times10^6$	1–50
HCO_3^-	8.5×10^6	100–300
CO_3^{2-}	3.9×10^8	1–10
Cl^-	4.3×10^9	1–700
SO_4^{2-}	4.7×10^5	10–200
Fe(II)	3.3×10^8	<1.0
NOM	$1.4–4.5\times10^8$	1–5 as C
Water source		**Chloride ($mg\,L^{-1}$)**
(B)		
Groundwater		≤700
Surface water		≤400
Municipal wastewater concentrate		~5000
Landfill leachate		~5000
Industrial wastewater		≤27000
Seawater		18900

However, a distinct advantage of EAOPs over AOPs is that the EAOPs develop an acidic diffusion boundary layer at the anode surface due to water oxidation, which can reach pH values < 1.0 [223]. This acidic diffusion layer was suggested as the primary reason for limited $OH^{\bullet}$ scavenging by HCO_3^- in experimental studies [224]. As HCO_3^- ions diffuse through the diffusion layer they are rapidly protonated to form H_2CO_3, which is effectively unreactive with $OH^{\bullet}$ (Table 1). It was shown that reaction rates of NDMA were not affected by the presence of a 5 mM HCO_3^- concentration [224]. However, NDMA oxidation rates declined significantly in the presence of HCO_3^- concentrations >1 mM when using the O_3/H_2O_2 AOP [225]. These results indicate that the acidic diffuse layer at the anode surface can provide a distinct advantage over other AOPs. This advantage is important because natural waters always contain carbonate species that are typically problematic to traditional AOPs.

Performance Comparison and Cost Analysis

Electrochemical water treatment technologies, specifically EAOPs, show promise as next-generation solutions to water treatment challenges. However, a relatively small amount of work has been dedicated to analyzing the cost of implementing these technologies. The most complete cost and performance comparison to date compares traditional oxidation processes (e.g., ozonation and Fenton oxidation) to the EAOP using BDD electrodes [226]. The general conclusion made from this study is that BDD electrodes have some key advantages that make them competitive with traditional oxidation processes [226]. Specifically, EAOP treatment achieved a more complete mineralization of organic compounds then either ozonation or Fenton oxidation. EAOP treatment was also found to have a higher efficiency at a higher contaminant concentration and did not accumulate refractory compounds during treatment. The electrical cost to power the electrodes was less for BDD electrodes then the cost of ozone generation, but had a comparable cost to Fenton oxidation. The oxidation of several organic compounds was assessed, and although the operating costs differed for each compound, ozonation had the highest operating cost and the Fenton process generally had the lowest. EAOPs were competitive with the Fenton process for the treatment of acidic waste streams and real wastewaters. The initial capital investment required for BDD electrodes was less than ozonation but higher than Fenton oxidation due to the current cost of BDD electrodes. Although the Fenton process was the most affordable oxidation method investigated, it has a tight pH regime and produces high volumes of Fe sludge that require further disposal [226].

Environmental impacts and cost analysis of EAOPs and AOPs using a life-cycle assessment (LCA) approach were recently investigated [227]. The study focused on the treatment of olive mill wastewater using electrochemical oxidation with BDD electrodes, wet air oxidation in a high-pressure reactor, and photocatalytic oxidation with a UV/TiO_2 reactor [227]. The impacts of the three treatment processes were normalized relative to the removal of $1\,g\,L^{-1}$ chemical oxygen demand (COD) and $1\,g\,L^{-1}$ of total phenol. General conclusions from this study were that the major environmental impact and cost of each technology was primarily due to the energy requirement for operation. The energy needed to remove $1\,g\,L^{-1}$ COD was 0.15kWh for EAOP, 0.8kWh for wet air oxidation, and 5.0kWh for UV/TiO_2. The energy needed to remove $1\,g\,L^{-1}$ total phenol was 1.2kWh for the EAOP, 2.9kWh for wet air oxidation, and 14.2kWh for UV/TiO_2. The environmental and human health impacts were primarily associated with the coal combustion process for electricity generation [227]. This study suggests that the use of a clean energy source would greatly limit environmental and human health impacts of the treatment processes.

FUTURE RESEARCH AND ENGINEERING NEEDS

It has been shown that EAOPs are promising technologies for water treatment, but several challenges still need to be overcome before they can be widely adopted. Future areas of research and development for EAOPs should be directed toward (1) improving materials design and engineering to allow for more efficient and competitive treatment systems, (2) improving the basic scientific understanding of EAOPs, (3) determining strategies to overcome toxic by-product formation, and (4) quantifying the economic costs and environmental impacts of EAOPs.

Materials Design and Engineering

The recent design of novel 3D electrodes and their use in flow-through mode allows for a more efficient EAOP treatment strategy. However, these electrodes still only utilize a small portion of their total available surface area due to ohmic loses in solution, and further improvements in reactivity are still necessary. Although a great improvement over traditional low surface area parallel plate electrodes, these 3D electrodes still have substantial room for improvement in order to avoid the use of a large number of electrochemical cells for a given application. The recent surge in electrode design, mainly in battery research, should be leveraged for the benefit of EAOPs. Materials science and engineering is needed to develop more active anodes with high available electroactive surface areas. Operating strategies and reactor design to incorporate these new materials in efficient electrochemical cells should also be a key focus of future work.

Applied research should also be directed at determining engineering strategies to allow for the adoption of EAOPs in water treatment. Example of such strategies are (1) placing EAOPs in the appropriate position in a treatment train approach, (2) coupling EAOPs to energy recovery strategies (e.g., fuel production), and (3) identifying target applications for EAOPs or the incorporation of them into hybrid treatment systems. Electrochemical cells have numerous benefits over existing technologies that should be exploited, including compact design, ease of automation, and self-generation of oxidizing chemicals. More focus should be directed in the development of point-of-use and point-of-entry treatment systems in order to take full advantage of these benefits.

Scientific Advancement

The vast majority of studies to date have focused on the effectiveness of EAOPs for the oxidation of synthetic and real waste streams. However, only a handful of studies focused on understanding the mechanisms of compound transformation. Therefore, a key challenge for the field is to develop experimental and theoretical techniques to further our understanding of EAOPs at the molecular scale. An example of future experimental work is the use of in situ spectroscopic techniques that can provide information on the state of the electrode or key reaction intermediates during electrode reactions. These experiments will allow structure-function relationships to be developed that can help in advancing scientific understanding and ultimately improve electrode design. These types of experiments provide information that can be coupled to theoretical DFT simulations that further enhance understanding. Advanced scanning electrochemical microscopy techniques are capable of providing spatially resolved information of electrode reactivity at the micron to submicron scale, which is useful in characterizing the reactivity of electrode materials and gaining mechanistic insight into contaminant transformation. Advancing the scientific understanding of EAOPs should lead to the development of more efficient and economical EAOP electrodes.

Toxic By-Product Formation

One of the primary challenges that hinder the adoption of EAOPs is the formation of halogenated by-products. Preventing ClO_4^- and BrO_3^- formation is especially important because they are terminal products of

Cl^- and Br^- oxidation, respectively, and the presence of these ions is ubiquitous in nearly all water streams. Furthermore, ClO_4^- is especially difficult to reduce back to Cl^- once it is formed [228–235]. Current research suggests that operational strategies and electrode modifications can limit ClO_4^- formation [180,205,206,236], but the formation of HOCs appears to be more difficult to control. Therefore, a combination of operating conditions, electrode modifications, and hybrid technologies needs to be investigated to address this critical challenge.

Economics and Environmental Impacts

More focus must be given to studying the cost and environmental impact of EAOP technologies. This area of research and development is quite lean and studies that provide information regarding these topics are important so that decision makers in industry are aware of the advantages of EAOPs for water treatment. Future work should focus on developing cost analyses for EAOPs in a variety of industries and comparing them to existing technologies. These types of studies should take advantage of complex economic and LCA models that are rapidly evolving. Adoption of EAOPs by industry will continue to be slow until a convincing argument from a technical, economical, and environmental standpoint can be made.

References

[1] H. Suty, C. De Traversay, M. Cost, Applications of advanced oxidation processes: present and future, Water Sci. Technol. 49 (2004) 227–233.

[2] J.J. Pignatello, E. Oliveros, A. MacKay, Advanced oxidation processes for organic contaminant destruction based on the Fenton reaction and related chemistry, Crit. Rev. Environ. Sci. Technol. 36 (2006) 1–84.

[3] K. Ikehata, N.J. Naghashkar, M.G. Ei-Din, Degradation of aqueous pharmaceuticals by ozonation and advanced oxidation processes: a review, Ozone-Sci. Eng. 28 (2006) 353–414.

[4] C. Comninellis, A. Kapalka, S. Malato, S.A. Parsons, L. Poulios, D. Mantzavinos, Advanced oxidation processes for water treatment: advances and trends for R&D, J. Chem. Technol. Biotechnol. 83 (2008) 769–776.

[5] E. Brillas, I. Sires, M.A. Oturan, Electro-Fenton process and related electrochemical technologies based on fenton's reaction chemistry, Chem. Rev. 109 (2009) 6570–6631.

[6] M. Umar, H.A. Aziz, M.S. Yusoff, Trends in the use of Fenton, electro-Fenton and photo-Fenton for the treatment of landfill leachate, Waste Manag. 30 (2010) 2113–2121.

[7] P.V. Nidheesh, R. Gandhimathi, Trends in electro-Fenton process for water and wastewater treatment: an overview, Desalination 299 (2012) 1–15.

[8] E. Rosales, M. Pazos, M.A. Sanroman, Advances in the electro-Fenton process for remediation of recalcitrant organic compounds, Chem. Eng. Technol. 35 (2012) 609–617.

[9] A. Yaqub, H. Ajab, Applications of sonoelectrochemistry in wastewater treatment system, Rev. Chem. Eng. 29 (2013) 123–130.

[10] K. Rajeshwar, J.G. Ibanez, G.M. Swain, Electrochemistry and the environment, J. Appl. Electrochem. 24 (1994) 1077–1091.
[11] T. Ochiai, A. Fujishima, Photoelectrochemical properties of TiO_2 photocatalyst and its applications for environmental purification, J. Photochem. Photobiol. C Photochem. Rev. 13 (2012) 247–262.
[12] G.H. Chen, Electrochemical technologies in wastewater treatment, Sep. Purif. Technol. 38 (2004) 11–41.
[13] M. Panizza, G. Cerisola, Electrochemical oxidation of aromatic sulphonated acids on a boron-doped diamond electrode, Int. J. Environ. Pollut. 27 (2006) 64–74.
[14] C.A. Martinez-Huitle, S. Ferro, Electrochemical oxidation of organic pollutants for the wastewater treatment: direct and indirect processes, Chem. Soc. Rev. 35 (2006) 1324–1340.
[15] A. Kraft, Doped diamond: A compact review on a new, versatile electrode material, Int. J. Electrochem. Sc. 2 (2007) 355–385.
[16] M. Panizza, G. Cerisola, Direct and mediated anodic axidation of organic pollutants, Chem. Rev. 109 (2009) 6541–6569.
[17] C.A. Martinez-Huitle, E. Brillas, Decontamination of wastewaters containing synthetic organic dyes by electrochemical methods: a general review, Appl. Catal. B Environ. 87 (3–4) (2009) 105–145.
[18] C.A. Martinez-Huitle, L.S. Andrade, Electrocatalysis in wastewater treatment: recent mechanism advances, Química Nova 34 (2011) 850–858.
[19] B.P. Chaplin, Critical review of electrochemical advanced oxidation processes for water treatment applications, Environ. Sci. Proc. Impact. 16 (2014) 1182–1203.
[20] F.C. Moreira, R.A.R. Boaventura, E. Brillas, V.J.P. Vilar, Electrochemical advanced oxidation processes: a review on their application to synthetic and real wastewaters, Appl. Catal. Environ. 202 (2017) 217–261.
[21] J. Radjenovic, D.L. Sedlak, Challenges and opportunities for electrochemical processes as next-generation technologies for the treatment of contaminated water, Environ. Sci. Technol. 49 (2015) 11292–11302.
[22] C.A. Martinez-Huitle, M.A. Rodrigo, I. Sires, O. Scialdone, Single and coupled electrochemical processes and reactors for the abatement of organic water pollutants: A critical review, Chem. Rev. 115 (2015) 13362–13407.
[23] G.V. Buxton, C.L. Greenstock, W.P. Helman, A.B. Ross, Critical review of rate constants for reactions of hydrated electrons, hydrogen atoms and hydroxyl radicals in aqueous solution, J. Phys. Chem. Ref. 17 (1988) 513–886.
[24] P. Canizares, C. Saez, A. Sanchez-Carretero, M.A. Rodrigo, Synthesis of novel oxidants by electrochemical technology, J. Appl. Electrochem. 39 (2009) 2143–2149.
[25] Y.D. Lan, C. Coetsier, C. Causserand, K.G. Serrano, On the role of salts for the treatment of wastewaters containing pharmaceuticals by electrochemical oxidation using a boron doped diamond anode, Electrochim. Acta 231 (2017) 309–318.
[26] J.M. Barazesh, C. Prasse, D.L. Sedlak, Electrochemical transformation of trace organic contaminants in the presence of halide and carbonate ions, Environ. Sci. Technol. 50 (2016) 10143–10152.
[27] J.T. Jasper, O.S. Shafaat, M.R. Hoffmann, Electrochemical transformation of trace organic contaminants in latrine wastewater, Environ. Sci. Technol. 50 (2016) 10198–10208.
[28] C.A. Martinez-Huitle, S. Ferro, A. De Battisti, Electrochemical incineration in the presence of halides, Electrochem. Solid. State Lett. 8 (2005) D35–D39.
[29] A. Tsitonaki, B. Petri, M. Crimi, H. Mosbaek, R.L. Siegrist, P.L. Bjerg, In situ chemical oxidation of contaminated soil and groundwater using persulfate: a review, Crit. Rev. Environ. Sci. Technol. 40 (2010) 55–91.

[30] A.M. Polcaro, A. Vacca, M. Mascia, S. Palmas, R. Pompei, S. Laconi, Characterization of a stirred tank electrochemical cell for water disinfection processes, Electrochim. Acta 52 (2007) 2595–2602.
[31] K.E. Carter, J. Farrell, Oxidative destruction of perfluorooctane sulfonate using boron-doped diamond film electrodes, Environ. Sci. Technol. 42 (2008) 6111–6115.
[32] Q. Zhuo, S. Deng, B. Yang, J. Huang, B. Wang, T. Zhang, G. Yu, Degradation of perfluorinated compounds on a boron-doped diamond electrode, Electrochim. Acta 77 (2012) 17–22.
[33] Q. Zhuo, S. Deng, B. Yang, J. Huang, G. Yu, Efficient electrochemical oxidation of perfluorooctanoate using a Ti/SnO_2-Sb-Bi anode, Environ. Sci. Technol. 45 (2011) 2973–2979.
[34] C. Comninellis, Electrocatalysis in the electrochemical conversion/combustion of organic pollutants for waste-water treatment, Electrochim. Acta 39 (1994) 1857–1862.
[35] S. Fierro, T. Nagel, H. Baltruschat, C. Comninellis, Investigation of the oxygen evolution reaction on Ti/IrO_2 electrodes using isotope labelling and on-line mass spectrometryle, Electrochem. Commun. 9 (2007) 1969–1974.
[36] C. Comninellis, A. DeBattisti, Electrocatalysis in anodic oxidation of organics with simultaneous oxygen evolution, J. Chim. Physique Et De Phys. Chim. Biol. 93 (1996) 673–679.
[37] O. Simond, V. Schaller, C. Comninellis, Theoretical model for the anodic oxidation of organics on metal oxide electrodes, Electrochim. Acta 42 (1997) 2009–2012.
[38] G. Foti, D. Gandini, C. Comninellis, A. Perret, W. Haenni, Oxidation of organics by intermediates of water discharge on IrO_2 and synthetic diamond anodes, Electrochem. Solid. State Lett. 2 (1999) 228–230.
[39] A. Kapalka, G. Foti, C. Comninellis, The importance of electrode material in environmental electrochemistry: formation and reactivity of free hydroxyl radicals on boron-doped diamond electrodes, Electrochim. Acta 54 (2009) 2018–2023.
[40] M.H.P. Santana, L.A. De Faria, J.F.C. Boodts, Electrochemical characterisation and oxygen evolution at a heavily boron doped diamond electrode, Electrochim. Acta 50 (2005) 2017–2027.
[41] Y.S. Hsu, S.K. Ghandhi, The effect of phosphorus doping on tin oxide-films made by the oxidation of phosphine and tetramethyltin, 1, growth and etching properties, J. Electrochem. Soc. 127 (1980) 1592–1595.
[42] R. Kotz, S. Stucki, B. Carcer, Electrochemical waste-water treatment using high overvoltage anodes, 1, physical and electrochemical properties of SnO_2 anodes, J. Appl. Electrochem. 21 (1991) 14–20.
[43] H. Liu, A. Vajpayee, C.D. Vecitis, Bismuth-doped tin oxide-coated carbon nanotube network: improved anode stability and efficiency for flow-through organic electrooxidation, ACS Appl. Mater. Interfaces 5 (2013) 10054–10066.
[44] C. Pulgarin, N. Adler, P. Peringer, C. Comninellis, Electrochemical detoxification of a 1,4-benzoquinone solution in waste-water treatment, Water Res. 28 (1994) 887–893.
[45] R. Cossu, A.M. Polcaro, M.C. Lavagnolo, M. Mascia, S. Palmas, F. Renoldi, Electrochemical treatment of landfill leachate: oxidation at Ti/PbO_2 and Ti/SnO_2 anodes, Environ. Sci. Technol. 32 (1998) 3570–3573.
[46] J.H. Grimm, D.G. Bessarabov, U. Simon, R.D. Sanderson, Characterization of doped tin dioxide anodes prepared by a sol-gel technique and their application in an SPE-reactor, J. Appl. Electrochem. 30 (2000) 293–302.
[47] S. Tanaka, Y. Nakata, T. Kimura, M.K. Yustiawati, H. Kuramitz, Electrochemical decomposition of bisphenol A using Pt/Ti and SnO_2/Ti anodes, J. Appl. Electrochem. 32 (2002) 197–201.
[48] C. Zanta, P.A. Michaud, C. Comninellis, A.R. Andrade, J.F.C. Boodts, Electrochemical oxidation of p-chlorophenol on SnO_2-Sb_2O_5 based anodes for wastewater treatment, J. Appl. Electrochem. 33 (2003) 1211–1215.

[49] M. Quiroz, S. Reyna, J. Sanchez, Anodic oxidation of pentachlorophenol at Ti/SnO_2 electrodes, J. Solid State Electrochem. 7 (2003) 277–282.
[50] X. Chen, F. Gao, G. Chen, Comparison of Ti/BDD and Ti/SnO_2–Sb_2O_5 electrodes for pollutant oxidation, J. Appl. Electrochem. 35 (2005) 185–191.
[51] C. Borras, C. Berzoy, J. Mostany, B. Scharifker, Oxidation of p-methoxyphenol on SnO_2-Sb_2O_5 electrodes: effects of electrode potential and concentration on the mineralization efficiency, J. Appl. Electrochem. 36 (2006) 433–439.
[52] Y.H. Wang, K.Y. Chan, X.Y. Li, S.K. So, Electrochemical degradation of 4-chlorophenol at nickel-antimony doped tin oxide electrode, Chemosphere 65 (2006) 1087–1093.
[53] B. Adams, M. Tian, A. Chen, Design and electrochemical study of SnO_2-based mixed oxide electrodes, Electrochim. Acta 54 (2009) 1491–1498.
[54] C. Borras, C. Berzoy, J. Mostany, J.C. Herrera, B.R. Scharifker, A comparison of the electrooxidation kinetics of p-methoxyphenol and p-nitrophenol on Sb-doped SnO_2 surfaces: concentration and temperature effects, Appl. Catal. B-Environ. 72 (2007) 98–104.
[55] L. Lipp, D. Pletcher, The preparation and characterization of tin dioxide coated titanium electrodes, Electrochim. Acta 42 (1997) 1091–1099.
[56] B. CorreaLozano, C. Comninellis, A. DeBattisti, Service life of Ti/SnO_2-Sb_2O_5 anodes, J. Appl. Electrochem. 27 (1997) 970–974.
[57] F. Montilla, E. Morallon, J.L. Vazquez, Evaluation of the electrocatalytic activity of antimony-doped tin dioxide anodes toward the oxidation of phenol in aqueous solutions, J. Electrochem. Soc. 152 (2005) B421–B427.
[58] F. Montilla, E. Morallon, A. De Battisti, S. Barison, S. Daolio, J.L. Vazquez, Preparation and characterization of antimony-doped tin dioxide electrodes. 3. XPS and SIMS characterization, J. Phys. Chem. B 108 (2004) 15976–15981.
[59] F. Montilla, E. Morallon, A. De Battisti, J.L. Vazquez, Preparation and characterization of antimony-doped tin dioxide electrodes, part 1, electrochemical characterization, J. Phys. Chem. B 108 (2004) 5036–5043.
[60] G. Li, Y.H. Wang, Q.Y. Chen, Influence of fluoride-doped tin oxide interlayer on Ni-Sb-SnO_2/Ti electrodes, J. Solid State Electrochem 17 (2013) 1303–1309.
[61] D.V. Smith, A.P. Watkinson, Anodic oxidation of phenol for wastewater treatment, Can. J. Chem. Eng. J. 59 (1981) 52–59.
[62] H. Sharifian, D.W. Kirk, Electrochemical oxidation of phenol, J. Electrochem. Soc. 133 (1986) 921–924.
[63] N.B. Tahar, R. Abdelhedi, A. Savall, Electrochemical polymerisation of phenol in aqueous solution on a Ta/PbO_2 anode, J. Appl. Electrochem. 39 (2009) 663–669.
[64] N.B. Tahar, A. Savall, Mechanistic aspects of phenol electrochemical degradation by oxidation on a Ta/PbO_2 anode, J. Electrochem. Soc. 145 (1998) 3427–3434.
[65] N.B. Tahar, A. Savall, A comparison of different lead dioxide coated electrodes for the electrochemical destruction of phenol, J. New Mater. Electrochem. Syst 2 (1999) 19–26.
[66] N.B. Tahar, A. Savall, Electrochemical degradation of phenol in aqueous solution on bismuth doped lead dioxide: a comparison of the activities of various electrode formulations, J. Appl. Electrochem. 29 (1999) 277–283.
[67] M. Panizza, G. Cerisola, Electrochemical degradation of methyl red using BDD and PbO_2 anodes, Ind. Eng. Chem.Res. 47 (2008) 6816–6820.
[68] L.L. Houk, S.K. Johnson, J. Feng, R.S. Houk, D.C. Johnson, Electrochemical incineration of benzoquinone in aqueous media using a quaternary metal oxide electrode in the absence of a soluble supporting electrolyte, J. Appl. Electrochem. 28 (1998) 1167–1177.
[69] F. Bonfatti, S. Ferro, F. Lavezzo, M. Malacarne, G. Lodi, A. De Battisti, Electrochemical incineration of glucose as a model organic substrate I. Role of the electrode material, J. Electrochem. Soc. 146 (1999) 2175–2179.
[70] G.H. Zhao, Y.G. Zhang, Y.Z. Lei, B.Y. Lv, J.X. Gao, Y.A. Zhang, D.M. Li, Fabrication and electrochemical treatment application of a novel lead dioxide anode with

superhydrophohic surfaces, high oxygen evolution potential, and oxidation capability, Environ. Sci. Technol. 44 (2010) 1754–1759.

[71] A.H. Ras, J.F. van Staden, Electrodeposition of PbO_2 and Bi-PbO_2 on Ebonex, J. Appl. Electrochem. 29 (1999) 313–319.

[72] Y. Song, G. Wei, R. Xiong, Structure and properties of PbO_2-CeO_2 anodes on stainless steel, Electrochim. Acta 52 (2007) 7022–7027.

[73] H. Liu, Y. Liu, C. Zhang, R. Shen, Electrocatalytic oxidation of nitrophenols in aqueous solution using modified PbO_2 electrodes, J. Appl. Electrochem. 38 (2008) 101–108.

[74] S. Abaci, U. Tamer, K. Pekmez, A. Yildiz, Performance of different crystal structures of PbO_2 on electrochemical degradation of phenol in aqueous solution, Appl. Surf. Sci. 240 (2005) 112–119.

[75] I. Sires, C.T.J. Low, C. Ponce-de-Leon, F.C. Walsh, The characterisation of PbO_2-coated electrodes prepared from aqueous methanesulfonic acid under controlled deposition conditions, Electrochim. Acta 55 (2010) 2163–2172.

[76] P.C.S. Hayfield, Development of a New Material - Monolithic Ti_4O_7 Ebonex® Ceramic, In Royal Society of Chemistry, Cambridge, 2002.

[77] D. Morris, Y. Dou, J. Rebane, C.E.J. Mitchell, R.G. Egdell, D.S.L. Law, A. Vittadini, M. Casarin, Photoemission and STM study of the electronic structure of Nb-doped TiO_2, Phys. Rev. B 61 (2000).

[78] G. Chen, S.R. Bare, T.E. Mallouk, Development of supported bifunctional electrocatalysts for unitized regenerative fuel cells, J. Electrochem. Soc. 149 (2002) A1092–A1099.

[79] L. Guo, Y. Jing, B.P. Chaplin, Development and characterization of ultrafiltration TiO_2 Magneli phase reactive electrochemical membranes, Environ. Sci. Technol. 50 (2016) 1428–1436.

[80] C. Kim, S. Kim, J. Choi, J. Lee, J.S. Kang, Y.E. Sung, W. Choi, J. Yoon, Blue TiO_2 nanotube array as an oxidant generating novel anode material fabricated by simple cathodic polarization, Electrochim. Acta 14 (2014) 1113–1119.

[81] C. Kim, S. Kim, J. Lee, J. Kim, J. Yoon, Capacitive and oxidant generating properties of black-colored TiO_2 nanotube array fabricated by electrochemical self-doping, ACS Appl. Mater. Interfaces 7 (2015) 7486–7491.

[82] Y. Yang, M.R. Hoffmann, Synthesis and stabilization of blue-black TiO_2 nanotube arrays for electrochemical oxidant generation and wastewater treatment, Environ. Sci. Technol. 50 (2016) 11888–11894.

[83] R.D. Shannon, Revised effective ionic-radii and systematic studies of interatomic distances in halides and chalcogenides, Acta Crystallogr. A 32 (SEP1) (1976) 751–767.

[84] H. Chhina, S. Campbell, O. Kesler, Ex situ and in situ stability of platinum supported on niobium-doped titania for PEMFCs, J. Electrochem. Soc. 156 (2009) B1232–B1237.

[85] A. Bauer, L. Chevallier, R. Hui, S. Cavaliere, J. Zhang, D. Jones, J. Roziere, Synthesis and characterization of Nb-TiO_2 mesoporous microsphere and nanofiber supported Pt catalysts for high temperature PEM fuel cells, Electrochim. Acta 77 (2012) 1–7.

[86] P. Hasin, M.A. Alpuche-Aviles, Y.G. Li, Y.Y. Wu, Mesoporous Nb-doped TiO_2 as Pt support for counter electrode in dye-sensitized solar cells, J. Phys. Chem. C 113 (2009) 7456–7460.

[87] S. Lee, J.H. Noh, H.S. Han, D.K. Yim, D.H. Kim, J.K. Lee, J.Y. Kim, H.S. Jung, K.S. Hong, Nb-doped TiO_2: a new compact layer material for TiO_2 dye-sensitized solar cells, J. Phys. Chem. C 113 (2009) 6878–6882.

[88] M. Yang, D. Kim, H. Jha, K. Lee, J. Paul, P. Schmuki, Nb doping of TiO_2 nanotubes for an enhanced efficiency of dye-sensitized solar cells, Chem. Commun. 47 (2011) 2032–2034.

[89] S.G. Kim, M.J. Ju, I.T. Choi, W.S. Choi, H.J. Choi, J.B. Baek, H.K. Kim, Nb-doped TiO_2 nanoparticles for organic dye-sensitized solar cells, RSC Adv. 3 (2013) 16380–16386.

[90] C. Das, P. Roy, M. Yang, H. Jha, P. Schmuki, Nb doped TiO_2 nanotubes for enhanced photoelectrochemical water-splitting, Nanoscale 3 (2011) 3094–3096.

[91] M.Z. Atashbar, H.T. Sun, B. Gong, W. Wlodarski, R. Lamb, XPS study of Nb-doped oxygen sensing TiO_2 thin films prepared by sol-gel method, Thin Solid Films 326 (1998) 238–244.

[92] A.M. Ruiz, A. Cornet, K. Shimanoe, J.R. Morante, N. Yamazoe, Effects of various metal additives on the gas sensing performances of TiO_2 nanocrystals obtained from hydrothermal treatments, Sens. Actuat. B Chem. 108 (2005) 34–40.

[93] M. Garcia-Mota, A. Vojvodic, H. Metiu, I.C. Man, H.Y. Su, J. Rossmeisl, J.K. Norskov, Tailoring the activity for oxygen evolution electrocatalysis on rutile TiO_2(110) by transition-metal substitution, ChemCatChem 3 (2011) 1607–1611.

[94] J.M. Kesselman, O. Weres, N.S. Lewis, M.R. Hoffmann, Electrochemical production of hydroxyl radical at polycrystalline Nb-doped TiO_2 electrodes and estimation of the partitioning between hydroxyl radical and direct Hole oxidation pathways, J. Phys. Chem. B 101 (1997) 2637–2643.

[95] A.C.M. Padilha, H. Raebiger, A.R. Rocha, G.M. Dalpian, Charge storage in oxygen deficient phases of TiO_2: defect Physics without defects, Sci. Rep. 6 (2016) 1–7.

[96] F.C. Walsh, R.G.A. Wills, The continuing development of Magneli phase titanium suboxides and Ebonex® electrodes, Electrochim. Acta 55 (2010) 6342–6351.

[97] L. Guo, K. Ding, K. Rockne, M. Duran, B.P. Chaplin, Bacteria inactivation at a substoichiometric titanium dioxide reactive electrochemical membrane, J. Hazard. Mater. 319 (2016) 137–146.

[98] C.L.K. Tennakoon, R.C. Bhardwaj, J.O. Bockris, Electrochemical treatment of human wastes in a packed bed reactor, J. Appl. Electrochem. 26 (1996) 18–29.

[99] G. Chen, E.A. Betterton, R.G. Arnold, W.P. Ela, Electrolytic reduction of trichloroethylene and chloroform at a Pt- or Pd-coated ceramic cathode, J. Appl. Electrochem. 33 (2003) 161–169.

[100] D. Bejan, E. Guinea, N.J. Bunce, On the nature of the hydroxyl radicals produced at boron-doped diamond and Ebonex® anodes, Electrochim. Acta 69 (2012) 275–281.

[101] D. Bejan, J.D. Malcolm, L. Morrison, N.J. Bunce, Mechanistic investigation of the conductive ceramic Ebonex (R) as an anode material, Electrochim. Acta 54 (2009) 5548–5556.

[102] A.M. Zaky, B.P. Chaplin, Porous substoichiometric TiO_2 anodes as reactive electrochemical membranes for water treatment, Environ. Sci. Technol. 47 (2013) 6554–6563.

[103] A.M. Zaky, B.P. Chaplin, Mechanism of p-substituted phenol oxidation at a Ti_4O_7 reactive electrochemical membrane, Environ. Sci. Technol. 48 (2014) 5857–5867.

[104] J.R. Smith, F.C. Walsh, R.L. Clarke, Reviews in applied electrochemistry, number 50-electrodes based on magneli phase titanium oxides: the properties and applications of Ebonex® materials, J. Appl. Electrochem. 28 (1998) 1021–1033.

[105] G. Chen, E.A. Betterton, R.G. Arnold, Electrolytic oxidation of trichloroethylene using a ceramic anode, J. Appl. Electrochem. 29 (1999) 961–970.

[106] K. Ellis, A. Hill, J. Hill, A. Loyns, T. Partington, The performance of Ebonex® electrodes in bipolar lead-acid batteries, J. Power Sources 136 (2004) 366–371.

[107] T. Ioroi, H. Senoh, S.I. Yamazaki, Z. Siroma, N. Fujiwara, K. Yasuda, Stability of corrosion-resistant magneli-phase Ti_4O_7-supported PEMFC catalysts at high potentials, J. Electrochem. Soc. 155 (2008) B321–B326.

[108] G.Y. Chen, C.C. Waraksa, H.G. Cho, D.D. Macdonald, T.E. Mallouk, EIS studies of porous oxygen electrodes with discrete particles—I, impedance of oxide catalyst supports, J. Electrochem. Soc. 150 (2003) E423–E428.

[109] C.C. Waraksa, G.Y. Chen, D.D. Macdonald, T.E. Mallouk, EIS studies of porous oxygen electrodes with discrete particles—II. Transmission line modeling, J. Electrochem. Soc. 150 (2003) E429–E437.

[110] S. El-Sherif, D. Bejan, N.J. Bunce, Electrochemical oxidation of sulfide ion in synthetic sour brines using periodic polarity reversal at Ebonex® electrodes, Can. J. Chem. Rev. Can. Chimi. 88 (2010) 928–936.
[111] M. Paulose, K. Shankar, S. Yoriya, H.E. Prakasam, O.K. Varghese, G.K. Mor, T. A. Latempa, A. Fitzgerald, C.A. Grimes, Anodic growth of highly ordered TiO_2 nanotube arrays to 134 μm in length, J. Phys. Chem. B 110 (2006) 16179–16184.
[112] H. Wu, D.D. Li, X.F. Zhu, C.Y. Yang, D.F. Liu, X.Y. Chen, Y. Song, L.F. Lu, High-performance and renewable supercapacitors based on TiO_2 nanotube array electrodes treated by an electrochemical doping approach, Electrochim. Acta 116 (2014) 129–136.
[113] W.P. Chen, Y. Wang, H.L.W. Chan, Hydrogen: a metastable donor in TiO_2 single crystals, Appl. Phys. Lett. 92 (2008) 112907–112909.
[114] J. Swaminathan, R. Subbiah, V. Singaram, Defect-rich metallic titania ($TiO_{1.23}$)-an efficient hydrogen evolution catalyst for electrochemical water splitting, ACS Cat. 6 (2016) 2222–2229.
[115] M.C. Santos, Y.A. Elabd, Y. Jing, B.P. Chaplin, L. Fang, Highly porous Ti_4O_7 reactive electrochemical water filtration membranes fabricated via electrospinning/electrospraying, AiCHE J. 62 (2016) 508–524.
[116] Y. Jing, L. Guo, B.P. Chaplin, Electrochemical impedance spectroscopy study of membrane fouling and electrochemical regeneration at a sub-stoichiometric TiO_2 reactive electrochemical membrane, J. Membr. Sci. 510 (2016) 510–523.
[117] Y. Jing, B.P. Chaplin, Electrochemical impedance spectroscopy study of membrane fouling characterization at a conductive sub-stoichiometric TiO_2 reactive electrochemical membrane: Transmission line model development, J. Membr. Sci. 511 (2016) 238–249.
[118] L. Hua, L. Guo, M. Thakkar, D. Wei, M. Agbakpe, L. Kuang, M. Magpile, B.P. Chaplin, Y. Tao, D. Shuai, X. Zhang, S. Mitra, W. Zhang, Effects of anodic oxidation of a substoichiometric titanium dioxide reactive electrochemical membrane on algal cell destabilization and lipid extraction, Bioresour. Technol. 203 (2016) 12–117.
[119] W. Haenni, P. Rychen, M. Fryda, C. Comninellis, Industrial applications of diamond electrodes, in: Thin-Film Diamond II, vol. 77, Elsevier, 2004, pp. 149–196.
[120] J. Ristein, Surface transfer doping of semiconductors, Science 313 (2006) 1057–1058.
[121] J. Xu, M.C. Granger, Q. Chen, J.W. Strojek, T.E. Lister, G.M. Swain, Boron-doped diamond thin-film electrodes, Anal. Chem. 69 (1997) 591A–597A.
[122] G.M. Swain, R. Ramesham, The electrochemical activity of boron-doped polycrystalline diamond thin-film electrodes, Anal. Chem. 65 (1993) 345–351.
[123] D. Gandini, P.A. Michaud, I. Duo, E. Mahe, W. Haenni, A. Perret, C. Comninellis, Electrochemical behavior of synthetic boron-doped diamond thin film anodes, New Diamond Front. Carbon Technol. 9 (1999) 303–316.
[124] Y. Show, M.A. Witek, P. Sonthalia, G.M. Swain, Characterization and electrochemical responsiveness of boron-doped nanocrystalline diamond thin-film electrodes, Chem. Mater. 15 (2003) 879–888.
[125] A. Ay, V.M. Swope, G.M. Swain, The physicochemical and electrochemical properties of 100 and 500 nm diameter diamond powders coated with boron-doped nanocrystalline diamond, J. Electrochem. Soc. 155 (2008) B1013–B1022.
[126] S. Wang, V.M. Swope, J.E. Bulter, T. Feygelson, G.M. Swain, The structural and electrochemical properties of boron-doped nanocrystalline diamond thin-film electrodes grown from Ar-rich and H_2-rich source gases, Diamond Relat. Mater. 18 (2009) 669–677.
[127] A.F. Azevedo, N.A. Braga, F.A. Souza, J.T. Matsushima, M.R. Baldan, N.G. Ferreira, The effect of surface treatment on oxidation of oxalic acid at nanocrystalline diamond films, Diamond Relat. Mater. 19 (2010) 462–465.

[128] B.P. Chaplin, I. Wylie, H. Zheng, J.A. Carlisle, J. Farrell, Characterization of the performance and failure mechanisms of boron-doped diamond ultrananocrystalline diamond electrodes, J. Appl. Electrochem. 41 (2011) 1329–1340.
[129] B.P. Chaplin, D.K. Hubler, J. Farrell, Understanding anodic wear at boron doped diamond film electrodes, Electrochim. Acta 89 (2013) 122–131.
[130] A.E. Fischer, Y. Show, G.M. Swain, Electrochemical performance of diamond thin-film electrodes from different commercial sources, Anal. Chem. 76 (2004) 2553–2560.
[131] M. Fryda, D. Herrmann, L. Schafer, C.P. Klages, A. Perret, W. Haenni, C. Comninellis, D. Gandini, Properties of diamond electrodes for wastewater treatment, New Diamond Front. Carbon Technol. 9 (1999) 229–240.
[132] Y. Xin, F. Liu, J. Wang, X. Li, Study on Ti/BDD Film Electrode with Ta Interlayer, Rare Met. Mater. Eng. 41 (2012) 247–249.
[133] I. Gerger, R. Haubner, H. Kronberger, G. Fafilek, Investigation of diamond coatings on titanium substrates for electrochemical applications, Diamond Relat. Mater. 13 (2004) 1062–1069.
[134] V. Fisher, D. Gandini, S. Laufer, E. Blank, C. Comninellis, Preparation and characterization of Ti/diamond electrodes, Electrochim. Acta 44 (1998) 521–524.
[135] Y. Tian, X.M. Chen, C. Shang, G.H. Chen, Active and stable Ti/Si/BDD anodes for electro-oxidation, J. Electrochem. Soc. 153 (2006) J80–J85.
[136] W.M. Haynes, CRC Handbook of Chemistry and Physics, 91st ed., New York, CRC Press, 2010.
[137] P.Y. Lim, F.Y. Lin, H.C. Shih, V.G. Ralchenko, V.P. Varnin, Y.V. Pleskov, S.F. Hsu, S. S. Chou, P.L. Hsu, Improved stability of titanium based boron-doped chemical vapor deposited diamond thin-film electrode by modifying titanium substrate surface, Thin Solid Films 516 (2008) 6125–6132.
[138] M.A. Lowe, A.E. Fischer, G.M. Swain, Formation of a crack-free and debonding-resistant boron-doped diamond thin film on titanium using a dual-coating strategy, J. Electrochem. Soc. 153 (2006) B506–B511.
[139] T. Kolber, K. Piplits, R. Haubner, H. Hutter, Quantitative investigation of boron incorporation in polycrystalline CVD diamond films by SIMS, Fresen. J. Anal. Chem. 365 (1999) 636–641.
[140] K.B. Holt, A.J. Bard, Y. Show, G.M. Swain, Scanning electrochemical microscopy and conductive probe atomic force microscopy studies of hydrogen-terminated boron-doped diamond electrodes with different doping levels, J. Phys. Chem. B 108 (2004) 15117–15127.
[141] H.V. Patten, S.C.S. Lai, J.V. Macpherson, P.R. Unwin, Active sites for outer-sphere, inner-sphere, and complex multistage electrochemical reactions at polycrystalline boron-doped diamond electrodes (pBDD) revealed with scanning electrochemical cell microscopy (SECCM), Anal. Chem. 84 (2012) 5427–5432.
[142] H.B. Martin, A. Argoitia, J.C. Angus, U. Landau, Voltammetry studies of single-crystal and polycrystalline diamond electrodes, J. Electrochem. Soc. 146 (1999) 2959–2964.
[143] C.H. Goeting, F. Marken, A. Gutierrez-Sosa, R.G. Compton, J.S. Foord, Electrochemically induced surface modifications of boron-doped diamond electrodes: an X-ray photoelectron spectroscopy study, Diamond Relat. Mater. 9 (2000) 390–396.
[144] H. Girard, N. Simon, D. Ballutaud, M. Herlem, A. Etcheberry, Effect of anodic and cathodic treatments on the charge transfer of boron doped diamond electrodes, Diamond Relat. Mater. 16 (2007) 316–325.
[145] N. Simon, H. Girard, D. Ballutaud, S. Ghodbane, A. Deneuville, M. Herlem, A. Etcheberry, Effect of H and O termination on the charge transfer of moderately boron doped diamond electrodes, Diamond Relat. Mater. 14 (2005) 1179–1182.
[146] M. Wang, N. Simon, G. Charrier, M. Bouttemy, A. Etcheberry, M.S. Li, R. Boukherroub, S. Szunerits, Distinction between surface hydroxyl and ether groups on boron-doped diamond electrodes using a chemical approach, Electrochem. Commun. 12 (2010) 351–354.

[147] M. Wang, N. Simon, C. Decorse-Pascanut, M. Bouttemy, A. Etcheberry, M.S. Li, R. Boukherroub, S. Szunerits, Comparison of the chemical composition of boron-doped diamond surfaces upon different oxidation processes, Electrochim. Acta 54 (2009) 5818–5824.
[148] O. Azizi, D. Hubler, G. Schrader, J. Farrell, B.P. Chaplin, Mechanism of perchlorate formation on boron-doped diamond anodes, Environ. Sci. Technol. 45 (2011) 10582–10590.
[149] H. Notsu, I. Yagi, T. Tatsuma, D.A. Tryk, A. Fujishima, Surface carbonyl groups on oxidized diamond electrodes, J. Electroanal. Chem. 492 (2000) 31–37.
[150] M.C. Granger, G.M. Swain, The influence of surface interactions on the reversibility of ferri/ferrocyanide at boron-doped diamond thin-film electrodes, J. Electrochem. Soc. 146 (1999) 4551–4558.
[151] D. Mishra, Z.H. Liao, J. Farrell, Understanding reductive dechlorination of trichloroethene on boron-doped diamond film electrodes, Environ. Sci. Technol. 42 (2008) 9344–9349.
[152] H. Notsu, I. Yagi, T. Tatsuma, D.A. Tryk, A. Fujishima, Introduction of oxygen-containing functional groups onto diamond electrode surfaces by oxygen plasma and anodic polarization, Electrochem. Solid State Lett. 2 (1999) 522–524.
[153] A. Forrester, S. Hepburn, Spin traps. A cautionary note, J. Chem. Soc. C (1971) 701–703.
[154] H. Chandra, M.C.R. Symons, Hydration of spin-trap cations as a source of hydroxyl adducts, J. Chem. Soc. Chem. Commun. 16 (1986) 1301–1302.
[155] L. Eberson, Formation of hydroxyl spin adducts via nucleophilic addition-oxidation to 5, 5-dimethyl-1-pyrroline N-oxide (DMPO), Acta Chem. Scand. 53 (1999) 584–593.
[156] V. Cerri, C. Frejaville, F. Vila, A. Allouche, G. Gronchi, P. Tordo, Synthesis, redox behavior and spin-trap properties of 2, 6-di- tert-butylnitrosobenzene (DTBN), J. Org. Chem. 54 (1989) 1447–1450.
[157] F. Leinisch, K. Ranguelova, E.F. DeRose, J.J. Jiang, R.P. Mason, Evaluation of the Forrester-Hepburn mechanism as an artifact source in ESR spin-trapping, Chem. Res. Toxicol. 24 (2011) 2217–2226.
[158] F. Leinisch, J.J. Jiang, E.F. DeRose, V.V. Khramtsov, R.P. Mason, Investigation of spin-trapping artifacts formed by the Forrester-Hepburn mechanism, Free Radic. Biol. Med. 65 (2013) 1497–1505.
[159] K. Ranguelova, R.P. Mason, The fidelity of spin trapping with DMPO in biological systems, Magn. Reson. Chem. 49 (2011) 152–158.
[160] Y. Jing, B.P. Chaplin, Mechanistic study of the validity of using hydroxyl radical probes to characterize electrochemical advanced oxidation processes, Environ. Sci. Technol. 51 (2017) 2355–2365.
[161] A.J. Bard, M.V. Mirkin, Scanning Electrochemical Microscopy, second ed., CRC Press, Boca Raton, FL, 2012, 659.
[162] K.D. Sugden, K.E. Wetterhahn, Reaction of chromium (V) with the EPR spin traps 5,5-dimethylpyrroline N-oxide and phenyl-N-tert-butylnitrone resulting in direct oxidation, Inorg.Chem. 35 (1996) 651–657.
[163] P.M.C.W. Hanna, P. Mason, Z. Health, P.O. Box, A. Hagi, M. Nishi, A. Murakami, When are metal ion-dependent hydroxyl and alkoxyl radical adducts of 5,5-dimethyl-1-pyrroline N-oxide artifacts? Arch. Biochem. Biophys. 296 (1992) 640–644.
[164] J.R. Feng, L.L. Houk, D.C. Johnson, S.N. Lowery, J.J. Carey, Electrocatalysis of anodic oxygen-transfer reactions—the electrochemical incineration of benzoquinone, J. Electrochem. Soc. 142 (1995) 3626–3632.
[165] Y. Yoon, Y. Jung, M. Kwon, E. Cho, J.W. Kang, Alternative electrode materials and ceramic filter minimize disinfection byproducts in point-of-use electrochemical water treatment, Environ. Eng. Sci. 30 (2013) 742–749.

[166] A.J. Bard, L.R. Faulkner, Electrochemical Methods: Fundamentals and Applications, second ed., John Wiley and Sons, Hoboken, NJ, 2001, 833.
[167] S.G. Bratsch, Standard electrode-potentials and temperature coefficients in water at 298.15-K, J. Phys. Chem. Ref. Data Monogr. 18 (1989) 1–21.
[168] L. Ouattara, M.M. Chowdhry, C. Comninellis, Electrochemical treatment of industrial wastewater, New Diamond Front. Carbon Technol. 14 (2004) 239–247.
[169] A. Anglada, R. Ibanez, A. Urtiaga, I. Ortiz, Electrochemical oxidation of saline industrial wastewaters using boron-doped diamond anodes, Catal. Today 151 (2010) 178–184.
[170] A. Anglada, A.M. Urtiaga, I. Ortiz, Laboratory and pilot plant scale study on the electrochemical oxidation of landfill leachate, J. Hazard. Mater. 181 (2010) 729–735.
[171] P. Canizares, B. Louhichi, A. Gadri, B. Nasr, R. Paz, M.A. Rodrigo, C. Saez, Electrochemical treatment of the pollutants generated in an ink-manufacturing process, J. Hazard. Mater. 146 (2007) 552–557.
[172] P. Canizares, L. Martinez, R. Paz, C. Saez, J. Lobato, M.A. Rodrigo, Treatment of Fenton-refractory olive oil mill wastes by electrochemical oxidation with boron-doped diamond anodes, J. Chem. Technol. Biotechnol. 81 (2006) 1331–1337.
[173] D. Haaken, T. Dittmar, V. Schmalz, E. Worch, Influence of operating conditions and wastewater-specific parameters on the electrochemical bulk disinfection of biologically treated sewage at boron-doped diamond (BDD) electrodes, Desalination Water Treat. 46 (2012) 160–167.
[174] J. Jeong, J.Y. Kim, J. Yoon, The role of reactive oxygen species in the electrochemical inactivation of microorganisms, Environ. Sci. Technol. 40 (2006) 6117–6122.
[175] J. Niu, H. Lin, C. Gong, X. Sun, Theoretical and experimental insights into the electrochemical mineralization mechanism of perfluorooctanoic acid, Environ. Sci. Technol. 47 (2013) 14341–14349.
[176] J. Niu, H. Lin, J. Xu, H. Wu, Y. Li, Electrochemical mineralization of perfluorocarboxylic acids (PFCAs) by Ce-doped modified porous nanocrystalline PbO_2 film electrode, Environ. Sci. Technol. 46 (2012) 10191–10198.
[177] T. Ochiai, Y. Iizuka, K. Nakata, T. Murakami, D.A. Tryk, A. Fujishima, Y. Koide, Y. Morito, Efficient electrochemical decomposition of perfluorocarboxylic acids by the use of a boron-doped diamond electrode, Diamond Relat. Mater. 20 (2011) 64–67.
[178] A. Anglada, A. Urtiaga, I. Ortiz, Pilot scale performance of the electro-oxidation of landfill leachate at boron-doped diamond anodes, Environ. Sci. Technol. 43 (2009) 2035–2040.
[179] M. Mascia, A. Vacca, S. Palmas, A.M. Polcaro, Kinetics of the electrochemical oxidation of organic compounds at BDD anodes: modelling of surface reactions, J. Appl. Electrochem. 37 (2007) 71–76.
[180] A. Donaghue, B.P. Chaplin, Effect of select organic compounds on perchlorate formation at boron-doped diamond film anodes, Environ. Sci. Technol. 47 (2013) 12391–12399.
[181] P. Cañizares, J. García-Gómez, I. Fernández de Marcos, M.A. Rodrigo, J. Lobato, Measrement of mass-transfer coefficients by an electrochemical technique, J. Chem. Educ. 83 (2006) 1204.
[182] H. Liu, C.D. Vecitis, Reactive transport mechanism for organic oxidation during electrochemical filtration: mass-transfer, physical adsorption, and electron-transfer, J. Phys. Chem. C 116 (2012) 374–383.
[183] R.B. Byrd, W.E. Stewart, E.N. Lightfoot, Transport Phenomena, second ed., Wiley, New York, 2007.
[184] A. Lasia, Electrochemical impedance spectroscopy and its applications, in: B. E. Conway, J.O.'.M. Bockris, R.E. White (Eds.), Modern Aspects of Electrochemistry, vol. 32, Kluwer Academic/Plenum Publishers, New York, 1999, pp. 143–248.
[185] J. Newman, K.E. Thomas-Alyea, Electrochemical Systems, third ed., Wiley, New York, 2004.

[186] E.J. Wilson, C.J. Geankoplis, Liquid mass transfer at very low Reynolds numbers in packed beds, Ind. Eng. Chem. Fund. (1966) 5.
[187] M.E.H. Bergmann, J. Rollin, T. Iourtchouk, The occurrence of perchlorate during drinking water electrolysis using BDD anodes, Electrochim. Acta 54 (2009) 2102–2107.
[188] M.E.H. Bergmann, J. Rollin, Product and by-product formation in laboratory studies on disinfection electrolysis of water using boron-doped diamond anodes, Catal. Today 124 (2007) 198–203.
[189] M.E.H. Bergmann, A.S. Koparal, T. Iourtchouk, Electrochemical advanced oxidation processes, formation of halogenate and perhalogenate species: a critical review, Crit. Rev. Environ. Sci. Technol. 44 (2014) 348–390.
[190] M.E.H. Bergmann, T. Iourtchouk, W. Schmidt, G. Nüske, M. Fischer, Perchlorate Formation in Electrochemical Water Disinfection, Perchlorates: Production, Uses and Health Effects, Nova Science Publishers, UK, 2011, pp. 11–141.
[191] M.E.H. Bergmann, T. Iourtchouk, J. Rollin, The occurrence of bromate and perbromate on BDD anodes during electrolysis of aqueous systems containing bromide: first systematic experimental studies, J. Appl. Electrochem. 41 (2011) 1109–1123.
[192] M.E.H. Bergmann, C. Comninellis, G. Chen (Eds.), Drinking Water Disinfection by In-line Electrolysis-Product and Inorganic By-Product Formation, Electrochemistry for the Environment, Springer, New York, 2009, pp. 163–205.
[193] H. Bergmann, A.T. Koparal, A.S. Koparal, F. Ehrig, The influence of products and by-products obtained by drinking water electrolysis on microorganisms, Microchem. J. 89 (2008) 98–107.
[194] A. Anglada, A. Urtiaga, I. Ortiz, D. Mantzavinos, E. Diamadopoulos, Boron-doped diamond anodic treatment of landfill leachate: evaluation of operating variables and formation of oxidation by-products, Water Res. 45 (2011) 828–838.
[195] J. Boudreau, D. Bejan, N.J. Bunce, Competition between electrochemical advanced oxidation and electrochemical hypochlorination of acetaminophen at boron-doped diamond and ruthenium dioxide based anodes, Can. J. Chem. 88 (2010) 418–425.
[196] H. Li, J. Ni, Electrogeneration of disinfection byproducts at a boron-doped diamond anode with resorcinol as a model substance, Electrochim. Acta 69 (2012) 268–274.
[197] A.Y. Bagastyo, D.J. Batstone, I. Kristiana, W. Gernjak, C. Joll, J. Radjenouic, Electrochemical oxidation of reverse osmosis concentrate on boron-doped diamond anodes at circumneutral and acidic pH, Water Res. 46 (2012) 6104–6112.
[198] A.Y. Bagastyo, D.J. Batstone, K. Rabaey, J. Radjenovic, Electrochemical oxidation of electrodialysed reverse osmosis concentrate on Ti/Pt-IrO_2, Ti/SnO_2-Sb and boron-doped diamond electrodes, Water Res. 47 (2013) 242–250.
[199] E.T. Urbansky, M.R. Schock, Issues in managing the risks associated with perchlorate in drinking water, J. Environ. Manage. 56 (1999) 79–95.
[200] E. Urbansky, Perchlorate as an environmental contaminant, Environ. Sci. Pollut. Res. 9 (2002) 187–192.
[201] Perchlorate in Drinking Water; R-16-04, California Department of Health Services, Sacramento, CA, 2006.
[202] Perchlorate Fact Sheet for Public Water Suppliers, Massachusetts Department of Environmental Protection, Boston, MA, 2006.
[203] Y.J. Jung, K.W. Baek, B.S. Oh, J.W. Kang, An investigation of the formation of chlorate and perchlorate during electrolysis using Pt/Ti electrodes: the effects of pH and reactive oxygen species and the results of kinetic studies, Water Res. 44 (2010) 5345–5355.
[204] D. Hubler, Modeling Electrochemical Water Treatment Processes, (Dissertation), The University of Arizona, Ann Arbor, 2012. Pro Quest/UMI. (Publication No. AAT 3548029.).
[205] W. Jawando, P. Gayen, B.P. Chaplin, The effects of surface oxidation and fluorination of boron-doped diamond anodes on perchlorate formation and organic compound oxidation, Electrochim. Acta 174 (2015) 1067–1078.

[206] P. Gayen, B.P. Chaplin, fluorination of boron-doped diamond film electrodes for minimization of perchlorate formation, ACS Appl. Mater. Interfaces 9 (2013) 27638–27648.
[207] C.R. Costa, F. Montilla, E. Morallon, P. Olivi, Electrochemical oxidation of acid black 210 dye on the boron-doped diamond electrode in the presence of phosphate ions: Effect of current density, pH, and chloride ions, Electrochim. Acta 54 (2009) 7048–7055.
[208] K.E. Carter, J. Farrell, Electrochemical oxidation of trichloroethylene using boron-doped diamond film electrodes, Environ. Sci. Technol. 43 (2009) 8350–8354.
[209] O. Scialdone, A. Galia, S. Randazzo, Oxidation of carboxylic acids in water at IrO_2-Ta_2O_5 and boron doped diamond anodes, Chem. Eng. J. 174 (2011) 266–274.
[210] H. Gallard, F. Pellizzari, J.P. Croue, B. Legube, Rate constants of reactions of bromine with phenols in aqueous solution, Water Res. 37 (2003) 2883–2892.
[211] D. Khamis, E. Mahe, F. Dardoize, D. Devilliers, Peroxodisulfate generation on boron-doped diamond microelectrodes array and detection by scanning electrochemical microscopy, J. Appl. Electrochem. 40 (2010) 1829–1838.
[212] C. Provent, W. Haenni, E. Santoli, P. Rychen, Boron-doped diamond electrodes and microelectrode-arrays for the measurement of sulfate and peroxodisulfate, Electrochim. Acta 49 (2004) 3737–3744.
[213] X.C. Gui, Z.P. Zeng, A.Y. Cao, Z.Q. Lin, H.Q. Zeng, R. Xiang, T.Z. Wu, Y. Zhu, Z. K. Tang, Elastic shape recovery of carbon nanotube sponges in liquid oil, J. Mater. Chem. 22 (2012) 18300–18305.
[214] S.D. Boyce, J.F. Hornig, Reaction pathways of trihalomethane formation from the halogenation of dihydroxyaromatic model compounds for humic-acid, Environ. Sci. Technol. 17 (1983) 202–211.
[215] L.M. Rebenne, A.C. Gonzalez, T.M. Olson, Aqueous chlorination kinetics and mechanism of substituted dihydroxybenzenes, Environ. Sci. Technol. 30 (1996) 2235–2242.
[216] W.A. Arnold, J. Bolotin, U. Von Gunten, T.B. Hofstetter, evaluation of functional groups responsible for chloroform formation during water chlorination using compound specific isotope analysis, Environ. Sci. Technol. 42 (2008) 7778–7785.
[217] J. Iniesta, P.A. Michaud, M. Panizza, G. Cerisola, A. Aldaz, C. Comninellis, Electrochemical oxidation of phenol at boron-doped diamond electrode, Electrochim. Acta 46 (2001) 3573–3578.
[218] J.F. Zhi, H.B. Wang, T. Nakashima, T.N. Rao, A. Fujishima, Electrochemical incineration of organic pollutants on boron-doped diamond electrode. Evidence for direct electrochemical oxidation pathway, J. Phys. Chem. B 107 (2003) 13389–13395.
[219] P. Canizares, J. Garcia-Gomez, J. Lobato, M.A. Rodrigo, Modeling of wastewater electro-oxidation processes part I. General description and application to inactive electrodes, Ind. Eng. Chem. Res. 43 (2004) 1915–1922.
[220] P. Canizares, J. Garcia-Gomez, J. Lobato, M.A. Rodrigo, Modeling of wastewater electro-oxidation processes part II. Application to active electrodes, Ind. Eng. Chem. Res. 43 (2004) 1923–1931.
[221] R. Vargas, C. Borras, D. Mendez, J. Mostany, B.R. Scharifker, Electrochemical oxygen transfer reactions: electrode materials, surface processes, kinetic models, linear free energy correlations, and perspectives, J. Solid State Electrochem. 20 (2016) 875–893.
[222] A.M. Polcaro, A. Vacca, M. Mascia, S. Palmas, J.R. Ruiz, Electrochemical treatment of waters with BDD anodes: kinetics of the reactions involving chlorides, J. Appl. Electrochem. 39 (2009) 2083–2092.
[223] B.P. Chaplin, Critical review of electrochemical advanced oxidation processes for water treatment applications, Environ. Sci.: Processes Impacts 16 (2014) 1182–1203.
[224] B.P. Chaplin, G. Schrader, J. Farrell, Electrochemical destruction of N-nitrosodimethylamine in reverse osmosis concentrates using boron-doped diamond film electrodes, Environ. Sci. Technol. 44 (2010) 4264–4269.

[225] C. Lee, J. Yoon, U. Von Gunten, Oxidative degradation of N-nitrosodimethylamine by conventional ozonation and the advanced oxidation process ozone/hydrogen peroxide, Water Res. 41 (2007) 581–590.
[226] P. Canizares, R. Paz, C. Saez, M.A. Rodrigo, Costs of the electrochemical oxidation of wastewaters: a comparison with ozonation and Fenton oxidation processes, J. Environ. Manage. 90 (2009) 410–420.
[227] E. Chatzisymeon, S. Foteinis, D. Mantzavinos, T. Tsoutsos, Life cycle assessment of advanced oxidation processes for olive mill wastewater treatment, J. Clean. Prod. 54 (2013) 229–234.
[228] M. Wasberg, G. Horanyi, The reduction of clo 4-ions on rh electrodes, J. Electroanal. Chem. 385 (1995) 63–70.
[229] C. Almeida, B.F. Giannetti, T. Rabockai, Electrochemical study of perchlorate reduction at tin electrodes, J. Electroanal. Chem. 422 (1997) 185–189.
[230] B.H. Gu, W.J. Dong, G.M. Brown, D.R. Cole, Complete degradation of perchlorate in ferric chloride and hydrochloric acid under controlled temperature and pressure, Environ. Sci. Technol. 37 (2003) 2291–2295.
[231] K.D. Hurley, J.R. Shapley, Efficient heterogeneous catalytic reduction of perchlorate in water, Environ. Sci. Technol. 41 (2007) 2044–2049.
[232] K.D. Hurley, Y.X. Zhang, J.R. Shapley, Ligand-enhanced reduction of perchlorate in water with heterogeneous Re-Pd/C catalysts, J. Am. Chem. Soc. 131 (2009) 14172–14173.
[233] X. Chen, X.C. Huo, J.Y. Liu, Y. Wang, C.J. Werth, T.J. Strathmann, Exploring beyond palladium: catalytic reduction of aqueous oxyanion pollutants with alternative platinum group metals and new mechanistic implications, Chem. Eng. J. 313 (2017) 745–752.
[234] J.Y. Liu, J.K. Choe, Y. Wang, J.R. Shapley, C.J. Werth, T.J. Strathmann, Bioinspired complex-nanoparticle hybrid catalyst system for aqueous perchlorate reduction: rhenium speciation and its influence on catalyst activity, ACS Catal. 5 (2015) 511–522.
[235] J.Y. Liu, M.W. Han, D.M. Wu, X. Chen, J.K. Choe, C.J. Werth, T.J. Strathmann, A new bioinspired perchlorate reduction catalyst with significantly enhanced stability via rational tuning of rhenium coordination chemistry and heterogeneous reaction pathway, Environ. Sci. Technol. 50 (2016) 5874–5881.
[236] A. Cano, P. Canizares, C. Barrera, C. Saez, M.A. Rodrigo, Use of low current densities in electrolyses with conductive-diamond electrochemical—Oxidation to disinfect treated wastewaters for reuse, Electrochem. Commun. 13 (2011) 1268–1270.

Further Reading

[1] P.A. Michaud, M. Panizza, L. Ouattara, T. Diaco, G. Foti, C. Comninellis, Electrochemical oxidation of water on synthetic boron-doped diamond thin film anodes, J. Appl. Electrochem. 33 (2003) 151–154.
[2] C. Comninellis, A. Kapalka, S. Malato, S.A. Parsons, L. Poulios, D. Mantzavinos, Advanced oxidation processes for water treatment: advances and trends for R&D, J. Chem. Technol. Biotechnol. 83 (6) (2008) 769–776.

CHAPTER

18

Integration of Electrochemical Advanced Oxidation With Membrane Separation and Biodegradation

Ane M. Urtiaga, Raquel Ibañez, María J. Rivero, Inmaculada Ortiz

University of Cantabria, Santander, Spain

OUTLINE

https://doi.org/10.1016/B978-0-12-813160-2.00018-3

INTRODUCTION

Despite the ability of EAOPs to effectively degrade and mineralize persistent and toxic organic pollutants, the large-scale implementation of the environmental applications of EAOPs is still hindered by several factors. The elevated operation costs due to energy consumption likely represents the main restriction at present [1]. This chapter evaluates process integration as a strategy to boost the effectiveness of EAOPs by combining them with other technologies in search of synergies and innovative solutions to overcome EAOP limitations. In our first case study, EAOPs will be shown as a powerful tool to increase the biodegradability of organic contaminants contained in recalcitrant landfill leachates, with the aim of degrading the initial organic load down to the point at which the improvement of the biodegradability potential makes the pretreated leachate suitable for the much more cost-effective aerated sludge treatment. Our second approach consists of using membrane separation aimed at the increase of the concentration of pollutants before the application of electrochemical degradation as a way to increase the kinetics of diffusion limited electro-oxidation. This concept is illustrated by two cases of study dealing with (i) the electrochemical treatment of reverse osmosis brines generated in water reclamation facilities, and (ii) the removal of perfluorinated compounds from industrial process streams.

ELECTROCHEMICAL TREATMENT AS A WAY TO INCREASE EFFLUENT BIODEGRADABILITY

Although chemical oxidation for complete mineralization of pollutants is usually costly, its combination with biological treatment has been reported to reduce the overall operating costs [2]. In many cases, not only the removal of the target compound but also its mineralization can involve long reaction times and energy costs that burden direct implementation of electrochemical oxidation as a single water treatment technology. A process design strategy can consider the effect of the electrochemical treatment on the biodegradability so that the most advantageous process from the economic and technical point of view can be obtained by the integration of technologies: electrochemical oxidation plus biological treatment.

During the initial stages of electrochemical oxidation, the organic contaminants are partially degraded, resulting in the formation of intermediate compounds. As a result, the biodegradability potential of the effluent changes during the oxidation process [3]. Therefore, it is necessary to determine the process conditions and time needed to turn a recalcitrant effluent into a biodegradable one.

This strategy can be considered with complex effluents such as the leachates obtained upon disposal in landfills of municipal solid wastes. The degradation of the organic fraction of the wastes in combination with percolating rainwater leads to the generation of a highly contaminated liquid. Landfill leachate usually contains a complex variety of materials and organic compounds such as fatty acids and many other hazardous chemicals [4]. Some of the compounds detected in the leachates are identified as xenobiotic organic pollutants, such as bisphenol A (BPA), *p*-cresol, and pharmaceutical products such as ibuprofen. Moreover, landfill leachates contain ammonium, metals, and inorganic salts, with chloride being the most abundant anion [5,6]. As a result, leachates generated in municipal waste landfills are characterized by their low biodegradability.

The literature shows several ways to assess the biodegradability of an effluent, which involves the quantification of the fraction of the total organic matter that microorganisms are able to assimilate. The ratio between the biological oxygen demand measured after 5 days and the chemical oxygen demand (BOD_5/COD) provides an initial picture of the biodegradation potential of wastewater. The Zahn-Wellens test, which allows the acclimation of the biomass to the aqueous waste composition, is also extended for measuring inherent biodegradability [7,8], although it requires very extended evaluation periods.

Respirometry assays can be implemented for a quick evaluation of biodegradability. These tests are based on the direct relationship between the specific oxygen uptake rate (OUR) and biomass activity. The respirometry technique measures the oxygen consumed by the biomass when degrading the substrates, thereby providing very precise information regarding microorganism activity and the rate at which the different substrates are consumed. In a respirogram, the values of oxygen consumption rate per unit volume and time are plotted versus time. The biodegradable chemical oxygen demand, COD_b, can be obtained after area integration. OUR can be directly correlated with food to volume (F/V) ratio because the amount of substrate to be decomposed is proportional to the amount of oxygen consumed by microorganisms, at a steady state. The respirometry bioassay simulates the typical operation conditions of an activated sludge unit, using a high biomass to COD ratio. Moreover, it allows a significant reduction of the testing time compared to BOD_5, BOD_{21}, and the Zahn-Wellens tests. The results of the biodegradability potential can be expressed in terms of COD_b/COD ratio.

The present application is focused on the analysis of the effect of electrochemical oxidation on the enhancement of the biodegradability potential of persistent and toxic pollutants, using respirometry assays to determine the biodegradability development. While the literature is profuse in analyzing the electrochemical oxidation of a single compound [9,10], works considering a mixture of compounds are rather scarce [11], a factor that introduces uncertainty in the direct extrapolation for the performance prediction of the electrochemical technology when dealing with real effluents.

In the first stage of this application, a mixture of compounds with estrogenic, hazardous, and pharmaceutical properties that included bisphenol A (BPA), *p*-cresol, and ibuprofen was electrochemically oxidized, with particular attention paid to the increase of the biodegradation potential. Later, the results were applied to the treatment of landfill leachates. A mixture of BPA, *p*-cresol, and ibuprofen in water was prepared, using ultrapure water (Millipore). The initial concentration of each compound was set at 100 mg/L and sodium chloride (2 g/L) was added as an electrolyte. Electrochemical experiments were conducted in a batch system composed of a feed tank (2 L) and a single compartment flow-by cell (DiaCell 106, Adamant Technologies), provided with a boron doped diamond (BDD) on a silicon anode and a stainless steel (316 L) cathode. Both electrodes were circular with a surface area of 70 cm^2 each and an electrode gap of 5 mm. A current density of 100 A/m^2 was applied, a value slightly higher than the limiting current density calculated as 87 A/m^2. The electrogenerated free chlorine was removed prior to respirometric analysis by adding the stoichiometric amount of sodium thiosulfate. Further details of the experimental set-up can be found elsewhere [12,13].

Fig. 1A shows the evolution of the concentration of BPA, *p*-cresol, and ibuprofen in the aqueous solution during the electrochemical treatment (concentration values are given as the fraction of the initial concentration C/C_0), together with the changes in biodegradability, expressed as the ratio COD_b/COD. The evolution of COD and dissolved organic carbon (DOC) are provided in Fig. 1B. It is remarkable that BPA and *p*-cresol were completely removed after 1 h of treatment while ibuprofen removal took about 2 h. Initially, before the complete removal of the parent contaminants, the effluent was not biodegradable as the COD_b/COD ratio was close to zero. However, just after the complete removal of the estrogenic BPA and the toxic *p*-cresol, the biodegradability potential started to increase. It was after approximately 2.5 h when the COD_b/COD reached a value of 0.4, a magnitude that is commonly used to define wastewater to become suitable for biological treatment. At the same time, DOC after 2.5 h of electrochemical treatment was only reduced by a factor of 35%, pointing to the transformation of the initial contaminants into other intermediates, which nevertheless could be further treated by means of a bioprocess to remove the remaining organic load. Therefore, it can be concluded that the electrochemical treatment enabled the increase of biodegradability even at low mineralization rates by the conversion of the initial contaminants into less toxic intermediate products.

The next step was focused at the treatment of landfill leachates. Fig. 2 shows the results for leachates generated in the sanitary landfill of

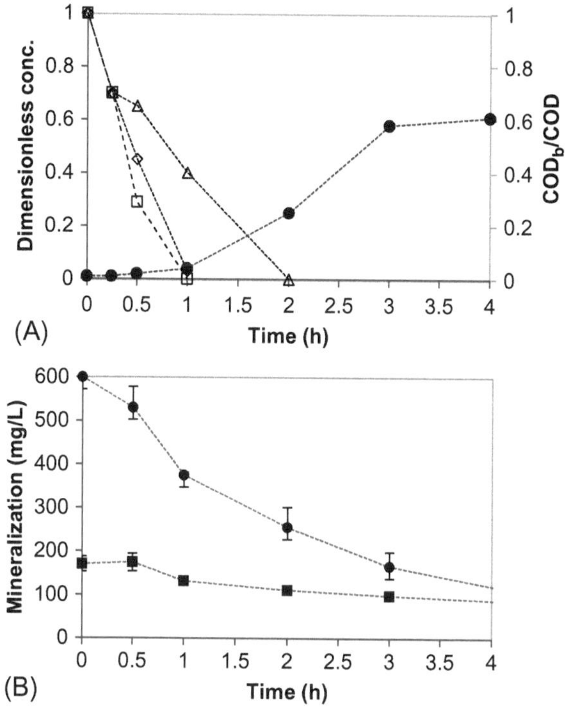

FIG. 1 (A) Bisphenol A (□), *p*-cresol (◇), and ibuprofen (Δ) in the left axis, and biodegradability evolution COD_b/COD (●) in the right axis, during the electrochemical oxidation using BDD anodes. (B) Mineralization of the sample, expressed as COD (●) and DOC (■). $J = 100\,A/m^2$. Electrolyte NaCl 2 g/L. Initial concentration of BPA, *p*-cresol, and ibuprofen 100 mg/L of each compound.

Meruelo (Cantabria, Spain). This landfill collects urban wastes generated in the region of Cantabria (northern Spain), and serves a total population of about 600,000. The electrochemical oxidation tests were performed at galvanostatic conditions and a current density of $900\,A/m^2$ was applied. Initial COD was $2228 \pm 702\,mg/L$ [14]. From respirometric assays, the landfill leachate was initially nonbiodegradable, as the COD_b/COD ratio was only 0.04. However, the recalcitrant character of the landfill leachate was completely modified after 1 h of EAOP treatment, when the COD_b/COD was increased to 0.40. Thus, the obtained results show that electrochemical oxidation can improve the biodegradability potential of landfill leachates in a short treatment time, and thus using low energy consumption. While COD reduction was only 50% in 1 h of EAOP, the increase of biodegradability makes the treated leachate suitable for subsequent aerated sludge treatment.

BPA, *p*-cresol, and pharmaceutical compounds are usually present in landfill leachates, contributing to their toxic and recalcitrant nature.

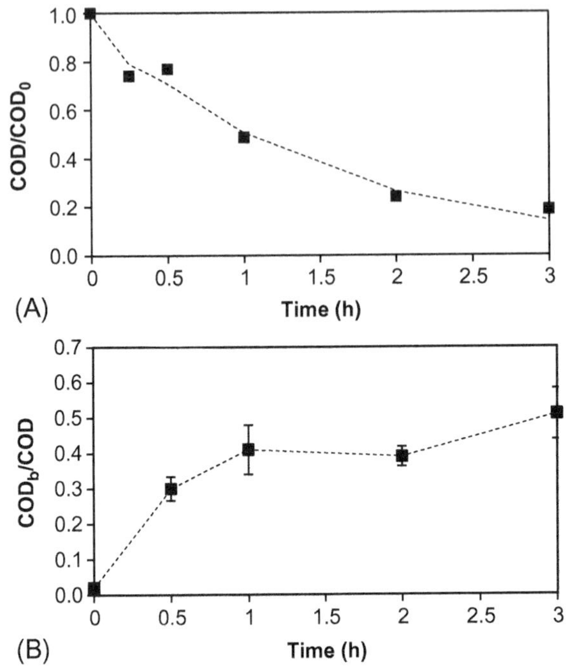

FIG. 2 Electrochemical oxidation of landfill leachate. (A) Dimensionless COD. (B) Biodegradability evolution. $J = 900\,A/m^2$. Sample: landfill leachates of Meruelo site (northern Spain).

Regardless of the AOP applied, the literature points out that the main degradation products of phenolic compounds, including BPA and *p*-cresol, come from the hydroxylation of the aromatic rings of the molecule that evolve to quinones and further oxidize to organic acids and eventually to CO_2 [15,16]. However, most of the AOP studies regarding phenol degradation pathways were carried out in the absence of chloride. The significant presence of chlorides in landfill leachates may lead to the formation of electrogenerated organochlorinated compounds in the degradation pathway [17]. Specific compounds can modify biodegradability during treatment. Considering, for instance, the electrochemical oxidation of 100 mg/L of BPA with 2000 mg/L of chloride ions, GC-MS analysis confirmed that during the first minutes of BPA electrooxidation, monochlorinated and dichlorinated compounds, such as *p*-clorophenol; benzofuran, 5-chloro-2, 3-dihydro-2-methyl; 2,4-diclorophenol; and benzofuran, 5, 7-chloro-2, 3-dihydro-2-methyl, were formed, as shown in Fig. 3 [18]. At longer oxidation times these compounds tended to disappear while trichloro compounds were formed and the appearance of phenol, 2,3,6-trichloro was also detected. Under the studied conditions, the formation of these compounds explains the low mineralization rate observed in the first hour of electrochemical treatment. Nevertheless, the rapid increase of biodegradability after 1 h of treatment points to the progressive

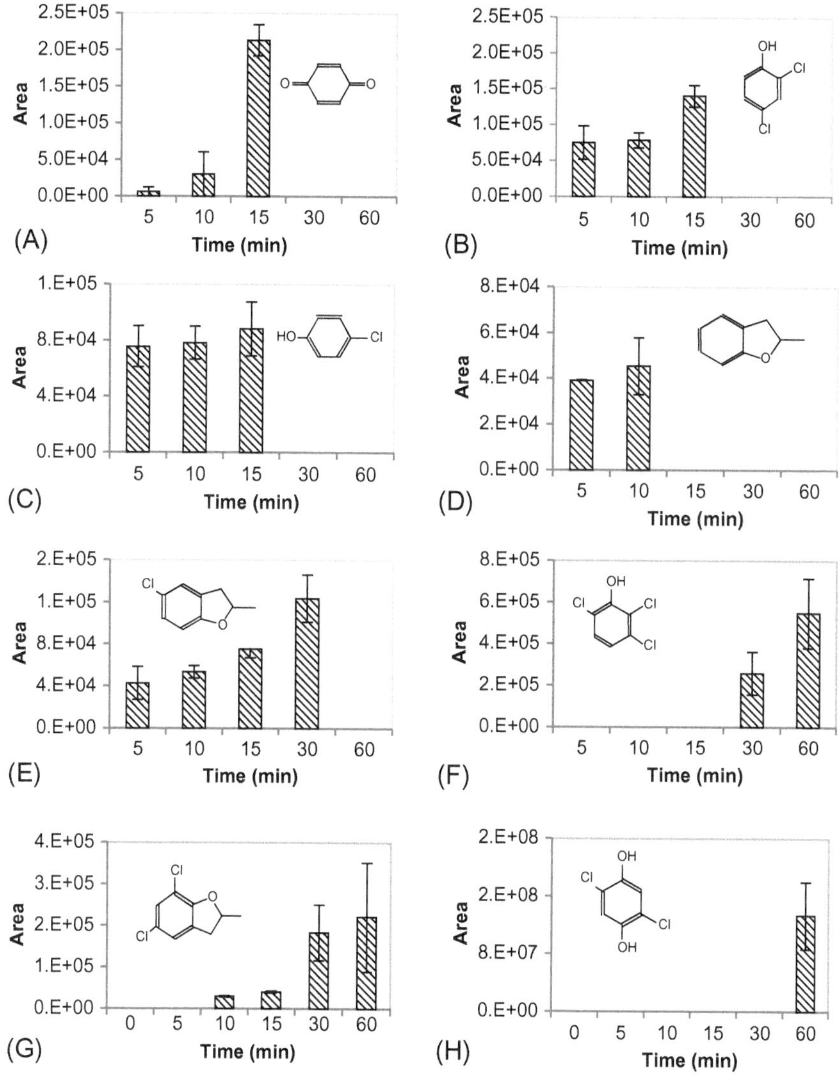

FIG. 3 BPA intermediates identified over 1h of electrochemical oxidation with BDD electrode (BPA, 100mg/L; 2000mg/L chloride as electrolyte). (A) *p*-Benzoquinone; (B) 2,4-diclorophenol; (C) *p*-clorophenol; (D) benzofuran, 2,3-dehydro-2-methyl; (E) benzofuran, 5-chloro-2,3-dihydro-2-methyl; (F) phenol, 2,3,6-trichloro; (G) benzofuran, 5,7-chloro-2,3-dihydro-2-methyl; and (H) hydroquinone, 2,5-dichloro. Results are present as the area of the chromatographic peaks.

oxidation and disappearance of chlorinated compounds. To conclude, these studies can be a useful tool to determine the time required for the electrochemical treatment to increase the biodegradability of the effluent to make it suitable for conventional treatment.

INTEGRATION OF ELECTROCHEMICAL TREATMENT WITH MEMBRANE SEPARATION

Membrane separation is one of the most extended technologies for producing clean and safe water. However, the use of membrane processes alone is not enough for the overall treatment of persistent organic pollutants (POPs) because the toxic and hazardous contaminants are retained in the concentrate stream, which should be treated before disposal. On the other hand, electrochemical treatment is highly effective for the degradation and mineralization of organic pollutants, although at the expense of high energy consumption, in particular when dealing with the low concentration range of emerging micropollutants and POPs. In this section, we report two applications derived from the concept of coupling membrane technology as a preconcentration step with EAOPs in the fields of water reclamation and reuse and for the treatment of industrial process waters polluted by perfluorinated compounds.

Increasing the Sustainability of Water Reuse

An illustrative example of the optimal performance of electrochemical oxidation of the effluents of wastewater treatment plants (WWTPs) in combination with membrane technologies was reported by Urtiaga et al. [19]. In this study, electrooxidation was applied to the treatment of the reverse osmosis rejection stream generated in a water reclamation unit. Using the effluent of a municipal WWTP as a nonconventional water source, the tertiary water treatment train consisted of an ultrafiltration (UF) unit, aimed at the removal of suspended solids and bacteria, and a reverse osmosis (RO) step for retention of salts, soluble organic matter, and colloids. The UF/RO system produced high quality de-ionized water suitable for most industrial applications. However, salts and organic pollutants accumulated in the RO rejection stream, which was the main waste effluent of the industrial water reclamation process [20]. Fig. 4 shows an overview of the pilot plant scheme that was applied in the water reclamation project, including the innovative electrochemical treatment of the RO concentrate.

The strategy for increasing the productivity of the water reclamation process consisted of the electrochemical treatment of the RO brine, with the aim of removing the retained organic compounds and ammonia. The physical-chemical composition of the RO rejection stream (Table 1) reveals its electrolyte character, with chloride and sulfate being the main contributors to the anions load together with moderate DOC content and ammonia. BDD anodes were selected for gaining two objectives: (i) the ammonia indirect oxidation by means of electrogenerated chlorine; and

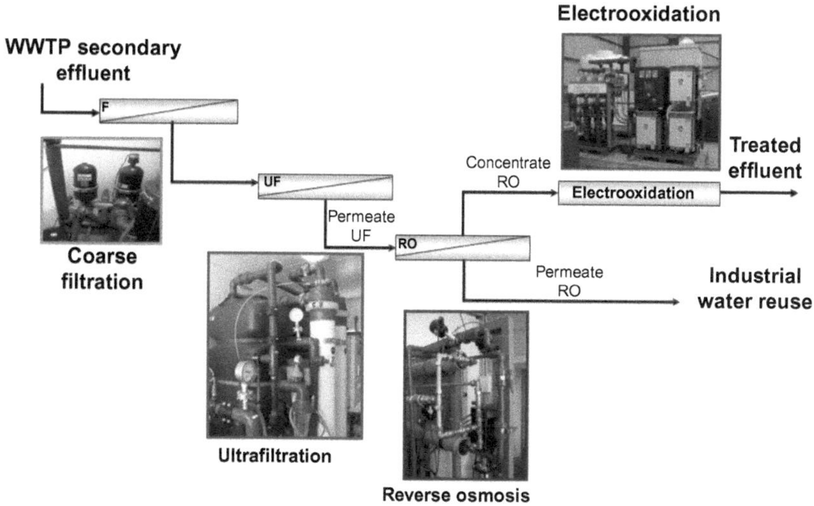

FIG. 4 Scheme of the advanced water treatment train, composed of ultrafiltration, reverse osmosis, and electrooxidation pilot plant units.

TABLE 1 Physical and Chemical Characteristics of the RO Concentrate Obtained in the UF/RO Water Reclamation Facility. Twelve Most Abundant Pharmaceutical Compounds Are Reported

Conductivity (mS/cm)	4.8
pH	7.9
DOC (mg/L)	26.6
NH_4^+ (mg/L)	207
Chloride (mg/L)	630
Sulfate (mg/L)	584
Caffeine (ng/L)	50,000
Nicotine (ng/L)	5683
4-AAA (ng/L)	15569
Atenolol (ng/L)	2779
Naproxen (ng/L)	9223
Fenofibric acid (ng/L)	1480
Gemfibrozil (ng/L)	9868
Ibuprofen (ng/L)	21,250
Furosemide (ng/L)	2522
Hidrochlorothiazide (ng/L)	27,000
Ofloxacin (ng/L)	2575
Benzafibrate (ng/L)	583

RO unit was operated at 70% water productivity.

(ii) the hydroxyl radical mediated oxidation of soluble organic compounds, with particular attention paid to the removal of emerging micropollutants. A group of 12 compounds representative of the most prevalent stimulants (caffeine, nicotine) and various therapeutic pharmaceutical categories (naproxen, ibuprofen, ofloxacin, furosemide, hydrochlorothiazide, gemfibrozil, bezafibrate, fenofibric acid, atenolol, and 4-AAA) was selected to study the effectiveness of the WWTP/UF/RO/EAOP treatment train [21,22].

Fig. 5 shows the excellent removal of the macropollutants achieved in the EAOP treatment, in an experiment performed at laboratory scale (BDD anode area = 70 cm^2, volume treated = 1 L). Both ammonia and COD were completely removed with the positive effect of increasing the applied current density on the removal kinetics of both parameters.

Fig. 6 illustrates the evolution of the pharmaceutical compounds along the sampling points of the treatment train. For each compound in Fig. 6, the central bar illustrates the concentration of the pollutant in the reclaimed water for industrial reuse. This can be observed that the permeate of the RO unit was virtually free of emerging micropollutants, although these compounds accumulated in the RO concentrate stream. Turning the attention to the electrochemical treatment, the two bars at the right side reveal the concentration of micropollutants at the inlet and outlet of the electrolysis step. Electrooxidation, under the conditions used in this study, removed more than 97% of hydrochlorothiazide, nicotine, atenolol, furosemide, and bezafibrate while 94% of ofloxacin, fenofibric acid, 4-AAA, naproxen, and gemfibrozil were eliminated. Ibuprofen

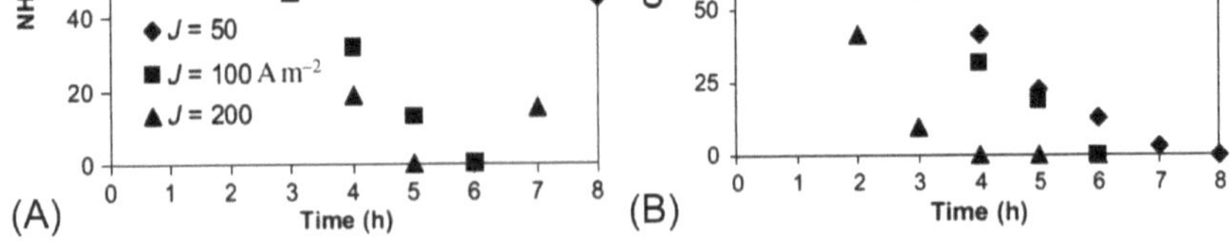

FIG. 5 Removal of ammonia (A) and organic pollutants (B) from the RO concentrate stream by BDD electrooxidation. Effect of the applied current density. No addition of electrolyte was needed. *Data from G. Pérez, A.R. Fernández-Alba, A.M. Urtiaga, I. Ortiz, Electro-oxidation of reverse osmosis concentrates generated in tertiary water treatment, Water Res. 44 (2010) 2763–2772. Copyright 2010. Elsevier.*

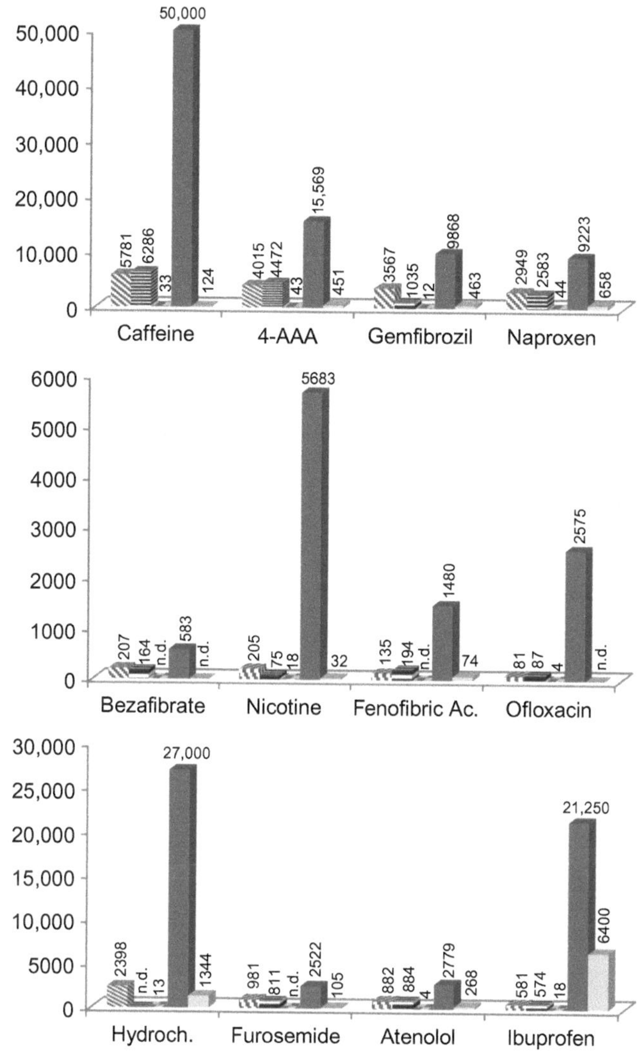

FIG. 6 Concentration of emerging pollutants in the different steps of the advanced treatment of the effluent of a municipal WWTP. WWTP effluent; UF permeate; RO permeate; RO concentrate; Elox effluent. n.d.: none detected. *Data from A.M. Urtiaga, G. Perez, R. Ibáñez, I. Ortiz, Removal of pharmaceuticals from a WWTP secondary effluent by ultrafiltration/reverse osmosis followed by electrochemical oxidation of the RO concentrate, Desalination 331 (2013) 26–34. Copyright 2013. Elsevier.*

was the most resistant to electrooxidation, with 70% of removal in the defined process conditions in accordance with the results reported in Fig. 6. Overall, the electrooxidation treatment reduced the total micropollutant content in the RO brine from 149 μg/L to 9.9 μg/L, although higher removal rates were expected for longer EAOP treatment time.

In conclusion, electrooxidation with BDD electrodes can provide more than 95% removal of most stimulant and pharmaceutical compounds from the RO concentrate effluent, demonstrating its excellent effectiveness for the mineralization of emerging contaminants in the waste concentrate stream generated during the reverse osmosis treatment of WWTP secondary effluents.

Increasing the Efficiency of Perfluorinated Compounds Removal

Perfluorocarboxilic acids (PFCAs) contain a fluorinated alkyl chain and a carboxylic hydrophilic end group. Long-chain PCASs are persistent, due to the extremely high energy of the C—F bond in the fully fluorinated alkyl chain, bioaccumulative, and toxic to laboratory animals and wildlife [23]. Hence, environmental protection institutions have established limits to PFCAs with eight or more fluorinated carbons [24]. For example, the United States Environmental Protection Agency has recently set health advisory levels for perflourooctanoic acid (PFOA) and PFOS in drinking water at 0.07 mg/L, both individually and combined [25]. As a result of the increasing concern about the effect of PFCAs in health and environment, nowadays industrial manufacturers have phased out the use of PFOA, PFOS, and longer chain homologues [26]. The alternatives are mostly short-chain PFCAs and different families of 6:2 fluorotelomers [23]. In contrast to PFCAs, 6:2 fluorotelomers are readily biodegradable, but they degrade into the persistent perfluorohexanoic acid (PFHxA) [27]. In general, shorter chain PFASs have been reported to be quickly eliminated in mammals, although PFHxA is equally persistent and cannot be degraded under biotic or abiotic conditions [28].

The present application was aimed at the development of a strategy for the removal and degradation of PFHxA from industrial process streams generated in the manufacturing of plastic resins [29]. PFHxA concentration in the industrial process water varied in the range of 80–200 mg/L. BDD was selected as anodic material, as the cleavage of the C-F will need the combination of hydroxyl radical meditated oxidation and direct electron transfer at the anode. A treatment train consisting of membrane nanofiltration (NF) followed by electrochemical degradation of the NF concentrate was proposed, as depicted in Fig. 7. This strategy brought several benefits, starting with the increase of the concentration of the target compound, that promoted diffusion from the liquid bulk toward the anode. At the same time, the electrolyte conductivity was increased, thus reducing the cell voltage to end up with the reduction of the feed volume to be treated by electrooxidation.

The DowFilm NF270 commercial membrane [30] provided PFHxA rejections as high as 99.4%; at the same time sulfate and calcium were also

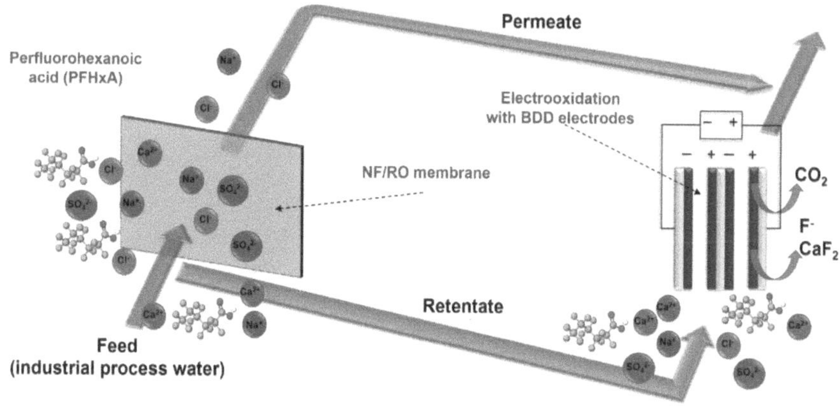

FIG. 7 Integration of NF and electrochemical degradation of perfluorohexanoic acid.

retained. The NF operation in concentration mode enabled a volume reduction factor of 5 and increased the PFHxA concentration in the retentate to 870 mg/L; at the same time, the conductivity was raised to 2.7 mS/cm. The NF retentate was treated in a commercial undivided electrochemical cell provided with two parallel flow-by compartments separated by a bipolar BDD electrode, and the same material for counter the anode and cathode. The electrochemical degradation rate of PFHxA reached 98% and was accompanied by its efficient mineralization, as the reduction of total organic carbon was higher than 95%. Energy consumption, which was 15.2 kWh/m^3 of treated NF concentrate, was minimized by operating the current density at 50 A/m^2. The possible adverse impact of fluoride ions obtained as the final degradation product was avoided by in situ precipitation as calcium fluoride, promoted by the incoming calcium contained in the industrial process stream. The observed kinetics of PFHxA oxidation were substantially faster than previously reported results using Ce-doped PbO_2 and BDD anodes [31], probably fostered by the optimal selection of operation variables achieved in this study (Fig. 8).

Overall, these results suggest that the process integration of nanofiltration separation and BDD electrochemical degradation is a promising alternative for the treatment of PFHxA that could be extended to the treatment of waters impacted by other PFASs, such as regulated PFOA and PFOS.

CONCLUDING REMARKS

Electrochemical oxidation using BDD anodes enables the degradation and mineralization of highly persistent pollutants, including landfill leachates, emerging micropollutants, and perfluorinated compounds. Still, this excellent technical performance is hindered by the high

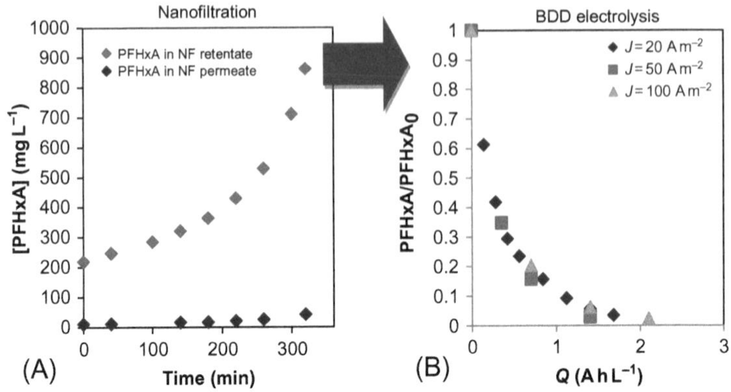

FIG. 8 (A) Nanofiltration. Change of PFHxA concentration in the retentate and permeate streams. Initial feed volume: 10L; volume reduction factor: 5; membrane area: 155 cm^2; feed pressure: 20 bar. (B) BDD electrooxidation. Change of the dimensionless PFHxA concentration with the specific charge passed. Two compartment bipolar cell, anode area: 140 cm^2. Initial PFHxA: 870 mg/L (final concentration of NF preconcentration stage). *Data from A. Soriano, D. Gorri, A. Urtiaga, Efficient treatment of perfluorohexanoic acid by nanofiltration followed by electrochemical degradation of the NF concentrate, Water Res. 112 (2017) 147–156. Copyright 2017. Elsevier.*

operation costs associated with the removal of substances in the low concentration range. The process integration strategy as described in this chapter allows the reduction of overall energy consumption, either by improving the biodegradability potential of the initially recalcitrant wastewaters or by a preconcentration stage by means of membrane separation.

Acknowledgments

The authors would like to acknowledge the funding received from MINECO and SPAIN-FEDER 2014-2020 for projects CTM2013-44081-R, CTM2014-57833-R, CTM2015-66078-R, CTM2016-75509-R, and to Network of Excellence E3TECH (CTQ2015-71650-REDT).

References

[1] A. Anglada, A. Urtiaga, I. Ortiz, Contributions of electrochemical oxidation to wastewater treatment: fundamentals and review of applications, J. Chem. Technol. Biotechnol. 84 (2009) 1747–1755.

[2] I. Oller, S. Malato, J.A. Sánchez-Pérez, Combination of advanced oxidation processes and biological treatments for wastewater decontamination—a review, Sci. Total Environ. 409 (2011) 4141–4166.

[3] J.A. Zazo, J.A. Casas, A.F. Mohedano, M.A. Gilarranz, J.J. Rodríguez, Chemical pathway and kinetics of phenol oxidation by Fenton's reagent, Environ. Sci. Technol. 39 (2005) 9295–9302.

[4] S. Renou, J.G. Givaudan, S. Poulain, F. Dirassouyan, P. Moulin, Landfill leachate treatment: review and opportunity, J. Hazard. Mater. 150 (2008) 468–493.

[5] A. Cabeza, A. Urtiaga, I. Ortiz, Electrochemical treatment of landfill leachates using a BDD anode, Ind. Eng. Chem. Res. 46 (2007) 1439–1466.
[6] O. Primo, M.J. Rivero, I. Ortiz, Photo-Fenton process as an efficient alternative to the treatment of landfill leachates, J. Hazard. Mater. 153 (2008) 834–842.
[7] Organisation for Economic Co-operation and Development OECD, Test No. 302B: Inherent Biodegradability: Zahn-Wellens/EMPA Test. OECD Guideline for Testing of Chemicals, 1992.
[8] International Standard Organisation (ISO), Water Quality. Evaluation of Ultimate Aerobic Biodegradability of Organic Compounds in Aqueous Medium. Static Test (Zahn-Wellens Method) (ISO 9888:1999), 1999.
[9] D. Mantzavinos, E. Psillakis, Enhancement of biodegradability of industrial wastewates by chemical oxidation pre-treatment, J. Chem. Technol. Biotechnol. 79 (2004) 431–454.
[10] S. Senthilkumar, C. Ahmed Basha, M. Perumalsamy, H.J. Prabhu, Electrochemical oxidation and aerobic biodegradation with isolated bacterial strains for dye wastewater: combined and integrated approach, Electrochim. Acta 77 (2012) 171–178.
[11] A. Urtiaga, A. Rueda, A. Anglada, I. Ortiz, Integrated treatment of landfill leachates including electrooxidation at pilot plant scale, J. Hazard. Mater. 166 (2009) 1530–1534.
[12] A. Cabeza, A. Urtiaga, M.J. Rivero, I. Ortiz, Ammonium removal from landfill leachate by anodic oxidation, J. Hazard. Mater. 144 (2007) 715–719.
[13] A. Anglada, A. Urtiaga, I. Ortiz, Laboratory and pilot plant scale study on the electrochemical oxidation of landfill leachate, J. Hazard. Mater. 181 (2010) 729–735.
[14] A. Anglada, A. Urtiaga, I. Ortiz, Pilot scale performance of the electro-oxidation of landfill leachate at boron doped anodes, Environ. Sci. Technol. 43 (2009) 2035–2040.
[15] S. Yoshihara, M. Murugananthan, Decomposition of various endocrine-disrupting chemicals at boron-doped diamond electrode, Electrochim. Acta 54 (2009) 2031–2038.
[16] Y.H. Cui, X.Y. Li, G. Chen, Electrochemical degradation of bisphenol A on different anodes, Water Res. 43 (2009) 1968–1976.
[17] A. Anglada, A. Urtiaga, I. Ortiz, D. Mantzavinos, E. Diamadopoulos, Boron-doped diamond anodic treatment of landfill leachate: evaluation of operating variables and formation of oxidation by-products, Water Res. 45 (2011) 828–838.
[18] E. Alonso, Contribución al tratamiento de lixiviados de vertedero de residuos sólidos urbanos mediante procesos de oxidación avanzada, (Tesis Doctoral), 2013.
[19] A.M. Urtiaga, G. Perez, R. Ibáñez, I. Ortiz, Removal of pharmaceuticals from a WWTP secondary effluent by ultrafiltration/reverse osmosis followed by electrochemical oxidation of the RO concentrate, Desalination 331 (2013) 26–34.
[20] A. Pérez-González, A.M. Urtiaga, R. Ibáñez, I. Ortiz, State of the art and review on the treatment technologies of water reverse osmosis concentrates, Water Res. 46 (2012) 267–283.
[21] G. Pérez, A.R. Fernández-Alba, A.M. Urtiaga, I. Ortiz, Electro-oxidation of reverse osmosis concentrates generated in tertiary water treatment, Water Res. 44 (2010) 2763–2772.
[22] G. Pérez, P. Gómez, R. Ibáñez, A.M. Urtiaga, I. Ortiz, Electrochemical disinfection of wastewater treatment plant (WWTP) effluent, Water Sci. Technol. 62 (2010) 892–897.
[23] ECHA, Candidate List of Substances of Very High Concern for Authorisation [WWW Document], https://echa.europa.eu/candidate-list-table, 2014.
[24] Stockholm Convention Secretariat, Decision SC-4/17: Listing of Perfluorooctane Sulfonic Acid, Its Salts and Perfluorooctane Sulfonyl Fluoride, Stockholm Convention Secretariat, 2009, pp. 1–4. http://chm.pops.int/TheConvention/Overview/TextoftheConvention/tabid/2232/.
[25] U.S. Environmental Protection Agency, Fact Sheet PFOA and PFOS Drinking Water Health Advisories, EPA 800-F-16-003, https://www.epa.gov/ground-water-and-drinking-water/supporting-documents-drinkingwater-health-advisories-pfoa-and-pfos, 2016.

[26] B. Gomez-Ruiz, S. Gómez-Lavín, N. Diban, V. Boiteux, A. Colin, X. Dauchy, A. Urtiaga, Boron doped diamond electrooxidation of 6:2 fluorotelomers and perfluorocarboxylic acids. Application to industrial wastewaters treatment, J. Electroanal. Chem. 798 (2017) 51–57.
[27] B. Gomez-Ruiz, S. Gómez-Lavín, N. Diban, V. Boiteux, A. Colin, X. Dauchy, A. Urtiaga, Efficient electrochemical degradation of poly- and perfluoroalkyl substances (PFASs) from the effluents of an industrial wastewater treatment plant, Chem. Eng. J. 332 (2017) 196–204.
[28] I. Fuertes, S. Gomez-Lavín, M.P. Elizalde, A. Urtiaga, Perfluorinated alkyl substances (PFASs) in northern Spain municipal solid waste landfill leachates, Chemosphere 168 (2017) 399–407.
[29] A. Soriano, D. Gorri, A. Urtiaga, Efficient treatment of perfluorohexanoic acid by nanofiltration followed by electrochemical degradation of the NF concentrate, Water Res. 112 (2017) 147–156.
[30] A. Pérez-González, R. Ibáñez, P. Gómez, A.M. Urtiaga, I. Ortiz, J.A. Irabien, Nanofiltration separation of polyvalent and monovalent anions in desalination brines, J. Membr. Sci. 473 (2015) 16–27.
[31] C.E. Schaefer, C. Andaya, A. Urtiaga, E.R. McKenzie, C.P. Higgins, Electrochemical treatment of perfluorooctanoic acid (PFOA) and perfluorooctane sulfonic acid (PFOS) in groundwater impacted by aqueous film forming foams (AFFFs), J. Hazard. Mater. 295 (2015) 170–175.

PART IV

RENEWABLE ENERGIES AND ELECTROCHEMICAL TECHNOLOGIES

CHAPTER

19

Prospective Applications of Renewable Energy-Based Electrochemical Systems in Wastewater Treatment

Vicente Montiel, David Valero, Francisco Gallud, Vicente García-García, Eduardo Expósito, Jesús Iniesta

University of Alicante, Alicante, Spain

OUTLINE

Electrochemical Water and Wastewater Treatment
https://doi.org/10.1016/B978-0-12-813160-2.00019-5

INTRODUCTION

Electrochemical technologies have been proved as quite useful approaches regarding wastewater treatment because they enable the transformation of organic and inorganic matter into less toxic and/or biodegradable products, for example through the electrooxidation or electroreduction processes. In addition, they enable the separation of those compounds either from the aqueous effluent to be purified or from the contaminated solid phase, for example, through electrocoagulation, electrodialysis or electroremediation methodologies.

The electrochemical technologies are supplied with direct current, which may come from a conventional alternating current electrical grid, by using a rectifier. This turns alternating current (AC) into direct current (DC), but electricity can also proceed from a renewable, environmental friendly energy source. In this regard, many of the scientific papers reported during the last 10–15 years show the results obtained from the combination of renewable energies with electrochemical treatment of wastewater and soils. The most recognized renewable energies nowadays are wind and solar photovoltaic. However, due to availability, accessibility, and ease of use, solar photovoltaic energy (PV) appears to be the most mature technology describing the more developed works about the use of renewable energy in combination with such electrochemical treatments (ET).

Therefore, this chapter focuses on revising the fundamentals of solar photovoltaic panels and how they can be connected as a power source to more efficiently carry out the electrochemical treatment of soils and wastewaters. The chapter also follows up the different works performed on the most used electrochemical wastewater treatments, emphasizing in each case the most successful combination approach of the renewable energy. In this regard, Fig. 1 sums up the target and scope of the chapter contents. Finally, this chapter provides an update of the combination of PV energy with the different electrochemical technologies toward environmental remediation applications as well as future perspectives on the topic.

SOLAR PHOTOVOLTAIC ENERGY AND ELECTROCHEMICAL TREATMENTS

Solar Panel

One of the most widespread and investigated renewable energy sources is solar photovoltaic. Solar photovoltaic panels (PV modules) convert solar irradiation into direct electric power. Among the advantages of

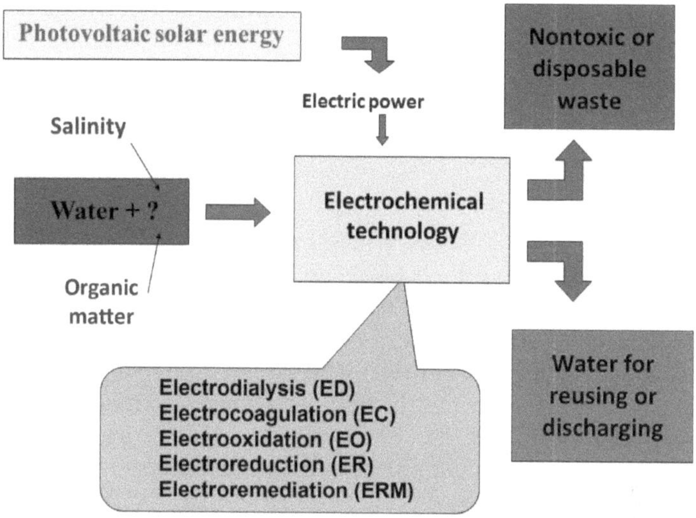

FIG. 1 Scheme of the combination of solar photovoltaic energy with electrochemical technologies in wastewater treatment.

solar energy, it is worth noting that solar energy is considered to be noncontaminant, silent, very abundant, not centralized, free, and infinite, so all the above added up to the fact that PV panels have a long lifespan and a low maintenance cost. Normally, a solar panel plant consists of PV panels, batteries (if energy storage is desired in the case daytime and energy demand are not simultaneous), a charge controller (it protects batteries from overvoltage), and a control and distribution system (it provides the required voltage and current intensity for achieving most efficient usage). Solar photovoltaic facilities spark great interest when used in remote locations as autonomous and reliable electric power supply systems. On the other hand, not only the electrochemical reactor does require energy to work, but also the pumping systems, instrumentation, and control. Such systems usually use electricity either from the electric grid power supply or through a DC power supply, which provides with constant current intensity or voltage similar to that required in electrochemical reactors. The former scenario is the easiest way to proceed because the electric grid is reliable and widely available. However, the use of the electricity grid is neither the only approach nor the best solution in every case. Alternatively, the combination of these two technologies such as electrochemistry and solar photovoltaic can provide a very profitable system because the supply of an electrochemical reactor with PV power presents a number of remarkable advantages, as follows:

- The system is portable, since it does not require an electric distribution infrastructure.
- The system can be the most inexpensive option when the system works discontinuously in a certain application, or when the needed interconnection between the PV panels and the electrochemical system can be direct.

Operating Principle

The photovoltaic generator is based on the photoelectric effect, that is, a material emits electrons when it is affected by electromagnetic radiation. In 1884, Heinrich Hertz first described the photoelectric effect; it was then explained by Albert Einstein in 1905. The explanation to this phenomenon stands on the assumption that in a material, the electrons cannot move freely until they reach a certain energy level, called Fermi energy. When a photon is absorbed by a material, its energy is transferred to one electron. Moreover, if the electron reaches the Fermi energy, it will be released from its atomic nucleus and be able to move freely. The photon's extra energy is then used to increase the electron's kinetic energy.

Mathematical Model

Unlike a conventional electric power supply, the photovoltaic generator does not provide a constant voltage or current intensity. Generated voltage and current depend on which and how many instruments are connected to the generator, the solar irradiance received at each moment, and the solar generator temperature. Fig. 2 shows the equivalent circuit of a photovoltaic generator [1,2].

As shown in Fig. 2, the photovoltaic generator is a power source where the value of the generated current intensity (I_L) is proportional to the solar power received from the solar generator. The rest of the components simulate a number of processes inside the generator and its connection. R_{SH} simulates a shunt resistance whereas R_S represents the resistive losses in the solar generator and the electric contact. The diode represents the recombination of electrons absorbed by the material before they get out

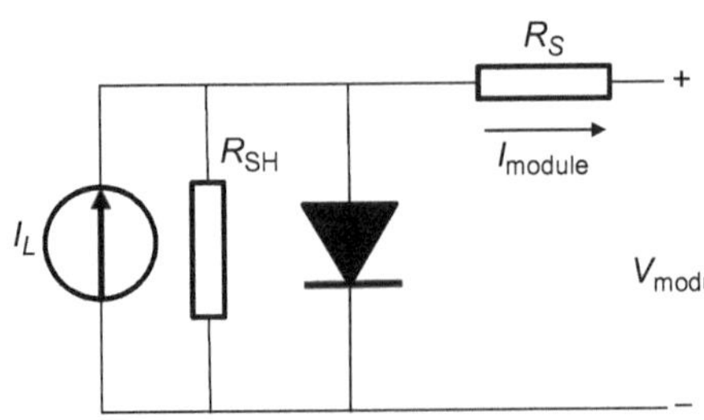

FIG. 2 Schematic diagram of a photovoltaic cell.

of the generator and do useful work. The relationship between voltage and current of the photovoltaic generator follows the equation:

$$I_{\text{module}} = I_L - I_0\left(e^{q\cdot\left(\frac{V_{\text{module}}+I_{\text{module}}\cdot R_S}{\gamma\cdot k\cdot T_{\text{module}}}\right)} - 1\right) - \frac{V_{\text{module}} + I_{\text{module}}\cdot R_S}{R_{SH}}$$

where:

V_{module} is the voltage at the photovoltaic generator.
I_{module} is the current at the photovoltaic generator.
I_L is the current generated by the solar energy in the photovoltaic generator, which is proportional to the solar radiation power onto the panel.
I_0 is the diode's reverse saturation current in the equivalent circuit.
q is the electron charge.
R_S is the resistance connected in series in the equivalent circuit.
k is the Boltzmann's constant.
γ is the shape factor that depends on the diode in the equivalent circuit of the photovoltaic generator.
T_{module} is the temperature inside the photovoltaic generator.

Fig. 3 shows the characteristic curve for one PV panel under solar radiation at a certain temperature and electric charge. The shadowed area indicates the power provided by the panel at different external charges. The same shadowed areas are also present either in typical intensity-voltage (*I*-*V*) curves of a photovoltaic generator or power-voltage curves.

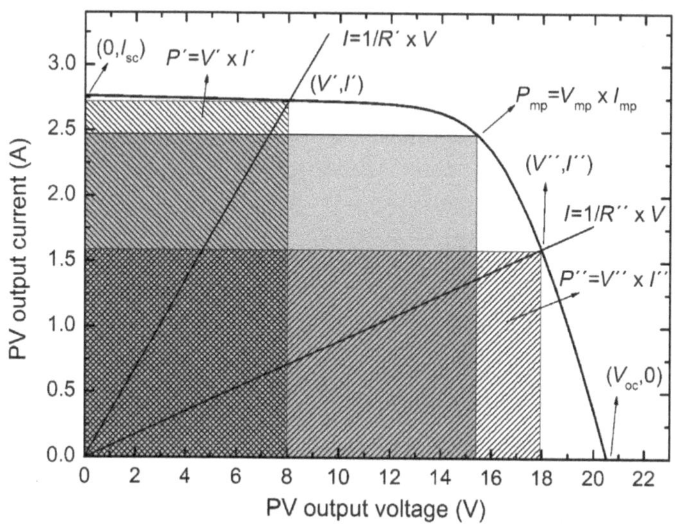

FIG. 3 Typical current intensity-voltage curve of a photovoltaic module.

For practical purposes, it is noteworthy that the usual parameters that define the PV module are the short circuit current (I_{sc}) and the open circuit voltage (V_{oc}). The short circuit current is the current generated by the PV panel when the voltage between the terminals is zero, that is, when the inner resistance of the reactor is zero ($R_{load} = 0$) and it depends mainly on solar irradiation. The open circuit voltage is the voltage measured in the absence of an electric charge ($R_{load} = \infty$) and represents the maximum voltage to be reached by the PV panel under certain conditions of radiation and temperature. The parameter V_{oc} depends mainly on the panel temperature. The maximum power (P_m) is the maximum electric power that a PV panel can provide for a certain solar irradiation (**G**) and panel temperature (T_{module}), and defines the point at the *I-V* curve where the product between current (I_{mp}) and voltage (V_{mp}) reaches its highest value. The peak power (P_p) represents the standard value used to classify the power that a panel can provide under irradiation conditions of 1,000 W m^{-2} at 298 K. The panel efficiency turn out to be another important parameter and is defined as the ratio value in percentage between the converted electric power and the total power received.

The *I-V* curve in Fig. 3 exhibits clearly two different zones. First, a plateau where the current values are close to I_{sc} within a high voltage values domain. The second region features an abrupt decrement in current intensity when voltage values approach V_{oc}. The ideal situation would be that working near the maximum power region in order to harvest all the power that the PV generator is capable of supplying. On one hand, if the working point deflects toward the plateau, higher current intensity will be attained but the measurements will be strongly influenced by solar irradiance changes, thereby varying I_{sc}. On the other hand, if the working point deflects to the zone close to V_{oc}, the current intensity value will be lower though more stable under instantaneous solar irradiation fluctuation. The above scenarios have important consequences regarding the usage of solar generators, that is, if a desired maximum power is required in the generator, then such a working point will be close to the voltage value that provides maximum power, also known as the point of maximum power (P_{mp}).

The Coupling of Solar Panels and Electrochemical Technologies

Several methods can connect a solar generator to an electrochemical system. This section describes those methods along with their advantages and disadvantages. An electrochemical system is divided into several subsystems that consume energy and are connected to a power supply for fully autonomous purposes. These sub-systems are as follows:

- Pumping devices consisting of several pumps supplied by a set voltage. Nonetheless, it is possible to reduce their energetic consumption under reduced flow rates.
- Instrumentation and control devices. The power consumption is constant and devices must be supplied by a set voltage.
- Electrochemical reactor. The working current density depends on the applied voltage. In many applications, the electrochemical reactor is considered as an electrical resistance.

From an energetic point of view, the target is to consume as much energy as possible for the reactor operation without overcoming its maximum working current and voltage as long as the pumping, instrumentation, and control devices are properly supplied with enough energy to be operated.

The easiest way of interconnecting the system is based on the direct mode. To do this, the solar generator is somehow designed so that its working voltage is necessary for all the elements that are part of the electrochemical system. However, because of its different nature, the pumping, instrumentation, control devices, and electrochemical reactor do not always need the same working voltage. In fact, the electrochemical reactor usually changes its required voltage as the reaction takes place. In addition, the solar generator loses efficiency, as the working point is not always that one with the maximum power generated (P_{mp}).

Another possible interconnection results from dividing the solar generator (i.e., solar plant) into a set of subfarms in order to make each one working in a different operation of the electrochemical system. In this way, the energetic requirements for each element of the subunit will be more easily resolved; nevertheless, though, the lack of efficiency still lingers. Once the solar plant is divided into subfarms, it is possible to design them in order to work as close to the maximum power point as possible for each of them. However, the different subplants would not be able to share energy between them so that the energy excess of a subplant is unable to be used by another element of the system. However, the above drawback can be solved by using tuneable solar subplants. This strategy is similar to the previous one, but each subplant will consist at the same time of several PV panels arrays, which can be connected either in parallel or in series. For example, a solar subplant made of six 12 V/1A solar panels produces a 12 V/6A generator at a single parallel array of the panels; three panels in a parallel array in series produces a 24 V/3A generator; two panels in a parallel array connected in series produces a 36 V/2A generator, and finally, all panels connected in series give a 72 V/1A generator. The main disadvantage of this technique is focused on the use of numerous relays so that the changes occur automatically. On top of that, there are still a limited number of possible configurations; hence, the problem of reusing energy from a subplant persists.

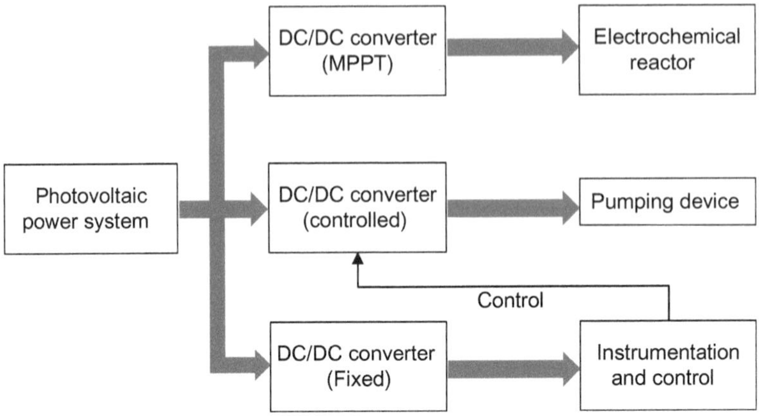

FIG. 4 Energy flow within the system.

A more efficient option of interconnecting the system regards the use of DC/DC converters. These elements are capable of converting a high direct voltage to another one with minimum energy loss. In this way, it is possible to have only one solar plant and, by means of these DC/DC converters, to supply all the different subsystems with different voltages. Furthermore, they can be configured so that the working point changes toward voltage and current intensity.

We lastly contemplate an evolution of the previous interconnection, which uses a DC/DC converter with maximum power point tracking (MPPT DC/DC) to the electrochemical reactor that makes possible the increase of the current density until its maximum value in the reactor. In conclusion, Fig. 4 sums up the different interconnections between the PV system and the different electrochemical devices.

Battery Usage

One of the main problems regarding with solar photovoltaic energy and even with another renewable energy is related to its availability and uncertainty. Particularly, solar energy is only available on sunny days, and during the daytime, power generation is far greater at central hours than near dawn or sunset periods. In order to overcome that obstacle, there are three possible strategies:

1. Batteries are used to store the excess energy during the central hours of the day for further use. This strategy is still disadvantageous due to its high cost and short lifespan of the battery along with its higher maintenance costs compared to solar panels.

2. Shutting down of the electrochemical system when the generated energy is not high enough. This solution requires the sizing of the electrochemical system in order to generate all the production during the central hours of the day.
3. Oversizing of the solar plant to provide that the electrochemical reactor works under maximum insolation condition.

Electrodialysis and Solar Photovoltaic Energy

Among all electrochemical technologies, electrodialysis (ED) is the most widely studied in combination with renewable energies (RE). The main limitations of the ED-RE coupling are—as in every RE coupling—the lack of stability and the fluctuation in the solar irradiation value, which leads to technical limitations when performing treatments. Even though this is a common problem of RE couplings, it can be easily overcome by simple shutdown and restart operations so that the systems can easily accommodate solar power fluctuations when the solar plant is not oversized.

ED contemplates several industrial applications related to solution purification or high salinity wastewater treatment [3]. However, the most studied application of this technology when coupled with an RE source is water desalination. In 1979, Lundstrom [4] launched for the first time a theoretical approach regarding the use of PV solar energy to directly power an ED system. Afterwards, laboratory scale as well as demonstrative plants were built. ED-PV also considered the use of batteries to store the energy that is employed to supply the ED reactor later. In this regard, Kuroda [5] proposed a two-stage desalination treatment system. The first stage consisted of the water treatment powered by PV during the highest solar irradiance hours while the second stage of desalination treatment was completed by the use of batteries during the lowest insolation conditions.

Adiga [6] described the usage of a system capable of treating up to 1 m^3 day^{-1} in Tanote, India, where for the first time the direct connection of a photovoltaic generator with an ED reactor was performed. Thanks to this work, the authors established some of the most important parameters that someone should take into account in order to perform the ED and PV connections, and proved that a solar tracing system allows the increase in efficiency and exploitation of the generator. Almadani [7] also conducted an investigation into the influence of some parameters such as flow rate or initial salt concentration on the operation of an ED-PV system using an electrodialysis reversal (EDR) with 24 pair of cells. Most of the studies undertaken for the desalination using ED-PV were performed using brackish water, although the literature describes some plants for desalination of seawater [8].

One of the main drawbacks of powering electrochemical systems using PV energy looks at the fluctuation in the irradiation value (*vide supra*) due to natural insolation variation during the day with a maximum value at noon, and those produced by sudden cloudy times and another weather phenomena. The first works regarding ED-PV were based on the usage of battery stacks in order to buffer the solar irradiation fluctuations and supply the reactor with a controlled voltage value [9]. Nevertheless, the direct connection of the generator and the reactor in the absence of battery usage is more feasible as well as inexpensive [10]. The works of Ortiz [2,11–14], performed in Alicante (Spain), described the procedure for carrying out a direct connection without battery usage, thereby reducing the cost of water production. These works also mentioned the development of a mathematical model that simulated the behavior of a PV generator according to the weather conditions and, on the other hand, the behavior of the ED reactor according to water salinity and the inlet electrical parameters. Accordingly, simulations results led the PV generator configuration as a function of the weather conditions, the number of panels available, and the electrical requirements of the system. These parameters were checked and proved by achieving drinking water at a pilot scale system with a cost of 0.17–0.32 € m^{-3}. The work performed by Cirez [15] continued Ortiz's work, concluding that the size and configuration of the PV generator must be suited to the ED reactor size in order to perform an optimized coupling.

Some studies, for example, that from Zhang [16], sought alternative routes for reducing the energy requirements to complete the treatment through a previous forward osmosis (FO) water treatment, which reduces both organic matter and salinity before reaching the ED unit. However, other authors stated that the higher competitiveness of the coupled systems must be fueled more efficiently along with a decrease of the panel price and operating costs. In this sense, Ghermandi [10] justified that the competitiveness of the ED-PV against another systems requires a PV price per electric power lower than 1.5–2 € W^{-1}. Bayod-Rújula [17] proposed the use of a slightly oversized PV plant in order to optimize the energy used for the treatment, owing to a greater versatility of the PV plant configuration.

As said before in this section, once ED-PV demonstrated the viability of the brackish water treatment, researcher's efforts focused on the optimization and efficiency improvement of the PV generator. With that target in mind, Uche [18] proposed the use of DC/DC converters to get a constant voltage at the ED reactor and to obtain as maximum power as possible from the PV modules. Other papers showed a hybrid energy supply [19] with a PV generator and batteries for the pumps and auxiliary equipment, and a secondary PV generator coupled directly to the ED reactor.

ED is increasing its presence in the water desalination sector for obtaining drinking water, such as the setting up of an ED plant in Hospitalet de

Llobregat, Spain, with a treatment capacity of 2.4 m^3 s^{-1} [20]. However the presence of systems powered by RE still represents only 5%, as Ghaffour [21] reflected in the work in which rules were established for a proper selection of a desalination treatment supplied by RE. Such rules reflected on every economical and technical aspect on the proper selection of renewable energy source combinations based on the final use of water and treatment requirements.

Even though most ED plants and scientific research in this field are devoted to brackish water and desalination, other alternative applications for ED exist such as wastewater treatment, which has great potentialities not yet fully exploited [22]. Moreover, several works have demonstrated the feasibility of ED for wastewater treatment when used as a tertiary treatment. In such cases, ED reaches the requirements for discharging into the public sewage system [3,23].

Future Perspectives

The challenge of desalinating water using solar photovoltaic technologies must cope with the competitiveness and improvements from an economic point of view. To do that, there are primarily two fundamental approaches:

(i) The PV generator. As the solar panel production costs decrease and the PV panel's efficiency is enhanced, the coupled systems will become more efficient and will reach prices that are more competitive.
(ii) The ion exchange membranes. The manufacture of cheaper membranes along with more ion separation selectivity will reduce ED costs and increase the efficiency of the treatment.

The ED-PV technology has the potential of settling down not only in isolated and remote locations, where its niche market is well recognized, but also in other geographical locations in which environmental and social-economic conditions are advantageous, as reported by Pugsley [24]. Setting up the ED-PV systems will allow drinking water for populations of up to 5,000 inhabitants, as has already begun in some locations where treatment plants of 1.6 m^3 h^{-1} have been installed [25]. It has been demonstrated that production costs of brackish water desalination through ED-PV turned out to be lower than using ED-RO [26].

In the coming decades, however, as the costs of all the main elements for manufacturing an ED-PV system decrease, there would be an increase in setting up ED toward either wastewater treatment acting as a tertiary process or simply industrial applications focused on the separation of products of high benefit present in wastewater and salt rejection streams. In addition, the installation of plants for obtaining potable or irrigation water coming from brackish water should be encouraged. In this sense, patents have already been registered for ED equipment elements powered by PV energy, which have demonstrated their effectiveness for the above applications [27]. Socioeconomic predictions estimate a great expansion of this technology between 2020 and 2050, once the price of water production by ED-PV is set in between 0.15 and 0.40 € m^{-3} [28].

Electrocoagulation and Solar Photovoltaic Energy

Electrocoagulation (EC) is an electrochemical technology closely related to chemical coagulation. EC is based on the application of an electric current to a sacrificial anode for example, Al or Fe. This produces the electrodissolution of the anodes, generating a clot formed by a wide variety of hydroxides, oxy-hydroxides, and polymeric hydroxides of Fe or Al during the wastewater treatment. This clot is capable of aggregating suspended particles as well as destabilizing and adsorbing dissolved or emulsified contaminants.

Other phenomena involved in the process of electrocoagulation are associated with the direct or indirect oxidation of organic matter and the generation of hydrogen bubbles in the cathode that helps the flotation of flocs formed. The formed sludge is usually separated by sedimentation and subsequent filtration. There are clear advantages in terms of (i) use of a smaller amount of chemical reagents, which does not increase the water salinity of treated wastewater, (ii) generation of a smaller quantity of sludge, and (iii) ease of operation, automation, and lower space requirements. Among the industrial sectors, the EC has demonstrated high effectiveness. The textile and galvanotechnical industries as well as a large number of industries in the agro-food sectors (dairy, meat, olive, etc.) are good examples. The coupling of electrocoagulation with photovoltaic solar energy is a recent topic. Valero [29] published the first work describing an EC system powered by a solar PV array. Since then, a vast number of studies have been published about the subject at laboratory scale systems. In that work, Valero once again demonstrated the feasibility of using an EC system working in continuous operation mode, directly powered by a PV array. The process was carried out on synthetic wastewater containing a textile dye, Remazol RB-133. The main challenge of electrocoagulation and solar photovoltaic energy (EC-PV) copes with the dependence of current intensity supplied by the solar plant on the solar irradiance and solar panel temperature. These parameters cannot be controlled, because of changing continuously during the day or even quickly on partly cloudy days. To give versatility to the EC-PV system against changes in solar irradiance, Valero and coworkers defined the parameter J_v as the current density per flow rate (C m^{-2} m^{-3}). Once the value of J_v for any wastewater treatment is determined, it must remain constant during the electrochemical process. Thus, when the current supplied by the solar plant changes, the flow rate must change to maintain the J_v constant; in other words, the volume of treated wastewater is directly related to solar irradiance. The same work emphasizes the importance of the configuration of the solar plant (number of panels connected in series, in parallel, or in a mixed configuration, and how each configuration can be a critical factor upon the efficiency of using PV power).

Most works found in the literature are carried out in a batch operation mode using EC-PV systems where the power supply of the EC system is controlled by batteries and a charge regulator. There are some reports applied to wastewater treatment from different industrial sectors, with the main objective of pointing out the electrical cost reduction associated with the EC process. In this way, Pirkarami [30] investigated the removal of textile dyes Acid Orange 2 and Reactive Blue 19 in both synthetic and real wastewaters using an EC-PV system in order to improve the economic efficiency of the treatment. The energy needed for the treatment was supplied by a PV module connected to a battery. The current provided to the electrocoagulator was controlled by a rheostat. From variables relative to the economic analysis of the treatment using an EC-PV, the authors considered the energy cost, electrodes, and electrolyte, evaluating the savings in 12% of the total costs. Furthermore, two very similar works performed under the same equipment and experimental system were described in [31,32]. Pirkarami's work, for example, studied the application of the EC-PV to real wastewater coming from a resin factory containing phenol, aldehydes, and textile water with Reactive Red 120, respectively. The authors were meant to provide a 17% decrease of the total treatment costs using PV. For real wastewater, mismatched results were obtained due to the complexity and heterogeneity of the studied waters. García-García [33] explored the elimination of COD, TOC, color, and turbidity in a more complex wastewater coming from an industrial wastewater treatment plant. The work was carried out by using EC and electrooxidation (EO) technologies supplied by PV modules. Again, the EC-PV was performed in a batch mode and power was supplied using a battery and a rheostat. Electrolysis at pH 4 using a Cu electrode achieved a decrease of 89% in COD, 97% in color, and 91% in turbidity.

Sharma [34] developed a SPEC system (solar photovoltaic electrocoagulation) for the treatment of synthetic natural waters containing kaolin and humic acid, with the purpose of being applied in remote communities in Australia. The study was done using a five-litter EC commercial reactor connected to a PV module of 65 W. The study evaluated the ability of supplying the EC reactor either directly to the PV module or using batteries and charge controllers. The treatment provided the removal of 87% turbidity and a decrease in the UV absorbance of 85%. When the PV panel was used in combination with the batteries and the charging control system, the system exhibited good stability. Hussin [35] also examined the use of an EC-PV system toward Pb elimination by means of EC in a batch operation. The energy supply of the EC reactor was carried out directly by a rheostat. The novelty of the above work lies in the use of a new sacrificial anode made of perforated Zn.

Novel advances were opened up by Valero et al. [29], who explored the features of the direct connection of EC and PV systems. In general, the above work aimed to understand the coupled operation of both systems to maximize the volume of wastewater treated upon the incident solar energy absorbed under different meteorological conditions. The merit of Valero's work was also founded on the possibility of working without batteries, thereby bringing about economic and environmental benefits.

Zhang [16] developed a direct EC-PV system that works in recirculating batch mode. They addressed the light-solar/electrical energy conversion and the energy consumption of the system applied to the removal of phosphates in a lagoon. The energy conversion efficiency (m) was calculated as the ratio between total photovoltaic energy and total electrical energy. The value (m) was also calculated under different meteorological conditions (sunny, fully, or partly cloudy days), and the authors concluded that, though the efficiency of phosphate removal was maximum on sunny days, the energy conversion efficiency was greater on partly cloudy days.

Dermentzis [36] also reported extensively on the use of EC directly coupled by a PV system in numerous works. Generally, the system works specifically under a continuous operation mode using the control parameter J_v developed previously by Valero et al. [29]. Thus, the system coped with the treatment of wastewater from the textile sector [37], chrome plating, and electroplating [36,37]. The most relevant conclusions noticed by the Greek group refer to the feasibility of coping with an EC-PV with a wide variety of wastewater, highlighting that the above applications are relevant in remote locations far from the electricity grid.

The innovative work from Naje [38] was based mainly on the development of a novel electrocoagulator by using an aluminum-rotating anode for the treatment of real textile water. A PV module powered the EC reactor through a rheostat. Results depicted the effects of several operational parameters such as speed of rotation, current density or recirculation speed on the efficiencies of COD and color elimination. It is noteworthy that the design of the reactor allowed the valorization of the hydrogen generated in the cathodic reaction (*vide infra*) with energy saving of 9.4% under optimal conditions. Similarly, Phalakornkule's work tackled the design and characterization of a new electrocoagulator [39]. This work included the development of a packed-bed reactor for the decoloration of water from the textile sector. Each anode was composed of an iron nut packed bed embedded in a plastic casing, with a gas bubbling tube placed in its center. The effect of the gas bubbling on the reactor operation is twofold: on one hand, it enhances the efficiency of the discoloration by improving mass transport conditions, and, on the other hand, it reduces the electrical resistance of the reactor and the energy consumption of the process. Those effects are especially interesting when the process is powered directly with a PV panel, as the authors vary the flow rate of water to be treated with instantaneous solar irradiation.

Future Perspectives

EC reactors are supplied by DC power such as that from PV panels. So the electric connection in a monopolar operation mode together with the operating power range of EC reactors are easily powered by a solar plant. However, the particularities of the design and operation of an EC system impose limits from both operational and economical points of view with respect to the development of autonomous EC-PV systems with high treatment capacities. Thus, it must be able to anticipate that the implantation of EC-PV systems is still incomplete and limited. In this regard, one of the main limitations for the development of EC-PV systems corresponds to the "complexity" of the whole EC devices. Apart from the electrocoagulator reactor, a complete EC system requires a series of coupled unitary operations, that is, pH and/or conductivity controllers, pumping and solution stirrers, sludge filtration and sensors, and electronic controllers. Thus, the accurate and simultaneous power supply of all the above unitary operations by a solar plant is quite complicated. Moreover, the difficulty is found on the different power supply characteristics of the unitary operations, such as DC or AC current, different operating voltages, and electrical supply stability.

Limitations regarding the expansion of EC-PV systems are found on its own market of this technology. The EC is mainly set in factories or purification stations where the energy supply by the electricity grid is available. Because EC systems tend to operate in a continuous mode, one of the greater limitations is found in direct coupling of EC-PV systems without the use of batteries.

Electrooxidation and Solar Photovoltaic Energy

The use of electrochemical technologies such as direct or mediated electrooxidation (EO) and electroreduction (ER) of organic or inorganic contaminants present in different types of aqueous effluents (urban or industrial wastewater) has been proven to be competitive compared to the most conventional technologies such as the use of activated sludge, oxidation by Fenton reagents, or ozonation. The electrooxidation treatment (EO) has shown its efficacy mainly in the treatment of recalcitrant pollutants in wastewater [40–44] at laboratory scale. Nevertheless, even though the introduction and development of electrochemical advanced oxidation processes (EAOP) has greatly contributed to the treatment and removal of a wide variety of substances (little or no biodegradable organic or inorganic compounds), the scale up of these processes as well as the evaluation of the economic costs compared to the more traditional methods settled in the market still remain a great handicap. In this sense, the decrease in energy costs of EO processes in the industry can be achieved by coupling RE [45]. Among the different RE for the generation of electricity, PV is considered for its advantageous benefits (*vide supra*). Mook and coworkers [45] have recently described in a review the possibilities and advantages of coupling PV power generators and electrochemical technologies referred to the treatment of wastewater containing high COD and dyes with low biodegradability. Another aspect to bear in mind for the reduction of energy costs refers to the possibility of generating hydrogen on the cathode using the coupling of the EO-PV process, as shown in the works reported by Navarro-Solis and Park [46,47]. The objective of this section is to show a global overview of the developments carried out on the coupling of PV energy with the EO-based electrochemical wastewater treatments (EO-PV). At the same time, the incorporation of PV into the electrochemical processes aims to reduce the economic costs and environmental impacts associated with the high demand of electrical energy that conventional electrochemical systems require for the treatment of wastewater [48]. In addition, this section dedicates a short view to hydrogen generation by means of water electrolysis [49,50] coupled with the electrochemical wastewater treatment by using EO. In the case of EO-PV systems, benefits from a significant reduction in CO_2 emissions and reduction of the costs associated with energy storage systems are formidably considered when compared to the use of batteries or supercapacitors.

Electrooxidative Process Coupled With PV Power

Wastewater treatment of either an urban or industrial nature by electrochemical methods is largely dominated by the use of advanced or nonelectrooxidative processes. However, the still-incipient research on the use of PV panels for wastewater treatment has taken three main roads: the development of new electrodic materials, the improvement of EO efficiencies, and the design and optimization of PV panels. Figueroa [51] was the first researcher to couple PV panels to the treatment of textile dyes Reactive Black and Acid Green 25 using Fenton reagents. The authors demonstrated that the generation of H_2O_2 and the decoloration of dyes through the Fenton process could be carried out by coupling a commercial PV module array to govern the voltage difference of the electrochemical flow reactor using stainless steel as the anode and a three-dimensional electrode based on a reticulated glassy carbon as the cathode. Although the electrogeneration of H_2O_2 from the electroreduction of O_2 can be exploited to perform the disinfection processes in wastewater, its efficacy is, for example, less than that when using ozone. In this respect, the work of Ciobotaru [52] described the use of an electrolyzer coupled with PV modules, which had an electrochemical configuration consisting of an electrochemical reactor with a polymeric electrolyte membrane (PEM) using identical Ti electrodes as anode and cathode, respectively, for the electrogeneration of O_3 in an acid medium. This Romanian group highlighted the possibilities of this technology for the disinfection of drinking water and the improvement in quality.

On the other hand, the work reported by Dominguez [53] was carried out using a batch mode configuration of the electrochemical reactor with a boron doped diamond anode (BDD) in order to perform the treatment of lignosulfonate in wastewater. Again, the authors connected the PV panels directly to the electrochemical reactor, obtaining a high TOC removal. Fujishima's group [54] opted again for the use of BDD anodes for the treatment of synthetic wastewater and real effluents from a river containing phenols, supplying electricity from PV energy. The oxidation of phenol was carried out simultaneously using photocatalysts based on TiO_2 in order to carry out the photooxidation of low phenol concentration in wastewater. The Fujishima group was not only able to couple the PV power to the electrolysis and photocatalysis units, but also the PV was capable of providing autonomy to the previous and final filtering units by using ion exchange membranes. The same authors concluded that it is possible to obtain potable water from contaminated river water provoked eventually by uncontrolled spills or accidents. Other use of BDD electrodes focused on the elimination and complete mineralization of the 2,4-Dichlorophenoxyacetic acid (2,4-D) herbicide using a 78 cm^2 electrochemical filter-press reactor coupling two PV panels with an effective area of 1.313 m^2 connected in series [55]. The above authors determined that one of the important variables to generate current corresponds to the irradiance power (>100 W m^{-2}). Montiel's group also demonstrated the feasibility of the EO-PV technology for the oxidation of the Remazol RB 133D dye at pilot scale [56]. In this work, a divided filter-press electrochemical reactor (geometric area of 3,300 cm^2) with a dimensionally stable anode for oxygen evolution and a carbon-felt cathode was used. In this work, a 40 PV module array (either connected in parallel or in series) was used and the results related to the elimination of color were compared with the EO and electrocoagulation processes, respectively. Furthermore, the same Spanish group incorporated the EO-PV technology for the anodic treatment of wastewater from the almond manufacturing sector, after a pretreatment. To this end, the group from the University of Alicante applied the EO using a filter-press electrochemical reactor with a geometric area of 3,300 cm^2. The most optimal anode for EO consisted of a chlorine DSA, working in alkaline conditions to minimize the formation of volatile or semivolatile halocompounds. The most interesting aspect of the Spanish group was the study of the influence of the PV array power and its configuration (80 PV modules connected either in series or in parallel) to reach a voltage required for the EO.

The efficacy and economic cost of the EO-PV process depends on the electrochemical process conditions as well as the effective area and configuration of the PV plant (*vide supra*). The solar irradiance, the location of the installation, and the tilt of the PV panels have a great influence. In this respect, Alvarez-Guerra [57] investigated and modeled the process called PSEO (photovoltaic solar-powered electrochemical oxidation) through a model that calculates the efficiency of electric energy from irradiance and the efficiency corresponding to wastewater mineralization [57]. From an engineering point of view, this model was able to predict the efficiency of the process at both laboratory and pilot scales. The authors concluded that the rate of mineralization is not related to the PSEO process, but that there is rather a direct relationship between the current intensity obtained from the PV panels and solar irradiance. This behavior is explained because the PV works at the plateau of the characteristic *I-V* curve (see Fig. 3). In addition, the model was capable of analyzing the electrodic area needed for the reactor as well as calculating the effective area of PV panels. The latter parameter depends mainly on the conductivity of the residual water and largely on the solar irradiance measured, which in turn is a function of the panel's location, according to the studies carried out in another paper by Alvarez-Guerra et al. [57].

Electrochemical Hydrogen Generation Coupled With PV Power

Most of the H_2 production in the world is generated via processes involving reforming and gasification of fossil fuels. However, the use of RE to carry out the H_2 generation from cleaner and environmentally friendly methods is now beginning. We can highlight, for example, electrolysis, thermolysis, photolysis, and photoelectrocatalysis, to name a few [58]. Among the different methodologies, the electrolysis of water is becoming an attractive technology for H_2 and/or O_2 generation [49]. In addition, the EO-PV technology can take advantage of the H_2 production by the electroreduction of water in the cathodic compartment. Specifically, the work performed by Navarro-Solis [47] describes how it might be possible to obtain H_2 as a by-product from the decoloration of dyes in wastewater through a Fenton process.

A subsequent work carried out by Kargi [59] also investigated the EO-PV for the elimination of organic matter, demonstrating the viability of the H_2 generation simultaneously to the EO process and exploring the effect of the cathodic material on the efficiency of H_2. In 2013, Kim [60] explored the oxidation of urea and urine in water using EO-PV technology, under simultaneous H_2 generation as a by-product in the cathodic process. The reports of both Kargi and Kim studied the effect of the presence of chlorides within the medium on the efficiency of H_2. In addition, Kim's work noticed that the EO-PV process along with H_2 generation would be viable without the addition of an external supporting electrolyte. Furthermore, economic calculations presented by the work of Kim and colleagues showed that H_2 generation could amortize 10% of the electrical energy necessary to carry out the treatment of urea. A pioneering work presented by Park [46] also carried out the simultaneous H_2 generation and the EO of organic matter with the presence of chlorides using a Bi-doped TiO_2 electrode as the anode and stainless steel as the cathode, using an undivided cell. More recently, Cho [61] followed up a similar study, demonstrating that the production of H_2 simultaneously with the treatment EO-PV of domestic wastewater using a TiO_2 functionalized with electroactive Bi (BiOx/TiO_2 anode).

Future Perspectives

With the purpose of increasing the efficiency of wastewater treatment by electrochemical processes, some of the variables relative to the design of EO-PV systems should be taken into account in the near future. First, the development of new anodic and cathodic materials that diminish prices over electrochemical reactors as well as the improvement of the EO and ER process efficiencies. Second, design and configuration of the panels. About the use of PV modules, the EO efficiency will depend on the effective area of PV panels, their configuration, and location.

As far as the industrial scale up of the EO-PV process is concerned, research and investment incentives on the installation of PV energy modules must be encouraged. Nonetheless, another aspect that fundamental research should bear in mind is focused on reducing the use of external supporting electrolytes into the aqueous effluents to be treated, so this research topic remains a vital challenge for EO-PV technology. In addition, the simultaneous EO-PV process and H_2 generation must be examined more deeply over the next few years. This research topic is of paramount importance from the fundamental and technological point of view [62]. Finally, the use of H_2 production in situ along with the use of batteries for the storage of electrical energy must be addressed shortly in order to gain the advantage of EO-PV systems to be energetically independent, continuous, cheap, and portable.

Electroremediation and Solar Photovoltaic Energy

Electrokinetic remediation or electroremediation (ERM) is a technology applied to the remediation of soils, sludge, and sediments. Generally, ERM is based on the application of a direct current between electrodes placed into the contaminated soil. The electric field originated causes the removal of contaminants present through physical, electrochemical, or electrokinetic processes, among which electromigration, electrophoresis, and electroosmosis are highlighted. It is noteworthy that several advantages of the above technology in terms of its removal effectiveness against a wide variety of contaminants (heavy metals, colloids, fluorides, etc.), its applicability to soils with a low permeability or the ability to work either ex situ or in situ.

The application of PV energy to ERM systems is very recent. Yuan [63] published the first work describing an ERM system coupled with a solar PV plant (ERM-PV). From that point on, few works have been published and all of them were performed at laboratory scale. Yuan used a PV module connected directly to the electrodes to carry out the electrokinetic remediation of soil contaminated with cadmium. The authors performed tests on sunny, cloudy, and rainy days. Such results were compared with others performed under a conventional procedure using a DC power supply. The authors also found that the applied voltage depended strongly on the weather conditions and that the current produced by the PV module was comparable to that from the DC power supply. Likewise, the removal efficiency achieved by solar cells was comparable with that from the DC power supply. Mohamedelhassan [64] studied the feasibility of ERM powered directly by solar panels for application as an electrokinetic barrier. The case under study prevented the contamination of a clayey soil by a hydraulic flow of cadmium from a raft of mine tailings. The study was conducted at laboratory scale using electrokinetic columns in 7-day experiments. Under such experimental conditions, solar panels provided energy and kept the Cd within the exclusion zone, even considering that the electrokinetic barrier worked only during periods of sunlight.

Bearing in mind that the energy cost accounts for approximately 25% of the total for an ERM treatment, Kim [60] explored the feasibility and viability of using an ERM system powered by PV modules with and without battery usage. The examined system was the regeneration of soil from an agricultural greenhouse with high nitrate content. However, the authors found that variations of soil pH at the vicinity of the electrodes could adversely affect crops. On the other hand, it was noticed that a careful control of the extent of treatment is mandatory to considerably reduce the saline concentration. Other examples of ERM-PV in similar works at laboratory scale are found by Zhou, Zhanag, Hassan, and Jeon [65–68], who displayed the removal of fluorides, Cr(VI), copper, and arsenic, respectively.

Future Perspectives

The characteristics of electroremediation and electrokinetic barrier technologies make them suitable candidates to be coupled with a solar plant owing to technical, logistical, and operational reasons. It is important to emphasize the simplicity of ERM-PV in terms of control and equipment required for operation, with minimal or no need for auxiliary devices powered by electric energy from either DC or AC current. Furthermore, implicit consequences about the simplicity of the ERM-PV technology resonate on the avoidance of DC-AC converters, the enhancement of the energy efficiency, investment reduction, and easier control. Such ERM features encourage applications in isolated contaminated areas far from the electric grid.

Even though the ERM-PV technology is little affected by discontinuities in the electricity supply, batteries can also be part of the set up when the solar plant is not connected directly to the electrodes. Nonetheless, research has to point out beneficial advantages derived from the technology regarding the scale up and the use of direct connections of PV panels to an ERM electrolytic cell. Such directions aim to:

- minimize the investment costs (e.g., batteries are one of the most expensive elements in a solar plant).
- simplify the control of distinct elements.
- prevent the generation of wastes due to exhausted batteries with subsequent detrimental environmental and economic consequences.

An Example of PV Coupled With Electrochemical Wastewater Treatments From the Almond Manufacture Sector

A figure of merit of this chapter is dedicated to the electrochemical wastewater treatment of aqueous effluents coming from the almond scalding and washing processes by coupling PV power. The almond manufacture sector is very important in Spain because it is the basis of peeled almond, nougat and ice cream production. The approach used was the application of different electrochemical processes in an orderly manner to reuse water obtained from the wastewater treatment. This work was deeply compiled by Valero on his doctoral thesis presented in 2014 at the Institute of Electrochemistry [23]. Valero's work elegantly demonstrates that electrochemical processes are able to be coupled with direct solar photovoltaic energy (without batteries) in order to carry out the electrochemical treatment of wastewater from the almond manufacture industry. Table 1 sums up the main characteristics and composition of wastewater to be treated:

The wastewater treatment was conducted using the EC, EO, and ED techniques. In all cases, the electric energy was supplied to the different electrochemical reactors using a solar plant from the University of Alicante. The work was carried out at a preindustrial plant scale so the results could be extrapolated to real water treatment from the same sector. Fig. 5 shows the proposed workflow for an optimum performance of wastewater treatment.

TABLE 1 Average Characteristics of an Aqueous Effluent From the Almond Manufacture Industry

Parameter	Value range
pH	4.6–5.7
Ionic conductivity (mS cm^{-1})	2.1–5.11
TOC (mg L^{-1})	1,400–2,800
COD (mg L^{-1})	4,800–6,300
Total suspended solids (mg L^{-1})	2,700–5,300
Apparent color (Pt—Co units)	16,000–30,000

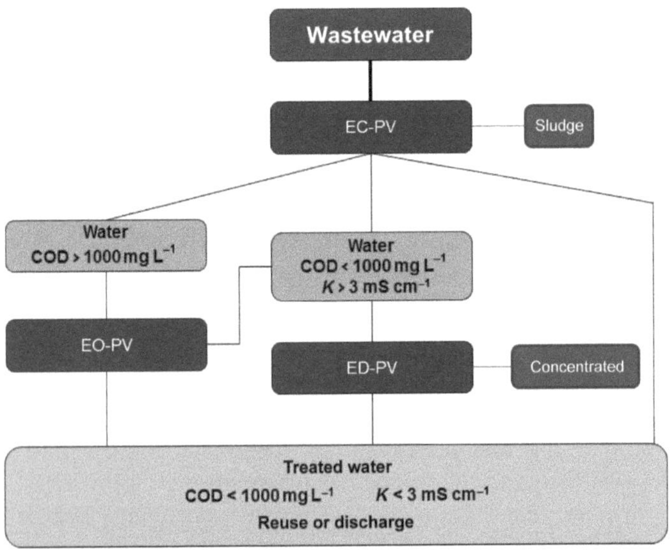

FIG. 5 General scheme displaying the workflow proposed for the selection of the electrochemical technique of wastewater treatment. *K* is the ionic conductivity. COD is the chemical oxygen demand expressed in mg O_2 L^{-1}.

It is noteworthy that the protocol proposed in Fig. 5 can be applied to different types of wastewater coming from another industrial sector featuring similar characteristics, as shown in Table 1. Valero addressed a wise strategy for optimizing the power supplied by the PV generator based on modifications in the PV array configuration (without changing the number of PV modules). Modification of the PV generator configuration when coupled with the electrochemical techniques involved changes in the working point of the plot *I-V* in Fig. 3. Most importantly, costs of wastewater treatment using direct coupling systems with photovoltaic generators were formidably lower, especially for the oxidation treatments when compared with the usage of conventional energy.

Coupling of Another Renewable Energy With Electrochemical Treatments

Electrodialysis and Wind Energy

Wind energy (WE) has also been introduced as other kind of RE that can be coupled with electrochemical technologies, although coupling with electrodialysis toward desalination or drinking water production is the most established technology. The reason for that is based on attractive advantages such as a low carbon footprint [69]. However, electrodialysis coupled with wind energy (ED-WE) exhibits an environmental impact associated with the management of brine production [70]. WE generated from a wind turbine based on a permanent magnetic generator is independent of wind conditions but is dependent on the resistance of the load connected directly to the system. Therefore, the energy efficiency of the wind turbine will vary with the internal resistance of the ED reactor. The main challenge WE copes with when coupled directly to ED is associated with the fluctuations and intermittent changes of wind, which can remarkably affect the current and voltage of the reactor. Nonetheless, excess energy due to strong winds can be stored as desalinated water.

There exist very few studies about the use of ED-WE. It is worth citing the work by Veza [71], who studied the operation of an electrodialysis reversal unit supplied directly with WE to treat brackish water ($5.0\ mS\ cm^{-1}$) on the island of Gran Canaria, Spain. WE was used to power both the reactor and pumping systems; the ionic conductivity of the produced water ranged between 0.2 and $0.5\ mS\ cm^{-1}$ with a production rate between 3 and $8.5\ m^3\ h^{-1}$ and an electrical consumption between 1.2 and $2.3\ kWh\ m^{-3}$. Strikingly, the ED-WE system adapted properly to wind fluctuations, even under strong wind conditions. Recently the work reported at laboratory scale by Malek [72] examined the influence of wind velocity, turbulence intensity, and oscillation period on the performance of an ED system to produce potable water from underground brackish water. Wind speed of up to $8\ m\ s^{-1}$ increased water production, and above that velocity water production tends to decrease. The moderate fluctuations in the wind speed did not appreciably influence the process; the higher intensity fluctuations lowered the water production. The increase in the internal resistance of the reactor associated with the decrease in ionic conductivity of produced water also influenced directly the angular velocity of the turbine rotor, thereby increasing its angular momentum and making the system less sensitive to extreme fluctuations of wind. The authors also stated that energy consumption was unaffected by wind fluctuations and/or oscillations, giving rise to an example of a particularly interesting robustness in isolated areas with difficult access to a conventional electricity grid.

Wind Energy Coupled With Another Electrochemical Technology

Unlike the coupling of ED-WE, the use of WE coupled with other electrochemical technologies toward wastewater treatment is very uncommon. Particularly, Jeon et al. (2016) explored for the first time the coupling of an EC system with a WE harvester at laboratory scale for the treatment of algae and dye-contaminated wastewater. High removal rates were obtained for algae (90%) and dye (97%) without the addition of other reagents after a 72 h treatment. Another study performed by Souza [73] compared the EO treatment of wastewater polluted with the pesticide 2,4-dichlorophenoxyacetic acid when conventional energy or WE were directly coupled with the reactor. Despite experiments being carried out under nonoptimal conditions and a gentle breeze, a complete mineralization was achieved, applying a total electric charge similar to that when supplied by conventional energy. Some differences are attributed to distinct oxidative mechanisms by the strong contribution of oxidizing agents generated at high current densities using WE. The same research group [73] recently compared the ERM treatment of soil polluted with the herbicide 2,4-dichlorophenoxyacetic acid during 15 days using conventional energy and WE coupled directly with the reactor. The continuous fluctuations of the wind speed, especially under calm conditions, provoked a decrease in the treatment efficiency (54% of herbicide removal) when compared with the efficiency obtained using conventional energy (90%).

Other Renewable Energies Coupled With Electrochemical Techniques

Generally, an electrochemical system will be coupled more effectively to an electromechanic RE source rather than to a thermal one. Among electromechanic energies, wave power (WP) allows us to obtain electric energy from the mechanic energy generated by the wave motion, making its application particularly attractive in coastal zones. With regard to thermal energy, geothermal energy (GE) was earlier coupled with electrodialysis (ED-GE) under high temperature conditions, as described by Boegli [74] using a 12 cell pair electrodialysis reactor (IONICS made) provided with Teflon-based membranes. The authors reported the desalinization of a brackish geothermal fluid (with salinity $<5\ g\ L^{-1}$) under temperatures between 60°C and 71°C. After 78 h of processing, they obtained a 25% salt removal with an 80% current efficiency for a current density of

23 mA cm^{-2}. Unfortunately, iron present in water fouled the membranes, doubling the resistance of the pair cell during the process.

There also exist several examples regarding the coupling of a triboelectric generator with an electrochemical process. For example, Fan [75] built a triboelectric generator that converted mechanic energy into electric energy. This technology has been coupled to different electrochemical processes (e.g., corrosion protection, toxic gas oxidation, or water splitting) [76]. The latter group used this technology for the electrochemical combustion of the azobencene dye present in wastewater from the pharmaceutical and cosmetic industries. They also used this technology for the electrochemical wastewater degradation from a degumming process [77], diminishing the amount of pollutants in 90%. They also combined this technology with an electrodialysis process, reducing the salinity in 98% of an 0.5 M NaCl solution [78]. Another example regards the coupling of an electrochemical sterilization process of wastewater, harvesting electric power from WP, and attaining complete sterilization [79].

Future Perspectives

The direct coupling of WE, WP, or GE renewable energies with electrochemical systems for wastewater treatment presents promising perspectives in regions where renewable energy sources are easy available and conventional energy supply is difficult. Coastal areas may particularly be beneficial due to a remarkable supply of wind and wave power. Currently, wind-powered commercial desalination plants are not working; only some wind turbine manufacturers offer turnkey wind desalination plants and even wind-diesel hybrids. Therefore, it becomes more important and evident that governments have to encourage the use of these renewable energies and to help reduce economic costs. Technological progress will continue improving the efficiency of those couplings, thereby diminishing implementation costs and enhancing their competitiveness.

References

[1] J.G. Eckstein, Detailed Modelling of Photovoltaic System Components, Solar Energy Laboratory, University of Wisconsin, Madison, 1990.
[2] J.M. Ortiz, E. Expósito, F. Gallud, V. García-García, V. Montiel, V.A. Aldaz, Desalination of underground brackish waters using an electrodialysis system powered directly by photovoltaic energy, Sol. Energy Mater. Sol. Cells 92 (2008) 1677–1688.
[3] D. Valero, V. García-García, E. Expósito, A. Aldaz, V. Montiel, Application of electrodialysis for the treatment of almond industry wastewater, J. Membr. Sci. 476 (2015) 580–589.
[4] J.E. Lundstrom, Water desalting by solar powered electrodialysis, Desalination 31 (1979) 469–488.

[5] O. Kuroda, S. Takahashi, K. Wakamatsu, S. Itoh, S. Kubota, K. Kikuchi, Y. Eguchi, Y. Ikenaga, N. Sohma, K. Nishinoiri, An electrodialysis sea water desalination system powered by photovoltaic cells, Desalination 65 (1987) 161–169.
[6] M.R. Adiga, S.K. Adhikary, P.K. Narayanan, W.P. Harkare, S.D. Gomkale, K. P. Govindan, Performance analysis of photovoltaic electrodialysis desalination plant at Tanote in Thar desert, Desalination 67 (1987) 59–66.
[7] H.M.N. AlMadani, Water desalination by solar powered electrodialysis process, Renew. Energy 28 (2003) 1915–1924.
[8] S.A. Kalogirou, Seawater desalination using renewable energy sources, Prog. Energy Combust. Sci. 31 (2005) 242–281.
[9] M.A. Lichtwardt, D. M. Williams, Solar-powered direct current electrodialysis reversal system, U.S.A. 6,042,701 (2000) 9.
[10] A. Ghermandi, S. Batayneh, Photovoltaic-powered membrane desalination in urban and rural areas, Ann. Arid Zone 49 (2010) 311–332.
[11] J.M. Ortiz, J.A. Sotoca, E. Expósito, F. Gallud, V. García-García, V. Montiel, A. Aldaz, Brackish water desalination by electrodialysis: batch recirculation operation modeling, J. Membr. Sci. 252 (2005) 65–75.
[12] J.M. Ortiz, E. Expósito, F. Gallud, V. García-García, V. Montiel, A. Aldaz, Photovoltaic electrodialysis system for brackish water desalination: modeling of global process, J. Membr. Sci. 274 (2006) 138–149.
[13] J.M. Ortiz, E. Expósito, F. Gallud, V. García-García, V. Montiel, A. Aldaz, Electrodialysis of brackish water powered by photovoltaic energy without batteries: direct connection behaviour, Desalination 208 (2007) 89–100.
[14] J.M. Ortiz Díaz-Guerra, Desalinización de aguas mediante sistemas de electrodiálisis alimentados directamente con energía solar fotovoltaica: estudio de viabilidad y modelización, Instituto Universitario de Electroquímica, Universidad de Alicante, Alicante, 2009.
[15] F. Círez, J. Uche, A.A. Bayod, A. Martíinez, Batch ED fed by a PV unit: a reliable, flexible, and sustainable integration, Desalin. Water Treat. 51 (2013) 673–685.
[16] S. Zhang, J. Zhang, W. Wang, F. Li, X. Cheng, Removal of phosphate from landscape water using an electrocoagulation process powered directly by photovoltaic solar modules, Sol. Energy Mater. Sol. Cells 117 (2013) 73–80. Y. Zhang, L. Pinoy, B. Meesschaert, B. Van Der Bruggen, A natural driven membrane process for brackish and wastewater treatment: photovoltaic powered ED and FO hybrid system, Environ. Sci. Technol. 47 (18) (2013) 10548–10555.
[17] A.A. Bayod-Rújula, A. Martínez-Gracia, Photovoltaic system for brackish water desalination by electrodialysis and electricity generation, Desalin. Water Treat. 7 (2009) 142–151.
[18] J. Uche, F. Círez, A.A. Bayod, A. Martínez, On-grid and off-grid batch-ED (electrodialysis) process: simulation and experimental tests, Energy 57 (2013) 44–54.
[19] B. Peñate, F. Círez, F.J. Domínguez, V.J. Subiela, L. Vera, Design and testing of an isolated commercial EDR plant driven by solar photovoltaic energy, Desalin. Water Treat. 51 (2013) 1254–1264.
[20] ATLL, Aigües del Ter Llobregat, From, www.atll.cat/es/page.asp?id=29, 2017.
[21] N. Ghaffour, J. Bundschuh, H. Mahmoudi, M.F.A. Goosen, Renewable energy-driven desalination technologies: a comprehensive review on challenges and potential applications of integrated systems, Desalination 356 (2015) 94–114.
[22] H. Sharon, K.S. Reddy, A review of solar energy driven desalination technologies, Renew. Sustain. Energy Rev. 41 (2015) 1080–1118.
[23] D. Valero, Tratamiento de las aguas residuales de la industria de la almendra mediante técnicas electroquímicas. Estudio de la alimentación de los sistemas mediante energía fotovoltaica, Institute of Electrochemistry, University of Alicante, Alicante, 2014.

[24] A. Pugsley, A. Zacharopoulos, J.D. Mondol, M. Smyth, Global applicability of solar desalination, Renew. Energy 88 (2016) 200–219.
[25] N.C. Wright, A.G. Winter, Justification for community-scale photovoltaic-powered electrodialysis desalination systems for inland rural villages in India, Desalination 352 (2014) 82–91.
[26] L. Karimi, L. Abkar, M. Aghajani, A. Ghassemi, Technical feasibility comparison of off-grid PV-EDR and PV-RO desalination systems via their energy consumption, Sep. Purif. Technol. 151 (2015) 82–94.
[27] V. Montiel, E. Expósito, V. García-García, F. Gallud, J. M. Ortiz and A. Aldaz (2016). Autonomous water treatment system, U. o. Alicante, Spain, WO2016181003 A1.
[28] C. Fernandez-Gonzalez, A. Dominguez-Ramos, R. Ibañez, A. Irabien, Sustainability assessment of electrodialysis powered by photovoltaic solar energy for freshwater production, Renew. Sustain. Energy Rev. 47 (2015) 604–615.
[29] D. Valero, J.M. Ortiz, E. Expósito, V. Montiel, A. Aldaz, Electrocoagulation of a synthetic textile effluent powered by photovoltaic energy without batteries: direct connection behaviour, Sol. Energy Mater. Sol. Cells 92 (2008) 291–297.
[30] A. Pirkarami, M.E. Olya, S. Tabibian, Treatment of colored and real industrial effluents through electrocoagulation using solar energy, J. Environ. Sci. Health A Tox. Hazard. Subst. Environ. Eng. 48 (2013) 1243–1252.
[31] M.E. Olya, A. Pirkarami, Electrocoagulation for the removal of phenol and aldehyde contaminants from resin effluent, Water Sci. Technol. 68 (2013) 1940–1949.
[32] A. Pirkarami, M.E. Olya, Removal of dye from industrial wastewater with an emphasis on improving economic efficiency and degradation mechanism, J. Saudi, Chem. Soc. 21 (2017) S179–S186.
[33] A. García-García, V. Martínez-Miranda, I.G. Martínez-Cienfuegos, P.T. Almazán-Sánchez, M. Castañeda-Juárez, I. Linares-Hernández, Industrial wastewater treatment by electrocoagulation-electrooxidation processes powered by solar cells, Fuel 149 (2015) 46–54.
[34] G. Sharma, J. Choi, H.K. Shon, S. Phuntsho, Solar-powered electrocoagulation system for water and wastewater treatment, Desalin. Water Treat. 32 (2011) 381–388.
[35] F. Hussin, F. Abnisa, G. Issabayeva, M.K. Aroua, Removal of lead by solar-photovoltaic electrocoagulation using novel perforated zinc electrode, J. Clean. Prod. 147 (2017) 206–216.
[36] K. Dermentzis, D. Marmanis, A. Christoforidis, A. Moumtzakis, Photovoltaic electrocoagulation process for remediation of chromium plating wastewaters, Desalin. Water Treat. 56 (2014) 1413–1418. K. Dermentzis, D. Marmanis, A. Christoforidis, in: Remediation of heavy metal bearing wastewater by photovoltaic electrocoagulation, 2nd International Conference—Water Resources and Wetlands, P. G. W. M. P. Bretcan, Tulcea, 2014, pp. 257–262.
[37] D. Marmanis, K. Dermentzis, A. Christoforidis, K. Ouzounis, A. Moumtzakis, Electrochemical treatment of actual dye house effluents using electrocoagulation process directly powered by photovoltaic energy, Desalin. Water Treat. 56 (2014) 2988–2993.
[38] A.S. Naje, S. Chelliapan, Z. Zakaria, M.A. Ajeel, K. Sopian, H.A. Hasan, Electrocoagulation by solar energy feed for textile wastewater treatment including mechanism and hydrogen production using a novel reactor design with a rotating anode, RSC Adv. 6 (2016) 10192–10204.
[39] C. Phalakornkule, T. Luanwuthi, P. Neragae, E.J. Moore, A continuous-flow sparged packed-bed electrocoagulator for dye decolorization, J. Taiwan Inst. Chem. Eng. 64 (2016) 124–133.
[40] J. Iniesta, E. Exposito, J. Gonzalez-Garcia, V. Montiel, A. Aldaz, Electrochemical treatment of industrial wastewater containing phenols, J. Electrochem. Soc. 14 (2002) D57–D62.

[41] M.S. Lucas, J.A. Peres, Decolorization of the azo dye reactive black 5 by Fenton and photo-Fenton oxidation, Dyes Pigm. 71 (2006) 236–244.
[42] L.S. Andrade, L.A.M. Ruotolo, R.C. Rocha, N. Bocchi, S.R. Biaggio, J. Iniesta, V. Garcia-Garcia, V. Montiel, On the performance of Fe and Fe,F doped Ti-Pt/PbO2 electrodes in the electrooxidation of the blue reactive 19 dye in simulated textile wastewater, Chemosphere 66 (2007) 2035–2043.
[43] L. Feng, E.D. van Hullebusch, M.A. Rodrigo, G. Esposito, M.A. Oturan, Removal of residual anti-inflammatory and analgesic pharmaceuticals from aqueous systems by electrochemical advanced oxidation processes. A review, J. Chem. Eng. 228 (2013) 944–964.
[44] C. Barrera-Diaz, P. Canizares, F.J. Fernandez, R. Natividad, M.A. Rodrigo Electrochemical, Advanced oxidation processes: an overview of the current applications to actual industrial effluents, J. Mex. Chem. Soc. 58 (2014) 256–275.
[45] W.T. Mook, M.K. Aroua, G. Issabayeva, Prospective applications of renewable energy based electrochemical systems in wastewater treatment: a review, Renew. Sustain. Energy Rev. 38 (2014) 36–46.
[46] H. Park, C.D. Vecitis, M.R. Hoffmann, Solar-powered electrochemical oxidation of organic compounds coupled with the cathodic production of molecular hydrogen, J. Phys. Chem. A 11 (2008) 7616–7626.
[47] I. Navarro-Solis, L. Villalba-Almendra, A. Alvarez-Gallegos, H-2 production by PEM electrolysis, assisted by textile effluent treatment and a solar photovoltaic cell, Int. J. Hydrogen Energy 35 (2010) 10833–10841.
[48] M. Panizza, P.A. Michaud, G. Cerisola, C. Comninellis, Electrochemical treatment of wastewaters containing organic pollutants on boron-doped diamond electrodes: prediction of specific energy consumption and required electrode area, Electrochem. Commun. 3 (2001) 336–339.
[49] M. Carmo, D.L. Fritz, J. Mergel, D. Stolten, A comprehensive review on PEM water electrolysis, Int. J. Hydrogen Energy 38 (2013) 4901–4934.
[50] M. Wang, Z. Wang, X. Gong, Z. Guo, The intensification technologies to water electrolysis for hydrogen production—a review, Renew. Sustain. Energy Rev. 29 (2014) 573–588.
[51] S. Figueroa, L. Vazquez, A. Alvarez-Gallegos, Decolorizing textile wastewater with Fenton's reagent electrogenerated with a solar photovoltaic cell, Water Res. 43 (2009) 283–294.
[52] I.E. Ciobotaru, F. Nenciu, D.I. Vaireanu, The electrochemical generation of ozone using an autonomous photovoltaic system, Rev. Chim. 64 (2013) 1339–1342.
[53] A. Dominguez-Ramos, R. Aldaco, A. Irabien, Photovoltaic solar electrochemical oxidation (PSEO) for treatment of lignosulfonate wastewater, J. Chem. Technol. Biotechnol. 85 (2010) 821–830.
[54] T. Ochiai, K. Nakata, T. Murakami, A. Fujishima, Y.Y. Yao, D.A. Tryk, Y. Kubota, Development of solar-driven electrochemical and photocatalytic water treatment system using a boron-doped diamond electrode and TiO2 photocatalyst, Water Res. 44 (2010) 904–910.
[55] F.L. Souza, J. Llanos, C. Sáez, M.R.V. Lanza, M.A. Rodrigo, P. Cañizares, Performance of wind-powered soil electroremediation process for the removal of 2,4-D from soil, J. Environ. Manage. 171 (2016) 128–132.
[56] D. Valero, V. García-García, E. Expósito, A. Aldaz, V. Montiel, Electrochemical treatment of wastewater from almond industry using DSA-type anodes: direct connection to a PV generator, Sep. Purif. Technol. 123 (2014) 15–22.
[57] E. Alvarez-Guerra, A. Dominguez-Ramos, A. Irabien, Design of the photovoltaic solar electro-oxidation (PSEO) process for wastewater treatment, Chem. Eng. Res. Des. 89 (2011) 2679–2685.
[58] S.Y. Tee, K.Y. Win, W.S. Teo, L.D. Koh, S. Liu, C.P. Teng, M.Y. Han, Recent progress in energy-driven water splitting, Adv. Sci. (2017).

[59] F. Kargi, Comparison of different electrodes in hydrogen gas production from electrohydrolysis of wastewater organics using photovoltaic cells (PVC), Int. J. Hydrogen Energy 36 (2011) 3450–3456.
[60] D.H. Kim, J.M. Jung, S.U. Jo, W.S. Kim, K. Baek, Photovoltaic powered electrokinetic restoration of saline soil, Sep. Sci. Technol. 47 (2012) 2235–2240.
[61] K. Cho, D. Kwon, M.R. Hoffmann, Electrochemical treatment of human waste coupled with molecular hydrogen production, RSC Adv. 4 (2014) 4596–4608.
[62] M. Bielmann, U.F. Vogt, M. Zimmermann, A. Zuttel, Seasonal energy storage system based on hydrogen for self sufficient living, J. Power Sources 196 (2011) 4054–4060.
[63] S. Yuan, Z. Zheng, J. Chen, X. Lu, Use of solar cell in electrokinetic remediation of cadmium-contaminated soil, J. Hazard. Mater. 162 (2009) 1583–1587.
[64] E. Mohamedelhassan, Res. J. Environ. Earth Sci. 3 (2011) 13.
[65] M. Zhou, S. Zhu, Y. Liu, X. Wang, Removal of fluorine from contaminated soil by electrokinetic treatment driven by solar energy, Environ. Sci. Pollut. Res. 20 (2013) 5806–5812.
[66] S. Zhang, J. Zhang, X. Cheng, Y. Mei, C. Hu, M. Wang, J. Li, Electrokinetic remediation of soil containing Cr(VI) by photovoltaic solar panels and a DC-DC converter, J. Chem. Technol. Biotechnol. 90 (2014) 693–700.
[67] I. Hassan, E. Mohamedelhassan, E.K. Yanful, Solar powered electrokinetic remediation of Cu polluted soil using a novel anode configuration, Electrochim. Acta 181 (2015) 58–67.
[68] E.K. Jeon, S.R. Ryu, K. Baek, Application of solar-cells in the electrokinetic remediation of as-contaminated soil, Electrochim. Acta 181 (2015) 160–166.
[69] R.G. Raluy, L. Serra, J. Uche, Life cycle assessment of desalination technologies integrated with renewable energies, Desalination 183 (2005) 81–93.
[70] Q. Ma, H. Lu, Wind energy technologies integrated with desalination systems: Review and state-of-the-art, Desalination 277 (2011) 274–280.
[71] J.M. Veza, B. Penate, F. Castellano, Electrodialysis desalination designed for off-grid wind energy, Desalination 160 (2004) 211–221.
[72] P. Malek, J.M. Ortiz, H.M.A. Schulte-Herbrüggen, Decentralized desalination of brackish water using an electrodialysis system directly powered by wind energy, Desalination 377 (2016) 54–64.
[73] F.L. Souza, C. Saez, J. Llanos, M.R.V. Lanza, P. Canizares, M.A. Rodrigo, Solar-powered CDEO for the treatment of wastewater polluted with the herbicide 2,4-D, Chem. Eng. J. 277 (2015) 64–69.
[74] W.J. Boegli, S.H. Suemoto, K.M. Trompeter, Geothermal desalting at the east mesa test site, Desalination 22 (1977) 77–90.
[75] F.R. Fan, Z.Q. Tian, Z. Lin Wang, Flexible triboelectric generator, Nano Energy 1 (2012) 328–334.
[76] S. Gao, J. Su, M. Wang, X. Wei, X. Zheng, T. Jiang, Electrochemical oxidation degradation of azobenzene dye self-powered by multilayer-linkage triboelectric nanogenerator, Nano Energy 30 (2016) 52–58.
[77] Z. Li, J. Chen, J. Zhou, L. Zheng, K.C. Pradel, X. Fan, H. Guo, Z. Wen, M.H. Yeh, C. Yu, Z. L. Wang, High-efficiency ramie fiber degumming and self-powered degumming wastewater treatment using triboelectric nanogenerator, Nano Energy 22 (2016) 548–557.
[78] Q. Jiang, Y. Han, W. Tang, H. Zhu, C. Gao, S. Chen, M. Willander, X. Cao, Z. Lin Wang, Self-powered seawater desalination and electrolysis using flowing kinetic energy, Nano Energy 15 (2015) 266–274.
[79] Q. Jiang, Y. Jie, Y. Han, C. Gao, H. Zhu, M. Willander, X. Zhang, X. Cao, Self-powered electrochemical water treatment system for sterilization and algae removal using water wave energy, Nano Energy 18 (2015) 81–88.

Index

Note: Page numbers followed by *f* indicate figures, and *t* indicate tables.

D

E

I

L

M

N

R

Kolibri 360 GmbH
Ferdinand-Jühlke-Straße 7
99095 Erfurt, Deutschland
produktsicherheit@kolibri360.de